Evolutionary Computation in Economics and Finance

Studies in Fuzziness and Soft Computing

Editor-in-chief
Prof. Janusz Kacprzyk
Systems Research Institute
Polish Academy of Sciences
ul. Newelska 6
01-447 Warsaw, Poland
E-mail: kacprzyk@ibspan.waw.pl
http://www.springer.de/cgi-bin/search_book.pl?series=2941

Further volumes of this series can be found at our homepage.

Vol. 81. V. Dimitrov and V. Korotkich (Eds.)
Fuzzy Logic, 2002
ISBN 3-7908-1425-3

Vol. 82. Ch. Carlsson and R. Fullér
Fuzzy Reasoning in Decision Making and Optimization, 2002
ISBN 3-7908-1428-8

Vol. 83. S. Barro and R. Marín (Eds.)
Fuzzy Logic in Medicine, 2002
ISBN 3-7908-1429-6

Vol. 84. L.C. Jain and J. Kacprzyk (Eds.)
New Learning Paradigms in Soft Computing, 2002
ISBN 3-7908-1436-9

Vol. 85. D. Rutkowska
Neuro-Fuzzy Architectures and Hybrid Learning, 2002
ISBN 3-7908-1438-5

Vol. 86. M.B. Gorzałczany
Computational Intelligence Systems and Applications, 2002
ISBN 3-7908-1439-3

Vol. 87. C. Bertoluzza, M.Á. Gil and D.A. Ralescu (Eds.)
Statistical Modeling, Analysis and Management of Fuzzy Data, 2002
ISBN 3-7908-1440-7

Vol. 88. R.P. Srivastava and T.J. Mock (Eds.)
Belief Functions in Business Decisions, 2002
ISBN 3-7908-1451-2

Vol. 89. B. Bouchon-Meunier, J. Gutiérrez-Ríos, L. Magdalena and R.R. Yager (Eds.)
Technologies for Constructing Intelligent Systems 1, 2002
ISBN 3-7908-1454-7

Vol. 90. B. Bouchon-Meunier, J. Gutiérrez-Ríos, L. Magdalena and R.R. Yager (Eds.)
Technologies for Constructing Intelligent Systems 2, 2002
ISBN 3-7908-1455-5

Vol. 91. J.J. Buckley, E. Eslami and T. Feuring
Fuzzy Mathematics in Economics and Engineering, 2002
ISBN 3-7908-1456-3

Vol. 92. P.P. Angelov
Evolving Rule-Based Models, 2002
ISBN 3-7908-1457-1

Vol. 93. V.V. Cross and T.A. Sudkamp
Similarity and Compatibility in Fuzzy Set Theory, 2002
ISBN 3-7908-1458-X

Vol. 94. M. MacCrimmon and P. Tillers (Eds.)
The Dynamics of Judical Proof, 2002
ISBN 3-7908-1459-8

Vol. 95. T.Y. Lin, Y.Y. Yao and L.A. Zadeh (Eds.)
Data Mining, Rough Sets and Granular Computing, 2002
ISBN 3-7908-1461-X

Vol. 96. M. Schmitt, H.-N. Teodorescu, A. Jain, A. Jain, S. Jain and L.C. Jain (Eds.)
Computational Intelligence Processing in Medical Diagnosis, 2002
ISBN 3-7908-1463-6

Vol. 97. T. Calvo, G. Mayor and R. Mesiar (Eds.)
Aggregation Operators, 2002
ISBN 3-7908-1468-7

Vol. 98. L.C. Jain. Z. Chen and N. Ichalkaranje (Eds.)
Intelligent Agents and Their Applications, 2002
ISBN 3-7908-1469-5

Vol. 99. C. Huang and Y. Shi
Towards Efficient Fuzzy Information Processing, 2002
ISBN 3-7908-1475-X

Shu-Heng Chen
Editor

Evolutionary Computation in Economics and Finance

With 110 Figures
and 66 Tables

Physica-Verlag
A Springer-Verlag Company

Professor Shu-Heng Chen
National Chengchi University
AI-ECON Research Center
Department of Economics
Taipei, Taiwan 11623
R.O. China
chchen@nccu.edu.tw

ISSN 1434-9922
ISBN 3-7908-1476-8 Physica-Verlag Heidelberg New York

Cataloging-in-Publication Data applied for
Die Deutsche Bibliothek – CIP-Einheitsaufnahme
Evolutionary computation in economics and finance: with 66 tables / Shu-Heng Chen, ed. – Heidelberg; New York: Physica-Verl., 2002
(Studies in fuzziness and soft computing; Vol. 100)
ISBN 3-7908-1476-8

Physica-Verlag Heidelberg New York
a member of BertelsmannSpringer Science+Business Media GmbH

http://www.springer.de

Hardcover Design: Erich Kirchner, Heidelberg

SPIN 10866945 88/2202-5 4 3 2 1 0 – Printed on acid-free paper

Foreword

For an editor of a book series there is a real pleasure when any book or volume is coming out. This pleasure is multiplied when he or she may write some introductory remarks. And this pleasure is further multiplied when the book or volume is a pioneering work, with great contributors, up to date material and comprehensive coverage.

This is exactly the case with this volume, and that is why it is a great honor and pleasure for me to be able to write these couple of sentences. First, let me mention that Professor Chen has somehow made my task difficult because in his introductory chapter he has both presented how the volume has been initiated, which topics have been chosen and why, including some historical perspective, and then has summarized the content of the papers. So, not much can be constructively added in these respects, and I will try to add some more general thoughts.

This volume can be viewed from various, at least two, perspectives, each of them being relevant to different groups of readers. On the one hand, one can say that the volume is concerned with economics and finance. Clearly, this is a very broad field and only some specific topics have been chosen. It is difficult to overestimate the relevance of these topics for all of us because our life depends so much on how effectively and efficiently financial means are acquired, circulated and utilized. Then, having formulated some problems, we apply for their solution tools of evolutionary computation. This may be viewed as a problem oriented approach.

On the other hand, one can say that the volume is concerned with evolutionary computation that is employed to solve financial and economic problems. Now, the focus is on evolutionary computation as a set of tools and techniques somehow emulating aspects of evolution processes in nature that is shown to provide effective and efficient means for the solution of relevant problems in a relevant field, economics and finance. This may be viewed as a tools and techniques oriented approach.

Clearly, which one of the above two views is relevant depends on the reader but this volume is excellent in both respects. Readers interested both in problems, and in tools and techniques will find a variety of interesting and useful information on both the state of the art and original and novel proposals.

One of the reasons is that the editor of the volume, a well known expert himself, has succeeded to gather top contributors whose research is opening new vistas at the crossroads of evolutionary computation and economics and finance. Their top expertise provides at the same time authoritative and critical surveys of past and other contributions. Therefore, the reader can have a complete view of what has been done, what is relevant, what is promising, what is still to be solved, etc.

To summarize, I wish to congratulate Professor Chen for his excellent work on this extraordinary volume. I think I can say on behalf of the whole scientific community interested in the topics of the volume that thanks to him we have now a comprehensive treatise that should provide us with the state of the art, inspire us, and provide a reference against which all next books and volumes related to those topics are to be compared.

November, 2001 Janusz Kacprzyk
Warsaw, Poland

Contents

Part I

An Overview

1 Evolutionary Computation in Economics and Finance: An Overview of the Book

Shu-Heng Chen

AI-ECON Research Center, Department of Economics, National Chengchi University, Taipei, Taiwan, 11623
chchen@nccu.edu.tw

Abstract. Being a start-off of the first volume on *evolutionary computation in economics and finance*, this chapter proposes an overview of the volume. The 20 contributed chapters of this volume are separated into three parts, namely, games, agent-based modeling and financial engineering. The reader will find himself/herself treading the path of the history of this research area, from the fledgling stage to the burgeoning era.

1.1 The Birth of this Volume

Evolutionary computation is generally considered as a consortium of *genetic algorithms*, *genetic programming*, *evolutionary programming* and *evolutionary strategies*. While it has been studied for almost 40 years by computer scientists, its application to economics and finance has a much shorter history. After a decade-long development, we believe that it is high time to have a special volume on this subject. The only questions are *who shall do this* and *how it can be done*.

In late 1998, I was invited by the chairs of the *First International Symposium on Intelligent Data Engineering and Learning* held in Hong Kong to organize a session on the application of genetic algorithms to computational finance. Just before a morning session on the second day of the conference, Professor Kacprzyk came to me and asked whether I was interested in contributing a volume to the book series "**Studies in Fuzziness and Soft Computing**", of which he serves as the series editor. This offer was not only a great honor but a precious opportunity to make come true what we had been dreaming about, a special volume devoted to *evolutionary computation in economics and finance*. The question immediately to follow is how this volume should be organized. Most editors would prefer to have a comprehensive coverage of the subject they are working with, and I am no exception.

To achieve this goal, this volume considers three areas of interest, namely, *game theory*, *agent-based economic modeling* and *financial engineering*. As a matter of fact, if we trace how the idea of evolutionary computation was applied to economics and finance, we would notice a path passing through these three areas chronologically. It originated in games ([13]), then extended to agent-based macroeconomic modeling ([47], [10]) and further to financial engineering ([15], [2], [16]). The more recent applications can be regarded as the enrichment of one of these three directions.

Having this framework in mind, we started to invite leading scholars from each of these areas to contribute a chapter. To guarantee the academic quality of the book, many of the papers, while invited, had been revised at the request of the editor before they were finally accepted. Two papers which failed be modified to a satisfactory degree were unfortunately rejected. In the end, there are totally 22 chapters, including this overview, published in this volume. These chapters are divided into three subdivisions according to the proposed framework: three chapters devoted to "**Game Theory**", ten chapters to "**Agent-Based Computational Economics**", and seven chapters to "**Financial Engineering**". The first chapter serves as the introduction to the book, and the last chapter documents research resources, including a bibliography. In the following, we will give a brief review of the three mainstays of the book, starting from game theory and moving on to agent-based computational economics and finally to financial engineering.

1.2 Playing Games: 1987

The earliest application of GAs to social sciences finds its area in games, which composes the first part of the book. There are three contributions presented in this part. Robert Marks' survey article "**Playing Games with Genetic Algorithms** " guides us through the development of GAs in game theory from the late 1980s to the present. This development can be roughly divided into three stages characterized by different but related types of games, namely, *the repeated prisoner's dilemma* (RPD) game, *the oligopoly game* and *the empirical game*. It dates back to 1987 when Robert Axelrod published the first paper of using the GA in games, more precisely, the RPD game. Behind this work and the follow-up work by John Miller, the first doctoral thesis on this subject, one can see the influence of the "Michigan School", initiated and led by John Holland. While the idea of GAs, *simulating natural selection*, was already known to computer scientists in the late 70s, its relevance to social sciences came almost a decade later. After describing this fledgling stage, Marks proceeds to the burgeoning stage and demonstrates how Axelrod's original piece of work has been enriched in the years to come. More than a dozen papers are included in his review. Some challenge the robustness of the strategy "Tit for Tat" by taking other factors, such as information processing costs and noise, into account. Some consider different coding schemes for game players, e.g., coding finite state automata. But the most important extension is to endow *all* players, not just a single one, with the capability to evolve, and in this case, GAs are used to simulate the interaction of players in a *co-evolutionary context*.

The second part of Marks' survey covers a lot of recent applications which can be closely related to industrial economics and policy issues. Since the

N-person RPD game and the oligopoly game are often considered close in spirit, it is only natural that GAs are applied to oligopoly games. Marks himself pioneered one of these applications. In a series of papers which he coauthored with Midgley and Cooper, Marks used *historical market data* of the canned ground coffee in the supermarket to breed artificial managers for the purpose of developing competitive marketing strategies. By conducting open-loop experiments, they concluded "that the historical brand managers could have improved their performance given insights from the response of the stimulus-response artificial managers". Henceforth, applications of GAs moved from stylized games to actual market interaction.

In contrast with Marks' paper, which shows how GAs can be helpful to game theorists, Riechmann's "**Genetic Algorithm Learning and Economic Evolution**" demonstrates how game theory can shed light on the foundation of GAs. Motivated by the recent successes of GAs as a learning mechanism when applied to some standard benchmark cases of economic theory, Riechmann is interested in knowing if there are *certain properties of genetic algorithms* which lead genetic learning models at least to the neighborhood of the results of mainstream economic models. One way to lay a theoretical foundation for GAs is to interpret GAs as a *Markov chain*, and apply the associated convergence analysis. This direction has been taken by GA theorists for years ([58]. [32]). What Riechmann do in this chapter and his earlier paper ([57]) is provide a game-theoretic interpretation, i.e., to interpret GAs as a *repeated N-person game*. With this new interpretation, Riechmann relates the *absorbing states* of a Markov chain to the *Nash Equilibra* of the repeated game and the *evolutionary stable states* of the repeated game. As a result, the convergence to "optimal" states and the stability of them can be analyzed in a Markovian framework. This work can be considered a continuation of the research line pioneered by [32], which attempted to make GAs transparent and accessible to economists.

In addition to the RPD game and the oligopoly game, the *repeated ultimatum game* is another type of game which has received extensive treatment among game theorists. It has also become a very active topic in experimental economics. Over the last two decades, a lot of laboratory experiments with human subjects have been conducted on this game. The ultimatum game is interesting because it shows the significance of fairness, perceived subjectively by the players and formed endogenously via their strategic interactions, on the final deal to come. John Duffy and Engle-Warnick's paper "**Using Symbolic Regression to Infer Strategies from Experimental Data**" contributes to the understanding of bargaining strategies involved in the ultimatum game. Using statistical models to infer from experimental data the decision rules human subjects may apply is nothing new. Reference [48], for instance, used linear regression to gauge the variables which human subject might use to form their price expectations. However, the expressive power of a prespecified parametric model may be too limited to represent the actual decision rules followed by human subjects. What Duffy and Engle-Warnick do in their chapter is to use *genetic programming* (**GP**), a tool with strong expressive power, to enhance the inference technique.

While this is not the only chapter on genetic programming in the book, it is distinguished by its emphasis on the *grammar*. The role of a grammar or *semantic restrictions* has been generally neglected by most economic and financial applications.[1] Most applications only introduce their function set and terminal set, the two key components of genetic programming, and let the decision rules be automatically generated without further restrictions. As a result, even though all decision rules can be syntactically correct, they may not be semantically valid.[2] Using the Backus-Nauer form grammar, Duffy et al. are able to add some semantic restrictions to the decision trees. Rules derived in this manner have a better chance of mimicking real (sensible) decision rules used by human subjects. This chapter is significant in that it provides a novel application of genetic programming to experimental economics. As a *data-mining tool*, genetic programming gives economists new insights into *what the subjects were doing during the experiments.*

1.3 Exploring Agent-Based Artificial Stock Markets: 1988

1.3.1 Agent-Based Artificial Stock Markets

The second part entitled **Agent-Based Computational Economics** (**ACE**) contains 10 chapters, and forms the largest proportion of the book. The size of this part acknowledges the great potential of this young but very active research area. While one can expect a special volume entirely devoted to agent-based computational economics[3], the ten chapters included in this part of the volume show how important evolutionary computation is for the development of agent-based computational economics.

The first four chapters concern one major subarea of ACE, namely, *the agent-based artificial financial market.* As we mentioned earlier, the first application area of evolutionary computation in economics is *game theory* ([13]). *Agent-based financial markets* are probably the second research arena where economists exploit evolutionary computation. In 1988, when John Holland and Brian Arthur established an economics program at the Santa Fe Institute, *artificial stock markets* were chosen as the initial research project. Two papers in this book are directly related to *agent-based artificial stock markets.* They can be regarded as an advancement of the early version of the SFI artificial stock market ([55], [62], [12], [44]).

---

[1] [18] is probably the only exception known to us.

[2] See [25] for a thorough discussion on this issue.

[3] In fact, there is already a special issue on this subject, which is guest edited by Leigh Tesfatsion and published by the Journal of Economic Dynamics and Control (JEDC), Volume 25, No. 3–4, March 2001.

The chapter by Jing Yang "**The Efficiency of an Artificial Double Auction Stock Market with Neural Learning Agents**" distinguishes itself from the conventional SFI artificial stock market in the *trading institutional design.* Trading institutional designs in the conventional SFI artificial stock market either follow the *Walrasian tatonnement scheme* ([55]) or the *rationing scheme* ([12], [44]). Yang, however, follows [23] and considers a *double auction* mechanism. This design narrows the gap between artificial markets and the real market, and hence makes it possible to compare the simulation results with the behavior of real data, e.g., tick-by-tick data. Furthermore, since stock market experiments with human subjects were also conducted within the double auction framework ([60]), it also facilitates the conversation between the experimental stock market and the agent-based artificial stock market.[4] In addition to a different trading institutional design, Yang also attempted a different approach to modeling the bounded rational agents. Instead of using genetic algorithms as exemplified by the SFI artificial stock market, she taps into neural networks, more precisely, the one-hidden-layer neural net.[5]

Despite these new set-ups, the basic result obtained by Yang is the same as what we learned from the SFI stock market, e.g., [44]. Yang's artificial market is also capable of generating what appears to be near equilibrium HREE (*homogeneous rational expectations equilibrium*) behavior. In fact, she shows that the same result holds regardless of using the Walrasian auctioneer or the double auction as the trading institutional design. Apart from the basic result, Yang also probes the possible force which may drive the market price persistently away from the HREE. The particular aspect she examines is the presence of *momentum traders*, i.e., traders who base their trading strategies upon the technical indicators, such as moving average (her MA(5)) or double moving average (her MA(10)). She then finds that the presence of momentum traders can not only drive the market price away from the HREE price, but also generate a lot of interesting phenomena, such as excess volatility, excess kurtosis (leptokurtotic), lack of serial independence of return, and high trading volume. This result, while interesting, is not surprising. The reason is that the neural nets employed by fundamentalists (value traders) are fed only with dividends. Therefore, they cannot possibly sense the presence of momentum traders and learn the relevance of other variables, e.g., the technical indicators. It would be interesting to know whether one can nullify the effect of the presence of momentum traders by endowing the value traders

[4] For an interesting discussion of the connection between the experimental approach and the agent-based computational approach to economics, one is referred to [33].

[5] Andrea Beltratti and Sergio Margarita pioneered this direction of research, i.e., applying artificial neural networks to modeling agents' learning behavior in the financial market. Many references can be found in [17], Chapter 7. A new version of the SFI market also considered this different learning model of learning agents. See, for example, [45].

with a more flexible framework to learn, e.g., using genetic programming, and that leads us to the next chapter.[6]

Shu-Heng Chen's chapter "**On AIE-ASM: Software to Simulate Artificial Stock Markets with Genetic Programming**", which he coauthored with Chia-Hsuan Yeh and Chung-Chih Liao, introduces a software, **AIE-ASM**, to run simulations of the artificial stock market based on a standard asset pricing model. One distinguishing feature of AIE-ASM is that agents' adaptive behavior is modeled with genetic programming rather than genetic algorithms. The software has been used to conduct a series of research, and this chapter is not intended to document any new results. Instead, it attempts to highlight an important, but always neglected, issue in agent-based computational economics, namely, *replicatability*. Authors of this chapter not only make their software available on the web, but also provide a detailed instructions on the implementation of it so that their published results can be verified. It is hoped that in the near future all researchers on ACE will make their source codes publicly available. Moreover, it should become a prerequisite for submission acceptance.

1.3.2 Artificial Foreign Exchange Markets

After the two papers on artificial stock markets, the next two chapters concern artificial foreign exchange markets. The first one entitled "**Exchange Rate Volatility in the Artificial Foreign Exchange Market**" is written by Jasmina Arifovic, who is probably the first person that finished a Ph.D thesis on applications of genetic algorithms on macroeconomics, and who turns out to be the most prolific researchers in this area. Arifovic published a series of journal articles on artificial foreign exchange markets([6], [9], and [7]). In these studies, she used genetic algorithms to replace the representative agent with artificial adaptive agents in a standard overlapping-generation model with two currencies ([39]), and was able to make two observations rarely seen in the standard OG model. First, [6] showed the *persistently endogenous fluctuation of the exchange rate*; second, by taking into account that governments of both countries finance their deficits via seignorage, [9] evidenced the equilibrium with the *collapse of one currency* (speculative attack).

The chapter differs from her earlier publications on the *implementation* of genetic algorithm learning. Depending on what to encode, GA learning can be implemented in two different ways, namely, *learning how to optimize* and *learning how to forecast.*[7] In terms of the foreign exchange market, what to optimize is the *saving* decision as well as the *portfolio decision.* Issues on either decision would not be that complicated if agents had *perfect foresight* on the exchange rate.[8] However, things can be difficult if the assumption of

[6] See also Duffy's chapter on this book.
[7] For details, see [20].
[8] If agents are perfect foresight on tho oxchange rate, then optimizing decisions requires only to solve a standard *intertemporal optimization problem.*

perfect foresight does not hold, and one must deal with *boundedly rational agents*. Therefore, an interesting application of GAs to boundedly rational agents should focus *directly* on learning how to forecast rather than *indirectly* on learning how to optimize.[9] What was implemented in [6], [9] and [7] is all *learning how to optimize*. This paper is probably Arifovic's first efforts to consider the alternative implementation of GAs, i.e., *learning how to forecast*.

In this chapter, the forecasting models of exchange rates employed by agents are *simple moving-average models*. They differ in the *rolling window size*, which are endogenously determined and can be time-variant. What is encoded by GAs is the size of the rolling window rather than the usual saving and portfolio decision. Simulations with this new coding scheme resulted in the convergence of the economies to *a single-currency equilibrium*, i.e., the collapse of one of the two currencies. This result was not found in [6]. The chapter, therefore, shows that different implementation of GA learning may have non-trivial effect on the simulation results. In one implementation, one can have *persistent fluctuation of the exchange rate* ([6]); in another case, one can have a single-currency equilibrium. Her second simulation reaffirms the robustness issue. In this simulation, she combines two different applications of GA learning. In addition to the original population of agents, who are learning how to forecast, she adds another population of agents, who are learning how to optimize. These two populations of agents undergo separate genetic algorithm updating. Simulations with these two separate evolving populations do not have the convergence to single currency equilibrium, but are characterized instead by persistent fluctuation.[10]

In retrospect, the ground-breaking project undertaken by the SFI economics program in 1988 was but a modest first step toward a more ambitious dream: *evolving the whole artificial economy*.[11] Over the last few years, a series of serious attempts has been made toward large-scale agent-based economic models, to name a few, Epstein and Axtell's *Sugarscape* ([34]), Pryor's *Aspen* ([14]) and Chris Barrett's *TRANSIMS* ([22]). The distinguishing feature of the large-scale agent-based economic models is that they are inspired by real-world objects rather than by standard theoretical models. Kiyoshi Izumi and Kazuhiro Ueda's "**Using an Artificial Market Approach to Analyze Exchange Rate Scenarios**" provides a good example. While the

[9] [29] showed that indirect coding can be considered as an *implicit* way to model *forecasting*.

[10] At this moment, it is hard to conclude how serious the robustness issue is. In some cases, these two different GA learning styles did not lead to significantly different results. For example, [5] and [20] both applied GAs to studying the inflation dynamics in the context of a two-period standard overlapping generation model. While one used the "how-to-optimize" implementation, and the other the "how-to-forecast" implementation, both their results supported the convergence to the low inflation stationary equilibrium.

[11] For a vivid description of this stage of development, see [64], p. 269.

persistent fluctuation of exchange rates is still at the core of the issue, their model of learning agents is based on the observations of *real dealers*.

Izumi and Ueda start their research with *fieldwork*. Strange as it may seen, fieldwork is neglected in most ACE studies despite its great potential for ACE modeling. This is probably because most ACE researchers care more about the emergent macro phenomena than about the rich microstructure of their ACE models.[12] To the question frequently asked in the literature, *whether genetic algorithms can represent a sensible learning process of humans*, Izumi and Ueda give a positive answer based on the findings from their fieldwork, including interviews and questionnaires.[13] Section 4 of the chapter spells out empirically a biological microfoundation of agent-based financial markets.

Their model of foreign exchange markets is called **AGEDASI TOF** (A GEnetic-algorithmic Double Auction Simulation in TOkyo Foreign exchange market), which was first proposed in [38].[14] AGEDASI TOF is also an agent-based extension of the standard asset pricing model. What makes AGEDASI TOF different from other extensions, such as the SFI and AIE artificial stock market, is that agents in AGEDASI TOF can get access to real-world data called *external data*. Artificial dealers in this chapter were trained by 17 real time series from January 1996 to December 1997. Using real-world data as a decision support for agents is not an entirely new idea in ACE; [49] is another example. Needless to say, there are certain advantages of using external data. It helps build more realistic ACE models, and hence facilitates more practical applications of ACE models, such as forecasting and policy analyses. But using external data makes the ACE model *half-open (exogenous) and half-close (endogenous)*. How to reconcile these two sub-systems so that there is no inconsistency between them is an issue yet to be addressed.[15]

The forecasting models used by artificial dealers in AGEDASI TOF are *perceptrons*. This architecture is similar to Yang's except that Yang considers one hidden layer. Weights of these perceptrons are encoded by strings of a finite number of alphabets rather than the typical binary strings. The

[12] However, Duffy ([33]) showed that the behavioral rules of the artificial agents can be suitably modeled on the basis of prior evidence from human subject experiments. Usefulness of fieldwork in ACE modeling is also demonstrated in [41].

[13] It is, however, interesting to note that the answer can vary case by case. For example, based on his observations from fieldwork, Duffy ([33]) did not even consider the necessity of genetic algorithms in modeling artificial agent behavior.

[14] While AGEDASI TOF is abbreviated for a double auction market, one has to keep in mind that the trading mechanism employed is simply the *Walrasian tatonnement scheme*, and is very similar to that of the SFI artificial stock market.

[15] Consider Izumi and Ueda's chapter as an example. The time series of GDP is exogenously given to artificial dealers, while the foreign exchange rate is endogenously generated by AGEDASI TOF. Therefore, in the ACE model, the exchange rate can not affect GDP, whereas in the real world it may not be the case.

canonical genetic algorithm is then applied to evolving this population of strings.[16]

Unlike the previous chapter, this chapter not only addresses the emergence of the fluctuation of exchange rates, it also addresses the policy of stabilizing exchange rates. Simulations of three policy scenarios are compared. It is shown that central bank's direct intervention in the foreign exchange market seems to have the strongest stabilizing effect than interest rate control and public announcement. This is probably the first application of the ACE model to the issue of economic stabilization. One may argue that while the ACE model can enhance our expressive power in economic modeling it may not be as useful as the *computational general equilibrium model* as far as policy analysis is concerned. This chapter dispels such skepticism.

1.3.3 Microeconomics

While the previous four chapters shows how EC techniques can help generate the rich dynamics of agent-based financial markets, the next four chapters extend their applications to issues like *pollution control*, *oligopolistic competition*, *the excess heterogeneity in wages* and *the role of institutions on social adaptability*. Bell and Beare's **Emulating Trade in Emissions Permits: An Application of Genetic Algorithms** concerns the effectiveness of different policies on pollution control. Pollution control is a very hot topic in microeconomics because it provides a perfect illustration on how the optimal level of pollution can potentially be achieved by using the *market mechanism* as a solution. However, when market participants are linked spatially through production externalities, the outcome of a competitive market is unclear. The main problem lies in the information revealed from the competitive market is not enough to capture the rents associated with reducing the net cost of externality ([35]). Even in a very simple setting, the problem can be too complex to be readily solved using conventional toolkits familiar to economists. Therefore, the effectiveness of the tradable permit scheme is ultimately an empirical question. The contribution of this chapter is to use a genetic algorithm to emulate an auction market for emissions permits and provides a basis for making both qualitative and quantitative assessments.

Three policy scenarios are evaluated. The benchmark is the *uniform emission tax*. The alternatives are an auction market and a *non-atomistic auction market*. The latter refers to an auction market with a central trading company in which each firm could hold shares. The simulation results show that the simple tradable emission permit scheme can be the most effective arrangement when the number of participants is small. However, as the number of participants increases, its performance declines dramatically and becomes

[16] In economics, people usually use the so-called *augmented GA*, i.e., the canonical GA plus the *election operator* (see Arifovic's chapter for more details), but the election operator is not used in AGEDASI TOF.

inferior to that of the uniform tax scheme and the non-atomistic auction market. The results imply that tax can be a more effective measure than the simple market mechanism to avoid an over-polluted environment. Nonetheless, the non-atomistic market mechanism still works.

A technical issue involved in this chapter is the fitness evaluation of a *multi-population genetic algorithm* (**MGA**).[17] Each firm in Bell and Beare's auction market is represented by a set of 20 strings corresponding to a range of potential trading strategy. Since the performance of each trading strategy depends on the strategies used by other firms, if there are n firms, we need at least 20^n rounds of trades to have a complete evaluation. This is not only computationally demanding, but also unrealistic. The solution proposed by the authors is to randomly take a sample from the population with a fixed sample size, k. The fitness evaluation is then made with this sample. When the number of firms, n, increases, the sample size becomes relatively smaller. The effect of sample size on fitness evaluation and the resultant impact on market performance have been widely studied in the literature ([46], [61], [21], [45]). These findings may help explain the negative impact of the number of firms on market efficiency as observed in the chapter.

Bell et al's paper shows that adaptive learning within a free competitive market does not necessarily lead to the most efficient results. In some cases, designing a non-atomistic market mechanism can be crucial to market efficiency. This chapter stresses the link between learning and institution arrangements. Our understanding of the effect of learning, or of institutional arrangements, for that matter, can be partial if they are studied separately.

Masayuki Ishinishi, Hiroshi Sato and Akira Namatame's **Cooperative Computation with Market Mechanism** provides another demonstration. The chapter discusses how cooperative solution can be obtained through the decentralized computation of economic agents with self-interest seeking behavior. An example used by the authors is a market composed of n firms who are competing with each other by selling a similar but not exactly the same product. Each firm faces a unique demand curve of its own product. Nonetheless, this demand curve is not entirely independent of the quantity supplied by other firms. In other words, demand curves are inter-related and hence are not necessarily fixed.

Firms in this interdependent market structure determines their quantities supplied via a simple adaptive learning mechanism under two institutional arrangements, namely, pure free competition and free competition with external influence. The external influence considered here is similar to a central company, discussed in the previous chapter, whose purpose is to maximize its own profits by collecting money from each firm. This central company, called a government in this chapter, also uses a simple learning algorithm to charge a non-uniform tax rate to firms. By redistributing profits to each firm, it can be shown that this simple learning algorithm, taken by both firms and the

[17] For a related discussion, see Vriend's chapter.

government, may drive the joint profits to a level higher than that of pure competition. This result particular applies to the case when the number of firms is increasing and the interdependent relations among them are *symmetric.* So, these two chapters in succession signifies the role of an (adaptive) institution arrangements in coordination with decentralized adaptive learning agents.

The simple learning algorithms considered here could have been genetic algorithms, as they are typically used in such kinds of application, but the authors did not make such a choice. Instead, they took something very similar to the Robbins-Monro algorithm ([59]). And that is one of the reasons why we include this paper in the volume. It draws readers' attention to the fact that EC is not the only model of adaptive learning, nor the obvious choice in some cases ([33]). To avoid an addiction to EC, it is important to justify the use of EC by reflecting upon its relative advantages to the alternatives.

The materials presented up to this point concern only the *macro* aspect of ACE models, such as the persistent fluctuation of the asset price and the aggregate level of pollution, but ACE modeling can be rich in microbehavior as well. Leigh Tesfatsion's "**Hysteresis in an Evolutionary Labor Market with Adaptive Search**" is a case in point. The specific issue addressed in the chapter is the famous *excess heterogeneity* problem in labor economics. Economists are generally puzzled by the empirical fact that *observationally equivalent work suppliers and employers have markedly different earning and employment histories* ([1]). In other words, factors such as schooling, age and experience can account for heterogeneity in wages only to a very limited extent. It implies that some *idiosyncratic factors* can be crucial to the wage decision. Conventional economics, however, can say very little about idiosyncrasies, except assuming that they are stochastic. This chapter centers on idiosyncrasies. Using an agent-based computational labor market, it shows that temporary shocks in the form of *idiosyncratic worksite interactions* can propagate up into sustained differences in earnings and employment histories for observationally equivalent workers and employers.

Essential elements of ACE modeling are explicitly depicted in the chapter. First, *object-oriented programming.* In her introductory article at the 2001 JEDC special issue on agent-based computational economics, Tesfatsion stated:

> what is new about ACE is its exploitation of powerful new computational tools, most notably *object-oriented programming.* These tools permit ACE researchers to extend previous work on *economic self-organization* and *evolution* in four key ways. (Ibid. p. 282, Italics added).

By using "**class agent**", her table 1 gives a good starting point to see the relation between agent-based computational economics and object-oriented programming. Second, *discovering and defining new concepts.* ACE models can generate important observations unknown to standard theoretical or empirical analysis. For example, terms such as *aggressive*, *persistently inactive* and *persistently nice* workers (employers) defined in her Section 3 are all new to labor economics. These concepts are very important to the ACE model and are relevant to efficient wage theories. Last, but not least, *source code.* Source code is equivalent to the *proof* seen in an analytical paper. Provision of source code facilitates replication.[18] The author gives a detailed account of the availability of the **Trade Network Game (TNG)**, Version 105b.

Genetic algorithms are involved because, in her agent-based computational market, worksite interactions between matched work suppliers and employers are modeled as a *two-person iterated prisoner's dilemma game*, and there is a long history of using GAs to study the 2-person IPD game. The 2-person IPD game is relevant because the relation between workers and employers described by the *efficient wage theories* is very similar to the cooperation or defection relation in the 2-person IPD game. Each worksite strategy is represented as a *finite state automaton* (**FSA**) with a fixed starting state, and each FSA is coded a bit string.[19] A genetic algroithm is then applied to each of the distinct trader type subpopulation, namely, buyers and sellers.

Unlike most applications which use the *augmented GA*, this paper applies the steady-state GA.[20] The difference between these two versions lies in *disruption avoidance.* In the augmented GA, disruption avoidance, via the election operator, is taken only at the end of the genetic operation step. It can guarantee that offspring are at least no worse than parents, but does not warrant that the best individuals will be retained. In the steady-state GA, disruption avoidance is enforced through the elitist operator right at the selection step. The elitist operator retains some number of the best individuals in each generation.[21] Economic studies comparing the performance of these two version of GAs are limited. [53] is the only paper that shows some advantages of the elitist operator over the election operator.

While the 2-person IPD game is a major part of her ACE labor market, some *complications* added to it make this study deviate from other familiar applications of GAs to 2-person IPD games. These complications, including a *deferred choice and refusal (DCR) mechanism* to determine trade part-

[18] See also the Chen et al's chapter for a similar discussion.

[19] How to effectively represent and encode the players' strategies in the IPD game is an issue drawing lots of attention. The coding scheme taken by Tesfatsion is very similar to the one initiated by John Miller in his Ph.D. thesis ([50]), which should be distinguished from the Axelrod's scheme. For other coding schemes, see also [67].

[20] A minor point which makes this application different from other economic applications is that the author uses two-point crossover rather than the most popular single-point crossover.

[21] The faction of new individuals in each generation has been called the *generation gap*.

ners, a *simple learning algorithm* to update expected utility assessment, and *job search costs* incurred by the workers make the simulation results more complex and *path-dependent.* In particular, when *job capacity* characterizing the economic fundamentals is tight, or loose for that matter, the substantial *network hysteresis* tends to arise that supports persistently heterogeneous earnings levels across employed work suppliers and across non-vacant employers even when each matched work supplier and employer pair expresses the same type of worksite behavior.

In contrast to the six ACE papers reviewed above which address learning in the context of *markets*, Francesco Luna's "**Computable Learning, Neural Networks and Institutions**" focus on the role of institutions in the learning process of economic agents. The simulation results substantiate the positive role played by institutions, but at the same time indicate that institutions can be a *nuisance* to the learning process.

This chapter is the only chapter in this volume about using *genetic algorithms* to *evolve a population of artificial neural networks* .[22] However, unlike many existing applications of *evolutionary artificial neural networks* (**EANNs**), which are basically engineering (optimization) oriented, this paper applies *genetic-algorithm neural nets* (**GANNs**) in a social-economic context, and in an innovative way. One can appreciate this novel applications from two different perspectives, one in terms of (optimization) perspective, and the other agent-based computational economics.

From the *engineering (optimization)* perspective, it can be considered a potential contribution to the literature on *evolutionary artificial neural nets.* According to [66], there are three different levels to evolve ANNs, namely, the connection weights, the architecture and the learning rules. Most studies in the literature concentrate on the first two categories, but this chapter puts emphasis on the last. More precisely, it employs a variation of the perceptron algorithm known as the *pocket algorithm*, and considers whether it would be advantageous to use the pocket algorithm, as a *learning rule*, to learn the weights of ANNs. The standard genetic algorithm is then used to evolve the learning rules of ANNs. The testbed includes the familiar XOR Boolean function and a *structural change* from the XOR function to $f(A, B) = A \vee (\overline{A} \wedge \overline{B})$. The author attempts to show that while the pocket algorithm is good for the first simulation, i.e., the environment without a structural change, it takes much longer to adapt to the new environment when there is a structural change.

The essence of this chapter is the analogy made between *institutions* and *neural nets* on the one hand, and *memory* (as a specific function of institutions) and the *pocket algorithm* (as a specific part of neural nets) on the other. The use of this metaphor provides an economic "translation" of the

22 While Yang's paper also deals with a population of artificial neural networks, each of the networks is treated *individually* and is updated via *conjugate gradient algorithms.*

well-known *no-free-lunch theorem* in the machine learning literature ([65]). At the end of this chapter, the author further pursues the significance of *mutation*, which is another hot topic in evolutionary computation. But, again, the author addresses its significance in the context of social learning, in lieu of optimization.

1.3.4 Technical Issues

Given the fact that there are so many user-supplied set-ups to be determined before conducting an experiment, one may wonder if variations of these set-ups would have major effect on the simulation results.[23] Even though a systematic study of the significance of these set-ups is beyond the scope of the book, this volume does offer two chapters to address some non-trivial aspects of these set-ups.

One is Nicholas Vriend's **On Two Types of GA-Learning**. This paper concerns the two styles of architecture commonly used in ACE models, namely, the *single-population GA* (**SGA**) and the *multi-population GA* (**MGA**). The distinction between these two types of architecture was made clearly in [36].

> Depending upon the model, an agent may be represented by a single string, or it may consist of a set of strings corresponding to a range of potential behaviors. For example, a string that determines an oligopolist's production decision could either represent a single firm operating in a population of other firms, or it could represent one of many possible decision rules for a given firm. (Ibid., p.367.)

[4] is the first study that compares different types of architecture in agent-based economic modeling. The author found that in converging to the rational expectations equilibrium (REE), the SGA and MGA behave the same. They both converged to the REE when the election operator was used, and failed to converge otherwise. Since no subsequent studies were attempted to verify the robustness of the results, to run GAs within either design turns out to be an arbitrary decision.

Vriend's paper is the first one to give a thorough analysis of the consequences of such a choice. To Vriend, the difference between the SGA and MGA is more than just a matter of *coding*. They can be different in the interaction *level* at which learning is modeled. For the MGA, learning is modeled at the *individual* level, i.e., agents learn exclusively on the basis of their own experience, whereas for the MGA, learning is modeled at the *population* level, i.e., agents learn from other agents' experiences as well. It is due to this distinction that the SGA is also called *social learning* and MGA *individual learning*. According to Vriend, there is an essential difference between individual and social learning, and the underlying cause for this is the so-called

[23] A long discussion of these user-supplied set-ups can be found in [25].

spite effect. The spite effect may occur in a social learning GA (SGA), but can never occur in an individual learning GA (MGA). To see how the spite effect can influence the outcome of the evolutionary process, Vriend uses the two different GAs to simulate the learning process of an oligopoly game. The simulation results show that while the individual learning GA moves close to the Cournot-Nash output level, the social learning GA converges to the competitive Walrasian output level. Contrary to what was observed in [4], the choice of the SGA and MGA can generally lead to non-trivial differences.

Vriend's finding pushes us to think harder of architecture selection. If the choice does matter, should one use the SGA, MGA, both of them or some others? This issue is not that difficult to solve because different types of architecture may have different implicit assumptions for learning, and some assumptions are not appropriate for a specific application. For example, the imitation process under the SGA obviously requires strategies to be observable; otherwise, they cannot be imitated. Unfortunately, in reality, strategies are usually secrets, which defy observations.[24] Therefore, the choice of architecture is to some extent empirically solvable.

The chapter by David Fogel, Kumar Chellapilla and Peter Angeline, "**Evolutionary Computation and Economic Models: Sensitivity and Unintended Consequences**, presents an even tougher choice, i.e, *representations* of learning agents. Fogel et al argue that the class of models used to represent learning agents can have an important effect on the resulting dynamics. To validate this argument, they provide two case studies. One is the the well-known *El Farol problem*, and the other the IPD game. For both cases, they show that by making slight adjustments to the representation of learning agents, wholly different behavior emerged.

The El Farol problem was originally proposed and studied by [11]. The problem concerns the attendance at the bar, El Farol, in Sante Fe. Agents' willingness to visit the bar at a specific night depends on their expectations of the attendance at that night. An agent will go to the bar if her expected attendance is less than the capacity of the bar; otherwise, she will not go. Reference [11] showed that the time series of attendance levels seems to always fluctuate around an average level of the capacity of the bar. However, agents in [11] reason with *a fixed set of models*, deterministically iterated over time. Discovering new models is out of the question in this set-up. Fogel et al. replace this fixed set of rules with a class of *auto-regressive* (**AR**) models. Furthermore, the number of lag terms and the respective coefficients are revised and renewed via *evolutionary programming* (**EP**).[25] The introduction of EP to the system of AR agents has a marked impact on the observed behav-

[24] See [30] for more detailed discussion of the observability assumption.

[25] From a technical viewpoint, this chapter is the only one in the book concerning the use of EP, and is one of the three applications of EANNs to ACE modeling. The other two are Yang's and Luna's chapters.

ior: the overall result is one of large oscillations rather than mild fluctuations around the capacity.

For the 2-person IPD games, the authors first review the simulations results based on the two most popular representations used in the game, namely, Axelrod's representation and the finite state automata. The main result that players tend to converge toward *mutual cooperation* is not sensitive to either representations. However, here, the authors consider a continuous version of the IPD game. Instead of defecting or cooperating, players can now choose a degree of defection or cooperation. To represent the continuum of strategies, they replace the finite state automata *with artificial neural nets*, and use EP to evolve these nets. Surprisingly, after these slight modifications, mutual (complete) cooperation is no longer the dominant pattern in the continuous IPD game. *Any cooperative behavior that did arise did not tend toward complete cooperation*, and the degree of mutual cooperation evolved is sensitive to the number of hidden nodes and the population size. In a specific case, *there is no tendency for cooperative behavior to emerge when only 2 hidden nodes are used.*

The striking result of the experiments with the El Farol problem and the iterated prisoner's dilemma is that small changes to simple models that adopt evolutionary dynamics can engender radically different emergent properties. The authors' concluding remakrs give us pause for thought: *that a certain behavior or property could not emerge from a complex system is an untenable conclusion because the unobserved behavior may indeed arise given only a minor modification to the model.*

1.4 Probing Econometrics and Financial Engineering: 1992

Five years after its first application to *games* and four years after its first contact with *the ACE models*, evolutionary computation ventured into another two territories: *econometrics* and *financial engineering.*

1.4.1 Econometrics

The first application of EC to econometrics is [42]. As Koza stated

> An important problem in economics is finding the mathematical relationship between the empirically observed variables measuring a system. In many conventional modeling techniques, one necessarily begins by selecting the size and shape of the model. After making this choice, one usually then tries to find the values of certain coefficients required by the particular model so as to achieve the best fit between the observed data and the model. But, in many cases, *the most important issue is the size and shape of the model itself.* (Ibid., p.57. Italics added.)

Econometricians offer no general solution to the determination of size and shape (the functional form). But, for Koza, finding the functional form of the model can be viewed as *searching a space of possible computer programs* for the particular computer program which produces the desired output for given inputs. Koza employed GP to elicit a very fundamental economic law, namely, the *quantity theory of money* or *exchange equation.*

However, unlike the conventional statistical foundation of econometrics, a well-established doctrine for the use of EC in econometric analysis is yet to be found. While some recent studies based on the *function approximation approach* may provide a unified principle in the future ([52]), for most practical applications tinkering and fiddling are still the order of the day. Given this background, the chapter by George Szpiro, **Tinkering with Genetic Algorithms: Forecasting and Data Mining in Finance and Economics**, gives a thorough discussion of a list of rules of thumb which can help boost the GP performance in econometric applications. Some of these rules are well noticed by the EC people, including using the time-variant mutation rate, using the elitist operator, reconciling the search space with population size, effectively reducing search space, and using least squares to fine-tune the values of constant coefficients. But some are not. Among them are the residual method, cross-breeding, and outliers avoidance. A number of examples are provided in the chapter to support the use of these rules of thumb.

1.4.2 Trading Strategies

In 1992 and the years to follow, partially because of the seminal work [19], which showed the significant prediction potential of the simple trading strategies, applications of evolutionary computation were extended further to the study of trading strategies. References [15] and [2] initialized this line of research. Reference [16] gave finance people the first systematic introductory book of genetic algorithms. Papers on this subject, however, did not find their way into prestigious journals until the late 1990s ([51], [3]).

There are four chapters on trading strategies included in this book. These chapters not only help us to see the enduring influence of the past, but also highlight some new directions to pursue in the future. Let us start the review of these four chapters by asking *why evolutionary computation is relevant to the study of trading strategies.* The answer is succinctly given in the chapter by Robert Pereira, "**Forecasting Ability But No Profitability: An Empirical Evaluation of Genetic Algorithm-Optimised Technical Trading Rules**". It starts with a review of two most basic classes of trading rules, namely, the *moving-average* (**MA**) rules and the *order-statistics* (**OS**) rules. Pereira works with a generalized version of the two classes and shows how these two classes of rules can be parameterized with three to four parameters. Conventional studies examining trading rules choose the corresponding parameter values *arbitrarily.* Following the base tone set by [16], Pereira reiterates that genetic algorithms can be used to search for the optimal parameter values for the generalized MA and OS trading rules.

The second question concerns *how to evaluate* the trading rules found by the EC techniques. Early applications were somewhat casual in this aspect, whereas recent advancements have laid out a set of elements for defining a rigorous application. These elements include the *profitability measure*, *benchmark*, and *statistical analysis*. The chapter has rich discussion on all these elements. For the profitability measure, the author acknowledges the difficulties in properly accounting for the *transaction costs*. In addition to the commonly used Sharpe ratio, the *break-even transaction cost* is reported for reference. For the benchmark, he introduces a *risk-adjusted buy and hold strategy* in place of the simple buy and hold strategy to make fair comparisons. For statistical analysis, he applies the *Cumby-Modest test* as well as the t test to examine the forecasting ability of trading signals. Moreover, the *bootstrap* method is used to test the significance of both the predictive ability and the profitability of technical trading rules.

This rigorous procedure enables us to take the third question seriously, i.e., *the effectiveness of trading strategies*. Using the daily closing All Ordinaries Accumulation index from the Australia stock market, Pereira found that the optimal rules outperform the benchmark (the risk-adjusted buy and hold strategy). The rules display some evidence of forecasting ability and profitability over the entire test period. However, this empirical evidence is subject to two qualifications. First, an examination of the results for the sub-periods indicates that the excess returns decline over time and are negative during the last couple of years. Second, the existence of thinly traded shares in the index can introduce a *non-synchronous trading bias*, and if the *non-synchronous trading bias* is taken into account, the rules display little, if any, evidence of profitability. While the first disadvantage may be solved by adding a re-learning scheme ([43]), there is no simple solution for the second one.

As mentioned in Pereira's chapter, *defining suitable profitability measure* is not that straightforward due to different notions of *risk* and *transaction costs*. The issue can be more complicated because during genetic search some profitability measure (fitness functions) may embody potentially *harmful biases*. Nevertheless, the effects of profitability measure on *genetic search* has not been well documented, and this is what the next chapter tries to make up for. In Chapter 17, **Evolutionary Induction of Trading Models**, Siddhartha Bhattacharyya and Kumar Mehta discuss several key aspects of *biased fitness evaluation*. One of them is the problematic bias caused by *compound returns*. Although continuously compounded returns are usually taken as a measure of ultimate trading model performance, they embody a selection bias towards *more active* trading rules, which may intensify the overfitting problem. To avoid such a bias, they propose profitability measure based on simple aggregate returns instead of compound returns.

The chapter also examines two technical aspects of trading applications of GAs. The first one concerns the *representation* of trading rules, particularly, the interpretation of the *action* of a trading rule. In most applications, a rule's condition evaluated as TRUE is considered a signal to buy, while a FALSE indicating a signal to sell. This interpretation takes the out-of-market position by default. But, what about taking the in-market position instead. Are they equivalent? This issue may be more complex than it appears. The seemingly logical equivalence between them can be blurred by the genetic search processes over a complicated landscape. In fact, in their empirical study with S&P 500, the authors show that while these two approaches are found to yield similar performance, the solutions can have dissimilar characteristics. Rules learnt by taking the out-of-market position by default are found to be of *greater complexity*, but with *lower specificity*.

The second issue concerns the *time horizons* of training samples. While conventional statistical analysis based on asymptotic theory suggests a large sample, financial time series rarely satisfy the required stationarity assumption. In this case, data over too long a length of time considered for training can confound the learning process with multiple, possibly overlapping or even conflicting patterns being present in the data. Given the dilemma, a medium sample is what one should be looking for. This point is well taken by the second experiment of the chapter.

Raymond Tsang and Paul Lajbcygier's chapter entitled **Optimizing Technical Trading Strategies with Split Search Genetic Algorithms** offers a welcome alternative to the often criticized standard genetic algorithms. The novel genetic algorithms they propose are *split search genetic algorithms* (**SSGAs**), a version of *distributed genetic algorithm* (**DGAs**) also called the *island model* of GAs. In DGAs, the population of chromosomes is partitioned into a number of isolated *subpopulations*, each one of which evolves *independently*, albeit with the same fitness function. A *migration operator* (*invasion operator*) periodically swaps individuals among subpopulations. One of the first works on this line of research is [56].

By adopting the multi-populations approach, one can differentiate subpopulations with different control parameters and have a more extensive and effective search. Tsang and Lajbcygier, for instance, use the *mutation rate* to differentiate two subpopulations, one with a high mutation rate and the other low. By this differentiation, the authors draw our attention to the role of the *mutation rate* in genetic algorithms, and even more importantly, to a possibly more effective way of implementing GAs.[26] The simulation results evidence the superiority of SSGAs over standard GAs, though the difference is not statistically significant.

The chapter by Tsang et al is also unique in its application domain. Eight futures, including commodities, foreign exchanges, bonds, and stocks

[26] The mutation rate was found important in some other economic applications ([8], [46], [53], [54], [63]).

are involved. In a similar vein to Pereira's chapter, GAs are used to optimize the parameterized trading rules, in this case, the filter rule and the moving filter rule. The performance of GAs is compared with that of the benchmark, the buy and hold strategy.

The works described in the previous three chapters demonstrate how EC can be directly applied to the development of trading strategies. Alternatively, one can first apply EC to financial forecasting, and then base the trading decisions on the resultant forecasts. The chapter by Mahmoud Koboudan, "**GP Forecasts of Stock Prices for Profitable Trading**", is a case in point. This chapter is composed of two parts. The first part centers around the question of *whether GP can in fact evolve equations that predict stock prices.* To answer this question, the author proposes a *predictability test* based on an *eta* statistic, and has most interesting findings: At the 5% level of significance, prices are at least 0.620 predictable, and returns are at most 0.280 predictable. This implies that return series are less predictable than price series.

The second part of the chapter focuses on the development of *single-day-trading-strategy* (**SDTS**) based on GP forecasts. An SDTS is a strategy whereby a trader buys a stock at a price close to the *daily low* and sells at a price close to the *daily high*, regardless of the order in which trading occurs. The main task of GP is to forecast the daily low and high. One unique feature of the SDTS is to require the investor to close every position by the end of the trading day. This is mainly because trading is based on a one-day forecast. Holding no position overnight helps avoid greater gambling losses in subsequent days for which no forecast is available to base a decision on. For an empirical illustration, stocks of Chase Manhattan and General Motors from NYSE and those of Dell Computer and MCI-Worldcom from NASDAQ are selected to apply the SDTS here.

In the appendix of the chapter, the author presents some best-fit equations for daily highs and lows. They are all very complicated. Perhaps this is a good place to mention a controversial issue in genetic programming. It is often argued that GP is unreliable because it tends to produce over-complex and insensible solutions.[27] But, as Kaboudan's paper shows, GP can be superior to a random walk in forecasting. For all four stocks, returns from trading based on the GP forecasts are higher than those based on a random walk.

1.4.3 Option Pricing

In addition to trading strategies, another active research area in financial engineering is *option pricing.* Evolutionary computation is not quiet on this area either. Early applications can be found in [26], [27], [31], [28], and [40]. By means of EC a data-driven approach is proposed for option pricing as opposed to the conventional analytical approach, e.g., the well-known Black-Scholes

[27] For a lengthy discussion of this issue, one is referred to [25].

model. The comparative advantages of the data-driven approach are determined by the tightness of the assumptions on which the analytical approach is based, and the historical data available for data mining. Early applications in general and the next two chapters in particular are motivated by these two determinants.

The chapter "**Option Pricing via Genetic Programming**" by Nemmara Chidambarab, Joaquin Triqueros and Chi-Wen Jevons Lee provides rich empirical evidence in support of GP in option pricing. They first examine how well GP can outperform the Black-Scholes model under controlled condition, and then employ GP to price real world options data, including S&P index options and five equity options. In the simulation study, they show that GP formulas beat the Black-Scholes model in ten out of ten cases where the underlying stock prices are generated using a *Possion jump-diffusion process* .[28] They work almost as well in pricing S&P Index options with genetic programs beating the Black-Scholes model in nine out of ten cases. For equity options, genetic programs beat or match the Black-Scholes model for four of the five stocks considered, though the results are not as strong as the case of S&P Index options. Also included in their performance comparison are neural networks and linear regression models. The performance evaluation is made based on the criterion *pricing errors*, in particular the *mean absolute pricing errors.*[29]

This chapter also addresses two important technical issues generally shared by EC. The first one is the *selection scheme.* As mentioned earlier, the selection scheme has received far less attention than it deserves in economic and financial applications. This chapter studies the performance of six selection schemes. Apart from roulette-wheel selection, tournament selection, steady-state selection, random selection (no selection), the authors consider a combination of random selection and roulette-wheel selection, called the *fitness-overselection selection.* In the literature, tournament selection was shown superior to roulette-wheel selection on many occasions, and the chapter also confirms this result; furthermore, it suggests that *tournament size* can play a role as well. Random selection performs worst, as expected. What is a little surprising is that, among the six schemes, the fitness-overselection scheme gives the lowest pricing errors.

---

[28] The authors pick the Possion jump-diffusion process because this process was empirically more plausible than the generally held normality assumption taken in the Black-Scholes model. This change, therefore, provides an opportunity to test the effectiveness of GP over the Black-Scholes model when one of the underlying assumptions is violated.

[29] By [37], a better criterion for performance comparison should be *hedging errors.*. While the authors also tested the hedging effectiveness of the GP model by constructing a *hedge portfolio* of the option, stock, and bond, detailed results are not available in this chapter.

The second issue is *the choice of control parameters.*[30] Instead of setting the parameter values arbitrarily, the authors follow *a two step process* to determine an optimal set of them. In the training and validation step, they test different parameter values. The parameters values that give the best results are then used in the next stage to develop GP formulas. The use of *a validation step* to solve the problem of parameter setting has become a standard procedure in EC, and is expected to be one to follow in economic and financial applications as well.

Another application related to option pricing is *implied volatility* . Measuring financial volatility is one of the toughest issues in current financial economics. Among the solutions, implied volatility, i.e., the volatility implied (inversely solved) from the option pricing model, is one extensively applied to estimating volatility, where one basically solves the inverse of the Black-Scholes function. Here, analytical approximations with different degrees of accuracy are available. However, difficulties arise when this idea is extended to *American put options*, where analytical exact pricing formulas have yet to be derived, and can only be priced via numerical procedures, such as the lattice approach and the finite-difference method. Unfortunately, solving implied volatility via numerical procedures can be very time consuming when they are used in connection with common numerical methods, such as the Newton–Raphson method. The chapter by Christian Keber, **Evolutionary Computation in Option Pricing: Determining Implied Volatilities Based on American Put Options**, shows how genetic programming can be used to derive accurate analytical approximations for determining implied volatility based on American put options.[31]

Keber first uses a randomly generated training sample of 1,000 American put options on non–dividend paying stocks to genetically derive an analytical approximation with GP. To ensure the accuracy of the assessment, the genetically derived approximation is applied to two data sets borrowed from the literature and two huge data sets artificially generated. To make a comparison, implied volatilities based on the Black-Scholes model are used as a benchmark. It is found that the genetically derived formula delivers better results, i.e., smaller deviations from true implied volatilities, than the benchmark. Moreover, the genetically derived formula provides accurate approximations for a wide range of parameter values of the underlying option and nearly perfect approximations for parameter values following specific assumptions which are consistent with current practice.

The promising results of the studies in these two chapters indicate that genetic programming can be a flexible instrument to derive valuation models for a broad variety of derivatives and associated stochastic price processes of the underlying asset. As Keber suggests, *exotic options* and derivatives where

---

30 See also Section 3.4, "Technical Issues".

31 Instead of estimating implied volatility, [24] proposed a direct approach to measure and estimate the financial volatility with genetic programming.

the stochastic price process of the underlying asset is only given implicitly should be a particular interesting application of GP.

1.5 Conclusions

While the goal of editing a single volume like this one is to give a comprehensive coverage of the field, there is a fundamental limit which one cannot transcend, and that is the size of the book. Normally speaking, 20 chapters already comes to the maximum capacity of a single volume. Anything beyond has to be reluctantly dropped out. Therefore, to compensate for what might be missing, we add a bibliography as a concluding chapter, which is composed of three parts: publications, useful web sites, and computer software. Complimentary guidance is also provided to enable readers to take full advantage of this bibliography.

References

1. Abowd J. M., Kramarz F., Margolis D. N. (1999) High Wage Workers and High Wage Firms. Econometrica **67**, 251–333
2. Allen F., Karajalainen F. (1993) Using Genetic Algorithms to Find Technical Trading Rules. Rodney L. White Center for Financial Research, The Wharton School, Technical Report, 20–93
3. Allen F., Karjalainen R. (1999) Using Genetic Algorithms to Find Technical Trading Rules. Journal of Financial Economics **51(2)**, 245–271
4. Arifovic J. (1994) Genetic Algorithms Learning and the Cobweb Model. Journal of Economic Dynamics and Control **18(1)**, 3–28
5. Arifovic J. (1995) Genetic Algorithms and Inflationary Economies. Journal of Monetary Economics **36(1)**, 219–243
6. Arifovic J. (1996) The Behavior of the Exchange Rate in the Genetic Algorithm and Experimental Economics. Journal of Political Economy **104**, 510–541
7. Arifovic J. (2001) Evolutionary Dynamics of Currency Substitution. Journal of Economic Dynamics and Control **25**, 395–417
8. Arifovic J., Eaton B. C. (1995) Coordination via Genetic Learning. Computational Economics **8(3)**, 181–203
9. Arifovic J., Gencay R. (2000) Statistical Properties of Genetic Learning in a Model of Exchange Rate. Journal of Economic Dynamics and Control **24**, 981–1005
10. Arthur W. B. (1992) On Learning and Adaptation in the Economy. Sante FI Economics Research Program Working Paper 92-07-038.
11. Arthur W. B. (1994) Inductive Reasoning and Bounded Rationality. American Economic Association Papers Proceedings **84**, 406–411
12. Arthur W. B., Holland J., LeBaron B., Palmer R., Tayler P. (1997) Asset Pricing under Endogenous Expectations in an Artificial Stock Market. In: Arthur W. B., Durlauf S., Lane D. (Eds.), The Economy as an Evolving Complex System II. Addison-Wesley, Reading, MA, 15–44

13. Axelrod R. (1987) The Evolution of Strategies in the Iterated Prisoner's Dilemma. In: Davis L. (Ed.), Genetic Algorithms and Simulated Annealing. Pittman, London, 32–41
14. Basu N., Pryor R. J. (1997) Growing a Market Economy. Sandia Report SAND-97-2093.
15. Bauer R. J. Jr., Liepins G. E. (1992) Genetic Algorithms and Computerized Trading Strategies. In: O'leary D. E., Watkins, R. R. (Eds.), Expert Systems in Finance, North Holland.
16. Bauer R. J. Jr. (1994) Genetic Algorithms and Investment Strategies. Wiley, New York.
17. Beltratti A., Margarita S., Terna P. (1996) Neural Networks for Economic and Financial Modelling. Thomson.
18. Bhattacharyya S., Pictet O., Zumbach G. (1998) Representational Semantics for Genetic Programming based Learning in High-Frequency Financial Data. In: Koza J., Banzhaf W., Chellapilla K., Deb K., Dorigo M., Foegl D., Garson N., Goldberg D., Iba H., Riolo R. (Eds.), Proceedings of the Third Annual Genetic Programming Conference. Morgan Kaufmann Publishers, San Francisco, 32–37
19. Brock W.A., Lakonishok J., LeBaron B. (1992) Simple Technical Trading Rules and the Stochastic Properties of Stock Returns. Journal of Finance **47**, 1731–1764
20. Bullard J., Duffy J. (1999) Using Genetic Algorithms to Model the Evolution of Heterogenous Beliefs. Computational Economics **13(1)**, 41–60
21. Cacho O., Simmons P. (1999) A Genetic Algorithm Approach to Farm Investment. Australian Journal of Agricultural and Resource Economics **43(3)**, 305–322
22. Casti J. (1997) Would-be Worlds: How Simulation is Changing the Frontier of Science. John Wiley & Sons.
23. Chan N. T., LeBaron B., Lo A. W., Poggio T. (1998) Information Dissemination and Aggregation in Asset Markets with Simple Intelligent Traders. Working Paper, MIT.
24. Chen S. -H. (1998) Modeling Volatility with Genetic Programming: A First Report. Neural Network World **8(2)**, 181–190
25. Chen S. -H. (2001) Fundamental Issues in the Use of Genetic Programming in Agent-based Computational Economics. In Namatame A. (Ed.), Proceedings of the First International Workshop on Agent-based Approaches in Economic and Social Complex Systems, 175–185
26. Chen S. -H., Lee. W. -C. (1997) Option Pricing with Genetic Algorithms: The Case of European-style Options. In: T. Back (Ed.), Proceedings of the Seventh International Conference on Genetic Algorithms. Morgan Kaufmann Publishers, San Francisco, CA, 704–711
27. Chen S. -H., Yeh C. -H., Lee W. -C. (1998) Option Pricing with Genetic Programming. In: Koza J., Banzhaf W., Chellapilla K., Deb K., Dorigo M., Foegl D., Garson M., Goldberg D., Iba H., Riolo R. (Eds.), Proceedings of the Third Annual Genetic Programming Conference. Morgan Kaufmann Publishers, San Francisco, CA, 32–37
28. Chen S. -H., Lee W. -C., Yeh C. -H. (1999) Hedging Derivative Securities with Genetic Programming. International Journal of Intelligent Systems in Accounting, Finance and Management **8(4)**, 237–251

29. Chen S. -H., Lin W. -Y., Tsao C. -Y. (1999) Genetic Algorithms, Trading Strategies and Stochastic Processes: Some New Evidence from Monte Carlo Simulations. In: Banzhaf W., Daida J., Eiben A. E., Garzon M. H., Honavar V., Jakiela M., Smith R. E. (Eds.), GECCO-99: Proceedings of the Genetic and Evolutionary Computation Conference. Morgan Kaufmann, 114–121
30. Chen S. -H., Yeh C. -H. (2001) Evolving Traders and the Business School with Genetic Programming: A New Architecture of the Agent-based Artificial Stock Market. Journal of Economic Dynamics and Control **25**, 363–393
31. Chidambaran N., Lee C., Trigueros J. (1998) An Adaptive Evolutionary Approach to Option Pricing via Genetic Programming. In: Koza J., Banzhaf W., Chellapilla K., Deb K., Dorigo M., Fogel D., Garzon M., Goldberg D., Iba H., Riolo R. (Eds.), Genetic Programming 1998: Proceedings of the Third Annual Conference. Morgan Kaufmann, San Francisco, CA, 187–192
32. Dawid H. (1996) Adaptive Learning by Genetic Algorithms. Springer, Berlin, Heidelberg, New York.
33. Duffy J. (2001) Learning to Speculate: Experiments with Artificial and Real Agents. Journal of Economic Dynamics and Control **25**, 295–319
34. Epstein J. M., Axtell R. (1996) Growing Artificial Societies: Social Science from the Bottom Up. MIT Press.
35. Green J. (1977) The Non-existence of Informational Equilibria. Review of Economic Studies **44**, 451–463
36. Holland J., Miller J. (1991) Artificial Adaptive Agents in Economic Theory. American Economic Review **81(2)**, 365–370
37. Hutchinson J., Lo A., Poggio T. (1994) A Nonparametric Approach to the Pricing and Hedging of Derivative Securities via Learning Networks. Journal of Finance **49 June**, 851–889
38. Izumi K., Okatsu T. (1996) An Artificial Market Analysis of Exchange Rate Dynamics. In: Fogel L. J., Angeline P. J., Back T. (Eds.), Evolutionary Programming V. MIT Press, 27–36
39. Kareken J., Wallace N. (1981) On the Indeterminacy of Equilibrium Exchange Rates. Quarterly Journal of Economics **96**, 207–222
40. Keber C. (2000) Option Valuation with the Genetic Programming Approach. In: Abu-Mostafa Y. S., LeBaron B., Lo A. W., Weigend A. S. (Eds.), Computational Finance – Proceedings of the Sixth International Conference. MIT Press, Cambridge, MA, 689–703
41. Kirman A. P., Vriend N. (2001) Evolving Market Structure: An ACE Model of Price Dispersion and Loyalty. Journal of Economic Dynamics and Control **25(3-4)**, 459–502
42. Koza J. (1992) A Genetic Approach to Econometric Modelling. In: Bourgine P., Walliser B. (Eds.), Economics and Cognitive Science. Pergamon Press, 57–75
43. Lanquillon C. (1999) Dynamic Aspects in Neural Classification. Intelligent Systems in Accounting, Finance and Managemnet **8(4)**, 281–296
44. LeBaron B., Arthur W. B., Palmer R. (1999) Time Series Properties of an Artificial Stock Market. Journal of Economic Dynamics and Control **23**, 1487–1516
45. LeBaron B. (2001) Evolution and Time Horizons in an Agent Based Stock Market. Forthcoming in Macroeconomic Dynamics.
46. Lettau M. (1997) Explaining the Facts with Adaptive Agents: The Case of Mutual Fund Flows. Journal of Economic Dynamics and Control **21**, 1117–1148

47. Marimon R., McGrattan E., Sargent T. J. (1990) Money as a Medium of Exchange in an Economy with Artificially Intelligent Agents. Journal of Economic Dynamics and Control **14**, 329–373
48. Marimon R., Spear S. E., Sunder S. (1993) Expectationally Driven Market Volatility: An Experimental Study. Journal of Economic Theory **61**, 74–103
49. Midgley D. F., Marks R. E., Cooper L. G. (1997) Breeding Competitive Strategies. Management Science **43(3)**, 257–275
50. Miller J. H. (1996) The Coevolution of Automata in the Repeated Prisoner's Dilemma. Journal of Economic Behavior and Organization **29**, 87–112
51. Neely C., Weller P., Ditmar R. (1997) Is Technical Analysis in the Foreign Exchange Market Profitable? A Genetic Programming Approach. Journal of Financial and Quantitative Analysis **32(4)**, 405–427
52. Nikolaev N. Y., Iba H. (2001) Genetic Programming of Polynomial Models for Financial Forecasting. Forthcoming in IEEE Transactions on Evolutionary Computation.
53. Novkovic S. (1998) A Genetic Algorithm Simulation of a Transition Economy: An Application to Insider-privatization in Croatia. Computational Economics **11(3)**, 221–243
54. Ostermark R. (1999) Solving Irregular Econometric and Mathematical Optimization Problems with a Genetic Hybrid Algorithm. Computational Economics **13(2)**, 103–115
55. Palmer R. G., Arthur W. B., Holland J. H., LeBaron B., Tayler P. (1994) Artificial Economic Life: A Simple Model of a Stockmarket. Physica D **75**, 264–274
56. Pettey C. B., Lutze M. R., Grefenstette J. J. (1987) A Parallel Genetic Algorithm. In: Proceedings of the Sceond International Conference on Genetic Algorithms, 155–161
57. Riechmann T. (1999) Learning and Behavioral Stability – An Economic Interpretation of Genetic Algorithms. Journal of Evolutionary Economics **9**, 225–242
58. Rudolph G. (1994) Convergence Properties of Canonical Genetic Algorithms. IEEE Transaction on Neural Networks **5(1)**, 96–101
59. Sargetn T. J. (1993) Bounded Rationality in Macroeocnomics. Oxford.
60. Smith V. L., Suchanek G. L., Williams A. W. (1988) Bubbles, Crashes, and Endogenous Expectations in Experimental Spot Asset Markets. Econometrica **56(6)**, 1119–1152
61. Szpiro G. (1997) The Emergence of Risk Aversion. Complexity **2(4)**, 31–39
62. Tayler P. (1995) Modelling Artificial Stocks Markets Using Genetic Algorithms. In: Goonatilake S., Treleaven P. (Eds.), Intelligent Systems for Finance and Business, Wiley, New York, NY, 271–288
63. Vallee T., Basar T. (1999) Off-line Computation of Stackelberg Solutions with the Genetic Algorithm. Computational Economics **13(3)**, 201–209
64. Waldrop M. M. (1992) Complexity: The Emerging Science at the Edge of Order and Chaos. Simon and Schuster.
65. Wolpert D. H., Macready W. G. (1997) No Free Lunch Theorem for Optimization. IEEE Transactions on Evolutionary Computation **1(1)**, 67–82
66. Yao X. (1993) A Review of Evolutionary Artificial Neural Netowrks. International Journal of Intelligent Systems **8(4)**, 539–567
67. Yao X., Darwen P. J. (1994) An Experimental Study of N-person Iterated Prisoner's Dilemma Games. Informatica **18**, 435–450

Part II

Games

2 Playing Games with Genetic Algorithms

Robert E. Marks

Australian Graduate School of Management, Universities of Sydney and New South Wales, Sydney 2052, Australia.
bobm@agsm.edu.au

Abstract. In 1987 the first published research appeared which used the Genetic Algorithm as a means of seeking better strategies in playing the repeated Prisoner's Dilemma. Since then the application of Genetic Algorithms to game-theoretical models has been used in many ways. To seek better strategies in historical oligopolistic interactions, to model economic learning, and to explore the support of cooperation in repeated interactions. This brief survey summarises related work and publications over the past thirteen years. It includes discussions of the use of game-playing automata, co-evolution of strategies, adaptive learning, a comparison of evolutionary game theory and the Genetic Algorithm, the incorporation of historical data into evolutionary simulations, and the problems of economic simulations using real-world data.

2.1 Introduction

Over the past twenty-five years, non-cooperative game theory has moved from the periphery of mainstream economics to the centre of micro-economics and macro-economics. Issues of information, signalling, reputation, and strategic interaction can best be analysed in a game-theoretic framework. But solution of the behaviour and equilibrium of a dynamic or repeated game is not as simple as that for a static or one-shot game. The multiplicity of Nash equilibria of repeated games has led to attempts to eliminate many of these through refinements of the equilibrium concept. At the same time, the use of the rational expectations assumption to cut the Gordian Knot of the intractability of dynamic problems has led to a reaction against the super-rational *Homo calculans* model towards boundedly rational models of human behaviour.

In the late 'eighties the Genetic Algorithm (GA) was first used to solve a dynamic game, the repeated Prisoner's Dilemma (RPD) or iterated Prisoner's Dilemma (IPD). Mimicking Darwinian natural selection, as a simulation it could only elucidate sufficient conditions for its Markov Perfect equilibria (MPE), rather than the necessary conditions of closed-formed solution, but over the past twenty-odd years, its use in economics has grown to facilitate much greater understanding of evolution, learning, and adaptation of economic agents. This brief survey attempts to highlight emergence of the marriage of GAs and game theory.

2.2 Deductive Versus Evolutionary Approaches to Game Theory

Ken Binmore and Partha Dasgupta [14] argued that the use of repeated games in economics can be viewed from two perspectives; first, deductively determining what the equilibrium outcome of a strategic interaction will be, using the Nash equilibrium concept as a solution method, and, second, evolutively, asking how the strategic players will learn as they play, which will also result in equilibrium. In general, the possible equilibria which follow from the second approach are a subset of those that follow from the first. Indeed, "learned" equilibria are one attempt to "refine" the possible Nash equilibria of the deductive approach. A good summary of the rationale and techniques of the "learning" approach to solving game equilibria is given in Fudenberg and Levine [39], who mention the GA as a means of exploring the space of strategies to play repeated games.

2.3 The Repeated Prisoner's Dilemma

Several years earlier, Robert Axelrod [9] had set up a tournament among computer routines submitted to play the RPD. Pairs of routines were matched in the repeated game, and their scores used to increase or reduce the proportion of each routine in the total population of potential players. Although Axelrod was a political scientist, he was influenced by William Hamilton, a biologist, and this simulated "ecology" mimicked in some way the natural selection undergone by species competing for survival in an ecological niche. It is well known that Anatol Rapoport's Tit for Tat emerged as a robust survivor among the many routines submitted. Foreshadowing the analytical tool of replicator dynamics (see below), however, the number of strategies in the game was fixed at the start: no new strategies could emerge.

2.4 Boundedly Rational Players

Axelrod's 1984 tournament pitted routines that were perforce boundedly rational. That one of the simplest, Tit for Tat, emerged as the best, as measured by its high score (although not always: see [57]), might only have been because of the bounded nature of its strategy (start off "nice", and then mimic your opponent's last move).

Deductive game theory has not assumed any limits to agents' abilities to solve complex optimisation problems, and yet Herbert Simon has been arguing for many years what applied psychologists and experimental economists are increasingly demonstrating in the laboratory: human cognitive abilities are bounded, and so descriptive models should reflect this. But one virtue of the assumption of perfect rationality is that global optima are reached, so that there is no need to ask in what manner our rationality is bounded. It may be no coincidence that two recent monographs on bounded rationality in economics (Thomas Sargent [79], and Ariel Rubinstein [78]) are written by authors who have previously been concerned with repeated games played by machines [77] and one of the first published uses of GAs in economics [56]. What is the link between these two areas? Some background will clarify.

2.5 Game-Playing Automata

The GA was developed by John Holland [47,48] and his students, including David Goldberg [42]. The original applications were in engineering and were predominantly optimisation problems: find $x = \arg\max f(x)$. The comparative advantage of GAs was that $f(.)$ was not required to be continuous or convex or differentiable or even explicitly closed-form – for a recent overview, see [70]. So the original focus was on optimisation of closed-form functions.

But the GA can search for more general optima. Axelrod [11] recounts how his colleague at Michigan, John Holland, mentioned that there was this new technique in Artificial Intelligence (or what has become known as Machine Learning) which, by simulating natural selection, was able to search for optima in extremely non-linear spaces. Axelrod [10], the first published research to use the GA with repeated games, was the result: against the same niche of submitted routines that he had used in his 1984 study, Axelrod used the GA to see whether it could find a more robust strategy in the RPD than Tit for Tat. I came across a reference to this then-unpublished work in Holland et al. [49] and obtained both the draft and the code in early 1988 from Axelrod. I had been interested in solutions to the RPD since a routine of mine had won the second M. I. T. competitive strategy tournament [34], an attempt to search for a generalisation for Tit for Tat in a three-person game with price competition among sellers of imperfect substitutes. What could this new technique tell me about my serendipitous routine?

Axelrod (with the programming assistance of Stephanie Forrest [36]) modelled players in his discrete RPD game as stimulus-response automata, where the stimulus was the state of the game, defined as both player's actions over the previous several moves, and the response was the next period's action (or actions). That is, he modelled the game as a state-space game [40,83], in which past play influences current and future actions, not because it has a direct effect on the game environment (the payoff function) but because all (or both) players believe that past play matters. Axelrod's model focused attention on a smaller class of "Markov" or "state-space" strategies, in which past actions influence current play only through their effect on a state variable that summarises the direct effect of the past on the current environment (the payoffs). With state-space games, the state summarises all history that is payoff-relevant, and players' strategies are restricted to depend only on the state and the time.

Specifically, Axelrod's stimulus-response players were modelled as strings, each point of which corresponds to a possible state (one possible history) and decodes to the player's action in the next period. The longer the memory, the longer the string, because the greater the possible number of states. Moving to a game with more than the two possible moves of the RPD will lengthen the string, holding the number of players constant. Increasing the number of players will also increase the number of states. Formally, the number of states is given by the formula a^{mp}, where there are a actions, p players, each remembering m periods [68].

Although the implicit function of Axelrod's 1987 paper is non-linear and open-form, in one way this pioneering work was limited: the niche against which his players interact is static, so the GA is searching a niche in which the other player's behaviour is determined at time zero: these routines do not learn. This type of problem is characterised as open-loop, and is essentially non-strategic [83].

2.6 Co-Evolution of Automata

The interest of game theory is in strategic interaction, in which all players respond to each other, and can learn how better to respond. I thought that GAs playing each other would be more interesting than Axelrod's static niche (which I replicated in [57]) and extended Axelrod's work to co-evolution of stimulus-response players in [58], later published as [60], although I termed the simultaneous adaptation of each player in response to all others' actions as "boot strapping", being unaware of the biologist's term.

Another approach to modelling players as co-evolving stimulus-response machines was taken by John Miller, in his 1988 Michigan PhD thesis (later published as [69]). He modelled the finite automaton explicitly, and let the GA derive new automata. This is explained further in [59]. Miller's approach has the advantage that offspring of the genetic recombination of the GA are functioning finite automata, whereas offspring in the Axelrod approach may include many irrelevant states which – via the curse of dimensionality [61] – dramatically increases the computation costs. Holland and Miller [50] argued for the merits of simulation in general (anticipating some of the arguments made by Judd [53]), and bottom-up artificial adaptive agents (such as the population of strategies playing the RPD to derive a fitness score for the GA) in particular.

As discussed at length in the two monographs, Sargent [79] and Rubinstein [78], modelling players as boundedly rational poses important questions for the modeller: which constraints are realistic? which ad hoc? Deterministic finite automata playing each other result in Markov perfect equilibria. But as soon as the researcher uses a (necessarily finite) automaton to model the player – or its strategy – the question of boundedness must be addressed.

Eight years ago, the literature on finite automata in repeated games could be categorised into two distinct branches: the analysis of the theoretical equilibrium properties of machine games, and the effect of finite computational abilities on the support of cooperative outcomes [59]. Since then, as discussed in the section, Empirical Games, below, there is a growing literature which uses such techniques as the GA and neural nets to explore historical market behaviour, especially in oligopolies.

Rubinstein [77] was using automata as players to explore some theoretical issues associated with the RPD. Others have followed, without necessarily using the GA to solve the strategy problem. Megiddo and Wigderson [67] used Turing machines (potentially of infinite size) to play the RPD; Chess [23] used simulations to generate best-response strategies in the RPD, and generated simple routines, but he did not use the GA; Cho [24–26] and Cho and Li [27] used "perceptrons," neural nets, to demonstrate support of the Folk Theorem in RPD games, with simple strategies, with imperfect monitoring, and with nets that are of low complexity. Fogel and Harrald [35] and Marks and Schnabl [65] compare neural nets and the GA for simulating the RPD, and Herbrich et al. [43] survey the use of neural networks in economics.

2.7 Learning

The earliest published paper to use the GA to solve the RPD is Fujiki and Dickinson [41], but instead of explicitly using automata, they modelled the players as Lisp routines, and allowed the GA to search the space of LisP routines for higher scoring code. (Lisp is a programming language often used in Artificial Intelligence applications.) The earliest publication in a peer-reviewed economics journal of a GA used in economics modelling was Marimon et al. [56], but it was not explicitly about using the GA to solve games. Instead, in a macro-economic model it used the GA to model learning, the first of many papers to do so.

I must admit that in an early 1989 conversation with Tom Sargent, I discounted the mechanisms of the GA – selection, crossover, and mutation – as models of the learning process, but Jasmina Arifovic, a student of Sargent's, in a continuing of series of papers over the past decade which model GA learning, has shown that this conclusion was wrong. Arifovic [2] uses the GA to simulate the learning of rational expectations in an overlapping-generations (OLG) model; her first published paper [3] was on learning in that canonical model of equilibrium determination: the Cobweb Model, in which she introduces the election operator, the first contribution by an economist to the practice of GAs, previously dominated by, first, engineers, and, second, mathematicians; Arifovic [4] shows that using the GA in an OLG model results in convergence to rational expectations and reveals behaviour on the part of the artificial adaptive agents which mimics that observed by subjects in experimental studies; [5] finds a two-period stable equilibrium in an

OLG model; Arifovic and Eaton [6] use the GA in a coordination game. The Cobweb Model has proved a popular subject of investigation using the GA: Dawid and Kopel [32] explore two models – one with only quantity choices and one with the prior decision of whether to exit or stay in the market; Franke [37] explores the stability of the GA's approximation of the moving Walrasian equilibrium in a Cobweb economy, and provides an interpretation and extension of Arifovic's election operator [3].

One of the traits of Tit for Tat as characterised by Axelrod [9] is that it is easily recognised. That is, other players can soon learn the type of player they are facing. A more complex strategy would not necessarily be as readily recognised. Dawid [29–31] argues that GAs may be an economically meaningful model of adaptive learning if we consider a learning process incorporating imitation (crossover), communication (selection), and innovation (mutation).

There are several other relevant papers on this topic. Bullard and Duffy [17] explore how a population of myopic artificial adaptive agents might learn forecasting rules using a GA in a general equilibrium environment, and find that coordination on steady state and low-order cycles occurs, but not on higher-order periodic equilibria. Bullard and Duffy [18] use a GA to update beliefs in a heterogeneous population of artificial agents, and find that coordination of beliefs can occur and so convergence to efficient, rational expectations equilibria. Riechmann [75] uses insights from the mathematical analysis of the behaviour of GAs (especially their convergence and stability properties) to argue that the GA is a compound of three different learning schemes, with a stability between asymptotic convergence and explosion. Riechmann [76] demonstrates that economic GA learning models can be equipped with the whole box of evolutionary game theory [89], that GA learning results in a series of near-Nash equilibria which then approach a neighbourhood of an evolutionary stable state. Dawid and Mehlmann [33], and Chen et al. [20] also consider genetic learning in repeated games. The recent volume edited by Brenner [16], on computational techniques of modelling learning in economics, includes several papers of interest, in particular Beckenbach [13] on how the GA can be re-shaped as a tool for economic modelling, specifically to model learning.

2.8 Replicator Dynamics

But the GA is not the only technique to come from biology into economics. Convergence of a kind was occurring as economics in general, and game theory in particular, was borrowing insights from a second source in biology. For the past fifty years there has been some interest in the Darwinian idea of natural selection as applied to firm survival in the (more or less) competitive marketplace, as summarised in Nelson and Winter [73]. Borrowing game theoretical ideas from economics, John Maynard Smith [66] introduced the concept of the evolutionarily stable strategy (ESS) to biology, and this in turn was used by economists eager to reduce the number of possible equilibria in repeated games. ESS is concerned with the invasion of a new species (or strategy) into an ecology (market) in which there is an existing profile of species (strategies). Axelrod [9] had been mimicking these changes as he allowed the proportions of his RPD-playing routines to alter in response to their relative success. The most widely discussed paper, and one of the earliest to use GA's in examining the changing profiles of strategies in the RPD, is Lindgren [54]. Two recent investigations of the RPD are by Ho [44], with information processing costs, and by Wu and Axelrod [90], with noise.

Axelrod [9] had argued that simple strategies such as Tit for Tat were collectively evolutionarily stable, and so could sustain cooperation in the finitely RPD. In an unpublished paper, Slade and Eaton [84] argue that Axelrod's conclusion is not robust to small deviations from the Axelrod setup, in particular, allowing agents to alter their strategies without announcement. Similarly, Nachbar [71,72] questions the robustness of Axelrod's findings, arguing that prior restrictions on the strategy set result in a convergence to cooperation with Tit for Tat, and that without these restrictions cooperation would eventually be exhausted.

Replicator dynamics [45,46] exhibit similar behaviour, as the profile of initial strategies changes, but unlike the GA as applied to stimulus-response machines (strategies) replicator dynamics cannot derive completely new strategies. Replicator dynamics have provided the possibility of closed-form investigation of evolutionary game theory [38,81,55], but evolutionary game theory in general, and replicator dynamics in particular, despite the word "evolutionary", are not directly related to GAs (see [29,30] for comparisons).

Chen and Ni [21] use the GA to examine Selten's [80] chain store game, in a variant due to Jung et al. [52], where there are "weak" and "strong" monopolists, in the context of answering questions of Weibull's on evolutionary game theory [89]. They characterise the phenomenon of coevolutionary stability, although it may be that the simulated behaviour observed are actually Markov Perfect equilibria [64] with a very long periodicity.

To what extent can interactions in oligopolies be modelled as the play of a repeated Prisoner's Dilemma? Yao and Darwen [91] examined the emergence of cooperation in an n-person iterated Prisoner's Dilemma, and showed that the larger the group size, the less likely its emergence. Chen and Ni [22] examine an oligopoly with three players, which they characterized as a time-variant, state-dependent Markov transition matrix, and conclude that, owing to the path dependence of the payoff matrix of the oligopoly game, this is not the case. Further, they argue that the rich ecology of oligopolistic behavior can emerge without the influence of changing outside factors (such as fluctuating income or structural changes), using GAs to model the adaptive behavior of the oligopolists.

2.9 Other Refinements

The literature on the RPD cited above restricts the player's decisions to its next action, but does not model an earlier decision: whether or not to play. This constraint is relaxed – players get to decide with whom to interact – in a series of papers by Leigh Tesfatsion and co-authors [85,8], culminating in her simulation of trading [87].

The GA has been used to search for the emergence of risk-averse players, in a simple model [86], and more recently in a robust model of the farmer's decision faced with high- and low-risk options [19]. Huck et al. [51] use replication and mutation (but not crossover or the GA) to search for the emergence of risk aversion in artificial players of repeated games.

Tomas Başar has been prominent in the mathematics of game theory: the book [12] solved many issues in dynamic, continuous, differential games before many economists had become aware of the problems. Recently, he has used GAs to solve for Stackelberg solutions, in which one player moves first [88].

Following in Başar's tradition, Özyildirim [74] applied the GA to search for approximations of the non-quadratic, non-linear, open-loop, Nash Cournot equilibria in a macro-economic policy game. She goes beyond many earlier studies in using historical data to explore the interactions between the Roosevelt administration and organised labour during the New Deal of the 'thirties. In a second paper [1], the GA is used to search for optimal policies in a trading game with knowledge spillovers and pollution externalities.

2.10 Empirical Games

The two papers by Özyildirim, discussed above [74,1], show the future, I believe: moving from stylised games, such as the RPD, to actual market interactions. Slade [83] provides a clear structure for this project, as well as reporting preliminary results. She finds that the Nash equilibrium of the one-shot game is emphatically not a reasonably approximation to real-world repeated interactive outcomes, and that most games in price or quantity appear to yield outcomes that are more collusive than the one-shot outcome, a finding which is consistent with the Folk Theorem.

Marks and his coauthors have been involved in a project to use GAs to solve for MPE using historical data on the interactions among ground-coffee brands at the retail level, where these players are modelled as Axelrod/Forrest finite automata in an oligopoly. This work is a generalisation of Axelrod [10] and Marks [60], and uses the ability of the GA to search the highly disjoint space of strategies, as Fudenberg and Levine [39] suggest.

The first two papers [62,68] build on a "market model" that was estimated from historical supermarket-scanner data on the sales, prices, and other marketing instruments of up to nine distinct brands of canned ground coffee, as well as using brand-specific cost data. The market model in effect gives the one-shot payoffs (weekly profits) for each brand, given the simultaneous actions of all nine (asymmetric) brands. There was natural synchrony in brands' actions: the supermarket chains required their prices and other actions to change together, once a week, at midnight on Saturdays.

Modelling the three most rivalrous historical brands as artificial adaptive agents (or Axelrod strings), the authors use the GA to simulate their coevolution, with the actions of the other six brands taken as unchanging, in a repeated interaction to model the oligopoly. After convergence to an apparent MPE in their actions, the best of each brand is played in open-loop competition with the historical data of the other six in order to benchmark its performance. The authors conclude that the historical brand managers could have improved their performance given insights from the response of the stimulus-response artificial managers, although this must be qualified with the observation that closed-loop experiments [83], allowing the other managers to respond, would be more conclusive.

Marks et al. [63] refine the earlier work by increasing the number of strategic players to four, by using four distinct populations in the simulation (against the one-string-fits-all single population of the earlier work), and by increasing the number of possible actions per player from four to eight, in order to allow the artificial agents to learn which actions (prices) are almost always inferior and to be avoided. Marks [61] explores some issues raised by the discrete simulations and the curse of dimensionality, and uses entropy as a measure of the loss of information in partitioning the continuous historical data (in prices etc.). Slade [82,83] considers two markets with similar rivalrous behaviour – gasoline and saltine crackers – but does not use a GA.

Curzon Price [28] uses a GA to model several standard industrial-organization models in order to demonstrate that the GA performs well as a tool in applied economic work requiring market simulation. He considers simple models of Bertrand and Cournot competition, a vertical chain of monopolies, and an electricity pool.

2.11 Conclusion

When John Holland [47] invented the GA, his original term for it was an "adaptive plan" [13], which looked for "improvement" in complex systems, or "structures which perform well." Despite that, most research effort, particularly outside economics, has been on its use as a function optimiser. But, starting with Axelrod [10], the GA has increasingly been used as an adaptive search procedure, and latterly as a model of human learning in repeated situations. In the 1992 second edition of his 1975 monograph, Holland expressed the wish that the GA be seen more as a means of improvement and less on its use as an optimiser.

This survey, although brief, has attempted to show that use of the GA in economics in general, and in game theory in particular, has been more and more focusing on its use as an adaptive search procedure, searching in the space of strategies of repeated games, and providing insights into historical oligopolistic behaviour, as well as into human learning processes.

Acknowledgements

The work in this study was partly supported by an Australian Research Council grant and the Australian Graduate School of Management. I wish to thank the editor, Shu-Heng Chen, for his encouragement in the face of my recent bereavement, and my colleague, Robin Stonecash, for her support.

References

1. Alemdar N. M., Özyildirim S. (1998) A Genetic Game of Trade, Growth and Externalities. Journal of Economic Dynamics and Control **22(6)**, 811–32
2. Arifovic J. (1990) Learning by Genetic Algorithms in Economic Environments. Santa Fe Institute Econ. Res. Prog. Working Paper 90–001.
3. Arifovic J. (1994) Genetic Algorithm Learning and the Cobweb Model. Journal of Economic Dynamics and Control **18(1)**, 3–28
4. Arifovic J. (1995) Genetic Algorithms and Inflationary Economies. Journal of Monetary Economics **36(1)**, 219–243
5. Arifovic J. (1998) Stability of Equilibria Under Genetic Algorithm Adaptation: An Analysis. Macroeconomic Dynamics **2(1)**, 1–21
6. Arifovic J., Eaton B. C. (1995) Coordination via Genetic Learning. Computational Economics **8(3)**, 181–203
7. Arifovic J., Eaton B. C. (1998) The Evolution of Type Communication in a Sender/Receiver Game of Common Interest with Cheap Talk. Journal of Economic Dynamics and Control **22(8–9)**, 1187–1207
8. Ashlock D., Smucker M. D., Stanley E. A., Tesfatsion L. (1996) Preferential Partner Selection in an Evolutionary Study of Prisoner's Dilemma. BioSystems **37(1–2)**, 99–125
9. Axelrod R. (1984) The Evolution of Cooperation. N. Y., Basic Books.
10. Axelrod R. (1987) The Evolution of Strategies in the Iterated Prisoner's Dilemma. In: Davis L. (Ed.) Genetic Algorithms and Simulated Annealing, London: Pittman, 32–41
11. Axelrod R. (1997) The Complexity of Cooperation. Princeton: P. U. P.
12. Başar T., Oldser G. J. (1982) Dynamic Non-Cooperative Game Theory. New York: Academic Press
13. Beckenbach F. (1999) Learning by Genetic Algorithms in Economics? In: Brenner (1999) op. cit., 73–100
14. Binmore K., Dasgupta P. (1986) Game Theory: A Survey. In: Binmore K., and Dasgupta P. (Eds.) Economic Organizations as Games, Oxford: B. Blackwell, 1–45

15. Birchenhall C. R. (1996) Evolutionary Games and Genetic Algorithms. In: Gilli M. (Ed.) Computational Economic Systems: Models, Methods & Econometrics, The series Advances in Computational Economics 5, Dordrecht: Kluwer Academic Publishers, 3–23
16. Brenner T. (Ed.) (1999) Computational Techniques for Modelling Learning in Economics, in the series Advances in Computational Economics 11, Dordrecht: Kluwer Academic Publishers
17. Bullard J., Duffy J. (1998) Learning and the Stability of Cycles. Macroeconomic Dynamics **2(1)**, 22–48
18. Bullard J., Duffy J. (1999) Using Genetic Algorithms to Model the Evolution of Heterogeneous Beliefs. Computational Economics **13(1)**, 41–60
19. Cacho O., Simmons P. (1999) A Genetic Algorithm Approach to Farm Investment. Australian Journal of Agricultural and Resource Economics, **43(3)**, 305–322
20. Chen S.-H., Duffy J., Yeh C. H. (1996) Equilibrium Selection via Adaptation: Using Genetic Programming to Model Learning in a Co-ordination Game. mimeo.
21. Chen S. -H., Ni C. -C. (1997) Coevolutionary Instability in Games: An Analysis Based on Genetic Algorithms. In: Proceedings of 1997 IEEE International Conference on Evolutionary Computation, Piscataway, N.J.: Institute of Electrical and Electronics Engineers, 703–708
22. Chen S. -H., Ni C. -C. (2000) Simulating the Ecology of Oligopolistic Competition with Genetic Algorithms, Knowledge and Information Systems, forthcoming.
23. Chess D. M. (1988) Simulating the Evolution of Behavior: The Iterated Prisoners' Dilemma, Complex Systems **2**, 663–670
24. Cho I. -K. (1995) Perceptrons Play the Repeated Prisoner's Dilemma, Journal of Economic Theory **67**, 266–284
25. Cho I. -K. (1996a) On the Complexity of Repeated Principal Agent Games. Economic Theory **7(1)**, 1–17
26. Cho I. -K. (1996b) Perceptrons Play Repeated Games with Imperfect Monitoring. Games and Economic Behavior **10(1)**, 22–53
27. Cho I. -K., Li H. (1999) How Complex are Networks Playing Repeated Games? Economic Theory **13(1)**, 93–123
28. Curzon Price T. (1997) Using Co-evolutionary Programming to Simulate Strategic Behaviour in Markets. Journal of Evolutionary Economics **7(3)**, 219–254
29. Dawid H. (1996a) Genetic Algorithms as a Model of Adaptive Learning in Economic Systems. Central European Journal for Operations Research and Economics **4(1)**, 7–23
30. Dawid H. (1996b) Learning of Cycles and Sunspot Equilibria by Genetic Algorithms. Journal of Evolutionary Economics **6(4)**, 361–373
31. Dawid H. (1999) Adaptive Learning by Genetic Algorithms: Analytical Results and Applications to Economic Models. Lecture Notes in Economics and Mathematical Systems, Vol 441. Heidelberg: Springer-Verlag, 2nd ed.
32. Dawid H., Kopel M. (1998) On Economic Applications of the Genetic Algorithm: A Model of the Cobweb Type. Journal of Evolutionary Economics **8(3)**, 297–315
33. Dawid H., Mehlmann A. (1996) Genetic Learning in Strategic Form Games. Complexity **1(5)**, 51–59

34. Fader P. S., Hauser J. R. (1988) Implicit Coalitions in a Generalised Prisoner's Dilemma, Journal of Conflict Resolution **32**, 553–582
35. Fogel D. B., Harrald P. G. (1994) Evolving Continuous Behaviors in the Iterated Prisoner's Dilemma In: Sebald A., Fogel L. (Eds.) The Third Annual Conference on Evolutionary Programming, Singapore: World Scientific, 119–130
36. Forrest S. (Ed.) (1991), Emergent Computation: Self-Organizing, Collective, and Cooperative Phenomena in Natural and Artificial Computing Networks, Cambridge: M.I.T. Press.
37. Franke R. (1998) Coevolution and Stable Adjustments in the Cobweb Model. Journal of Evolutionary Economics **8(4)**, 383–406
38. Friedman D. (1991) Evolutionary Games in Economics. Econometrica **59(3)**, 637–666
39. Fudenberg D., Levine D. K. (1998) The Theory of Learning in Games. Cambridge: M.I.T. Press
40. Fudenberg D., Tirole J. (1991) Game Theory. Cambridge: M.I.T. Press
41. Fujiki C., Dickinson J. (1987) Using the Genetic Algorithm to Generate Lisp Source Code to Solve the Prisoner's Dilemma. In: Grefenstette J. J. (Ed.) Genetic Algorithms and their Applications. Proceedings of the 2nd. International Conference on Genetic Algorithms. Hillsdale, N. J.: Lawrence Erlbaum
42. Goldberg D. E. (1988) Genetic Algorithms in Search, Optimization, and Machine Learning. Reading, Mass.: Addison-Wesley
43. Herbrich R., Keilbach M., Graepel T., Bollmann-Sdorra P., Obermayer K. (1999) Neural networks in economics. In: Brenner (1999), op. cit., 169–196
44. Ho T. H. (1996) Finite Automata Play Repeated Prisoner's Dilemma with Information Processing Costs. Journal of Economic Dynamics and Control **20(1–3)**, 173–207
45. Hofbauer J., Sigmund C. (1988) The Theory of Evolution and Dynamic Systems. Cambridge: C. U. P.
46. Hofbauer J., Sigmund C. (1998) Evolutionary Games and Population Dynamics, Cambridge: C. U. P.
47. Holland J. H. (1975) Adaptation in Natural and Artificial Systems. Ann Arbor: Univ. of Michigan Press. (A second edition was published in 1992: Cambridge: M.I.T. Press.)
48. Holland J. H. (1992) Genetic Algorithms. Scientific American July, 66–72
49. Holland J. H., Holyoak K. J., Nisbett R. E., Thagard P. R. (1986) Induction: Processes of Inference, Learning, and Discovery, Cambridge: M.I.T. Press.
50. Holland J. H., Miller J. H. (1991) Artificial Adaptive Agents in Economic Theory. American Economic Review: Papers and Proceedings **81(2)**, 365–370
51. Huck S., Müller W., Strobel M. (1999) On the Emergence of Attitudes Towards Risk. In: Brenner (1999), op. cit., 123–144
52. Jung Y. J., Kagel J. H., Levin D. (1994) On the Existence of Predatory Pricing: An Experimental Study of Reputation and Entry Deterrence in the Chain-store Game. Rand Journal of Economics 25(1), 72–93
53. Judd K. L. (1998) Numerical Methods in Economics, Cambridge: M.I.T. Press
54. Lindgren K. (1991) Evolutionary Phenomena in Simple Dynamics. In: Langton C., Taylor C., Farmer J., Rasmussen S., (Eds.), Artificial Life II, Vol. 10, Reading: Addison-Wesley, 295–324
55. Mailath G. J. (1998) Do People Play Nash Equilibrium? Lessons from Evolutionary Game Theory. Journal of Economic Literature **36(3)**, 1347–74

56. Marimon R., McGrattan E., Sargent T. J. (1990) Money as a Medium of Exchange in an Economy with Artificially Intelligent Agents. Journal of Economic Dynamics and Control **14**, 329–373
57. Marks, R. E. (1989a) Niche Strategies: The Prisoner's Dilemma Computer Tournaments Revisited. AGSM Working Paper 89–009. `<http://www.agsm.edu.au/~bobm/papers/niche.pdf>`
58. Marks R. E. (1989b). Breeding Optimal Strategies: Optimal Behavior for Oligopolists. In: Schaffer J. David (Ed.), Proceedings of the Third International Conference on Genetic Algorithms, San Mateo, CA.: Morgan Kaufmann, 198–207
59. Marks R. E. (1992a) Repeated Games and Finite Automata. In: Creedy J., Borland J., Eichberger J. (Eds.) Recent Developments in Game Theory. Aldershot: Edward Elgar, 43–64
60. Marks R. E. (1992b) Breeding Hybrid Strategies: Optimal Behaviour for Oligopolists. Journal of Evolutionary Economics, **2**, 17–38
61. Marks R. E. (1998) Evolved Perception and Behaviour in Oligopolies. Journal of Economic Dynamics and Control **22(8–9)**, 1209–1233
62. Marks R. E., Midgley D. F., Cooper L. G. (1995) Adaptive Behavior in an Oligopoly. In: Biethahn J., Nissen V. (Eds.) Evolutionary Algorithms in Management Applications, Berlin: Springer-Verlag, 225–239
63. Marks R. E., Midgley D. F., Cooper L. G. (1998) Refining the Breeding of Hybrid Strategies. Australian Graduate School of Management Working Paper 98–017, Sydney. `<http://www.agsm.edu.au/~bobm/papers/wp98017.html>`
64. Marks R. E., Midgley D. F., Cooper L. G. (2000) Breeding Better Strategies in Oligopolistic Price Wars. Submitted to the IEEE Transactions on Evolutionary Computation, special issue on Agent-Based Modelling of Evolutionary Economic Systems, INSEAD Working Paper 2000/65/MKT, `<http://www.agsm.edu.au/~bobm/papers/ieee-marks.html>`
65. Marks R. E., Schnabl H. (1999) Genetic Algorithms and Neural Networks: A Comparison Based on the Repeated Prisoner's Dilemma. In: Brenner (1999), op. cit., 197–219
66. Maynard Smith J. (1982) Evolution and the Theory of Games. Cambridge: C. U. P.
67. Megiddo N., Wigderson, A. (1986) On Play by Means of Computing Machines. In: Halpern J. Y. (Ed.) Reasoning About Knowledge, Los Altos: Kaufmann, 259–274
68. Midgley D. F., Marks R. E., Cooper L. G. (1997) Breeding Competitive Strategies. Management Science **43(3)**, 257–275
69. Miller J. H. (1996) The Coevolution of Automata in the Repeated Prisoner's Dilemma. Journal of Economic Behavior and Organization **29**, 87–112
70. Mitchell M. (1996) An Introduction to Genetic Algorithms. Cambridge: M.I.T. Press
71. Nachbar J. H. (1988) The Evolution of Cooperation Revisited. mimeo, Santa Monica: RAND Corp., June.
72. Nachbar J. H. (1992) Evolution in the Finitely Repeated Prisoner's Dilemma, Journal of Economic Behavior and Organization **19(3)**, 307–326
73. Nelson R. R., Winter S. G. (1982) An Evolutionary Theory of Economic Change. Cambridge: Belknap Press of Harvard University Press
74. Özyildirim S. (1997) Computing Open-loop Noncooperative Solution in Discrete Dynamic Games. Journal of Evolutionary Economics **7(1)**, 23–40

75. Riechmann T. (1999) Learning and Behavioural Stability: An Economic Interpretation of Genetic Algorithms. Journal of Evolutionary Economics **9(2)**, 225–242
76. Riechmann T. (2001) Genetic Algorithm Learning and Evolutionary Games, in this volume.
77. Rubinstein A. (1986) Finite Automata Play the Repeated Prisoner's Dilemma. Journal of Economic Theory, **39**, 83–96
78. Rubinstein A. (1998) Modeling Bounded Rationality. Cambridge: M.I.T. Press
79. Sargent T. J. (1993) Bounded Rationality in Macroeconomics. Oxford: O. U. P.
80. Selten R. (1978) The Chain Store Paradox. Theory and Decision **9**, 127–59
81. Selten R. (1991) Evolution, Learning, and Economic Behavior. Games and Economic Behavior **3**, 3–24
82. Slade M. E. (1992) Vancouver's Gasoline-price Wars: An Empirical Exercise in Uncovering Supergame Strategies. Review of Economic Studies **59**, 257–274
83. Slade M. E. (1995) Empirical Games: The Oligopoly Case. Canadian Journal of Economics **28(2)**, 368–402
84. Slade M. E., Eaton B. C. (1990) Evolutionary Equilibrium in Market Supergames. University of British Columbia Department of Economics Discussion Paper: 90-30.
85. Stanley E. A., Ashlock D., Tesfatsion L. (1994) Iterated Prisoner's Dilemma with Choice and Refusal of Partners. In: Langton C. (Ed.) Artificial Life III, Vol. 17, Santa Fe Institute Studies in the Sciences of Complexity, Redwood City: Addison-Wesley, 131–175
86. Szpiro G. (1997) The Emergence of Risk Aversion. Complexity **2**, 31–39
87. Tesfatsion L. (1997) A Trade Network Game with Endogenous Partner Selection. In: Amman H. M., Rustem B., Whinston A. B. (Eds.) Computational Approaches to Economic Problems, Dordrecht: Kluwer Academic Publishers, 249–269
88. Vallee T., Başar T. (1999) Off-line Computation of Stackelberg Solutions with the Genetic Algorithm. Computational Economics **13(3)**, 201–209
89. Weibull J. W. (1995) Evolutionary Game Theory, Cambridge: M.I.T. Press
90. Wu J., Axelrod R. (1995) How to Cope with Noise in the Iterated Prisoner's Dilemma, Journal of Conflict Resolution **39(1)**, 183–189
91. Yao X., Darwen P. J. (1994) An Experimental Study of n-person Iterated Prisoner's Dilemma Games. Informatica **18**, 435–450

3 Genetic Algorithm Learning and Economic Evolution

Thomas Riechmann

Institut für Volkswirtschaftslehre, Universität Hannover, Königsworther Platz 1, 30167 Hannover, Germany
riechmann@vwl.uni-hannover.de

Abstract. This paper tries to connect the theory of genetic algorithm (GA) learning to evolutionary game theory. It is shown that economic learning via genetic algorithms can be described as a specific form of evolutionary game. It will be pointed out that GA learning results in a series of near Nash equilibria which during the learning process build up to finally reach a neighborhood of an evolutionarily stable state. In order to clarify this point, a concept of evolutionary stability of genetic populations will be developed. Thus, in a second part of the paper, it becomes possible to explain both, the reasons for the specific dynamics of standard GA learning models and the different kind of dynamics of GA learning models which use extensions to the standard GA.

3.1 Introduction

Genetic Algorithms (GAs) have been frequently used in economics to characterize a well defined form of social learning.[1] They have been applied to mainstream economics problems and mathematically analyzed as to their specific dynamic and stochastic properties.[2] But, although widely seen as conducting a rather evolutionary economic line of thought, up to now there is no piece of work explicitly focusing on what it is that makes GA learning an evolutionary kind of behavior.

This paper aims to clarify the scientific advantage of viewing genetic algorithms as evolutionary processes. Evolutionary game theory delivers some very useful tools, which can help explaining why economic GAs behave the way they do. In more detail, it can be found out what are the reasons for the special, non–converging dynamics of the standard GA, and — more than this — it can be explained why certain changes to the GA, like the introduction of the election operator[3], can change the GA–dynamics to an as dramatic extent as they do.

As genetic algorithms have been well introduced into economic research, this paper will not explicitly review the specific structure and working principles of GAs. The reader will be provided to have a basic notion of genetic

[1] Some frequently cited papers are [2], [3], [4], [5], [6], [11], and [12].
[2] [14], [30].
[3] See [3].

algorithms, which can be gained from e.g. [18] or [27]. The kind of genetic algorithms found in these books will be called 'basic' or 'standard' genetic algorithm. In this paper, the standard GA will mainly be dealt with. Nevertheless, the analysis carried out will give rise to the opportunity of analyzing more difficult variants of genetic algorithms. Thus, in a further part of this paper, even enhanced or augmented genetic algorithms will be briefly dealt with, clarifying the reasons why some of these variants behave differently from the standard GA.

This paper will face the question what is evolutionary in GA learning. First, it will be shown that there is a close connection between evolutionary game theory and genetic algorithm learning. Using this notion it will be pointed out that evolving genetic populations can be interpreted as a series of near Nash equilibria. Then, in a second step, the well–known concept of evolutionary stability will be transferred to the field of GA research. In a third step, it will be shown that under the regime of the market genetic algorithm learning leads to a kind of evolutionary dynamics, which can be described as economic progress rather then just economic change. A further part of the paper makes clear why some frequently used modifications to the basic GA result in kinds of dynamics which are very different from the basic dynamics. At last, the paper ends with a summary.

3.2 The Standard Genetic Algorithm

As there is a growingly large number of different variants of genetic algorithms in economic research, this paper will mainly deal with the most basic GA, the standard GA, which is described in detail in [18]. The standard GA is the GA all other variant GAs derive from.

More precisely, most of the findings of this paper will only apply to standard, one–population, economic GAs. In addition to the standard GA being the simplest one, there are two more decisive characteristics of genetic algorithms which are dealt with in this paper. The first one is the fact that this paper does not focus two– or more–population–GAs, like those used in e.g. [4] or [11]. The second one is the fact, that this paper will only face 'economic' GAs, which means genetic algorithms that model processes of *social* learning via interaction within a single population of economic agents. The above sentence contains an implicit definition of social learning: Social learning means all kinds of processes, where agents learn from one another. Examples for social learning are learning by imitation or learning by communication. Learning by experiment, on the contrary, is no form of social learning. It is a form of isolated *individual* learning. In the following, the terms 'social learning' and 'economic learning' will be used as synonyms.[4]

[4] This means that GA models of e.g. the traveling salesman problem, which surely have an economic subject, are nevertheless no 'economic' GAs in the above meaning.

The standard GA is a stochastic process which repeatedly turns one population of bit strings into another. These bit strings are called genetic individuals. In economic research they normally stand for some economic strategy or routine in the sense of [28]. These routines are said to be used by economic agents.[5]

Each repeated 'turn' of the genetic algorithm essentially consists of two kinds of stochastic processes, which are variety generating and variety restricting processes. *Variety generating* processes are processes during which new economic strategies are developed by the economic agents. In the standard GA these processes are reproduction, which is interpreted as learning by imitation, crossover, which is interpreted as learning by communication, and mutation, which is interpreted as learning by experiment. All these processes, or genetic operators, take some old economic strategies and use them to find new ones, thus enhancing the variety of strategies within the current population. In the standard GA, there is one *variety restricting* process, which is the genetic operator of selection. Selection decreases the number of different economic strategies. It first evaluates the economic success of each strategy, thus often being interpreted as playing the role of the market as an information revealing device.[6] Then it selects strategies to be part of the next population. The selection operator of the standard GA does so by applying a kind of biased stochastic process: Selection for the next generation is done by repeatedly drawing with replacement strategies from the pool of the old population to be taken over into the next one. The chance of each strategy to be drawn is equal to its relative fitness, which is the ratio of its market success to the sum of the market success of all strategies in the population. Thus, the number of different strategies within a population is reduced again.

3.3 Genetic Algorithms as Evolutionary Processes

Close relationships between economic learning models and models of evolutionary theory have been recognized before. [25] gives a clear notion of the similarities of learning models on the one hand and evolutionary processes on the other. As genetic algorithms, too, have been broadly interpreted as models of economic learning[7], this section will show that they can also be regarded as evolutionary processes.

At a first glance, it is the structure of genetic algorithms and evolutionary models that seems to suggest a close relationship between GAs and evolutionary economic theory: Both face the central structure of a population of

[5] It is important to stress the following point: A genetic individual is not interpreted as an economic agent, but as an economic strategy *used by* an economic agent. This interpretation allows for several agents employing the same strategy.

[6] Note, that this has already been described in [21].

[7] For such an interpretation see e.g. [15].

economic agents interacting within some well defined economic environment and aiming to optimize individual behavior.

As the aim of this paper is to describe genetic algorithms as evolutionary processes, the first question to be answered is the question, if GAs are evolutionary processes at all. In the following, it will be argued that GAs are a specific form of evolutionary processes, i.e. evolutionary games.

[17, p. 16] gives three characteristics for an evolutionary game:

> 'By an evolutionary game, I mean any formal model of strategic interaction over time in which (a) higher payoff strategies tend over time to displace lower payoff strategies, but (b) there is some inertia, and (c) players do not systematically attempt to influence other players' future actions.'

Prior to checking these three points, it is important to notice that economic GAs are in fact models of strategic interaction. In the interpretation as models of social learning, GAs deal with a number of economic agents, each trying to find a behavioral strategy which, due to her surrounding, gives her the best payoff possible. GAs are models of *social* learning, which in fact is a way of learning by interaction.[8] Thus, it can be recognized, that GAs are in fact models of 'strategic interaction'.

Moreover, GAs are dynamic processes which reproductively prefer higher payoff strategies over lower payoff ones. It has been shown that in the standard GA the probability of a strategy i to be reproduced from its current population $\overline{n}$ into next period's population, $P(i|\overline{n})$, only depends on its relative fitness $R(i|\overline{n})$, which is the strategy's payoff or market success relative to the aggregate payoff of the population $\overline{n}$.[9] Higher relative fitness leads to a higher reproduction probability:

$$\frac{dP(i|\overline{n})}{dR(i|\overline{n})} > 0 . \tag{3.1}$$

Thus, Friedman's condition (a) is fulfilled.

Secondly, according to Friedman, inertia means that changes in behavior do not take place too abruptly.

Looking at the genetic or game theoretic population, it is mutation, or learning by experiment, which causes the most abrupt changes. Whereas the strategy of a single economic agent might be changed more dramatically by imitation or communication, this is not true for the population as a whole. At

[8] There are, on the contrary, ways of individual, i.e. non–social learning as e.g. statistical forms of learning or neural network learning.

[9] For a more precise description of this, refer to [30].
Note, that this only applies to GA models which do not implement agents having some kind of memory of their own, as this would turn the GA Markov process into a time variant one. Thus, this is another restriction of the class of GA learning models this paper is able to derive concluions about.

the level of the population, only learning by experiment is able to introduce strategies or at least parts of strategies which have not occured in society before. In fact, learning by imitation or communication can only lead to the adoption of strategies, which — at least partly — have been used by members of the population before. Thus, at the level of the population, only mutation can cause real innovation or truly abrupt changes.

As mutation, or learning by experiment, is the source of the most abrupt changes in GA learning, it should be proved that small changes by mutation are more likely than big ones. For the standard GA, using binary representation of genetic individuals, the mutation operator is quite simple. Mutation randomly alters ('flips') single bits of the bit string by which an economic strategy (i.e. a genetic individual) is coded. Each bit of the string has a small probability μ to be changed. μ, which is called the mutation probability, is the same for every bit within every genetic individual of every population. Figure 3.1 shows an example of the mutation operator. The fact that small

0 0 1 0 0 0 0 0 0 0 0 0 1 0 0 0 1

↓

0 0 1 0 0 0 0 0 0 1 0 0 1 0 0 0 1

Fig. 3.1. Mutation (example)

changes by mutation are more likely than big ones can be shown as follows: The probability of an economic strategy i to be turned into strategy j by mutation, $P_m(i,j)$, depends on the length of the genetic individuals' bit strings, L, the mutation probability μ, and the number of bits that have to be flipped in order to turn i into j, which is called the Hamming distance between i and j, $H(i,j)$:

$$P_m(i,j) = \mu^{H(i,j)} (1-\mu)^{L-H(i,j)} \quad . \tag{3.2}$$

Simple differentiation gives

$$\frac{\partial P_m(i|j)}{\partial H(i,j)} = \mu^{H(i,j)} (1-\mu)^{L-H(i,j)} [\ln \mu - \ln (1-\mu)] \begin{cases} <0 & \text{for} \quad \mu < \frac{1}{2} \\ =0 & \text{for} \quad \mu = \frac{1}{2} \\ >0 & \text{for} \quad \mu > \frac{1}{2} \end{cases} . \tag{3.3}$$

For the normal parameter value of μ,[10] this means the obvious: Small changes in strategy are more likely than big changes.[11] Thus, it becomes evident that GA learning processes are processes that contain some inertia.

[10] Normally, the mutation probability μ ranges somewhere between 1/100 and 1/1 000.

[11] The result given in (3.2) deserves two further remarks. First, the fact that for $\mu > 1/2$ big changes are more likely than small ones explains why for relatively

Friedman's third point ('players do not systematically attempt to influence other players' future actions') can be proved more verbally. The agents that are modeled by an economic GA have very restricted knowledge. By the time an economic agent forms her latest economic strategy she does not know anything about the plans of the other agents in her population. All an economic agent in a GA model can do is to try her best to adopt to her neighbors' *past* actions, for the near past is all an economic agent can remember. Taking into account these very limited individual abilities, it is easy to conclude, that there is no room for *systematic* influences on other agents' actions.

From the above it can be concluded, that models of economic GA learning are in fact models that can be interpreted as evolutionary games as well.

3.4 Populations as Near Nash Equilibria

The main structure in genetic algorithm learning models is the genetic population. It can be noticed that a population is nothing more than a distribution of different economic or behavioral strategies.[12] This is true for genetic populations as well as for populations in the game theoretic meaning of the word. Thus, it can be said that a genetic population *is* a game theoretic population.

Genetic algorithms are in fact describing a repeated economic game. Imagine a genetic algorithm using a population of M genetic individuals with the length of each individual's bit string of L. This GA is able to deal with every economic strategy in S, the set of all available strategies. S has the size $N = |S| = 2^L$. This means that this GA can be interpreted as a repeated symmetric one population M person game with up to N possible pure strategies. But, compared to 'normal' evolutionary games, within most economic GA learning models, the rules are different. Whereas in evolutionary games most of the time a strategy is repeatedly paired with single competing strategies, in genetic algorithm learning, each strategy plays against the whole aggregate rest of the population.[13] There is no direct opponent to a single strategy. Instead, every economic agent aims to find a strategy $i \in S$ that performs as

---

large values of μ, GA results seem to become very similar to random walks. Secondly, the result yields an interesting interpretation for the field of economic learning. If mutation is interpreted as learning by experiment, (3.2) shows that a little experimenting is a good thing to do, while too many experiments will disturb the generation of valuable new economic strategies. If mutation is interpreted as making mistakes in imitation or communication (see e.g. [1]), (3.2) simply means that you should not make too many of those mistakes.

[12] For an in–depth discussion of this, refer to [13] or [30].

[13] While this notion is true for most of the economic GA models, for some it is not, including [6] and [2], which make use of different mechanisms of matching the agents.

good as possible relative to its environment, which is completely determined by the current population $\overline{n}$ and the objective function $R(\cdot)$. [14]

This means that every economic agent i faces problem (3.4):[15]

$$\max_{i \in S} R(i|\overline{n}) \ . \tag{3.4}$$

This directly leads to the concept of Nash equilibria. A Nash strategy is defined as the best strategy *given the strategies of the competitors.* Thus, a Nash strategy is exactly what every economic agent, alias genetic individual, is trying to reach. As a first step of analysis, a genetic algorithm can be seen as modeling a system of economic agents, each trying to play a Nash strategy. In economic terms this means that every agent tries to coordinate her strategy with the other agents' ones, for this is the best way of maximizing her profit (or utility or payoff or whatever the model wants the agent to maximize). A genetic population can be interpreted as a population of agents, each trying to play Nash.

3.5 Evolutionary Stability of Genetic Populations

While a genetic population represents a primarily static concept, learning processes are of course genuinely dynamic processes. Thus, in order to analyze GA learning as an evolutionary learning process, the dynamics and concepts of stability have to be analyzed.[16]

This paper will make use of the concept of evolutionary stability, especially the notion of evolutionarily stable strategies or evolutionarily stable states (ESS).[17] In short, a strategy is evolutionarily stable if, relative to its population, it performs better than any new, 'invading' strategy. Though widely used in economic dynamics, the concept of ESS has two weaknesses which make this concept seem to be of only limited suitability for the analysis

[14] The exact mathematical formulation can be found in equation [30, equation (7)], which, in game theoretic terms, gives the payoff to agent i playing against the rest of population $\overline{n}$. It should be noted, that in games a player's payoff depends on his action and the action of every opponent, so that the best formulation of fitness or payoff is $R(i|\overline{n})$.

[15] Although (3.4) looks a bit complicated, even compared to most of the mainstream economic models, it is in fact remarkably simple. All it says is 'Do the best you can with respect to your neighborhood!'

[16] Replicator dynamics (see e.g. [34], [22], or [23]), which have often been used to characterize evolutionary dynamics, seem to be unsuited for some economic problems. (In [24, p. 286] it is even suggested that 'There is nothing in economics to justify replicator dynamics'.) Applied to the analysis of GA learning, replicator dynamics, not directly accounting for stochastics, are simply not precise enough to cover the whole GA learning process.

[17] See [33], [22], [23], [32], [34], [26], or [24] (to mention just a few of various pieces of work on this topic).

of genetic algorithms. The first weakness lies in the fact that the concept of ESS is based on symmetric two person games only. As mentioned above, this is not the form of games usually played in GA learning. Most GAs have each genetic individual playing against the aggregate rest of the population. Secondly, there is no explicit formulation of the selection process underlying the concept of evolutionary stability. ESS are based on the notion that invading 'mutant' strategies are somehow rejected or eliminated from the population. It is not clear how this rejection will be carried out. Genetic algorithms, in contrast, present a clear concept of rejection: Every strategy will be exposed to a test, which is best described as a one–against–the–rest game. Then it will be reproduced or rejected with a probability depending on its performance (i.e. market performance) in the game. GA reproduction or rejection has two main features, it selects due to performance and it selects due to probability, which means that a bad strategy will be rejected almost surely but not with probability one.

Thus, a refined concept of evolutionary stability for genetic algorithms is presented. An attempt to set up a concept of evolutionary stability for genetic algorithms which is keeping the spirit of the ESS is the following: A genetic population is evolutionarily stable if the process of the genetic algorithm rejects an invasion by one or more strategies from the genetic population. Invasion itself can either take the form of a totally new strategy entering the population or it can simply mean a change in the frequency of the strategies already contained within the population. Thus, a clearer definition of an evolutionarily stable population might be: A population is evolutionarily stable if it is resistant against changes in its composition.

More formally, a genetic population $\overline{n}$ will be called *evolutionarily superior* to population $\overline{m}$, (denoted as $\overline{n} \overset{es}{>} \overline{m}$) if it exhibits two characteristics[18]:

(a) Every strategy i contained within population $\overline{n}$ has at least the same fitness in the basic population $\overline{n}$ as it has in the invaded population $\overline{m}$, while at least one strategy has even more fitness in $\overline{n}$ than in $\overline{m}$.
(b) The invading strategies $k \in \{\overline{m} \setminus \overline{n}\}$ are the worst performing strategies contained in $\overline{m}$, so that they will be most surely rejected.

[18] Note that the following characterizes a kind of weak dominance concept. The fact, that this concept is called *superiority* has, in earlier drafts of this paper, led to misconception: The name is chosen in order to resemble the similarity to the concept of Pareto superiority. But, neither the concept of evolutionary superiority nor the concept of Pareto superiority make any statements about some kind of welfare. Originally, Pareto superiority is just a means to order points within a highly dimensioned space. Applying evolutionary superiority analogously to the original meaning of the Pareto criterion has no welfare implication at all. It is just used in order to make genetic populations weakly comparable with respect to the process of the genetic algorithm and GA's way of turning one population into another.

In mathematical terms, a genetic population $\overline{n}$ is evolutionarily superior to $\overline{m}$, if

$$R(i|\overline{n}) \geq R(i|\overline{m}) \; \forall \, i \in \overline{n} \, , \tag{3.5}$$

$$\wedge \; \exists j \text{ with } R(j|\overline{n}) > R(j|\overline{m}) \; , \tag{3.6}$$

$$\wedge \quad R(k|\overline{m}) < R(i|\overline{m}) \; \forall \, i \in \overline{n}; \; \forall k \in \{\overline{m} \setminus \overline{n}\} \; . \tag{3.7}$$

Equations 3.5 and 3.6 reflect characteristic (a) given above, equation 3.7 is the formalization of characteristic (b). For full validity, a further remark is necessary, even if it is a little beyond the scope of this paper: Within genetic algorithms, invading strategies can only result from reproduction ('imitation'), crossover ('communication') or mutation ('experiment') within the population itself. This means that the final outcome of GAs without mutation (i.e. processes with learning by imitation and communication only), which are always uniform populations, may have other populations being superior to them, but — without mutation — better populations simply cannot arise.[19]

Note that (3.5) to (3.7) induce a partial ordering on the set of genetic populations S'. A population is called stable in the concept of (3.5) to (3.7) if there is no other population that is superior to it: n is an evolutionarily stable population, if

$$\nexists \overline{m} \in S' \quad \text{with} \quad \overline{m} \overset{\text{es}}{>} \overline{n} \, . \tag{3.8}$$

Condition (3.8) is in fact a generalization of the concept of evolutionary stability.[20]

Due to the fact that genetic algorithm selection is a probabilistic rather than a deterministic process, invading strategies, even in a evolutionarily stable population, may not be rejected within a single round of the algorithm. It can only be stated that the invader will be driven out of the population within finite time. That is to say: If a genetic population is evolutionarily stable, it will recover from an invasion within a finite number of steps of the GA, which means that *in the long run* the population will not lastingly be changed. Nevertheless, once an evolutionarily stable population is invaded, there may show up a number of evolutionarily inferior populations within the next few rounds of the GA. These populations represent transitory states of the process of rejecting the invader. [30] shows that there is in fact more than one population that will occur in the long run. These may be transitory populations as well as different populations which are evolutionarily stable, too.

19 See [30] for the restrictions different learning techniques put on the set of available strategies.

20 Note e.g. the similarity to the definition in [34, pp. 36].

3.6 Evolutionary Dynamics

As a consequence of what has been developed in the preceding parts, dynamics of genetic algorithms can be characterized in a more evolutionary manner. First of all it can be noticed that every population describes a game theoretic outset which is a near Nash equilibrium. The genetic algorithm as a process of turning one population into another can be viewed as at least an approximation of the moving Nash equilibria process. More than this, turning to the criterion of evolutionary superiority ((3.5) to (3.8)), the GA always selects in favor of the superior population. This notion can be used to characterize genetic learning dynamics: The stochastic process GA continuously discards populations in favor of better ones in the sense of criterion (3.8). This only describes the direction of the process, not the exact path that is taken in time. In fact, due to the stochastic properties of genetic algorithms, the exact path of the process highly depends on the starting point, i.e. the composition of the very first genetic population. And although the path up to an evolutionarily stable equilibrium may differ, Markov chain theory shows that one such state will be reached, and specifically that it will be reached irrespectively of the starting conditions. There may be path dependence, lock–ins, or related phenomena, but in the case of genetic algorithm learning these will only be of temporary nature. In the long run, genetic algorithm theory promises, the 'best' state will be reached.[21]

Knowing the special form of the dynamic process of the GA and the direction in which this process will lead, a few more words can be said about the role of heterogeneity for the dynamics.

It seems important to notice the way economic changes take place. Starting with some population, genetic operators (i.e. learning mechanisms) cause changes in the population. New types of behavior are tested. The test is performed by exposing the strategies to the market. The market reveals the quality of each tested strategy relative to all the other strategies within the population. Then selection lets economic agents give up poorly performing strategies and adopt better ones (imitation) or even create new ones by communication (crossover) or experimentation (mutation). After that, again,

[21] This may be regarded as a weakness of the concept of genetic algorithm learning, as it neglects the possibility of modeling path dependence or lock–ins. So it may be worthwhile to mention two further points, which are mainly beyond the scope of this paper. First, depending on the underlying (economic) problem, some GAs spend long times supporting populations which are not evolutionarily stable. Some keywords pointing to this topic are 'deceptiveness' of genetic algorithms and the problem of 'premature convergence'. Secondly, the lack of ability to model lasting lock–ins or path dependence applies to the basic genetic algorithm. There are variations of genetic algorithms which are capable of modeling these phenomena. One keyword pointing into this direction of research may be 'niching mechanisms'. Again, a good starting point for more descriptions of all of the special cases and variants of GAs is [18].

strategies are tested and evaluated by the market, by that way coordinating the agents' strategies.

There are in fact two crucial points to this repeated process: First, it is the diversity of strategies that drives economic change, i.e. the succession of populations constantly altering their composition. Under the regime of genetic algorithm learning, this change in individual as well as in social behavior heavily (while not entirely) relies on learning by imitation and learning by communication. As was pointed out in greater detail in [30], these kinds of learning can only take place within heterogeneous populations. Thus, in a way, it can be said that it is heterogeneity that is the main force behind economic change.

The second crucial point to the process of genetic algorithm learning is the role of selection, which can be interpreted as the role of the market. While the act of learning will be enough to achieve economic *change*, economic *development* can only be reached by the cooperation of learning and selection. In order to turn the succession of different populations into the succession of constantly improving populations (in the sense of evolutionary superiority), a device is needed that makes it possible to distinguish successful strategies from less successful ones. Having at hand such a device, it is possible to decide which strategies shall live and grow and which ones shall die. This device is the market in economics as it is the selection operator within genetic algorithms. It is the market and only the market that turns economic change into economic development.[22]

Summarizing, under the regime of the market, evolutionary dynamics of genetic algorithm learning is mainly driven by two forces: Heterogeneity, which constantly induces behavioral (and by that, economic) change, and the market as a coordination device, revealing information about the quality of each type of behavior and ruling out poorly performing strategies, thus turning economic change into economic development.

Finally, looking at genetic algorithm learning from an evolutionary point of view, one more point has to be added. It has been shown that, as long as possible, genetic algorithm learning and market selection improve individual and with that social behavior. Yet, once an evolutionarily stable state of behavior has been reached, there certainly is no room for further improvement. But, due to the special structure of genetic algorithms, this does not mean that in this state economic agents stop changing their behavior. Learning, or what has above been called change, still continues and will not cease to continue. Still, there will arise new ways of individual behavior within a population. Now it is the role of the market (i.e. selection) to drive these strategies out of the population again. Due to the probabilistic nature of the GA, this process may take more than one period, thus producing one or even more transitory populations until an evolutionarily stable population

[22] This reflects a rather classical economic thought, given, e.g., in [20] (usually quoted as [21]).

is regained. To put it in different words: Even after an evolutionarily stable state is reached, evolutionary stability is continuously challenged by new strategies. While in the first phase of the GA learning process some of these new strategies are integrated into the population, in the second phase all of the invaders will be replaced again. So there is an ongoing near equilibrium movement resulting from the continuous rejection of invading strategies.

In fact, genetic algorithm learning leads to an

> 'interplay of coordinating tendencies arising from competitive adaptions in the markets and de–coordinating tendencies caused by the introduction of novelty'

([36, p. xix]), which has often been regarded as a key feature of evolutionarily economic analysis of the market.[23]

Long run dynamics of GA learning processes have in mathematical terms been characterized as a state of a stable distribution of genetic populations ([30]). With the help of evolutionary game theory, a clear economic reason can be found, why this state shows up: It is a process of near equilibrium dynamics, caused by the continuously ongoing challenge of the ESS by new strategies and the rejection of these strategies that prevents social behavior from total convergence but still keeps it near enough to some stable state.

3.7 Modified Genetic Operators and Their Impact on Stability

Within this paper, only the most basic type of genetic algorithms has been looked at. In economic research, various forms of GAs are used which employ modifications of the operators described in this paper. Some modifications of genetic operators have a major impact on the dynamics and accordingly on the stability properties of the genetic algorithm containing them. Two such modifications will briefly be mentioned.

3.7.1 Selection

Within this paper the standard form of the selection operator has been analyzed. While this 'biased roulette wheel' selection used in the standard genetic algorithm leads to the reported results, there are different types of selection which lead to algorithms with different behavior.[24]

Above all, there is a group of elitist selection operators, including the selection within evolution strategies[25] and the election operator introduced in [3].[26] Elitist selection ensures that at least the first best genetic individual of a

---

[23] See [35].

[24] An overview of various selection schemes can be gained from [19].

[25] For a survey, refer to [9].

[26] For an interpretation and an extension of the election operator, see [16].

population will become a member of the next generation's population. These differing selection operators show up a much stronger tendency of leading to strict asymptotic convergence and uniformity of genetic populations.[27] This tendency can easily be explained. In contrast to roulette wheel selection, elitist selection ensures that invading strategies which turn out to be the worst strategies throughout the population will be replaced at once. This means that there will be no room for transitory populations. Bad strategies, i.e. strategies obeying condition (3.7), will be ruled out before they can even enter a population. This certainly leads to asymptotic behavioral stability.

Which selection operator to choose for a genetic algorithm in an economic model heavily depends on the economic interpretation of the operators and on its relevance to the problem to be modeled. This interpretation, in turn, mainly depends on the role, the author of the model wants to assign to e.g. chance, mass phenomena, network externalities and related topics. The problem concentrates upon the question whether the best behavior in a certain period will inevidently find its way into next period's pool of behavioral strategies (elitist selection) or if there will be any forces that can prevent this (roulette wheel selection).

3.7.2 Mutation

The analysis performed above shows that mutation is a strong force behind economic change. Yet, used as a metaphor of economic learning by experiment, mutation in its simplest form may be seen as underestimating economic agents' rational capacity. Why should agents not notice if their repeated experiments cease to gain improvements? A modified mutation operator could be thought of, endogenizing each agents propensity to experiment. A modification of this type, based on earlier papers by [7], [8] and [10], has been analyzed in [29], where this change in mutation is found to smooth but not to totally remove the resulting near equilibrium dynamics of the GA. The reasons for this finding can be found in the fact that once a relatively good state is reached, mutation probability is reduced. Learning by experiment decreases if there is not much left to learn. Thus, there are less invading strategies producing less transitory populations which leads to a slow down in economic fluctuations.

Again, which kind of mutation operator to choose mainly depends on the economic interpretation which should be applied to it.

[27] A convergence analysis for genetic algorithms with elitist selection has been carried out in [31]. In this piece of work, Rudolph proves that genetic algorithms with elitist selection *will* converge to a uniform population.

3.8 Summary

Learning by genetic algorithms is a specific form of a repeated evolutionary game. This fact gives you at hand the whole range of analytical tools evolutionary game theory offers for the analysis of dynamic processes.

This paper proves that GA learning in fact is an evolutionary game. It uses the notion of Nash equilibria and a transferred concept of evolutionary stability to describe in detail the dynamics of genetic algorithm learning both in its standard form and in some of its variants. Though this is just the beginning of some more pieces of work still to be done, the results are rather enlightening and help explain why GA learning works the way it apparently does.

References

1. Alchian A. A. (1950) Uncertainty, Evolution, and Economic Theory. Journal of Political Economy **58**, 211–221
2. Andreoni J., Miller J. H. (1995) Auctions with Artificial Adaptive Agents. Games and Economic Behavior **10**, 39–64
3. Arifovic J. (1994) Genetic Algorithm Learning and the Cobweb–model. Journal of Economic Dynamics and Control **18**, 3–28
4. Arifovic J. (1995) Genetic Algorithms and Inflationary Economies. Journal of Monetary Economics **36**, 219–243
5. Arifovic J. (1996) The Behavior of the Exchange Rate in the Genetic Algorithm and Experimental Economies. Journal of Political Economy **104**, 510–541
6. Axelrod R. (1987) The Evolution of Strategies in the Iterated Prisoner's Dilemma. In: Davis L. (Ed.) Genetic Algorithms and Simulated Annealing. Pitman, London, 32–41
7. Thomas Bäck. (1992) The Interaction of Mutation Rate, Selection, and Self–Adaption within a Genetic Algorithm. In: Männer R., Manderick B. (Eds.) Parallel Problem Solving from Nature 2. Elsevier Science, Amsterdam
8. Bäck T. (1992) Self–Adaption in Genetic Algorithms. In: Proceedings of the First European Conference on Artificial Life. Cambridge, MA, London, MIT Press
9. Bäck T., Hoffmeister F.,Schwefel Hans-Paul (1991) A Survey of Evolution Strategies. In: Belew R. K., Booker L. B. (Eds.) Proceedings of the 4th International Conference on Genetic Algorithms, San Mateo, California, 13.–16. Juli 1991. Morgan Kaufmann, 2–9
10. Bäck T., Schütz M. (1996) Intelligent Mutation Rate Control in Canonical Genetic Algorithms. In: Ras W., Michalewicz M. (Eds.) Foundation of Intelligent Systems 9th International Symposium, ISMIS 96, Berlin, Heidelberg, New York, Springer, 158–167
11. Birchenhall C. (1995) Modular Technical Change and Genetic Algorithms. Computational Economics **8**, 233–253
12. Bullard J., Duffy J. (1998) A Model of Learning and Emulation with Artificial Adaptive Agents. Journal of Economic Dynamics and Control **22**, 179–207

13. Davis T. E., Principe J. C. (1993) A Markov Chain Framework for the Simple Genetic Algorithm. Evolutionary Computation **1**, 269–288
14. Dawid H. (1994) A Markov Chain Analysis of Genetic Algorithms with a State Dependent Fitness Function. Complex Systems **8**, 407–417
15. Dawid H. (1996) Adaptive Learning by Genetic Algorithms. Springer, Berlin, Heidelberg, New York
16. Franke R. (1997) Behavioural Heterogeneity and Genetic Algorithm Learning in the Cobweb Model. Discussion Paper 9, IKSF—Fachbereich 7—Wirtschaftswissenschaft. Universität Bremen
17. Friedman D. On Economic Applications of Evolutionary Game Theory. Journal of Evolutionary Economics **8**, 15–43
18. Goldberg D. E. (1989) Genetic Algorithms in Search, Optimization, and Machine Learning. Addison–Wesley, Reading, Massachusetts
19. Goldberg D. E., Deb K. (1991) A Comparative Analysis of Selection Schemes Used in Genetic Algorithms. In: Foundations of Genetic Algorithms. San Matoe, California, Morgan Kaufmann, 69–93
20. Hayek F. A. von (1969) Freiburger Studien, chapter 15, Der Wettbewerb als Entdeckungsverfahren, 249–265, J. C. B. Mohr (Paul Siebeck), Tübingen
21. Hayek F. A. von (1978) New Studies in Philosophy, Politics, Economics and the History of Ideas. Chapter 12, Competition as a Discovery Process. Routledge & Kegan Paul, London, 179–190
22. Hofbauer J., Sigmund K. (1988) The Theory of Evolution and Dynamical Systems. Cambridge University Press, Cambridge, UK
23. Hofbauer J., Sigmund K. (1998) Evolutionary Games and Population Dynamics. Cambridge University Press, Cambridge, UK
24. Mailath G. J. (1992) Introduction: Symposium on Evolutionary Game Theory. Journal of Economic Theory **57**, 259–277
25. Marimon R. (1993) Adaptive Learning, Evolutionary Dynamics and Equilibrium Selection in Games. European Economic Review **37**, 603–611
26. Marks R. E. (1992) Breeding Hybrid Strategies: Optimal Behaviour for Oligopolists. Journal of Evolutionary Economics **2**, 17–38
27. Mitchell M. (1996) An Introduction to Genetic Algorithms. MIT Press, Cambridge, MA, London
28. Nelson R. R., Winter S. G. (1982) An Evolutionary Theory of Economic Change. MIT Press, Cambridge, MA, London
29. Riechmann T. (1998) Learning How to Learn. Towards an Improved Mutation Operator within GA Learning Models. In: Computation in Economics, Finance and Engeneering: Economic Systems. Cambridge, England, 1998. Society for Computational Economics
30. Riechmann T. (1999) Learning and Behavioral Stability – An Economic Interpretation of Genetic Algorithms. Journal of Evolutionary Economics **9**, 225–242
31. Rudolph G. (1994) Convergence Analysis of Canonical Genetic Algorithms. IEEE Transactions on Neural Networks **5**, 96–101
32. Samuelson L. (1997) Evolutionary Games and Equilibrium Selection. MIT Press Series on Economic Learning and Social Evolution. MIT Press, Cambridge, MA, London
33. Smith J. M. (1982) Evolution and the Theory of Games. Cambridge University Press, Cambridge, UK
34. Weibull J. (1995) Evolutionary Game Theory. MIT Press, Cambridge, MA, London

35. Witt U. (1985) Coordination of Individual Economic Activities as an Evolving Process of Self–organization. Economic Appliquée, XXXVII, 569–595
36. Witt U. (1993) Introduction. In: Witt U. (Ed.) Evolutionary Economics. Edward Elgar, Aldershot, England, xiii–xxvii

4 Using Symbolic Regression to Infer Strategies from Experimental Data

John Duffy[1] and Jim Engle–Warnick[2]

[1] University of Pittsburgh, Pittsburgh PA 15260, USA
jduffy+@pitt.edu
[2] Nuffield College, Oxford University, Oxford OX1 1NF, UK

Abstract. We propose the use of a new technique–symbolic regression–as a method for inferring the strategies that are being played by subjects in economic decision-making experiments. We begin by describing symbolic regression and our implementation of this technique using genetic programming. We provide a brief overview of how our algorithm works and how it can be used to uncover simple data generating functions that have the flavor of strategic rules. We then apply symbolic regression using genetic programming to experimental data from the repeated "ultimatum game." We discuss and analyze the strategies that we uncover using symbolic regression and conclude by arguing that symbolic regression techniques should at least complement standard regression analyses of experimental data.

4.1 Introduction

A frequently encountered problem in the analysis of data from economic decision-making experiments is how to infer subjects' strategies from their actions. The standard solution to this inference problem is to make some assumptions about how actions might be conditioned on or related to certain strategically important variables and then conduct a regression analysis using either ordinary least squares or discrete dependent variable methods. A well-known difficulty with this approach is that the strategic specification that maps explanatory variables into actions may be severely limited by the researcher's view of how subjects ought to behave in the experimental environment. While it is possible to experiment with several different strategic specifications, this is not the common practice, and in any event, the set of specifications chosen remains limited by the imagination of the researcher.

In this paper, we propose the use of a new technique – symbolic regression using genetic programming – as a means of inferring the strategies that are being played by subjects in economic decision-making experiments. In contrast to standard regression analysis, symbolic regression involves the breeding of simple computer programs or functions that are a good fit to a given set of data. These computer programs are built up from a set of model primitives, specified by the researcher, which include logical if–then–else operations, mathematical and Boolean operators (and, or, not), numerical constants, and current and past realizations of variables relevant to the problem that is being solved. These programs can be generated for each subject in a population and may be depicted in a decision tree format that facilitates their interpretation as individual strategies.

The genetic programming algorithm that we develop for breeding and selecting programs is an automated, domain–independent process that involves large populations of computer programs that compete with one another on the basis of how well they predict the actions played by experimental subjects. These computer programs are selected for breeding purposes based on Darwin's principle of survival of the fittest and they also undergo naturally occurring genetic operations such as crossover (recombination) that are appropriate for genetically mating computer programs. Following several generations of breeding computer populations, the algorithm evolves programs that are highly fit in terms of their ability to predict subject actions. The directed, genetic search process that genetic programming embodies, together with the implicit parallelism of a population–based search process has proven to be a very powerful tool for function optimization in many other applications [4].

The advantage of symbolic regression over standard regression methods is that in symbolic regression, the search process works simultaneously on both the model specification problem and the problem of fitting coefficients. Symbolic regression would thus appear to be a particularly valuable tool for the analysis of experimental data where the specification of the strategic function used is often difficult, and may even vary over time. We begin by describing genetic programming and how it can be used to perform symbolic regression analysis. We then explain how our algorithm is capable of uncovering simple data generating functions that have the flavor of strategic rules. We apply our symbolic regression algorithm to experimental data from the repeated ultimatum game. We discuss and analyze the strategies that we uncover using symbolic regression and we conclude by arguing that symbolic regression should at least complement standard regression analyses of experimental data.

4.2 Symbolic Regression Using Genetic Programming

The use of genetic programming for symbolic regression was first proposed by John Koza [4] as one of several different applications of genetic programming. In addition to symbolic regression, genetic programming has been successfully applied to solving a large number of difficult problems such as pattern recognition, robotic control, the construction of neural network architectures, theorem proving, air traffic control and the design of electrical circuits and metallurgical processes. The genetic programming paradigm, as developed by Koza and other artificial intelligence researchers, is an approach that seeks to automate the process of program induction for problems that can be solved on a computer i.e. for problems that are computable. The basic idea is to use Holland's [3] genetic algorithm to search for a computer program that constitutes the best (approximate) solution to a computable problem given appropriate input data and a programming language. A genetic algorithm is a stochastic, directed search algorithm based on principles of population genetics that artificially evolves solutions to a given problem. Genetic algorithms operate on populations of finite length, (typically) binary strings (patterned after chromosome strings) that encode candidate solutions to a well–defined problem. These strings are decoded and evaluated for their fitness, i.e. for how well each solution comes to solving the problem objective. Following Darwin's principle of survival of the fittest, strings with relatively higher fitness values have a relatively higher probability of being selected for mating purposes to produce the succeeding 'generation' of candidate solutions. Strings selected for mating are randomly paired with one another and, with certain fixed probabilities, each pair of 'parent' strings undergo versions of such genetic operations as crossover (recombination) and mutation. The strings that result from this process, the 'children', become members of the next generation of candidate solutions. This process is repeated for many generations so as to (artificially) evolve a population of strings that yield very good, if not perfect solutions to a given problem. Theoretical work on genetic algorithms, e.g. [2] reveals that these algorithms are capable of quickly and efficiently locating the regions of large and potentially complex search spaces that yield highly fit solutions to a given problem. This quick and efficient search is due to the use of a population–based search, and to the fact that the genetic operators ensure that highly fit substrings, called schema, (or subtrees in genetic programming) increase approximately exponentially in the population. These schema constitute the "building blocks" used to construct increasingly fit candidate solutions. Indeed, Holland [3] has proven that genetic algorithms optimize on the trade–off between searching for new solutions (exploration) and exploiting solutions that have worked well in the past.

Genetic programming is both a generalization and an extension of the genetic algorithm that has only recently been developed by Koza [4] and others. In genetic programming the genetic operators of the genetic algorithm e.g. selection, crossover and mutation operate on a population of variable rather than fixed length character strings that are not binary, but are instead interpretable as executable computer programs in a particular programming language, typically LISP. In LISP (or in similar LISP–like environments), program structures, known in LISP as Symbolic expressions, or 'S–expressions' can be represented as dynamic, hierarchical decision trees in which the non–terminal nodes are functions or logical operators, and the terminal nodes are variables or constants that are the arguments of the functions or operators. The set of non–terminal functions and operators and the set of terminal variables and constants are specified by the user and are chosen so as to be appropriate for the problem under study.

In our application, each decision tree (self executing computer program) is viewed as a potential strategy for one subject, playing a particular role in a particular economic decision-making game. The fitness of each candidate decision tree is simply equal to the number of times the program makes the same decision as the experimental subject over the course of the experimental session, given the same information that was available to the subject at the time of the decision. Koza has termed the problem of finding a function, in symbolic form, that fits a finite sample of data as symbolic regression. While there may be other ways of performing a symbolic regression, we know from the work of Holland that a genetic-algorithm-based search will be among the most efficient, hence, the use of genetic programming for symbolic regression analysis. As mentioned in the introduction, the major advantage of symbolic regression using genetic programming over standard regression methods is that one does not have to prespecify the functional form of the solution, i.e., symbolic regression is data–to–function regression. Instead, one simply specifies two sets of model primitives: (1) the set of non–terminal functions and operators, N, and (2) the set of terminal variables or constants, T. The dynamical structure of the player's strategy (in our application) is then evolved using genetic operations and the grammar rules of the programming language (LISP).

4.3 An Illustration

Our application of symbolic regression using genetic programming is perhaps best illustrated by an example. We will consider the well-known two player, repeated ultimatum game, using data from an experiment conducted by Duffy and Feltovich [1] where subjects played this game for 40 periods. The symbolic regression technique that we illustrate here can easily be applied to other experimental data sets as will (hopefully) become apparent from the description that follows.

In the ultimatum game, two players, A and B, sometimes referred to as proposer and responder, must decide how to divide a \$10 pie. The proposer (player A) proposes a split of the \$10 pie and player B can either accept or reject the proposed split, with acceptance meaning implementation of the offer and rejection resulting in nothing for either player. In the Duffy-Feltovich experiment, the proposers (player As) could propose only integer dollar amounts, e.g. a split of \$6 for A and \$4 for B. This version of the ultimatum game has many Nash equilibria but the unique subgame perfect equilibrium is for Player A to demand \$9 and for player B to accept this proposal earning \$1. The well-known finding from many ultimatum game experiments is that the subgame perfect equilibrium prediction fails to hold; the most commonly observed outcome is for the proposer to propose a nearly equal split of the \$10 pie and for the responder to accept this proposal. In the Duffy-Feltovich experiment, players of both types were randomly paired with one another for 40 periods. Duffy and Feltovich report that the modal proposal by proposers in their baseline (control) treatment is \$6 for player A and \$4 for player B, and that this proposal is frequently accepted by responders.

Our focus here is on understanding the strategic behavior of the responders (the player Bs) in Duffy and Feltovich's control treatment of the ultimatum game. The discreteness of the proposal space in the Duffy-Feltovich experimental design greatly simplifies our implementation of symbolic regression using genetic programming. Furthermore, the large number of observations 40) for each subject allows us to run separate regressions for each subject, in an effort to uncover individual strategies. We can therefore look for heterogeneity in the strategic behavior of subjects who were assigned to play the same role in all rounds of the experiment. In data sets with a smaller number of observations per subject one could use the symbolic regression technique to search for the strategy that best characterizes a *population* of players of a given type.

4.4 The Regression Model

The first step in conducting a symbolic regression is to specify a grammar for the programming language that will be used to evolve the structures (computer programs) that characterize the play of the game. A generative grammar for a programming language simply specifies the rules by which the set of non–terminal symbols and terminal symbols (model primitives) may be combined. In particular, the grammar guarantees that, however we assemble the primitives into a structure, the result is always a valid structure. Non–terminal symbols are those requiring further input, and terminal symbols are those that do not require any further input. For example, the non–terminal symbolic logic operator "if" requires three additional inputs, denoted in brackets { }: if {condition} then {do something} else {do other thing}. The inputs to the non–terminal "if" symbol may themselves be either non–terminals or terminals. An example of a terminal is a variable, constant or action requiring no further input. For example, in modeling the behavior of responders in the ultimatum game, the input for {do something} in the if expression above might be the terminal action accept; the input for do other thing might be the terminal action reject.

As in spoken languages, the grammar of a programming language is intended to be extremely general, admitting a wide variety of different symbolic operators and expressions. Rather than constructing the grammar of a programming language from scratch, the practice in genetic programming is to make use of the grammar of an existing, high-level programming language like LISP or APL. In this paper we make use of grammar of LISP. The advantage of LISP is that the input structures are all symbolic text arrays which are readily converted into programs (and vice versa). Furthermore, parse tree manipulations are easily implemented and program structures are free to vary in size up to some maximum length.

Our implementation of LISP is simulated using C++, but other programming languages can also be used, including, of course, LISP itself.

4.4.1 Grammar

We use the Backus–Nauer form grammar as described in Geyer–Schulz.[2] This grammar consists of non–terminal and terminal nodes of a tree, and a structure with which to build the tree. An example grammar that allows for nested "if" statements, and which we use for the ultimatum game is given in Fig. 4.1. The grammar we use for the ultimatum game specifies

```
Node                          Possible Derivations

<fe> = ( <f6><f0><fe><fe> ) or ( <f6><f0><f2><fe> ) or
                  ( <f6><f0><fe><f2> ) or ( <f6><f0><f2><f2> )
<f0> = ( <f7><f0><f0> ) or ( <f8><f0> ) or ( <f1> ) or ( <f10> )
<f1> = ( <f9><f3><f3> ) or ( <f9><f4><f5> )
<f2> = ( <f10 >)
<f3> = ( <f11> )
<f4> = ( <f12> )
<f5> = ( <f13> )
<f6> = "if"
<f7> = "or" or "and"
<f8> = "not"
<f9> = "<" or ">" or "="
<f10> = "0" or "1" or "b1" or "b2" or "b3"
<f11> = "4" or "5" or "6" or "7" or "8" or "a0" or "a1" or "a2" or "a3"
<f12> = "T"
<f13> = "5" or "10" or "15" or "20" or "25" or "30" or "35
```

Fig. 4.1. "Nested If Statement" grammar for ultimatum game

that the set of non–terminals includes nested if–then statements, logical and, or, and not statements, and the mathematical operators $<$, $>$, and $=$. The set of terminals includes the past 3 proposals made by the player As that a player B has met (a1–a3), along with the player B's own past 3 responses (b1–b3). Also included is player A's current proposal, denoted a0. In our application, the internal representation of player A proposals (a0–a3) is an integer from 0-9 which denotes the amount of the $10 prize that a player A proposes to keep for him or herself (thus $10 - the player A's proposal is the amount to be received by player B). The internal representation of a player B's response is either a 0 or a 1 with a 0 representing reject, and a 1 representing accept. The set of terminals also includes the set of integers from 4–8, which player Bs may use to condition their decisions; we chose this set of integers since player A proposals were restricted to be integer amounts and since most proposals were for amounts in this range. Finally, we include time, T, as an additional terminal symbol, along with integer values for 5-period intervals of play. If T is chosen, then the number referenced comes from $\langle f13\rangle$ as indicated by the grammar $(\langle f9\rangle\langle f4\rangle\langle f5\rangle)$. If one of the mathematical operators, $<$, $>$, or $=$ are chosen, the two numbers compared come from $\langle f11\rangle$ as indicated by the grammar $(\langle f9\rangle\langle f3\rangle\langle f3\rangle)$. The nodes $\langle f2\rangle - \langle f5\rangle$ simply add parentheses to nodes $\langle f10\rangle - \langle f13\rangle$, so that our algorithm understands these symbols to be terminal nodes. We can summarize the textual aspects of the grammar by noting that the set of non–terminals, $\mathcal{N} = \{if,\ or,\ and,\ not,\ <,\ >,\ =\}$ and the set of terminals, $\mathcal{T} = \{a0,\ a1,\ a2,\ a3,\ b1,\ b2,\ b3,\ 4,\ 5,\ 6,\ 7,\ 8,\ T,\ 5,\ 10,\ 15,\ 20,\ 25,\ 30,\ 35\}$.

In addition to specifying the set of non–terminals and terminals, the grammar also specifies how operations may be performed on the set $\mathcal{N}(\mathcal{T})$. The starting node, as specified in Fig. 4.1, allows for four different initial derivations of a decision tree, (individual strategy) all of which begin with the non–terminal logical operator "if" = node $\langle f6\rangle$. As noted above, this operator requires three inputs, and the grammar in Fig. 4.1 specifies restrictions on these inputs. For instance, the second input, which is the condition statement that the if operator evaluates, must always come from node $\langle f0\rangle$, which in turn requires either a Boolean operator from nodes $\langle f7\rangle - \langle f8\rangle$, (*and*, *or* , *not*), or a mathematical operator from node $\langle f9\rangle$ or a terminal from node $\langle f10\rangle$. The other two inputs in $\langle fe\rangle$ are designed to be as general as possible, with either node $\langle fe\rangle$ or node $\langle f0\rangle$ possible for each input position. Similarly, the rules for non–terminal nodes $\langle f0\rangle$ take account of the input needs of the operators in the first position. For instance, an "and" or "or" operator requires two inputs, whereas a "not" operator requires only one, and terminals from node $\langle f10\rangle$ require no inputs. In addition to defining the structure of if–then statements and logical statements the $\langle fe\rangle$ and $\langle f0\rangle$ nodes also call on their own nodes, thus allowing for nested versions of both types of statements.

When constructing a tree we first choose uniformly from one of the four given derivations for $\langle fe\rangle$. Note that each of these derivations begins with the conditional if statement. Given a particular choice for $\langle fe\rangle$, we then proceed from left to right and choose uniformly from the possible derivations for each non–terminal node until only terminal nodes remain. To illustrate how this is done, we will derive a rule using the grammar of Fig. 4.1 for the ultimatum game.

4.4.2 Deriving a Rule

Consider the following rule for player B, the responder, in the ultimatum game: reject if the proposer's current offer is to keep more than \$5, otherwise accept. In the syntax of LISP, the symbolic expression is written as:

$$(if((> (a0)(5)))(0)(1).$$

Parentheses are used to control the evaluation of the expression, with the expression in the innermost set of parentheses being evaluated first. Figure 4.2 shows how the construction of this rule proceeds starting with a random choice for $\langle fe \rangle$. At each step we take the first non–terminal node, working

Symbol	Derivation	Resulting Rule
start	---	<fe>
<fe>	4th	(<f6><f0><f2><f2>)
<f6>	1st	(if<f0><f2><f2>)
<f0>	3rd	(if(<f1>)<f2><f2>)
<f1>	1st	(if((<f9><f3><f3>))<f2><f2>)
<f9>	2nd	(if((><f3><f3>))<f2><f2>)
<f3>	1st	(if((>(<f11>)<f3>))<f2><f2>)
<f11>	6th	(if((>(a0)<f3>))<f2><f2>)
<f3>	1st	(if((>(a0)(<f11>)))<f2><f2>)
<f11>	2nd	(if((>(a0)(5)))<f2><f2>)
<f2>	1st	(if((>(a0)(5)))(<f10>)<f2>)
<f10>	1st	(if((>(a0)(5)))(0)<f2>)
<f2>	1st	(if((>(a0)(5)))(0)(<f10>))
<f10>	2nd	(if((>(a0)(5)))(0)(1))

Fig. 4.2. Derivation of a decision rule

from left to right through the rule, and replace it with one of its derivations. The process continues until the only remaining nodes are terminal. The result is a valid, interpretable decision rule.

4.4.3 Tree Representation

What we have really derived (and what provides for a better interpretation) is a dynamic, hierarchical decision tree which consists of non–terminal and terminal nodes. The genetic program begins with a randomly generated population of these decision trees. Over many generations, the program creates trees of variable length up to a certain maximum depth using the non–terminal nodes to perform genetic operations in a search to find the best fit tree. The tree for the above rule is given in Fig. 4.3. The depth of the tree is defined as the number of non–terminal nodes, in this case 13. When generating rules, a maximum allowable depth is chosen, and whenever a random tree is generated that is larger, it is thrown out and replaced. The depth can be thought of as a measure of the complexity of the tree (or strategy).

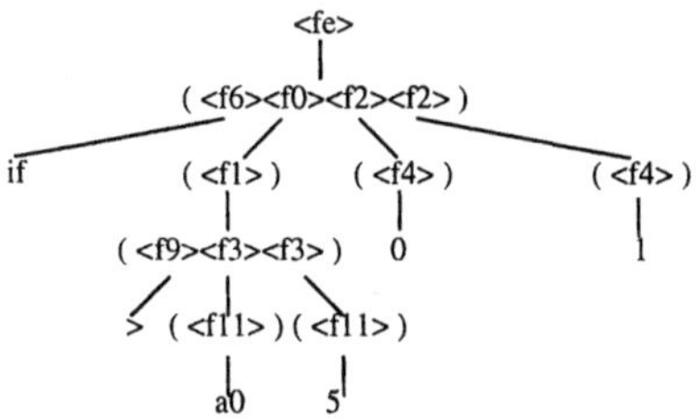

Fig. 4.3. Tree for rule: (if((>(a0)(5)))(0)(1))

4.4.4 Genetic Operation – Crossover

The crossover operation first selects two rules to be parents. It then randomly chooses one of the non–terminal nodes in the first parent, finds all identical nodes in the second parent and uniformly chooses one of these. It then cuts the two subtrees at these nodes, swaps them and recombines the subtrees with the parent trees. By cutting and swapping at the same nodes, the crossover operation ensures that the resulting recombined trees are always syntactically (and semantically) valid programs. If the crossover operation results in a tree that exceeds the maximum depth, the tree is discarded and crossover is repeated until two valid trees result or until a maximum number of attempts is exceeded.

As an example, consider again the rule derived above, and also consider another rule for the second parent, say, if T is less than 30, then accept if the proposer's current offer is to keep less than \$7 and reject otherwise, else reject. The trees are shown in Figs. 4.4 and 4.5.

To illustrate crossover, suppose we randomly choose the first node $\langle f0 \rangle$ in the second level of the first parent tree. We then have to choose an $\langle f0 \rangle$ in the second parent, so suppose we take the $\langle f0 \rangle$ in the third level of the second parent tree. Both nodes are highlighted in the Figs. 4.4 and 4.5. Next, follow these non–terminal $\langle f0 \rangle$ nodes in each parent until their paths terminate at terminal nodes. These are the subtrees that will be swapped between the parents to create new offspring, or children. One of these children is illustrated in Fig. 4.6: Note that the new strategy is quite different from the parent strategy (parent 1) from which it came, as this new strategy instructs the responding player to reject if the proposer's current offer is to keep less than \$7.

A few genetic programs also include a mutation operation that is in addition to the crossover operation. However, as Koza[4] has pointed out, the

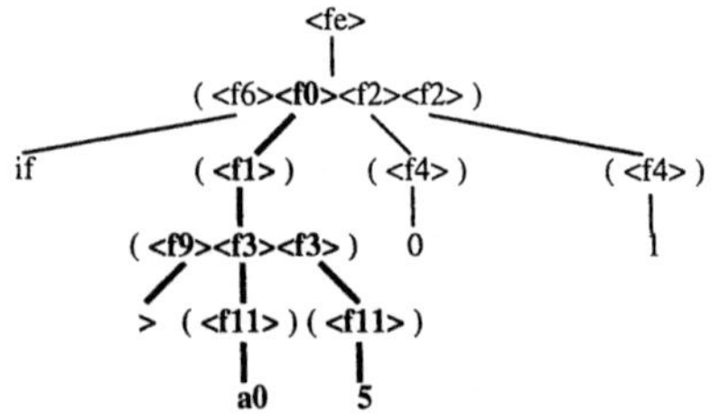

Fig. 4.4. Parent 1 for crossover: (if((>(a0)(5)))(0)(1))

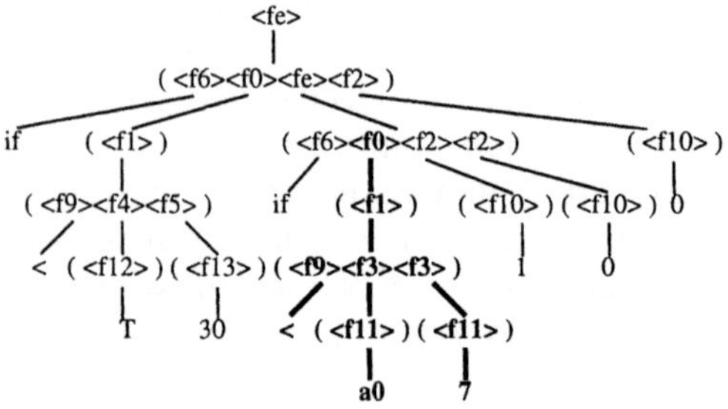

Fig. 4.5. Parent 2 for crossover: (if((<(T)(30)))(if((<(a0)(7)))(1)(0))(0))

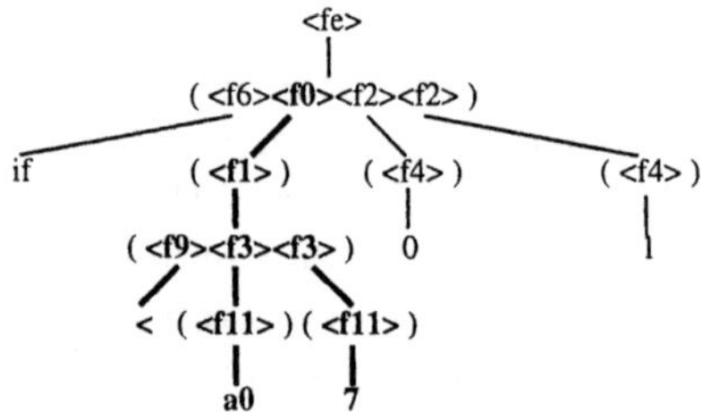

Fig. 4.6. Child 1 from crossover: (if((<(a0)(7)))(0)(1))

position-independence of subtrees in genetic programming would seem to obviate the need for a separate mutation operation; in effect, the crossover operation by itself serves as a kind of macromutation. Indeed, Koza[4] shows that the addition of a separate mutation operator does not lead to any substantial improvement in the performance of genetic programming. Following Koza and most other genetic programming researchers, we chose not to use a mutation operation in our application of genetic programming.

4.5 The Algorithm

The following algorithm is used in our application of symbolic regression to the ultimatum game data.

(a) Randomly generate a population of n rules.
(b) Play the rules against the same opponents that each experimental subject faced and evaluate their fitness. If stopping criterion is met, then stop.
(c) Choose k rules to survive to the next generation, with the probability of survival proportional to relative fitness.
(d) Choose $n - k$ parent rules on which to perform crossover with probability of being chosen proportional to relative fitness. The resulting recombined decision rules are the parent's offspring or "children."
(e) The next "generation" of rules is changed to include only the k survivors and the $n - k$ children. Go to step 2 and repeat steps 2–5.
(f) End the algorithm after a maximum number of generations, or if a perfect fitness score is achieved.

Note that a genetic program can be seen as the piecing together of building blocks, or sub–trees, which contribute a greater than average fitness to the complete rule, and which are small enough to survive crossover. The fact that reproduction and crossover are performed based on relative fitness increases the probability that a better than average fit subtree is used in future generations, and thus help to direct the search, as in Holland's genetic algorithm.

4.6 Parameters and Fitness Specification

In our regression analysis of the ultimatum game data we used the grammar as described in Fig. 4.1. We considered population sizes of $n = 200$ trees, with a maximum depth of 150. The number of trees that were selected for copying intact into the next generation was $k = 20$ or 10%. The remaining 180 trees were created through crossover alone. The selection of trees to be copied into the next generation as well as the selection of trees for crossover purposes was based on an adjusted and normalized fitness criterion. First, each decision rule was decoded to determine its raw fitness, which is simply the number of actions out of 37 that it correctly predicted for a particular player B. (Recall we are allowing for 3 lagged values and we have 40 observations). Let us denote the raw fitness score of decision rule i at generation t by $f(i,t)$. The adjusted fitness measure of rule i at generation t is given by:

$$af(i,t) = \frac{1}{1+f(i,t)}.$$

This transformation converts the raw fitness value into the $[0,1]$ interval, and also ensures that small fitness differences are sufficiently exaggerated, which becomes an important issue as the population of decision rules becomes increasingly fit over time. Note too, that this adjustment implies that the most fit rules are now those with the lowest adjusted fitness values. The normalized fitness value takes the adjusted fitness value, and makes it a relative, population–wide measure. The normalized fitness of rule i at generation t is defined by

$$nf(i,t) = \frac{af(i,t)}{\sum_{j=1}^{n} af(i,t)}.$$

The sum of all normalized fitness values is 1. When selection and crossover decisions are made they are made on the basis of this normalized fitness value. In particular, the 10% of strings selected for reproduction as well as the strings selected for crossover purposes are selected randomly (with replacement) with probability that is inversely proportional to normalized fitness values. These fitness measures and methods for determining the next generation from the current generation are standard in the genetic programming literature (see, e.g. [4]).

4.7 Regression Results for the Ultimatum Game

We selected 8 individual player Bs from the ultimatum game experimental data. The criterion for selecting a particular player B was that the player rejected a Player A offer at least twice in a 40–round game; we focused on these cases, as other cases had too little variation in actions played to detect any meaningful strategy other than "always accept". We present the data for each of these 8 player B subjects in the 8 examples of Table 4.1 below. In each example, the first line of data reveals the integer amount that a player A proposed to keep for him/herself. The second line represents the player B's actual response: 0 = reject, 1 = accept.

Table 4.1. Symbolic regressions on player B actions in the ultimatum game: 8 examples

Example 1

Data/rule comparisons in generations 1, 10, 20, & 30

subject		best rule fitness	mean fitness
A	6658766656766776677667886676566666656666		
B	1010011110011001000110001001110110111000		
rule			
1	1011000011001101110011101100010010001	25	17.62
10	1011000011001101110011101100010011100	26	19.20
20	0001000001000100110001100100000000000	23	19.67
30	1011100011101101111011101110011011001	25	19.71

Rule expressions (minimum depth among best fit in generations 1, 10, 20, & 30)

```
 1 (IF(B2)(0)(1))
10 (IF((<(T)(35)))(IF(B2)(0)(1))(B1))
20 (IF(NOT(AND(NOT(NOT(B2)))(NOT(NOT(NOT((=(T)(15))))))))(IF(B3)(0)(1))(IF((>(5)(8)))(0)(0)))
30 (IF(NOT(NOT(NOT(B2))))(1)(IF(NOT(B3))(B1)(IF(B3)(0)(IF(B2)(0)(1)))))
```

Minimum depth best fit rule interpretations in generations 1, 10, 20, & 30

```
 1    if accepted at t-2, reject
      else accept
10    if t < 35
            if accepted at t-2, reject
            else accept
      else repeat action at t-1
20    if rejected at t-2 and t is not 15
            if accepted at t-3, reject
            else accept
      else reject
30    if rejected at t-2, accept
      else if rejected at t-3, repeat action at t-1
      else if accepted at t-3, reject
      else accept
```

Results of expanded search with population size 500

(if(not((>(7)(a0))))(if(((>(8)(4))(0)(0))(1)), fitness = 29, generation 21

Example 2

Data/rule comparisons in generations 1, 10, 20, & 30

subject		best rule fitness	mean fitness
A	6546577776665666656576666556666676666656		
B	1110110001111001111101100110011101010011		
rule			
1	1011011111101111110110111011110111011	24	17.55
10	1110000111111111111111111011111111111	25	20.44
20	111001111111111111111101101111110101011	25	21.56
30	0010011111101111100110110010100101011	27	22.49

Rule expressions (minimum depth among best fit in generations 1, 10, 20, & 30)

1 (IF(OR(NOT((<(T)(10))))(NOT(NOT(NOT((=(7)(5)))))))(IF(0)(B1)(IF((<(A0)(A3)))(1)
(IF(B2)(B1)(1))))(IF(B3)(IF((>(T)(20)))(IF(1)(B2)(1))(B2))(0)))

10 (IF((>(A0)(A2)))(B3)(IF(AND(1)((<(T)(15))))(IF((=(T)(30)))
(IF(NOT(AND((=(A1)(7)))(OR((<(A0)(A3)))(AND((=(T)(10)))(1)))))(1)(B2))(B1))(1)))

20 (IF((>(A0)(A2)))(B3)(IF(NOT(NOT(B2)))(IF((=(T)(30)))(IF(NOT((>(A2)(A0))))(1)(B2))(B1))(1)))

30 (IF((=(T)(30)))(IF(AND((=(T)(10)))((<(T)(20))))(B1)(B3))(IF(B2)(IF(B2)
(IF(NOT((>(A0)(A2))))(B1)(0))(B1))(IF((<(T)(5)))
(IF(AND(NOT(NOT(NOT(B3))))(AND((<(A3)(A2)))(OR(NOT(0))(AND(B3)(B3)))))(0)(B1))(1))))

Minimum depth best fit rule interpretations in generations 1, 10, 20, & 30

1 if offer at t-0 > offer at t-3, accept
else if accepted at t-2, repeat action at t-1
else accept

10 if offer at t-0 > offer at t-2, repeat action at t-3
else if t < 15, repeat action at t-1
else accept

20 if offer at t-0 > offer at t-2, repeat action at t-3
else if accepted at t-2
    if t = 30
        if offer at t >= offer at t-2, accept
        else repeat action at t-2
    else repeat action at t-1

30 if t = 30, repeat action at t-3
else if accepted at t-2
    if not offer at time t > offer at time t-2, repeat action at t-1
    else reject
else if t < 5
    if rejected at t-3 and offer at t-3 < offer at t-3, reject
    else repeat action at t-1
else accept

Results of expanded search with population size 500

no improvement over original search

Example 3

Data/rule comparisons in generations 1, 10, 20, & 30

subject		best rule fitness	mean fitness
A	56679777297776375777888877817778868787 77		
B	11000111001111111111100011101111001010111		
rule			
1	0001110011111111111000111111111110010101	27	20.32
10	00111100111111111110011110111100111111	28	23.52
20	00111101111111111110011110111100111111	29	24.50
30	01111101111111111110111110111101011111	29	25.63

Rule expressions (minimum depth among best fit in generations 1, 10, 20, & 30)

1 (IF((>(5)(A0)))(1)(IF(NOT((>(T)(30))))(IF((=(T)(25)))(B1)(IF(1)(B1)(0)))(B2)))
10 (IF(B3)(B1)(1))
20 (IF(OR(0)(B3))(IF((<(A2)(6)))(IF(1)(IF(AND(B1)((<(A0)(A1))))(1)(1))(1))(B1))
(IF(0)(0)(IF((>(7)(7)))(IF((=(T)(20)))(B2)(0))(1))))
30 (IF(B2)(IF(B3)(IF(AND(0)(B3))(B3)(B1))(1))(IF((=(A1)(6)))(B3)(1)))

Minimum depth best fit rule interpretations in generations 1, 10, 20, & 30

1 if offer at t-0< 5, accept
else if t <=30, repeat action at t-1
else repeat action at t-2
10 if accepted at t-3, repeat action at t-2
else accept
20 if accepted at t-3
    if offer at t-2 < 6, accept
    else repeat action at t-1
else accept
30 if accepted at t-2
    if accepted at t-3, repeat action at t-1
    else accept
else if offer at t-1 = 6, repeat action at t-3
else accept

Results of expanded search with population size 500

higher fitnesses with more complicated "randomizing" rules

Example 4

Data/rule comparisons in generations 1, 6, & 7

subject		best rule fitness	mean fitness
A	9576766977866797778477788877778767467786		
B	0111111011011101110111100011110111111101		
rule			
1	1111111111111111111111111111111111111	28	22.18
6	1011000111100110111000011110111111001	30	25.95
7	1111011011101110111100011110111111101	37	26.96

Rule expressions (minimum depth among best fit in generations 1, 6, & 7)

1 (IF(NOT(0))(1)(1))
6 (IF((>(A0)(A3)))(0)(1))
7 (IF((>(A0)(7)))(0)(1))

Minimum depth best fit rule interpretations in generations 1, 6 & 7

1 always accept
6 if offer at time 0 > offer at time t-3, reject
else accept
7 if offer at time 0 > 7, reject
else accept

Example 5

Data/rule comparisons in generations 1, 10, 20, & 30

subject		best rule fitness	mean fitness
A	667659899667777876677874778778677877777		
B	1111100001111110111110111101101110111111		
rule			
1	11110001111111111111111111111111111111	29	23.11
10	11100001111111111111111111111111111111	30	27.08
20	11100001111111111111111111111111111111	30	27.62
30	11100001111111111111111111111111111111	30	27.65

Rule expressions (minimum depth among best fit in generations 1, 10, 20, & 30)

1 (IF(NOT(B1))(IF(B1)(B1)(B2))(IF(B1)(1)(1)))
10 (IF(NOT(0))(IF(NOT(B1))(IF((>(7)(A3)))(B1)(B3))(1))(IF(NOT(B1))(B1)(1)))
20 (IF(0)(1)(IF((<(T)(10)))(B1)(1)))
30 (IF((<(T)(10)))(IF(OR(OR(1)((=(T)(25))))(NOT(B1)))(B1)(1))(1))

Minimum depth best fit rule interpretations in generations 1, 10, 20, & 30

1 if rejected at t-1, repeat action at t-2
else accept
10 if rejected at t-1
if offer at t-3 > 7, repeat action at t-1
else repeat action at t-3
else accept
20 if t < 10, repeat action at t-1
else accept
30 if t < 10, repeat action at t-1
else accept

Results of expanded search with population size 500

(if(not((>(8)(a0))))(0)(1)), fitness = 37, generation 10

Example 6

Data/rule comparisons in generations 1, 10, 20, & 30

		best rule fitness	mean fitness
subject			
A	5769777776698396887777667778767757877758		
B	101000001110010100111111111111110000010		
rule			
1	000000011000000001111111111111000000	28	22.79
10	000000111011010011111111111111000011	29	24.91
20	000000011010010111111111111111010000	29	25.42
30	000000110001000011111111111111000000	30	26.02

Rule expressions (minimum depth among best fit in generations 1, 10, 20, & 30)

```
 1  (IF((>(A2)(A3)))(B2)(IF(B1)(B2)(B1)))
10  (IF((=(A0)(7)))(IF((<(4)(A3)))(B1)(B3))(IF(B2)(IF(1)(IF(B2)(B1)(0))(B1))(IF((<(T)(5)))
    (IF((>(T)(10)))(IF((=(T)(30)))(IF(NOT(1))(1)(IF(NOT(1))
    (IF((>(T)(25)))(B1)(B3))(B3)))(0))(0))(1))))
20  (IF((=(8)(A3)))(IF((<(4)(A3)))(B1)(B3))(IF(B2)(IF(B2)(IF(B2)(B1)(0))(B1))(IF(NOT((>(T)(10))))
    (IF(B2)(B1)(0))(IF(NOT((<(T)(10))))(B3)(B3)))))
30  (IF((<(T)(20)))(IF(NOT(AND(NOT(1))(AND((=(T)(5)))(OR(AND((>(7)(5)))(1))(AND(B3)(B2))))))
    (IF(B3)(IF((>(6)(5)))(0)(B2))(B1))(B2))(IF(B2)(IF(B2)(B1)(0))(IF(B2)(B1)(0))))
```

Minimum depth best fit rule interpretations in generations 1, 10, 20, & 30

```
 1    if offer at t-2 > offer at t-3, repeat action a t -2
      else if accepted at t-1, repeat action at t-2
      else repeat action at t-1 (reject)
10    if offer at t = 7
              if offer at t-3 < 4, repeat action at t-1
              else repeat action at t-3
      else if accepted at t-2, repeat action at t-1
      else if t < 5, reject
      else accept
20    if offer at t-3 = 8, repeat action at t-1
      else if accepted at t-2, repeat action at t-1
      else if t <= 10
              if accepted at t-2, repeat action at t-1
              else reject
      else repeat action at t-3
30    if t < 20
              if accepted at t-3, reject
              else repeat action at t-1
      else if accepted at t-2, repeat action at t-1
      else reject
```

Results of expanded search with population size 500

no improvement over original search

Example 7

Data/rule comparisons in generations 1, 10, 20, & 30

subject		best rule fitness	mean fitness
A	5478669677977762778875778877087787787877		
B	11101101110111111110011110011101101101011		
rule			
1	111111111111111111111111111111111111	26	20.39
10	111111111111111111111111111111111111	26	24.26
20	111111111111111101111101111111111111	28	25.12
30	111111111111111111111111111111101111	27	25.28

Rule expressions (minimum depth among best fit in generations 1, 10, 20, & 30)

1 (IF(0)(B1)(1))
10 (IF(1)(1)(1))
20 (IF((<(A0)(8)))(1)(IF((<(8)(5)))(1)(IF(NOT((=(5)(8))))(B1)(B1))))
30 (IF(1)(IF(AND(NOT(NOT(AND(NOT(NOT(AND((=(T)(35)))(B1))))(B1))))((>(T)(5))))
(IF((<(A3)(A1)))(1)(IF(NOT((<(A0)(4))))(IF((<(8)(5)))(B1)(IF((<(T)(15)))(B1)(B3)))(B2)))(1))
(IF(AND(NOT(AND(1)(NOT((<(T)(10))))))(NOT((<(8)(5)))))(1)(B3)))

Minimum depth best fit rule interpretations in generations 1, 10, 20, & 30

1 always accept
10 always accept
20 if offer at t < 8, accept
else repeat action at t-1
30 if t = 35 and accepted at t-1
    if offer at t-3 < offer at t-1, accept
    else if offer at t >= 5, repeat action at t-3
    else repeat offer at t-2
else accept

Results of expanded search with population size 500

(if((>(a0)(7)))(if((<(7)(a0)))(0)(b1))(if(0)(b1)(1))), fitness = 37, generation 6

Example 8

Data/rule comparisons in generations 1, 10, 20, & 30

subject		best rule fitness	mean fitness
A	56967465865465566565756767666666665666666		
B	10011111011101111111011010111110001100100		
rule			
1	11111111111111111111111111000000000	28	20.28
10	11111111111111111111111111111111100	27	23.60
20	11111111111111111111111111111110000	27	23.82
30	11111111111111111111111111111110000	27	24.22

Rule expressions (minimum depth among best fit in generations 1, 10, 20, & 30)

1 (IF((>(T)(30)))(0)(1))
10 (IF((>(T)(35)))(B3)(1))
20 (IF((>(T)(35)))(0)(1))
30 (IF((>(T)(35)))(IF(OR(NOT(NOT((<(T)(5)))))((<(A3)(7))))
(IF(NOT(OR(NOT(NOT(1)))(OR(NOT((>(8)(A1))))(OR(NOT(NOT((>(7)(4)))))
(OR(1)((<(A3)(7))))))))(1)(0))(1))(1))

Minimum depth best fit rule interpretations in generations 1, 10, 20, & 30

1 if t > 30, reject
else accept
10 if t > 35, repeat action from t-3
else accept
20 if t > 35, reject
else accept
30 if t > 35
if offer at t-3 < 7, reject
else accept
else accept

Results of expanded search with population size 500

no improvement over original search

For each player B history, we report the results of our symbolic regression in several ways. First, we provide the binary string strategy that was generated by the best–of–generation rule, along with this rule's raw fitness value and the mean raw fitness value of the population of 200 rules at various generations (iterations of the algorithm). The symbolic regression was stopped after a maximum of 30 generations, or earlier if a perfect raw fitness score was achieved. If there was more than one best of generation rule at each generation that we report, we chose to report the most parsimonious rule, i.e. the rule with the minimum depth. We also provide the best of generation rules themselves, in both symbolic (computer program) form, and in a decoded and reduced ("plain English") form. Finally, we report whether an expanded search over 30 generations with a larger population size of 500 rules yielded any improvement in fitness, and if so, we report the best rule obtained.

Recall that a perfect raw fitness score in this application is 37. This perfect score was achieved (within 30 generations) in three of the 8 examples, examples 4, 5 and 7. In example 4 for instance, player B's strategy was uncovered to be of the simple form: reject any proposed split that gives player A more than \$7, otherwise accept. This strategy was uncovered by our genetic program algorithm with 200 rules after only 7 generations. Of course, such threshold strategies are fairly easy to uncover by carefully examining the history of play for these player Bs. However, the point of our exercise is that we can use symbolic regression to automate this inference process. Furthermore, inference may not be so easy if the player's strategy is not perfectly deterministic or is time–varying due to learning.

Indeed, time dependence is shown to matter in Example 5, where the minimum depth, best–of–generation rule at the end of 30 generations yielded a raw fitness score of 30. This rule had player B playing the same action played in the previous period for the first 10 periods of the game. Following period 10, the strategy was to always accept any player A proposal. This rule, of course, misses the player B's rejections of player A offers of 8 following period 10, but it is able to detect that a change has occurred in the proposals being made by player As starting around period 10 (in this case, it is a change in the magnitude of player A proposals, which become smaller). Of course, if we allowed for a more continuous dimension for time T this rule might be further improved. Note that the $T < 10$ substring survives in the best of generation string for some time (the last 10 generations). The survival of this substring illustrates the notion that highly fit building blocks are more likely to be chosen over time. The actual strategy of the player B in example 5, like the rule in example 4, is a deterministic rule: a close study of the history of play of player B in example 5 reveals that this player always rejected proposals that were greater than 7. While our algorithm was not able to detect this rule in 30 generations with a population of 200 strings, when we increased the population size to 500 strings, we were able to perfectly uncover this player B's deterministic rule in 10 generations. Similarly, the expanded search with 500 rules led to a perfect fit rule in example 7 after just 6 generations.

The other 5 examples in Table 4.1 yield best–of–generation rules that are more difficult to interpret, with ending raw fitness scores of 30 or less, and no improvement from expanded searches with 500 rules. We see in the inferred rules from these examples that time conditioning is of some importance. For instance, in Example 6, the best of generation rule at the end of the simulation run (30 generations) is able to detect a break in player B behavior around period 20, and the rule conditions its actions on this break–point. We also see the survival over many generations of highly fit substrings in the best of generation rules. For example, the substring "if accepted at $t-3$" appears in the minimum depth, best fit rule of generations 10, 20 and 30 in Example 3 as does the substring "if $t > 35$" in Example 8. The survival of such highly fit substrings would be more apparent if we reported all of the rules, not just the minimum depth, best–of–generation rule. Finally, we see a lot of strategies that condition on the actions played by the player B in the past, as well as on the proposals made by player As in the past, rather than on the current $t = 0$ proposal. While the responders in the experiment may have been conditioning on this type of information, – they did have the past history of 10 rounds of play on their computer screens – we suspect that what the genetic program is actually doing is using conditioning on past actions as a means of implementing probabilistic choices. In our current implementation, there is no randomization mechanism; therefore, the genetic program may have to invent such a mechanism in order to advance the search for more fit strings. We plan to implement the possibility of randomized choices in future work, by including in the set of terminals a floating point constant chosen randomly from the set of real numbers on the unit interval.

4.8 Summary and Conclusions

In this paper we have discussed how genetic programming can be used to conduct a symbolic regression analysis. We have then applied this procedure to experimental data from the repeated ultimatum game. Our aim was to uncover player's strategies from the actions these players played, a perennial inference problem in the analysis of experimental data. We find that our algorithm is frequently capable of uncovering simple, deterministic strategies, but has more difficulty with strategies that appear to be random or time-varying. While simple rules may be detected by careful inspection of the data, the fact that our algorithm can automate this process is an appealing feature. We believe that with further work, as well as with longer simulation run–times, we will be able to make further progress on the recovery of random or time–varying rules.

The main advantage of our approach over standard regression analyses is that we do not have to prespecify the structure of the regression equation. Instead, we only need provide a set of terminals and nonterminals as part of the grammar of some language that is then used to derive decision trees (e.g. LISP). Finally, our technique can be used to assess the degree of homogeneity of strategic play among subjects assigned to play the same role in economic decision-making experiments. If strategic behavior is not sufficiently homogeneous, as we have found in our simple examples, then it may not be reasonable to pool data across subjects as is frequently done in analyses of experimental data.

References

1. Duffy J., Feltovich N. (1999) Does Observation of Others Affect Learning in Strategic Environments? An Experimental Study. International Journal of Game Theory **28**, 131–152
2. Geyer–Schulz A. (1997) Fuzzy Rule–Based Expert Systems and Genetic Machine Learning. 2nd edn. Physica–Verlag
3. Holland J. (1997) Adaptation in Natural and Artificial Systems. University of Michigan Press
4. Koza J. R. (1992) Genetic Programming: On the Programming of Computers by Means of Natural Selection. MIT Press

Part III

Agent-Based Computation Economics

5 The Efficiency of an Artificial Double Auction Stock Market with Neural Learning Agents*

Jing Yang

Department of Economics, Concordia University
Financial Market Department, Bank of Canada
jyang@bank-banque-canda.ca

Abstract. The goal of this paper is to investigate the convergence property in a double auction market and test the robustness of the results. We construct an artificial equity market where agents trade a risky asset that pays a stochastic dividend each period. Artificial Neural Networks take on the role of traders, who form their expectations about the future return and place orders based on their expectations. Market prices are set endogenously by trading among agents in a double auction market. The efficiency of this artificial market is measured by the convergence of the price to the Rational Expectations Equilibrium (REE). We find that market dynamics under double auction converge to the REE in in some experiments. This convergence, however, is sensitive to the deviation from rationality among the agents. In the experiment where we introduce noise trading, convergence becomes unattainable. Minimal rationality is not sufficient to generate convergence in a double auction market when the market price is endogenous.

5.1 Motivation and Introduction

The continuous double auction under the open-outcry floor trading has been the principle trading system in equity markets for more than 140 years. Technological advances have now introduced a new form of market trading system: the computerized double auction mechanism (widely used in electronic trading). What are the features of this market? How do these features affect the price formation process? How efficient is this type of market system compared to those with alternative trading systems? Experimental studies answer these questions by examining two different types of artificial markets: a laboratory

* This paper is drawn from a chapter in my doctoral dissertation at Concordia University. I am grateful to my thesis supervisors Professor Bryan Campbell, Lawrence Kryzanowski and Anastas Anastapoulos for their insightful advice and valuable support. I have also benefited from helpful comments by seminar participants of the CEF 99 conference at Boston. Special thanks forward to my Bank collegues Nicolas Audet, Toni Gravelle and Peter Thurlow. Obviously, any errors are the responsibility of the author. The views expressed in this paper do not represent the official views of Bank of Canada.

market where human subjects trade with each other and generate market dynamics; or a computer simulated market where artificial traders replace the human traders. The main weakness of the laboratory market is that some parameters are unobservable or uncontrolled, like a trader's risk preference, learning behaviour and trading strategies. In this sense, a computer simulated artificial market improves upon the laboratory market, since those parameters as well as the trading system can be carefully controlled and modified to test many different hypotheses. On the other hand, this computational tool brings in many new untested parameters. Many laboratory market are conducted under continuous double auction. References [28] and [16] introduced laboratory asset markets and feature the markets with insider traders and uninformed traders and conclude that the equilibrium prices do reveal insider information. The lesson from the double auction asset market experiments of [32] seem to be rather different. They report frequent large bubbles-episodes where transaction prices rise well above the Rational Expectations Equilibrium (REE) values for an extended time period, which usually ends in a sudden price crash to one that is close or below the fundamental value. Reference [8] studied a market where a stochastically lived asset pays dividend each period.

This is a more complicated market where token valuations are not specified exogenously by the experimenter but are determined endogenously and must be inferred by traders over the course of the trading process. They concluded that " the double auction does not fail completely at generating convergence to competitive equilibrium prices for stochastically lived asset, but the auction mechanism performs much more slowly and erratically than in simpler settings." Computer simulated markets extend the experimental study by permitting a manipulation of the agents' trading strategies and a trading mechanism. After conducting of many real trading experiments in laboratories, [18] were interested in just how much "intelligence" was necessary to generate the results they were seeing. They ran a computer experiment with "Zero Intelligence" (ZI) agents who are almost completely random in their behavior. They used a double auction market similar to those used in many laboratory experiments. Their results showed that the budget constraint is critical to achieve the market efficiency. Further examples of a simple double auction can be found in [29]. They focused on the performance of different trading strategies. Experiments involving simulated markets frequently appear in the literature. Reference [10] construct an double auction asset market experiment to investigate information dissemination and aggregation. They find in tests of information dissemination, prices can converge to the fundamental value in almost all cases. But convergence is difficult to achieve in tests of price aggregation in many cases. To test the robustness of the results, technical traders and noise traders are included. The Santa Fe stock market ([2] and [23]) investigate market efficiency and price convergence with a rational expectations asset pricing model, a genetic algorithm learning and Walrasian tatonnement as the market clearing mechanism. Reference [24] provides a very useful survey of some portions of this literature. This paper addresses the market efficiency question by measuring the price convergence to a rational expectations equilibrium price. In particular, we consider the price deviation and trading volumes. We conduct this study with 3 purposes in mind: (1) Simulate a computerized double auction platform in order to help us better understand the price formation process in this trading structure and to test other hypotheses on market design and trading behaviour; (2) Replace the human trader with the artificial neural network traders to observe whether they are able to learn to form the rational expectations and how they adjust their expectations within each period; (3) Compare this double auction market outcome to theoretical benchmarks and to the results from laboratory markets. It is worth to mention that in most of the laboratory market studies where double auctions are used, the "redemption value" (return) of the risky asset is exogenously set by the designers and the underlying distribution of this value is discrete and uniform . By contrast, in our simulated market, the market price, a major part of the return, is endogenously determined by the trading process. Dividend, the other part of the return, follows an exogenous stochastic process. The remainder of the paper is organized as follows. Section 5.2 presents the market structure. In this section, our basic asset pricing model, the agents' expectation formation, their trading strategies and a double auction trading mechanism are described. Section 5.3 provides the design of all the experiments. The results of these computational experiments are presented in Sect. 5.4. The market simulation results are evaluated, tested and compared with the findings in the literature. Section 5.5 concludes and suggests future extensions to the work reported herein.

5.2 Market Structure

There are three building blocks in a market simulation: environment; expectations formation and trading strategies; and trading mechanism.

5.2.1 Basic Framework

We set up a model of an asset market along the lines of [19] and [21]. Consider a market in which N traders, indexed by j. Traders are assumed to have the constant absolute risk aversion (CARA) utility function. Time is indexed by t and the horizon is indefinite. Traders decide on their desired asset composition between a risky stock and a risk-free bond. The risk free bond is in infinite supply and pays a constant interest rate r. Stock is issued in units, and pays a stochastic dividend, D_t, which follows an exogenous stochastic process $\{D_t\}$ as follows

$$D_t = \bar{d} + \rho D_{t-1} + \xi_t, \tag{5.1}$$

where $\xi_t \sim N(0, \sigma_\xi^2)$.

Each trader attempts to maximize next period's wealth, by optimizing the allocation between the risky asset and the riskfree asset as follows:

$$Max \quad \hat{E}_t^j[-exp(-\lambda W_{t+1}^j)] \tag{5.2}$$

$$s.t \ W_{t+1}^j = (1 + r_t)W_t^j + Q_t^j[P_{t+1} + D_{t+1} - (1 + r_r)P_t] \tag{5.3}$$

where λ is the constant coefficient of risk aversion, P_t is the stock price and E_t^j denotes the expectation of trader j conditional on the set of information Ω_t available to j at time t, including historical information on market prices and dividends.

It is well known that under CARA utility and Gaussian distributions for predictions, the agent's desired demand, Q_t^{j*}, for holding risky asset is linear in the expected excess return:

$$Q_t^{j*} = \frac{\hat{E}_t^j[P_{t+1} + D_{t+1} - (1 + r)P_t]}{\lambda \hat{\sigma}^2_{j, P_{t+1}+D_{t+1}}|\Omega_t} \tag{5.4}$$

Risk aversion and the existing forecast error prevent traders from taking an infinite position based on expected return differentials.

Market clearing price can be find by balancing aggregate demand to aggregate supply.

$$\sum_{j=1}^{N} Q_t^{j*}(P_t) = \sum_{j=1}^{N} \frac{\hat{E}_t^j[P_{t+1} + D_{t+1} - (1 + r)P_t]}{\lambda \hat{\sigma}^2_{j, P_{t+1}+D_{t+1}}} = \bar{Q} \tag{5.5}$$

Under the *full information* and homogeneous expectations, every trader observes each of these components in the dividend process and holds identical expectations, the index j can be omitted from th expectations operator. Using equation (1) and (5), it is not difficult to solve for a homogeneous rational expectation equilibrium (REE) under full information as

$$P = fD_t + g \tag{5.6}$$

where $f = \frac{\rho}{1+r-\rho}$ and $g = \frac{1}{r}(1 + f)[\bar{d} - \lambda(1 + f)\sigma_\xi^2(\frac{\bar{Q}}{N})]$.

Finally, as a benchmark for evaluating the equilibria in subsequent experiments, it is useful to find the analytical expression for the expected return in equilibrium. Given equation (5) and the direct relationship between price and dividend in equation (6), the optimal forecasts in the full revealing rational expectation equilibrium is easy to find. One form of these forecasts is

$$E_t(P_{t+1} + D_{t+1}) = \rho(1 + f)D_t + [(1 + f)\bar{d} + g] \tag{5.7}$$

This is the homogeneous rational expectation equilibrium(h.r.e.e.) forecasting we seek. This forecasting equation is one of the crucial items that traders need to estimate. It is used as a benchmark for our agents to learn. When the underlying components of the dividend process are not directly observable, the key question is how the next period price and dividend expectations might be formed by each trader. We will consider this issue in the Sect. 5.2.2.

5.2.2 Heterogeneous Traders and Their Trading Strategies

We assume this market is populated by two types of traders: value traders (Artificial Neural Network traders) and momentum traders. Among them, the value traders follow the rational expectation model and try to learn the state of nature and the rational expectations forecasting given by equation (7) simultaneously. Their existence is expected to keep market prices close to the rational expectations equilibrium. Momentum traders are chartists and do not follow the rational expectations model. They act according to some technical indicators. The presence of this type of traders is to test the robustness of the results.

ANN traders with their trading strategies For value traders, they need to learn all the parameters in equation (7) without having any information about the distribution of the dividends. We assume value traders employ artificial neural networks (ANN)[1] to learn the parameters.

Artificial neural networks provide a conceptually simple procedural model of how agents might learn to approximate their true but unknown conditional expectation function and hence form "boundedly rational" expectations. This structure will offer better desirable properties comparing to traditional learning method (say, least squares): It avoid any presumption on beliefs and require no ex ante knowledge of the structure. And it will better mirror actual cognitive process, in which different individual might well "cognize" different patterns and arrive at different forecast from the same market information.

Feedforward neural networks are employed by ANN traders with one input, one hidden layer (with three neurons) and one output(for more details see Appendix). There are two reasons for choosing this relatively simple network design. First, the single hidden layer feedforward network possesses the universal approximation property that it can approximate any nonlinear function to an arbitrary degree of accuracy with a suitable number of hidden units ([36]). Second, since the rational expectation mapping in equation (7) is a simple linear one, a more complex architecture is unnecessary. The motivation for keeping one hidden layer is to provide value traders with

[1] Reference [31] is a good introductory book that concentrates on one type of neural network; namely, feedforward neural networks, one of the most commonly used NNs. Reference [20] provide a clear and concise description of the theoretical foundations of neural networks using a statistical mechanical framwork.

the possible learning structure for out-of-equilibrium behavior. The homogeneous rational expectation equilibrium(h.r.e.e.) ,represented by the equation (7), rely on some very strong assumptions such as homogeneous beliefs among agents;common ex ante knowledge on the distribution of the state variable, and a Walransian tatonnement. While in our market simulation, since all of those assumptions are relaxed, value traders shall expect to encounter some out-of-equilibrium behaviour, especially when momentum traders participate on the market.

Each value trader possess an ANN model, which have similar structure but there are differences in the initial value of the parameters. Traders train their model according to newly revealed market information on return at the end of each period, and then use those updated model to forecast the next period return.

We now consider the question of when a trader will choose to trade via a market order or limit order, or not seek to trade at all. The problem is structured as follows. Let the trader consider to rebalance his portfolio at the beginning of every trading round (Each trading period contains multiple trading rounds.) To simplify the analysis, we assume that all orders are for a fixed number of shares($\triangle Q$) and when any market or limit order is executed, it is satisfied fully at the stated price. We assume that an unfulfilled limit order is canceled prior to the next decision point.

At any of the time t decision point, the trader j is faced with the five possible options:

(a) Post a limit buy order at price $P^b_{j,t}$
(b) Post a limit sell order at price $P^a_{j,t}$
(c) Post a market order to buy shares at the current best ask price, a
(d) Post a market order to sell shares at the current best bid price, b
(e) Do nothing.

The formal way for trader to find the optimal order placement strategies is maximizing his expected utility of wealth, $Max(U_1, U_2, U_3, U_4, U_5)$, where U_i is the expected utility of choosing above option (i). In this simulation, we use a simpler way to approximate traders' optimal trading strategies.[2]

In order to choose his action, an ANN trader will find his reservation price first. At this price his optimal holding equal to his current holdings and he has no incentive to trade, which implies he will leave the market.

A risk averse trader finds his optimal demand by maximizing his utility in terms of next period wealth. His reservation price can be solved from the optimal demand function given by equation (4). Let $Q^{j*}_t = Q^j_t$, where Q^j_t is the current holding and the Q^{j*}_t is the desired holding in current period, the

[2] We should keep in mind that finding the formal solution for optimal strategy is very important. Even the double auction appears to be very complex a game to yield a clear game theory solution. This will be studied in the future.

reservation price for trader j is

$$P^{R,j} = \frac{\hat{E}_t^j(P_{t+1} + D_{t+1}) - \lambda\hat{\sigma}^2_{j,P_{t+1}+D_{t+1}} Q_t^j}{1+r} \tag{5.8}$$

As can be seen, this reservation price is negatively related to his current holding level, and positively related to the expectation value of next period return.

Given this reservation price, he will never buy higher or sell lower than this price. Let a denote the best ask, and b the best bid $(a > b)$in the current trading round, the value trader will

(a) Post a market buy order , if $a < P^R$
(b) Post a market sell order, if $b > P^R$
(c) Do nothing, If a(or b)$=P^R$
(d) Otherwise, post a limit order to bid or ask shares, the order size, at the price

$$P_{j,t} = P^{R,j} - SB \tag{5.9}$$

B is an indicator variable where $B = +1$ for an ask order and $B = -1$ for a bid order. S is a spread between the reservation price and the price quoted. For a given reservation price, SB picks up half of the bid and ask spread. S is an important parameter that reflects price discreteness. It is a linear function[3] of order size $\triangle Q$, but the parameter for this linear function could be different across traders if they have different forecasting variances,

$$S^j = \left(\frac{\lambda\hat{\sigma}^2_{j,P_{t+1}+D_{t+1}}}{1+r}\right)\triangle Q \tag{5.10}$$

The reservation price here serves as a mean-reverting device for the market. When traders hold very similar expectations, the difference between their reservation prices come from the difference between their current holding in the risky asset. The low-reservation-price trader (who has large position) posts a low bid and ask price; while the high-reservation-price trader (who has small position) posts a high bid and ask price. As a result, the low-reservation-price trader could only become a seller since his ask order will have a high probability of being executed. Therefore, he can reduce his position and keep selling until it reaches his optimal holding.

Momentum traders and their trading strategies Momentum traders are chartists who believe that future price movements can be determined by

[3] It is important to keep in mind that the spread should also be a function of time, the closer to the end of a trading day, the higher bid price should be, which reflects the probability of the order execution. This aspect of the market will be studied in the future.

examining patterns in past price movements as represented by various moving averages (MAs). Several studies support the notion that the price history can be used to predict market movements,[4] which is partly due to the "self-fulfilling" nature of expectations. The moving average trading rule states that when the short-term (usually 1–to–5 day) moving average is greater than the long-term moving average (usually more than 50 days) [7], a rising market is indicated. Thus, this trading rule would generates a buy signal. Based on such market trends, the momentum trader decides to enter or exit the market.

The momentum traders are divided into two groups according to their choice of trading rules. The first group of momentum traders compares the current market price P_t with MA(5). That is,

- If P_t >MA(5), they buy ΔQ shares.
- If P_t =MA(5), they hold their current position.
- If P_t <MA(5), they sell ΔQ shares.

The second group of momentum traders identifies a trading opportunity by comparing MA(5) with MA(10). Specifically:

- If MA(5)>MA(10), they buy ΔQ shares.
- If MA(5)=MA(10), they hold their current position.
- If MA(5)<MA(10), they sell ΔQ shares.

5.2.3 The Trading Institution: Double Auction

We chose double auction mechanism for two reasons. First, major stock, commodity and many other markets are organized as double auctions. A classic example of a double auction market is the NYSE. Second, laboratory double auctions with human traders are known to yield data that approximate the competitive equilibrium in a variety of environments. [5]

Any trading mechanism can be viewed as a type of trading game in which the players meet at some venue and act according to some rules. These rules dictate who can trade, when and how orders can be submitted, who may see or handle the orders, how orders are processed, and how prices are eventually set. It has been recognized that these rules are very important because they

[4] In a study by [5] which uses bootstrapping techniques, two of the most popular technical trading rules are used to predict the movements of the Dow Jones Industrial Average (DJIA).

[5] It is worth to mention that most of the laboratory double auction markets set the redemption values of the risky asset exogenously. When these values become endogenous, as [17] point out: 'we start to see how the nice properties of DA markets can start to break down in more complicated enviroments where token valuations are not specified exogenously by the experimenter but are determined endogenously and must be inferred by traders over the course of the trading process.'

can affect bidding incentives, and therefore the terms and the efficiency of an exchange.

Rational expectations equilirium(REE) rely on a Walransian tatonnement auction, which has long served the needs of general equilibrium price theory. In this type of auction, the time at which the orders arrive is not relevant. It is a path-independent process. In this auction, the Walrasian auctioneer, who can observe all the traders' demand functions, adjusts prices in various markets to equilibrate aggregate supply and aggregate demand, which reflects a *competitive equilibrium (CE).* The key feature of these process is that agents are not permitted to trade "out of equilibrium", until the adjustment process has determined a CE price vector.

While in a double auction (known by experimentalists as the *continuous double auction* and by practitioners as an *open-outcry market*) we used in this paper, the price is path-dependent. The time at which the orders arrive is relevant. In a double auction (DA), transaction prices arises from a two-sided auction where buyers improve (raise) bids and sellers improve (lower) asks until one of the buyers and one of the sellers reach agreement.

In our simplified double auction market[6], traders can either submit a bid(ask), or accept an existing bid(ask). If a bid(ask) exists for the stock, any subsequent bid(ask) could be higher or lower than the current one, but only the best bid(ask) price can be obeserved by traders. There is no role of market maker. A transaction occurs when an existing bid (ask) is accepted. Given the absence of opportunity to borrow or sell short, traders have to trade subject to their budget constraints. A trader leaves the market when either the market price equals his reservation price or he runs out of money.

Each trading period contains 40 trading rounds. At the beginning of each trading round, we assume a random permutation of the traders, which determines the subsequent order of traders. Initially, the value traders come to the market with their own reservation prices and attempt to post or accept a bid (ask) order by comparing their reservation prices with the existing best ask (best bid). They can only observe the best bid and the best ask price. Momentum traders go to the market with their "buy", "sell" or "hold" signals and submit market orders according to these signals. At the end of each trading period, the last transaction price, *closing price*, is recorded as a market price of this period. Agents trade with each other according to the following scenarios:

- Scenario 1. If the best bid, b, and the best ask, $a (a > b)$, exist on the market
  - If $P^{R,j} > a$, he will post a market order, buy at this ask price;
  - If $P^{R,j} < b$, he will post a market order, sell at this bid price;
  - If $b < P^{R,j} < a$ and $P^{R,j} < (a + b)/2$, he will post a sell order at a price of $(P^{R,j} + S^j)$;

[6] It is similar to the one in [10].

  - If $b < P^{R,j} < a$ and $P^{R,j} > (a+b)/2$, he will post a buy order at a price of ($P^{R,j} - S^j$)
- Scenario 2. If only the best ask, a, exists
  - If $P^{R,j} > a$, he will post a market order, and buy at this ask price;
  - IF $P^{R,j} < a$, he will post a buy order at a price of $(P^{R,j} - S^j)$
- Scenario 3. If only the best bid, b, exist
  - If $P^{R,j} < b$, he will post a market order, sell at this bid price;
  - If $P^{R,j} > a$, he will post a sell order at a price of $(P^{R,j} + S^j)$;
- Scenario 4. If no bid or ask exist,
  - he will has an equal chance to post a buy or a sell order at price of $(P^{R,j} - S^j)$ or $(P^{R,j} + S^j)$ respectively.

5.3 Experiment Design

The rational expectations equilibrium price, given by equation (6), will be used as our bench mark for the four experiments. The traders are assumed to work in the following environment:

(a) The riskfree interest rate, r, is fixed to be 0.1; risk aversion parameter, λ, is set to be 0.5.
(b) Throughout the experiments, we specify the dividend process as equation (1) where $\bar{d} = 0.5$, $\rho = 0.7$, and $\sigma_\xi^2 = 0.25$.
(c) Traders are assumed to hold an endowment of 2 units of shares and 2000 dollars in cash.
(d) Value traders are assumed to be equipped with an ANN learning mechanism. The neural network is programmed to learn the value of the parameters in rational expectation equation. The initial starting value of weights is randomly draw from an uniform distribution between [-1 1].
(e) Momentum traders are initially endowed with "hold" signals until they get enough information to form their trading rules. Moving averages of the stock price are calculated with equal weighted parameters. Momentum traders using MA(5) or MA(10) wait for 5 or 10 periods respectively to enter the market. They enter the market with their "buy" or "sell" signals generated from their trading rules.
(f) A minimum unit of time corresponds to a trading round. Each trading period contains 40 trading rounds. Each trading round begins with a permutation of the traders to determine the order of the participation in the market. The trading round is completed when each trader is considered in turn.
(g) The market is run for 1000 periods and 25 times for same experiment.

- Experiment 0. 10 value(ANN) traders with a Walrasian Auction
  The purpose of this experiment is to check whether the value ANN trader is able to learn the rational expectations mapping between next period's return and the lagged dividend, which is given by equation (7).

- Experiment 1. 10 value(ANN) traders with a Double Auction (Experienced traders)
  The purpose of this experiment is to test the convergence of prices to the REE with a double auction trading institution. Value traders in this experiment are trained on the presample price and dividend data for 40 periods before they enter the market.
- Experiment 2. 10 value(ANN) traders with a Double Auction (Inexperienced traders)
  The value traders in this experiment are not trained by the presample data before they enter the market. This experiment is designed to check the effect of traders' experience on the convergence property.
- Experiment 3. 10 value traders, 10 momentum traders with a Double Auction
  The purpose of this experiment is to check the sensitivity of the convergence on this double auction market to the deviation from rationality. The presence of momentum traders should add some noise to the market. In this experiment, it would be interesting to observe whether the value traders can maintain rational expectation equilibrium given the existence of some erroneous signals caused by the presence of "irrational" momentum traders.

5.4 Computational Results

The efficiency of a market in this paper is measured by how close and how fast the prices converge to REE. The variables we consider are price deviation and trading volume. The diminishing trading volume suggests that the market is approaching to its equilibrium.

In general, ANN traders are able to learn rational expectation parameters collectively. Convergence to the REE occurs rapidly in Experiment 1, the case of experienced traders and identical trading strategies. Such convergence occurs more slowly in experiment 2, where value agents are not trained before they enter the market. In experiment 3, with the presence of momentum traders, convergence is unattainable. This can be explained by the fact that in the first experiment, all the traders are fundamental traders. Their forecasts are supervised by the fundamental values of the stock. In the case that includes momentum traders, no price convergence occurs. The reason is that momentum traders only obey their own trading rules (e.g. MA(5) or MA(10)), which are not consistent with rational expectation. They disregard the views of value traders, and value traders are not able to learn those technical trading rules either. The value traders may not even realized the presence of momentum traders. In other words, the fundamental value of our stock is not common knowledge.

Can ANN learn the rational expectations?

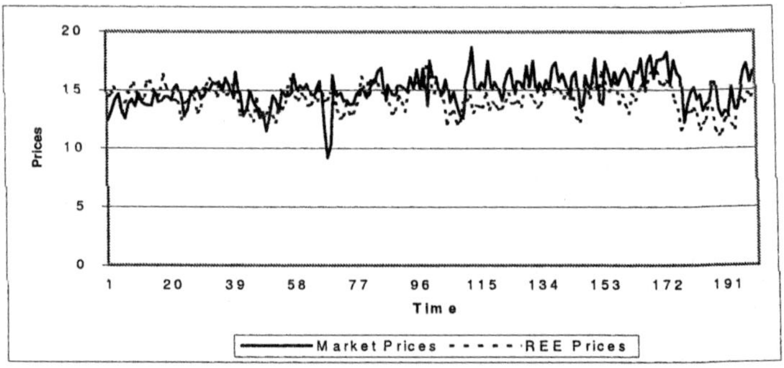

Fig. 5.1. Experiment 0: Market price vs. REE price

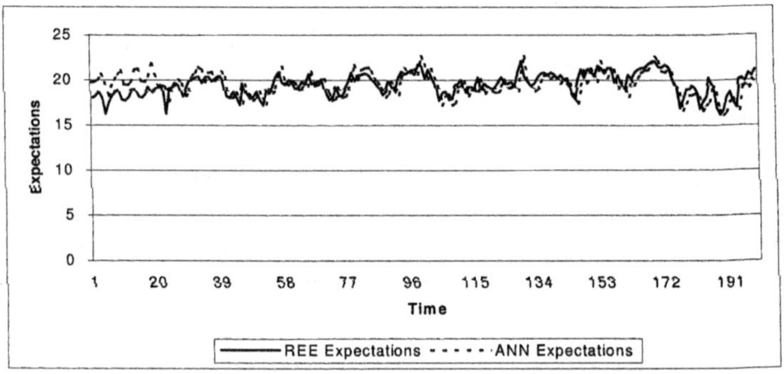

Fig. 5.2. Experiment 0: Rational expectation vs. ANN expectation

In experiment 0, value traders use ANN to form their expectations and submit their demand functions to a "Walrasian auctioneer", which implies this auctioneer can observe every trader's demand function. The auctioneer calculates the market price which represents the point where aggregated demand equals supply. As is seen from Fig. 5.1, the price series appears to be nearly identical to the price in the rational expectations regime, the absolute deviations between ANN expectations and rational expectations are very small (Fig. 5.2). Under linear homogeneous REE, the price and dividend is a linear function of lagged dividend. For a further test, the market price and dividend is regressed on a lag of the dividend and a constant term. The estimated term is very close to the analytical solution of value of parameters in equation equation (7).

Do prices converge to a fully revealed rational expectations equilibrium?

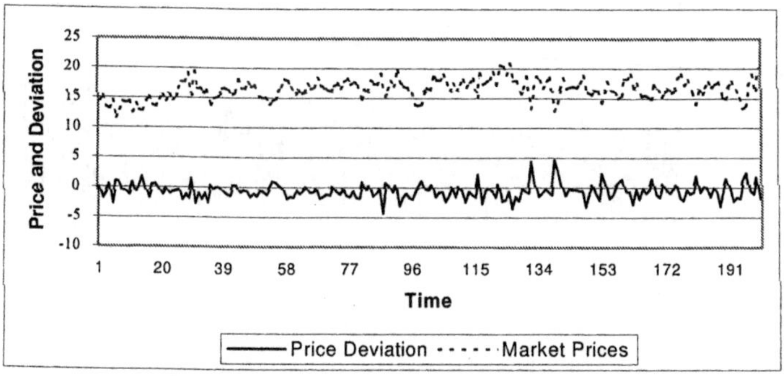

Fig. 5.3. Experiment 1: Market price vs. price deviation

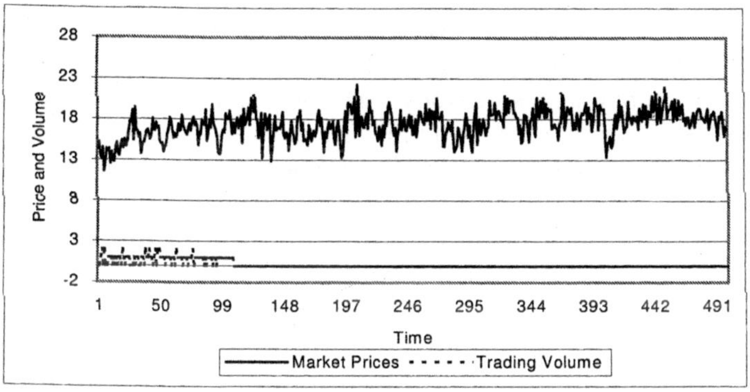

Fig. 5.4. Experiment 1: Market price vs. trading volume

In Experiment 1, 10 value traders use an ANN learning mechanism to forecast the next period's price and dividend. They are identical in terms of their information set, information processing ability, but heterogeneous in terms of their network parameters. All of the networks are trained with rational expectations before the traders enter the market. After a certain period of learning so that homogeneous rational expectations are reached, trading ceases (Fig. 5.4). This confirms Tirole's famous "No-trade Theorem" [33]. As also shown in Fig. 5.3, in experiment 1, the deviation of market price from REE is bounded and close to zero.

In experiment 2, however, in the early periods, the market shows relatively high price deviations and compare with the first experiment, convergence is slower and the price deviation is bounded in a wider range (Fig. 5.5). The reason for this is that the ANNs in this market are not pretrained by any market data. The difference between experiment 1 and 2 can be interpreted

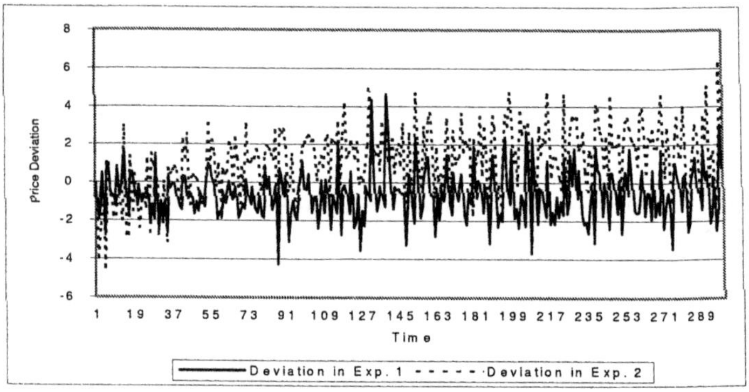

Fig. 5.5. Price deviation in experiment 1 and experiment 2

as the difference between experienced traders and inexperienced traders. In the inexperienced traders case, the convergence to REE is slow. This result is consistent with [8]. Another interpretation for this difference is the difference between new public issues of stocks and long outstanding stocks. For new public issues of stocks, there is no historical data available to train traders. In this case, we observe high volatility in market prices comparing with experiment 1.

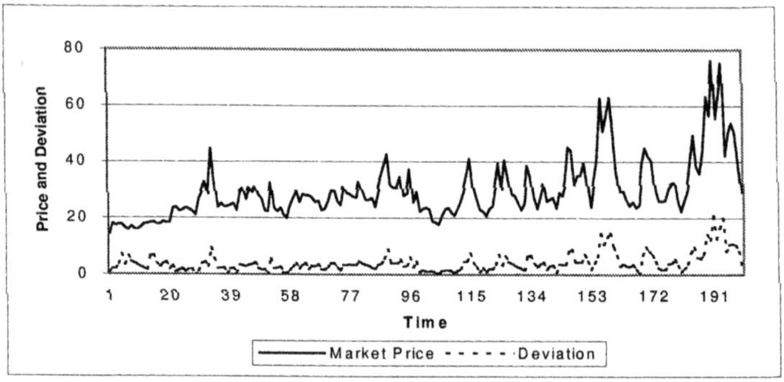

Fig. 5.6. Experiment 3: Market price vs. price deviation

In experiment 3, with the presence of irrational momentum traders, convergence is unattainable. This result is contradict to the finding in [18] "zero intelligence" (ZI) traders case, but it is consistent with the results in Camerer and Weighelt's laboratory market [8]. The difference in the results comes from the way of the market setting. In ZI case, the return of the risky asset ("redemption value") is exogenous, set by the designer, but in our experiments, this end-of-period return is determined endogenously and must be inferred by traders over the course of the trading process. In this case, convergence is more difficult to archive than in simpler setting.

Under null hypothesis of linear homogeneous REE, the price and dividend is a linear function of lag dividend as shown in equation (7). Follow [23], we use the following regression and analyze the estimated residual series.

$$P_{t+1} + D_{t+1} = c_0 + c_1 D_t + \varepsilon_t \tag{5.11}$$

In the homogeneous REE this estimated residual series should be independent and identically distributed, N(0, 1.89). This is done for experiment 1, 2 and 3. Results are given in Table 5.1.

Table 5.1. Basic summary statistics for the residuall and trading volumes for each of the three experiments

Experiment	Variance	Kurtosis	ρ	Trading Volume
Experiment 1	1.91	0.33	0.036	0.25
	(0.12)	(0.31)	(0.007)	(0.03)
Experiment 2	2.93	0.71	0.078	0.96
	(0.65)	(0.56)	(0.012)	(0.05)
Experiment 3	4.23	4.76	0.351	2.42
	(0.97)	(1.01)	(0.098)	(0.11)

1. The experiments are run for 1000 periods and 25 times for each. The statistics shown in this table are the average over 25 runs. Numbers in parenthesis are standard errors estimated using the 25 runs.
2. The kurtosis reported is excess kurtosis.
3. The trading volume is the average value for each period over all trading periods.

The first column shows the variance of the residual from all three experiments. The parameters are chosen to give the theoretical value of 1.89. All three cases show a higher variability but to different degrees. The first two experiments are very close. The next column gives the excess kurtosis, which should be zero under normal distribution. In the three experiments, the third one shows a significant amount of excess kurtosis. The third column presents the autocorrelation in the residual. In all three experiments, the autocorrelation are small, in the first two the autocorrelation is close to zero, which is comparable to the low auto correlations obeseved on the real market. The last column presents the average trading volume in each trading round in the three experiments. The last experiment shows very high trading activity.

In general, the case of heterogeneous preferences and trading strategies generates a market that exhibits richer "psychology". It is consistent with the fast learning case in Santa Fe artificial stock market of [2]. It is not consistent, however, with the attainment of a REE.

Where does the price volatility come from? Is it from learning or the auction institution?

The above results in experiment 3 identify richer and more complex market dynamics. Prices diverge substantially from theoretical (fundamental) prices. The differences between the two price series provide systematic evidence of temporary price bubbles and crashes. This appearance of bubbles and crashes suggests that, momentum traders have affected the market.

In this experiment, a large proportion of the variance in the price and dividend is left to be explained by the homogeneous rational expectation hypothesis in equation (11). What are the other potential explanatory variables? To answer this question, two other technical indicators used by momentum traders are added to the simple linear regression of price and dividend on the lagged dividend,

$$P_{t+1} + D_{t+1} = c_0 + c_1 D_t + c_2 I_{t,MA(5)} + c_3 I_{t,MA(10)} + \xi_{t+1} \tag{5.12}$$

Here the first indicator variable is a 5-period moving average indicator, which equals to one when price is above 5-period moving average or zero when price is below that. The second indicator variable represents whether the 5-period moving average is above 10 period moving average, which takes value of one or zero for 5-period moving average is above or below 10 period moving average respectively. the regression results on equation (11) and (12) for only experiment 3. The numbers presented in the Table are the estimated parameters and R^2 values. The standard errors are given in parenthesis. The results show that both two technical indicators give significant extra predictability. The parameters are small but statistically significant. The R^2 values in the last column confirms the fact that adding technical trading indicators explains a higher proportion of the variance in price an dividend.

Table 5.2. Regression on fundamental and technical indicators

Regression equation	constant	D	$I_{MA(5)}$	$I_{MA(10)}$	R^2
Eq.(11)	4.65	1.85	NA	NA	0.43
	(0.08)	(0.54)			(0.05)
Eq.(12)	4.21	2.02	0.054	0.086	0.57
	(0.13)	(0.87)	(0.023)	(0.034)	(0.08)

* The numbers in the parenthesis are the standard errors estimated in the 25 runs.

It is obvious that in the presence of momentum traders, ANN traders cannot drive the market price to the rational expectations equilibrium. They cannot pick up the part of the variation which can only be explained by the technical indicators. To understand this question more clearly, we need to look at the market microstructure aspect in terms of order flow.

A concrete example will help. In the experiment with only value traders, the individual reservation price serve as a mean-reverting device to drive the market price to equilibrium by adjusting their positions. Consider a two trader case. Suppose for certain periods, Trader 1 keep bidding high and buying stocks from Trader 2 and this buying pressure drove up the market price. As a result, Trader 1 has a large position of stock and hence, a relatively low reservation price. Trader 2, however, has a higher reservation price. Since agents will not buy above or sell below these price, it is impossible for Trader 1 to buy shares from Trader 2, since the limit buy order that Trader 1 posted is always below Trader 2's reservation price. Therefore, the low-reservation-price trader could only possibly be a seller, by which he could reduce his holding level until it reaches the optimal holding. This selling pressure will drive the market price to its equilibrium. With the presence of the momentum traders, however, the link between the reservation price and the market price is broken. Since even in the case where Trader1 posts a low bid quote, it is still very possible for a momentum trader to "sell at the market", since they are not rational traders. As a result, a high-position trader still keeps increasing his holding level, and this will drive the market price a long way away from it's equilibrium level. Convergence to the REE fails. In this case, minimal rationality seems insufficient to produce convergence to REE prices.

To check whether the large divergence in experiment 3 is from the trading institution, it would be interesting to replace the double auction in this experiment with a Walrasian auction, and then compare the market dynamics from these two different trading institutions. Unfortunately, we cannot perform this test since the 10 momentum traders have no demand functions to submit to the Walrasian auctioneer. In this case, it is impossible for the Walrasian auctioneer to find a market clearing price since the link between the individual demand and aggregate demand is broken. Therefore, it is still too early to draw strong conclusions regarding where this large divergence comes from.

5.5 Conclusions and Directions for Future Research

Our artificial double auction market is capable of generating behaviour close to REE under certain circumstances. The speed of convergence varies across trader's experience. When irrational traders emerge, the market is driven by more noisy factors than just fundamental trading. The convergence is unattainable in this case. The value trader cannot learn the chartists' strategies: they may not even realize the existence of the chartists. But it is still not clear whether the observed market inefficiencies should be attributed to the learning effects, or to other differences in trading institution design.

When the intrinsic value of the risky asset becomes endogenous, the convergence is sensitive to the deviation of rationality and the minimal rationality in this case is not sufficient.

Some interesting hypotheses can be tested on this market are informational efficiency, the role of market transparency and market microstructure. Another interesting extension to this initial work is related to the choice of market mechanism. In this paper, we use a double auction. It would be appealing to simulate different types of auctions or to endow some participants with more market power to observe whether and how they could corner a market.

A Appendix

We assume each value trader possess a ANN(1-3-1) model. They have similar structure but there are differences in the initial values of the parameters. Five steps are taken to form their prediction of next period return and risk.

(a) Initialize a network with K hidden nodes($K = 3$), J inputs($J = 1$) and one output. The parameters start with initial values drawn from a uniform distribution whose range is set at $[-1, 1]$;

(b) Forecast with the current values of the parameters. An input at time t is presented to the network and forwarded through the network as follows:

$$Y = h(\sum_{k=1}^{K} \alpha_k f({\beta_k}' X + \gamma) + a_0), \tag{5.13}$$

where $\beta_k = [\beta_{k1}\beta_{k2} \dots \beta_{kJ}]'$, $X = [X_1 X_2 \dots X_J]'$ and $\gamma = [\gamma_1\gamma_2 \cdots \gamma_k]'$. $f()$ is the logistic function, $f(u) = [1 + exp(u)]^{-1}$, and $h()$ is a linear function, $h(s) = s$. The signals, X, containing only one element, the dividend, which is received from the outside are transmitted to the hidden neurons by a logistic function. The outputs of the hidden units is then weighted again, and are then summed and transformed by $h(\cdot)$. At the beginning of each period, we include the most recent dividend in the input layer. Each trader invokes his own network to forecast the return in next period.

(c) Compute the forecasting error. At the end of each trading period, when the market price and dividend are revealed, the sum of squared deviations between the output of the network and the target are computed as:

$$E_t = \frac{1}{2} \sum_{t=1}^{T} (Y_t - (\sum_{k} \alpha_k f({\beta_k}' X + \gamma_k) + a_0))^2 \tag{5.14}$$

(d) Estimate the parameters. Given an input and output pair X_t, Y_t, the estimation of all the parameters in this network is done by minimizing the sum of the squared error above. The optimization is done by a conjugate gradient algorithm.

(e) Update the forecasting errors. At the end of each trading period, agents update their estimated conditional variances according to an exponentially weighted average of squared forecasting error,

$$\sigma^2_{j,P_{t+1}+D_{t+1}} = \theta(P_t + D_t - \hat{E}_{t-1}(P_{t-1} + D_{t-1}))^2 + (1-\theta)\sigma^2_{j,P_t+D_t} \quad (5.15)$$

This variance will be used in traders' demand function(4), the reservation price function(8) and the spread function(10). We keep these values fixed inside each trading period, updating them according to equation (15) at each invocation of ANN. In our experiment, θ is fixed to be 0.013. If the expectations are unbiased, forecasts on the return will be upheld on average by the market, and therefore the price sequence will be a rational expectations equilibrium. Thus, the price fluctuates as the information Ω_t fluctuates over time, and it reflects the "correct" or "fundamental" value.

References

1. Arifovic J. (1996) The Behavior of the Exchange Rate in the Genetic Algorithm and Experimental Economics. Journal of Political Economy **104**, 510–541
2. Arthur W. B., Holland J., LeBaron B., Palmer R., Tayler P. (1997) Asset Pricing Under Endogenous Expectations in an Artificial Stock Market. In: Arthur W. B., Durlauf S., Lane D. (Eds.) The Economy as an Evolving Complex System II, Addison-Wesley, Reading, MA, 15–44
3. Belttratti A., Margarita S., Terna P.(1996) Neural Networks for Economic and Financial Modeling. International Thomson Computer Press, London, UK
4. Bray M. (1982) Learning, Estimation, and the Stability of Rational Expectations. Journal of Economic Theory **26**, 318–339
5. Brock W. A., Lakonishok J., LeBaron B. (1992) Simple Technical Trading Rules and the Stochastic Properties of Stock Returns. Journal of Finance **47**, 1731–1764
6. Brock W. A., Dechert W. D., Scheinkman J. A., LeBaron B. (1996) A Test for Independence Based on the Correlation Dimension. Econometric Reviews **15(3)**, 197–235
7. Brown, Goetzmann, Kumar (1998) The Dow Theory: William Peter Hamilton's Track Record Reconsidered. Journal of Finance **53**, 1311–1333
8. Camerer C., Weigelt K. (1992) Convergence in Experimental Double Auctions for Stochastically Lived Assets. In: Friedman D., Rust J., (Eds.) The Double Auction Maket: Institutions, Theories, and Evidence. AddisonWesley, Reading, MA, 355–390
9. Campbell J. Y., Lo A. W., MacKinlay A. C. (1996) The Econometrics of Financial Markets. Princeton University Press, Princeton, NJ
10. Chan, LeBaron, Lo, Poggio (1998) Information Dissemination and Aggregation in Asset Markets with Simple Intelligent Traders. Working Paper, MIT
11. Chen S. H., Yeh C. H. (1999) Evolving Traders and the Facualty of the Business School: A New Architecture of the Artificial Stock Market, This volume

12. Cliff D., Bruten J. (1997) Zero is Not Enough: On the Lower Limit of Agent Intelligence for Continous Double Auction Markets. Technical Report HPL97141, Hewelett Packard Laboratories, Bristol, UK
13. De Long J. B., Shleifer A., Summers L. H. (1990) Noise Trader Risk in Financial Markets. Journal of Political Economy **98**, 703–738
14. Engle R. F. (1982) Autoregressive Conditional Heteroskedasticity with Estimates of the Variance of United Kingdom Inflation. Econometrica **50**, 987–1007
15. Fogel D. B. (1995) Evolutionary Computation: Toward a New Philosophy of Machine Intelligence, IEEE Press, Piscatway, NJ.
16. Forsythe R., Palfrey T.R., Plott C.R. (1982) Asset Valuation in an Experimental Market. Econometrica **50**, 537–567
17. Friedman D., Rustin J. (1992) Preface. In: Friedman D., Rust J. (Eds.) The Double Auction Market: Institutions, Theories, and Evidence, AddisonWesley, Reading, MA, 155–198
18. Gode, Sunder (1993) Allocative Efficiency of Markets with Zero-Intelligence Traders: Market as a Partial Substitute for Individual Rationality. Journal of Political Economy **101(1)**, 119–137
19. Grossman S., Stiglitz J.(1980) On the Impossibility of Informationally Efficient Markets. American Economic Review **70**, 393–408
20. Hertz J., Krogh A., Palmer R. (1991) An Introduction to the Theory of Neural Computation, Addison Wesley Publishing Company, Reading, Mass
21. Hussman J.(1992) Market Efficiency and Inefficiency in Rational Expectations Equilibria. Journal of Economic Dynamics and Control **16**, 655-680.
22. LeBaron B. (1998) Technical Trading Rules and Regime Shifts in Foreign Exchange. In: Acar E., Satchell S. (Eds.) Advanced Trading Rules, Butterworth-Heinemann, 5–40
23. LeBaron B., Arthur W. B., Palmer R. (1999) Time Series Properties of an Artificial Stock Market. Journal of Economic Dynamics and Control **23**, 1487–1516
24. LeBaron B. (2000) Agent-Based Computational Finance: Suggested Reading and Early Research. Journal of Economic Dynamics and Control, 24(5-7), 679-702
25. Margarita S., Beltratti A. (1993) Stock Prices and Volume in an Artificial Adaptive Stock Market. In: New Trends in Neural Computation: International Workshop on Artificial Networks. SpringerVerlag, Berlin, 714–719
26. O'Hara (1995) Market Microstructure Theory. Blackwell Publishers, Cambridge, USA
27. Plott C. R., Sunder S. (1982) Efficiency of Experimental Security Markets with Insider Information: An Application of Rational Expectations Models. Journal of Political Economy **90**, 663–698
28. Plott C. R., Sunder S. (1988) Rational Expectations and the Aggregation of Diverse Information in Laboratory Settings. Econometrica **56**, 1085–1118
29. Rust J., Miller J., Palmer R. (1992) Behavior of Trading Automota in a Computerized Double Auction Market, In: Friedman D., Rust J. (Eds.) The Double Auction Market: Institutions, Theories, and Evidence. Addison Wesley, Reading, MA, 155–198
30. Sargent T. (1993) Bounded Rationality in Macroeconomics. Oxford University Press, Oxford, UK
31. Smith, Murray (1993) Neural Networks for Statisticians. Van Nostrand Reinhold, New York, N.Y.

32. Smith V. L., Suchanek G. L., Williams A. W. (1988) Bubbles, Crashes, and Endogenous Expectations in Experimental Spot Asset Markets. Econometrica **56(6)**, 1119–1152
33. Tirole J. (1982) On the Possibility of Speculation under Rational Expectations. Econometrica **50**, 1163–1181
34. Wasserman P. (1993) Advanced Methods in Neural Computing. Van Nostrand Reinhold, New York, N.Y.
35. White H. (1989) Connectionist Nonparametric Regression: Multilayer Feedforward Networks can Learn Arbitrary Mappings. Neural Networks **3**, 535–549
36. White H. (1992) Artificial Neural Networks: Approximation and Learning Theory. Blackwell Publishers, Cambridge, MA

6 On AIE-ASM: Software to Simulate Artificial Stock Markets with Genetic Programming

Shu-Heng Chen[1], Chia-Husan Yeh[2], and Chung-Chih Liao[3]

[1] AI-ECON Research Center, Department of Economics, National Chengchi University, Taipei, Taiwan
chchen@nccu.edu.tw
[2] Department of Information Management, I-Shou University, Kaohsiung, Taiwan
imcyeh@saturn.yzu.edu.tw
[3] Graduate Institute of International Business, National Taiwan University, Taipei, Taiwan
liao@aiceon.org

Abstract. Agent-based computational economic modeling requires demanding work on computer programming. Publications of agent-based computational economic modeling usually do not provide readers with adequate information to permit replication of the experiments reported in the papers. Such failure makes the findings from the agent-based simulations hard to verify and defies technical improvement. To facilitate the growth of this research area, it is necessary for authors to make their source codes available in a public domain. This paper is a documentation accompanying the software AIE-ASM, which is available on the website. The software is designed to simulate the agent-based artificial stock market based on a standard asset pricing model. Genetic programming, as part of the software, is used to drive the learning dynamics of traders. An example based on the version of single-population genetic programming is demonstrated in this paper.

6.1 Introduction

One of the earliest and most productive applications of genetic algorithms to economics is the agent-based modeling of artificial stock markets. This research agenda originated from Brain Arthur and John Holland about a decade ago, when an economic research unit was established at the Santa Fe Institute in New Mexico([2]). It can be considered as the legacy of John Holland in economics in that it replaces the representative, perfect-foresight agent in the *standard asset pricing model* with Holland's *artificial adaptive agents* ([11]). The first journal publication on the agent-based artificial stock market was [14]. This line of research progressed slowly in the beginning, and has eventually reached its burgeoning stage over the last few years. Amid the excitement about its prospect, researchers, however, are encountered with a problem which is generally not shared by conventional economic approaches, but is very common in the agent-based computational approach, i.e. *replicatability.*

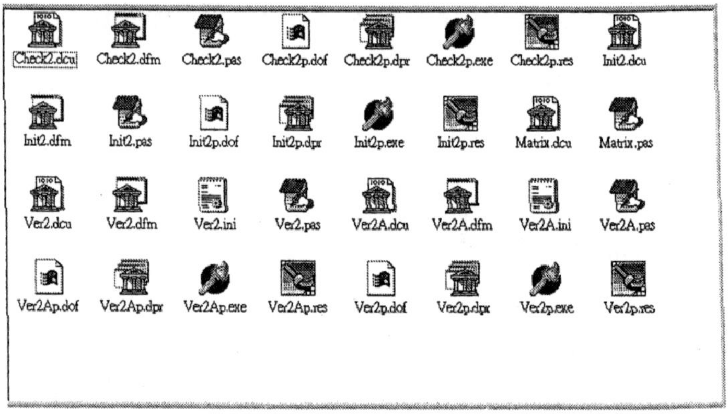

Fig. 6.1. An overview of the software: AIE-ASM

Agent-based computational economic modeling requires demanding work on computer programming. Usually, the publications as outcomes of running these programs do not provide readers with adequate information to permit replication of the experiments reported in the papers. This may generally make the findings or conclusions from the agent-based simulations hard to verify, which in turn defies technical improvement. To facilitate the growth of this research area, it is necessary for authors to make their source codes available in a public domain. This paper provides documentation accompanying the software AIE-ASM, which is available on the website. The software is designed to simulate the agent-based artificial stock market based on a standard asset pricing model. *Genetic programming* (**GP**), as part of the software, is used to drive the learning dynamics of traders. An example based on the version of *single-population genetic programming* is demonstrated in this paper. Other references related to the software can be found in [7] and [8].

6.2 AIE-ASM, Version 2: A User's Guide

One of the formidable tasks for the artificial stock market is the *design issue*. As [13] pointed out: "The computational realm has the advantages and disadvantages of a wide open space in which to design traders, and new researchers should be aware of the daunting design questions that they will face. Most of these questions still remain relatively unexploited at this time. (p.696)" Nevertheless, one should notice that this issue is not confined to the artificial stock market, and is widely shared by all research in *bounded rationality*. For example, [16] stated "This area is wilderness because the research faces so many choices after he decides to forgo the discipline provided by equilibrium theorizing."(p.2)

```
BinarySet=+\-\*\%\
UnarySet=Exp\Log\Sin\Cos\Abs\Sqrt\
X_Var=X00\X01\X02\X03\X04\X05\X06\X07\X08\X09\X10\
Y_Var=Y00\Y01\Y02\Y03\Y04\Y05\Y06\Y07\Y08\Y09\Y10\
Extra_Var=Z00\Z01\Z02\Z03\Z04\
Const_Item=R

[Stock_Market]
H0=1.0
M0=100.0
Irate=0.1
Beta=0.001

[Traders]
Reproduct=50
Mutation=100
MutProb=3.3
MutMode=1
LeaveProb=5
DepthLimit=17
MaxExp=1700.0
LibraryProb=100
Lambda=0.5
G_Gap=1
TryNo=5
Theta1=0.5
Theta2=0.0001
Theta3=0.0133
Gen_No=20000
```

Fig. 6.2. The file "Ver2A.ini" in the software "AIE-ASM"

Too many choices is a tough problem for us. There are *lots of parameters* in GP, and the results we obtained using the GP method might be sensitive to the particular parameter values chosen. We invite interested researchers to try what they may think sensitive, and provide a link below to the source code we used in [7].

The software **AIE-ASM** is available from the following web address:
http://www.aiecon.org/software.htm
The software is composed of 32 files as shown in Fig. 6.1. Among them,

there are four executable files, namely, Ver2p.exe, Ver2Ap.exe, Init2p.exe and Check2p.exe. The parameters declared in the Table 1 of [8] are all stored in the file **Ver2A.ini** (Fig. 6.2).

The dynamics of the stock market are initialized by using the initial values of the stock prices $\{P_{0-i}\}_{i=0}^{11}$ and the stock prices plus dividends $\{P_{0-i} + D_{0-i}\}_{i=0}^{11}$ specified in the file **p.txt**. To rerun a simulation with different initial values, one simply inputs a series of new data into the data file, p.txt. The time series of dividends $\{D_t\}$ follows a stochastic process which is exogenously given. Therefore, it is generated before the simulation and is stored in the file **Dstar.txt**. If the end user would like to test a different stochastic process, such as AR(1), she can simply generate a new series with the stated stochastic property and store it in the file Dstar.txt, and then rerun the program.

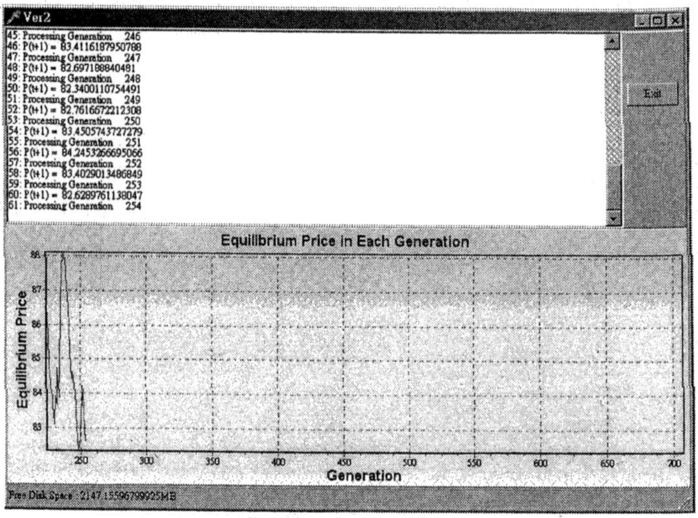

Fig. 6.3. The results of a sample run shown on the screen

Before running the program, all executable files (.exe files) and parameter files (.ini files) have to be put under the same directory, and the place to store the data files (.txt files) should be declared in the parameter files. To run the program, simply use *mouse* to *click* the file Init2p.exe followed by clicking the file Ver2p.exe. One second after the execution of Ver2p.exe, Ver2Ap.exe will be executed automatically. Depending on the number of generations to evolve, the execution can take a lot of memory space. To see whether the execution is terminated unintendedly due to insufficient memory size, the user may like to click Check2p.exe after clicking Ver2p.exe. What Check2p.exe does is to check whether the execution is unintendedly terminated. If so, it will

Table 6.1. Time series outputs generated from AIE-ASM: Data1.txt

Variables	Name in Data1.txt
Stock price	P_{t+1}
Dividends	D_{t+1}
Trading volumes	V_t
Number of buyers	BuyNo
Number of sellers	SellNo
Number of non-participants	BeSNo
Shares to buy	TotalBid
Shares to sell	TotalOffer
Average of depth of trees	AveDepth
Average of nodes of trees	AveNode
Average of shares held	AveHi
Mean of traders' expectations	AveVijs
Variance of depth of tress	VarDepth
Variance of nodes of trees	VarNode
Variance of shares held	VarHi
Variance of traders' expectations	VarVijs
Number of trades with successful search	SucceedNo
Number of martingale believers	MartingaleNo
Number of traders who decide to search	VisitGenNo
Number of traders who decide to search	VisitLibNo

VisitGenNo is available only in the case without the B-School, whereas No VisitlibNo is available only in the case with the B-school.

restart Ver2p.exe and continue the execution right from the point it was terminated so that you need not rerun the program, and the data generated before the termination will be kept. Generally speaking, we recommend the user to click this file if the number of generations to evolve is more than 2000. Once Ver2p.exe is executed, a time series plot of the stock price will be immediately shown on the screen (Fig. 6.3).

In addition to the price series directly shown on the screen, there is a complete output file named **Data1.txt**. What has been stored in this file is the time series variables listed in Table 6.1. These variables are aggregate results. It is also possible to trace the whole population in an individual manner. The outputs about microstructure are stored in **Data2.txt**. To have a feeling of the evolution of traders' expectations and decisions, you can find in Figures 6.4 and 6.5 snapshots of the histogram of $E_{i,t}(P_{t+1} + D_{t+1})$ and $h_{i,t}$ of a single simulation by using the parameters in the Table 1 of [8].

Depending on the computing environment, running genetic programming can be very time-consuming. For example, for a typical run of SGP with business school, it takes 5 days for 20,000 generations on **AMD K6 233** and **Pentium II 233** with **128 MB RAM**. The computational time required is not a linear function of the number of iterations due to the fact that the

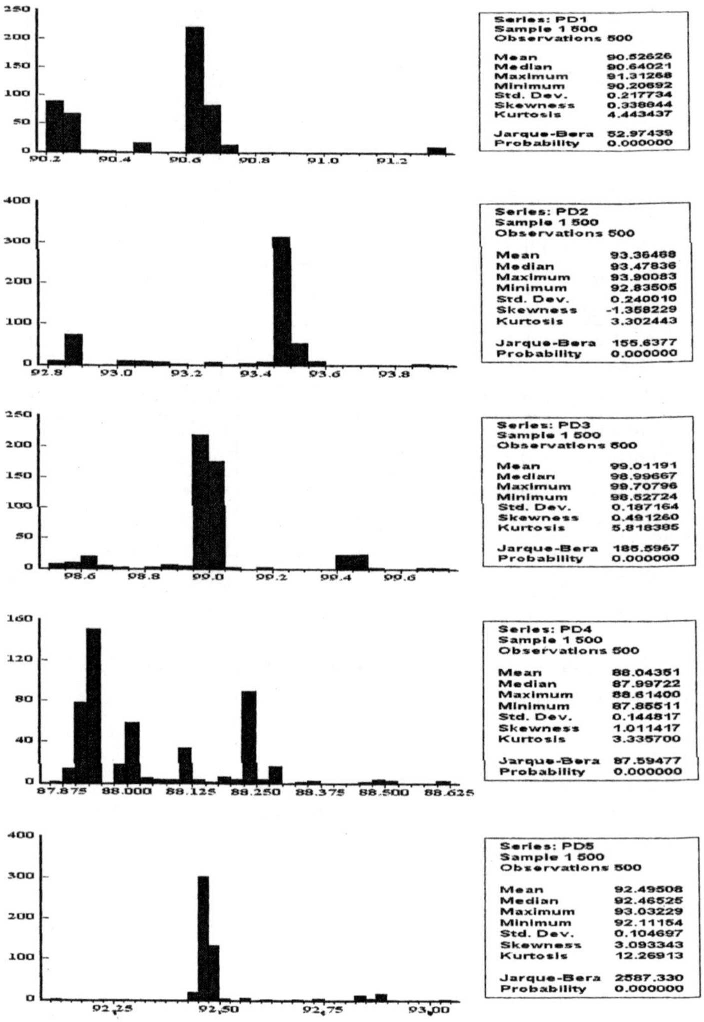

Fig. 6.4. Histogram of traders' expectations of the price and dividends

From the top to the bottom are the five snapshots of the histogram of traders' expectations on the 100th, 200th,...,500th trading day of a simulation based on the Table 1 of [8].

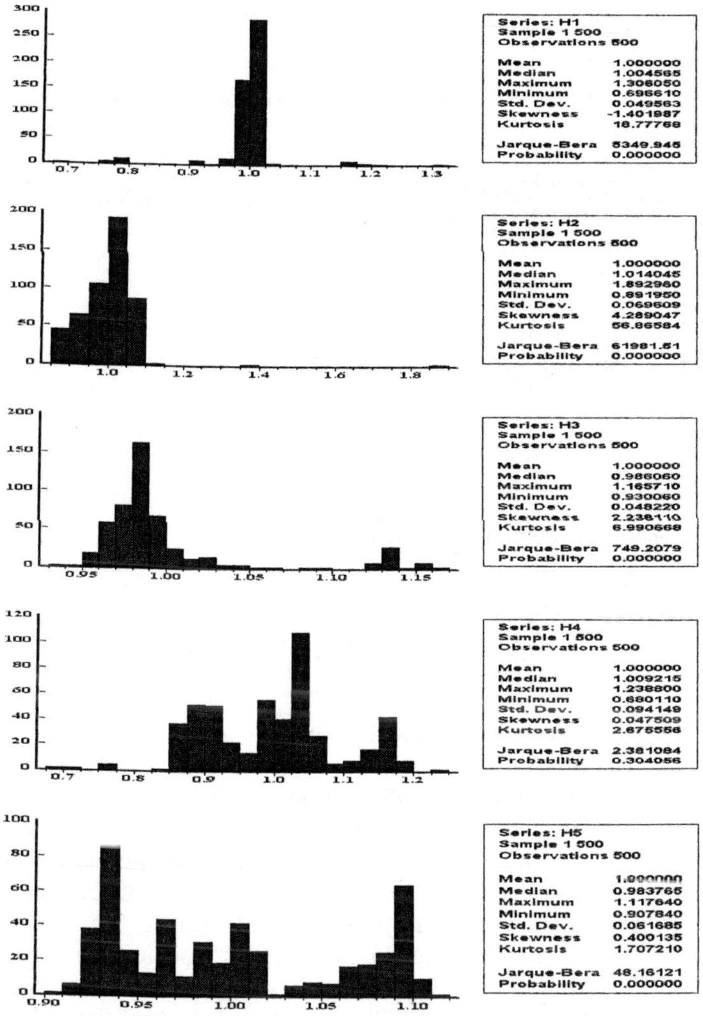

Fig. 6.5. Histogram of traders' portfolios

From the top to the bottom are the five snapshots of the histogram of traders' portfolios on the 100th, 200th,...,500th trading day of a simulation based on the Table 1 of [8].

tree size of the LISP program may increase with number of iterations (generations). This run-time speed limits the possibility of large-scale simulation. However, it is our belief that *one of the main contributions of this paper is to provide a public platform to facilitate more extensive research in this di-*

```
        op := GetOp(empty,Value);
      end;      (* end of while not empty *)

        if Top^.next <> nil then
      begin
        MessageDlg('Error: Data compute not match in T' ,mtWarning,[mbOK],0);
        exit;
      end;
      Vij := Top^.Value;

    if Abs(Vij) <= MaxExp then
        Vij := (Exp(Theta2*Vij)-Exp(-Theta2*Vij))/(Exp(Theta2*Vij)+Exp(-Theta2*Vij))
//  if Abs(Vij) <= 10000.0 then
//      Vij := Vij/10000.0
      else
      begin
        if Vij > 0.0 then
            Vij := 1.0
        else
            Vij := -1.0;
      end;
      Vij := DimC[u-1] * (1 + Theta1 * Vij);
//    Vij := (DimB[u-1] + 10.0) * (1 + Theta1 * Vij);

      FreeStack(Top);
      except
        FreeStack(Top);
      end;
end;


procedure MakeFullTree(var t:TwoPoints;d,depth:integer);
var
    p          : TwoPoints;
    i,j,sign : byte;
    val        : real;
```

Fig. 6.6. Traders' forecasting equation in the file Ver2.pas

rection. Those interested in conducting further experiments are encouraged to download the software and give it a try.

AIE-ASM is quite user-friendly because most of the parameters can be directly modified through the file **Ver2A.ini**. For example, to simulate artificial stock market with and without the b-school[1], one can make a choice between these two versions, simply modify the parameter value of **LibraryProb** (Fig. 6.2), which determines the probability of visiting the b-school if the trader decides to search. Setting this value at "0" means there is no chance of visiting the b-school, and hence the b-school is not included. Setting this value at "100" means that direct imitation from traders is infeasible, and that learning and adaptation is restricted to the b-school only. Anything between "0" and "100" makes a compromise between these two extremes.

The major files of the source code are **Ver2.pas** and **Ver2A.pas**. In some cases, the user needs to modify these two files to conduct her own experiments. For example, if the user wants to model traders' forecast in terms of level rather than increments (Equation (6.1)),

$$E_{i,t}(P_{t+1} + D_{t+1}) = (P_t + \mu)(1 + \theta_1 tanh(\theta_2 \cdot f_{i,t})), \tag{6.1}$$

[1] See "Search Processes" in the next section. For the details of the b-school, the interested reader is referred to [7].

then she needs to change the equation on line 796 contained in the procedure **ComputeVij** (line 680) and the one on line 631 in the procedure **ComputeHi** (line 511) (See Fig. 6.6).[2]

6.3 Search Process without Business School

The search process to be detailed below will determine the outcome of the trader's search. Basically, it describes how *promising ideas* (*forecasting rules*) are popularized or how *new ideas* are discovered during traders' search and adaptive process, i.e., *the evolution of a population of ideas.* Recently, *genetic algorithms* (GAs) and *genetic programming* (GP) have been extensively used to substantiate processes like this. However, the straightforward application of single-population GAs or GP, which rests on the assumption that strategies are observable and imitable, has been seriously criticized by [10]. [7] considered a modified version of single-population GP by introducing a mechanism called *business school.* In the following, we will, however, describe the standard style of GP, i.e., the version without business school. For making a distinction, one is called *a search process with business school* and the other *a search process without business school.*

The search process without business school goes as follows. At each single search, with probabilities p_r, p_c, and $p_m(= 1 - p_r - p_c)$, the trader will *randomly* choose a *genetic operator* to search for an idea, where p_r is the probability of choosing the *reproduction* operator, p_c the probability of choosing *crossover*, and p_m the probability of choosing *mutation.* In the case where the reproduction operator is chosen, she will first randomly select two GP trees, say, $gp_{j,t}$ and $gp_{k,t}$ from GP_t, *the collection of all traders' forecasting rules at time t.* These rules are compared with each other based on their performance on accumulated profits over the last n_2 days. The one with higher increments is selected and is denoted by $gp^r_{i,t}$.[3] In the case of mutation, we

[2] For a lengthy discussion of using Equation (6.1), see [8]. Other possibilities are given as follows.

$$E_{i,t}(P_{t+1} + D_{t+1}) = P_t(1 + \theta_1 tanh(\theta_2 \cdot f_{i,t})) + \mu_d \tag{6.2}$$

$$\begin{aligned} E_{i,t}(P_{t+1} + D_{t+1}) = & \; P_t(1 + \theta_1 tanh(\theta_2 \cdot f^p_{i,t})) \\ + & \; D_t(1 + \theta_3 tanh(\theta_4 \cdot f^d_{i,t})) \end{aligned} \tag{6.3}$$

$$E_{i,t}(P_{t+1}+D_{t+1}) = \begin{cases} f_{i,t}, \text{if } \; (1-\theta_2)(P_t + D_t) \le f_{i,t} \le (1+\theta_1)(P_t + D_t), \\ (1+\theta_1)(P_t + D_t), \text{if } \; f_{i,t} > (1+\theta_1)(P_t + D_t), \\ (1-\theta_2)(P_t + D_t), \text{if } \; f_{i,t} < (1-\theta_2)(P_t + D_t) \end{cases} \tag{6.4}$$

[3] This is just the standard tournament selection style with tournament size 2.

follow the same procedure as reproduction except that $gp^r_{i,t}$ has a chance of being perturbed by *tree mutation* (See [12]), and the result is denoted by $gp^m_{i,t}$.

In the case of crossover, trader i first randomly selects two pairs of trees, say, $(gp_{j_1,t}, gp_{j_2,t})$ and $(gp_{k_1,t}, gp_{k_2,t})$ from GP_t. Then, as with the reproduction process, the better one of each pair is selected and further paired as parents. Two offspring, say $(gp_{j,t}, gp_{k,t})$, are born from the parents by the standard GP crossover. One of the two offspring is randomly selected and is denoted by $gp^c_{i,t}$. In sum, depending on the genetic operator used, $gp^*_{i,t}$ $(* = r, c,$ or $m)$ defines the outcome of a single search by trader i.

But, traders are not supposed to take whatever comes out of their search. Instead, trader i will first compare the performance of the new model with the old one, i.e., the one trader used at time t. If the accumulated profits of the new one are indeed better, then it will replace the old one.[4] In this case,

$$gp_{i,t+1} = gp^*_{i,t} \tag{6.5}$$

Otherwise, trader i will give up this newly-found rule and start another round of search, and she will do it again and again until either she finds a better one or she continuously fails I^* times. For the latter,

$$gp_{i,t+1} = gp_{i,t} \tag{6.6}$$

To wrap it up, a pseudo program is provided as follows (also see Fig. 6.7).

Procedure [Trader's Search]

0. **begin**
   1. Calculate $\Delta W(gp_{i,t})$
   2. A = **Random**(R,C,M) with (p_r, p_c, p_m)
   3. If A = C, go to step (11).
   4. (gp_1, gp_2) = (**Random**(GP_t), **Random**(GP_t))
   5. Calculate $\Delta W(gp_1)$ and $\Delta W(gp_2)$.
   6. gp_{new} = **Tournament Selection** $(\Delta W(gp_1), \Delta W(gp_2))$
   7. If A = R, go to step (19).
   8. $gp_{new} \leftarrow$ **Mutation** (gp_{new})
   9. Calculate $MAPE(gp_{new})$
   10. Go to step (19)
   11. (gp^1_1, gp^1_2) = (**Random**(GP_t), **Random**(GP_t))
   12. (gp^2_1, gp^2_2) = (**Random**(GP_t), **Random**(GP_t))
   13. Calculate $\Delta W(gp^1_1)$ and $\Delta W(gp^1_2)$.
   14. Calculate $\Delta W(gp^2_1)$ and $\Delta W(gp^2_2)$.
   15. gp_1 = **Tournament Selection**$(\Delta W(gp^1_1), \Delta W(gp^1_2))$
   16. gp_2 = **Tournament Selection**$(\Delta W(gp^2_1), \Delta W(gp^2_2))$

[4] Such a procedure, known as the *election operator*, is first introduced by [1].

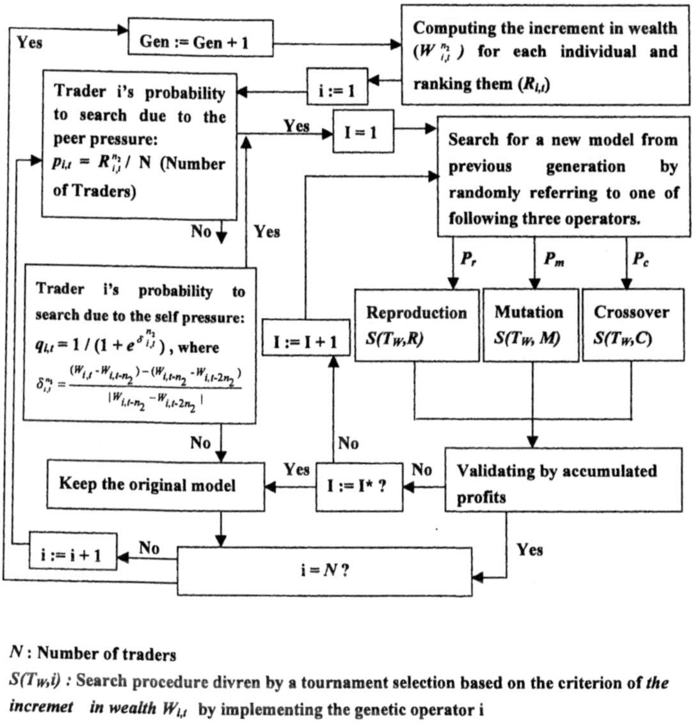

Fig. 6.7. The flowchart of the GP-driven search process

$(gp_1, gp_2) \leftarrow$ **Crossover** (gp_1, gp_2)

$gp_{new} =$ **Random**(gp_1, gp_2)
$gp_{i,t+1} =$ **Tournament Selection** $(\Delta W(gp_{i,t}), \Delta W(gp_{new}))$
end

6.4 An Example

In this section, we presents a few experimental designs based on the style of single-population genetic programming described in the previous section. In this case, the single-population genetic programming is directly operated on the society of traders. Two different designs are considered, which are identical in all aspects (control parameters) except for parameter n_2.

Table 6.2. Parameters of the GP-based artificial stock market: SGP without the B-school

The Stock Market	
Shares of the stock (H) per capita	1
Initial money supply (M_1) per capita	100
Interest rate (r)	0.1
Stochastic process (D_t)	IID ~ Normal(10,4)
Price adjustment function	*tanh*
Price adjustment (β_1)	0.2×10^{-4}
Price adjustment (β_2)	0.2×10^{-4}
Traders	
Number of traders (N)	500
Degree of RRA (λ)	0.5
Criterion of fitness	Increment in wealth
Sample size of $\sigma^2_{t\|n_1}$ (n_1)	10
Evaluation cycle(n_2)	1, 10
Search intensity (I^*)	5
($\theta_1, \theta_2, \theta_3$)	(0.5, 10^{-4}, 0.0133)
Genetic Programming	
Number of trees created by the full method	50
Number of trees created by the grow method	50
Function set	$\{+, -, \times, /, Sin, Cos, Exp, Rlog, Abs, Sqrt\}$
Terminal set	$\{P_t, P_{t-1}, \ldots, P_{t-10}, P_{t-1} + D_{t-1}, \ldots, P_{t-10} + D_{t-10}\}$
Selection scheme	Tournament selection
Tournament size	2
Probability of creating a tree by reproduction	0.10
Probability of creating a tree by crossover	0.70
Probability of creating a tree by mutation	0.20
Probability of mutation	0.0033
Probability of leaf selection under crossover	0.5
Mutation scheme	Tree Mutation
Maximum depth of tree	17
Number of generations	20,000
Number of trading days	20,000
Maximum number in the domain of Exp	1700
Criterion of fitness	Increment in Wealth

Parameter n_2 can be considered as *the time horizon of evolution*. In the evolutionary design proposed above, individual rules are assigned a fitness value at each round which is directly used to determine the probability of their being passed down to the next generation, via reproduction, crossover, or mutation. When $n_2 = 1$, the rules are functions whose validity is estimated in a single shot (only one point). In this case, the number of generations is also the time scale of simulation. In other words, we are simultaneously evolving the population of traders while deriving the price P_t. Alternatively, the fitness may be collected over several rounds, rather than on one single trial, so that the "genetic" of the model is applied only after several rounds, during which the cumulative strength of the rules is the profits accumulated over several rounds. Here, for the single-shot case, n_2 is set at 1, while for the multi-shot case, it is set at 10. All parameters are summarized in Table 6.2.

For each value of n_2, we ran two experiments, and each experiment lasted for 20,000 periods. Figure 6.8 is time series plot of the stock price of each run. Under *full information* and *homogeneous expectations*, the *homogeneous rational expectations equilibrium price* (**HREE**) is[5]

$$P_t = \frac{1}{r}[\mu - \lambda\sigma_\xi^2(\frac{H}{N})] = \frac{1}{r}[\mu - \lambda\sigma_\xi^2 h]. \tag{6.7}$$

[5] See [3], pp. 40-41.

Table 6.3. Linear dependence of the series

Periods	Series 1-1		Series 1-2		Series 2-1		Series 2-2	
	PSC	R^2	PSC	R^2	PSC	R^2	PSC	R^2
1-2000	(0,2)	0.28	(1,0)	0.09	(0,0)	0.00	(0,0)	0.00
2001-4000	(1,0)	0.21	(1,0)	0.28	(2,3)	0.07	(1,0)	0.13
4001-6000	(1,0)	0.22	(1,0)	0.22	(2,2)	0.19	(0,1)	0.14
6001-8000	(1,0)	0.14	(1,0)	0.21	(2,2)	0.16	(1,0)	0.15
8001-10000	(1,0)	0.14	(1,0)	0.11	(3,3)	0.26	(1,0)	0.13
10001-12000	(1,0)	0.18	(1,0)	0.10	(3,3)	0.39	(1,0)	0.09
12001-14000	(1,0)	0.14	(1,0)	0.09	(6,6)	0.69	(1,0)	0.09
14001-16000	(1,0)	0.23	(1,0)	0.14	(0,0)	0.00	(1,0)	0.22
16001-18000	(1,0)	0.23	(1,0)	0.21	(3,4)	0.19	(0,2)	0.29
18001-20000	(0,1)	0.16	(1,0)	0.16	(3,3)	0.25	(2,2)	0.23

Each series has 20,000 observations. They are equally divided into ten non-overlapping periods, and the PSC algorithm is run for each period to select the linear ARMA order, i.e., (p, q). Once the order is selected, the R^2 is derived based on the model with the selected order.

Since in our experiments (Table 1), $(\mu, \sigma_{\xi}^2, r, \lambda, h) = (10, 4, 0.1, 0.5, 1)$, the **HREE** price is 80. It seems difficult to predict from this reference price what actually happened in our artificial stock market. The dynamics of our markets are very wild. For example, in both experiments of $n_2 = 1$, we experienced a sequence of bubbles followed by crashes. The medians of these price series are 104.23, 102.88, 182.92, and 98.87 respectively, which are all higher than the **HREE** price.

To see whether these series satisfy *the efficient market hypothesis* , we first examined the *linear predictability* of the return series by using the Rissanen's *predictive stochastic complexity* (**PSC**) ([15]). The **PSC** criterion is a model selection criterion. It selects the model with the minimum PSC. By the PSC filtering algorithm, we can identify the linear ARMA model (p, q) of a series. If a series satisfies the **EMH**, then both p and q should be 0, i.e., there is no linear dependence, and hence the series is not linearly predictable. We applied the PSC algorithm to the ten equally-divided subseries of all four series, and the results are shown in Table 6.3.

From Table 6.3, almost all series are *linearly predictable* with a strictly positive p and q. Furthermore, if we run a regression based on the model with the selected order, the derived R^2 is non-trivial for most series. As a result, these series can hardly satisfy the *efficient market hypothesis* (**EMH**), and hence we cannot use these series to show the **EMH** as an emergent property. In fact, the feature of highly linear dependence is also rarely observed in real return series. Therefore, in addition to Harrald's criticism, the high R^2 of the return series also casts doubts on whether running single-population GP directly on a society of traders is a proper design for agent-based financial markets.

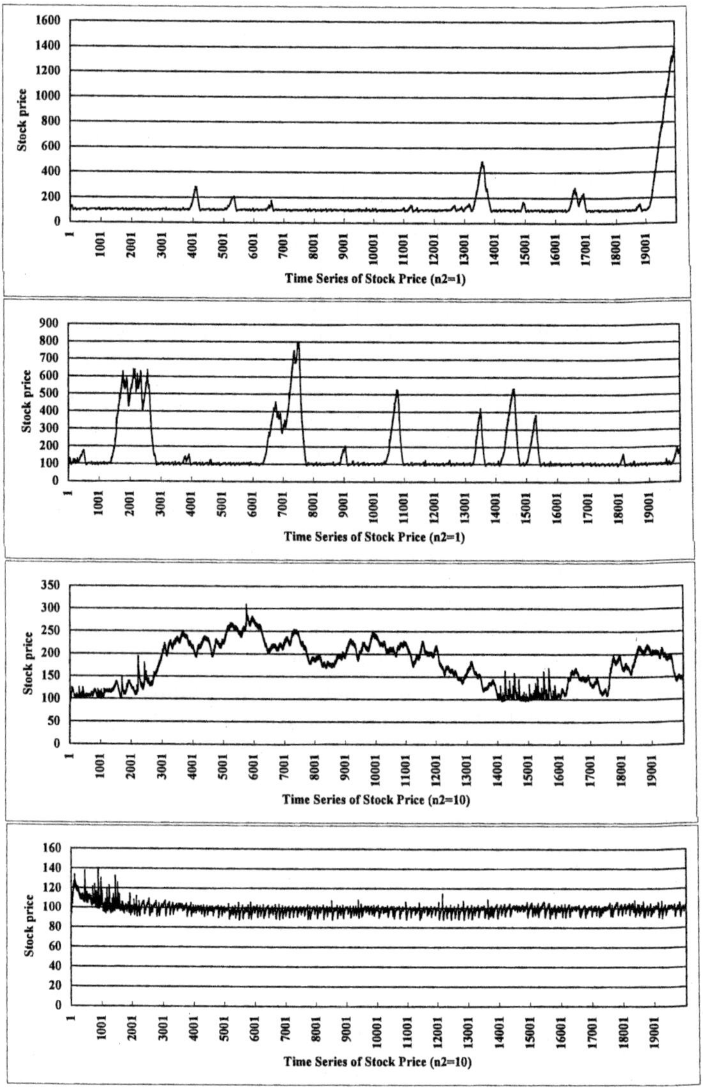

Fig. 6.8. Time series plot of the stock price: SGP without B-school

Earlier, we mentioned that SGP (GA) is not suitable for the architecture design due to Harrald's criticism. But, this does not rule out the attractiveness of this design because it may still generate series *behaving reasonably close* to the real return series. The evidence we have here, however, is not in favor of that possibility.[6]

6.5 A Summary of AIE-ASM Publications

In addition to the two journal articles ([7], [8]), simulations of this software can also be found in a few proceedings papers. They are available in the website:
http://econo.nccu.edu.tw/ai/staff/shc/vita.htm
or
http://aiecon.org/staff/shc/vita.htm

Reference [5] studied the effect on market efficiency of the search intensity resulting from *psychological pressure*. They found that increase in search intensity reduced the degree of heterogeneity of traders and made the chance of a successful search even more remote. On the other hand, it also led to a reduction in market expectations error.

Reference [9] examined the possible explanations for the presence of the causal relation between stock returns and trading volume. They found that the bi-directional causality between trading stock returns and trading volume ubiquitously existed in all their four different experiments. The implication of this result is that the presence of the stock price-volume causal relation does not require any explicit assumptions like information asymmetry, reaction asymmetry, noise traders, or tax motives. It can be a generic property in a market modeled as evolving decentralized system of autonomous interacting agents.

Reference [6] examined the significance of the *election operator*. By including or not including this operator, they considered two kinds of traders: the ones who are "prudent" will validate a now idea before putting it into practice, and the ones who are "casual" will take whatever suggested (*follow the herd*). They found that markets with prudent traders and markets with casual traders could exhibit non-trivial differences in the size of speculative bubbles and price efficiency.

Finally, [4] studied the behavior of price discovery within a context of an *agent based stock market*. Via their agent based simulations, it was found that, except for some extreme cases, the mean prices generated from these artificial markets deviated from the *homogeneous rational expectations equilibrium* (**HREE**) *prices* no more than by 20%. Furthermore, while the HREE price should be a *deterministic* constant in all of their simulations, the artificial price series generated exhibited quite wild fluctuations, which may be connected to the well-known *excess volatility* in finance.

[6] Of course, to show that *direct imitation* not only lacks behavioral foundation, but is also empirically implausible, one needs to run more experiments. We leave this as a future direction for research.

References

1. Arifovic J. (1994) Genetic Algorithm Learning and the Cobweb Model. Journal of Economic Dynamics and Control **18**, 3–28
2. Arthur B. (1992) On Learning and Adaptation in the Economy. SFI Working Paper, 92-07-038
3. Arthur W. B., Holland J., LeBaron B., Palmer R. and Tayler P. (1997) Asset Pricing under Endogenous Expectations in an Artificial Stock Market. In: Arthur W. B., Durlauf S., Lane D. (Eds.), The Economy as an Evolving Complex System II, Addison-Wesley, 15–44
4. Chen S. -H., Liao C. -C. (2000) Price Discovery in Agent-Based Computational Modeling of Artificial Stock Markets. In: Proceedings of the Second Asia-Pacific Conference on Genetic Algorithms and Applications. Global Link Publishing Company, Hong Kong, pp.380–387
5. Chen S. -H., Yeh C. -H. (2000a) On the Role of Intensive Search in stock Markets: Simulations Based on Agent-Based Computational Modeling of Artificial Stock Markets. In: Proceedings of the Second Asia-Pacific Conference on Genetic Algorithms and Applications. Global Link Publishing Company, Hong Kong, 397–402
6. Chen S. -H., Yeh C. -H. (2000b) On the Consequence of "Following the Herd": Evidence from the Artificial Stock Market. In: Arabnia H. R. (Ed.) Proceedings of the International Conference on Artificial Intelligence, Vol. II, CSREA Press, 388–394
7. Chen S. -H., Yeh C. -H. (2001a) Evolving Traders and the Business School with Genetic Programming: A New Architecture of the Agent-Based Artificial Stock Market. Journal of Economic Dynamics and Control **25**, 363–393
8. Chen S. -H., Yeh C. -H. (2001b) On the Emergent Properties of Artificial Stock Markets: The Efficient Market Hypothesis and the Rational Expectations Hypothesis. Forthcoming in *Journal of Economic Behavior and Organization*.
9. Chen S. -H., Yeh C. -H., Liao C. -C. (2000), Testing for Granger Causality in the Stock-Price Volume Relation: A Perspective from the Agent-Based Model of Stock Markets. In: Wang P. (Ed.) Proceedings of the Fifth Joint Conference on Information Sciences, Vol. II, 950–956
10. Harrald P. (1998) Economics and Evolution. Panel Paper Given at the Seventh International Conference on Evolutionary Programming, March 25–27, San Diego, U.S.A.
11. Holland J. H., Miller J. H. (1991) Artificial Adaptive Agents in Economic Theory. American Economic Review: Papers and Proceedings **81(2)**, 365–370
12. Koza J. (1992) Genetic Programming: On the Programming of Computers by Means of Natural Selection. The MIT Press.
13. LeBaron B. (2000) Agent-Based Computational Finance: Suggested Readings and Early Research. Journal of Economic Dynamics and Control **24**, 679–702
14. Palmer R. G., Arthur W. B., Holland J. H., LeBaron B., Tayler P. (1994) Artificial Economic Life: A Simple Model of a Stockmarket. Physica D **75**, 264–274
15. Rissanen J. (1989) Stochastic Complexity in Statistical Inquiry. World Science, Singapore.
16. Sargent T. J. (1993) Bounded Rationality in Macroeconomics, Oxford

7 Exchange Rate Volatility in the Artificial Foreign Exchange Market

Jasmina Arifovic

Simon Fraser University, Canada and
California Institute of Technology, U.S.
arifovic@sfu.ca

Abstract. This paper studies co-evolution of different decision rules in an artificial foreign exchange market. The behavior of the exchange rate depends on the type of decision rules that agents use. Evolution of the moving average and least squares forecasting techniques results in a speculative attack on one of the currencies and that currency's eventual collapse. Addition of the rules that evolve the portfolio fractions directly brings in persistent volatility of the exchange rate that resembles the actual exchange rates time series.

7.1 Introduction

Persistent fluctuations have characterized the behavior of the exchange rates ever since the flexible exchange rate system was introduced. Theoretical models, based on fundamentals like money supplies, real income, interest rates, inflation rates, and current account balances, have not been successful in capturing a high percentage of the variation in the exchange rate at short-or-medium-term frequencies. An alternative way to try to model the behavior exhibited under the flexible exchange rates system is to address explicitly the issue of agents' beliefs and the way in which they change over time.

This paper examines the exchange rate behavior in an artificial foreign exchange market (AFEM). The paper studies co-evolution of different decision rules in a simple general equilibrium monetary model where there are no changes over time in terms of fundamentals.[1] Agents can use different rules in making their portfolio decisions, i.e. how much of their savings to allocate to each currency. Portfolio decisions in turn affect the level of prices. Thus nominal prices, rates of return and exchange rates are endogenously determined. Their dynamics are influenced solely by changes in agents' portfolio decisions.

As a specific application of the AFEM framework, the results of simulations where moving average forecasting techniques evolve are presented. Binary strings encode the size of the sample used in the computation of a forecast and indicate which of the availabe techniques will be used. Rules

[1] In the rational expectations equilibrium, the exchange rate is constant and, due to the perfect substitutability of the two currencies that are traded, the equilibrium exchange rate is also indeterminate.

are updated using the genetic algorithm. These simulations result in the collapse of one of the two currencies and the convergence of the economy to a single-currency equilibrium. The addition of least squares to the population of evolving forecasting techniques brings about even faster collapse. These speculative attacks imply greater volatility of the exchange rate, a feature that also characterizes actual exchange rate time series. However, a complete collapse of one of the currencies that is due purely to shifts in agents' beliefs is not necessarily final outcome of speculative attacks that occur in the actual foreign exchange markets.

The addition of populations of binary strings that encode portfolio fractions generates volatility of the exchange rate that persists over time. Speculative attacks on both currencies occur, but they come to an end before the investment into the currency that is under the attack can be driven down to zero. Thus, the volatility of the exchange rate behavior observed in this AFEM environment captures some of the behavior observed in the actual foreign exchange markets.

It is worthwhile pointing out that, unlike the other artificial stock market models that require a sequence of exogenous shocks to one of the fundamentals (usually a dividend is assumed to follow a stochastic, AR(1) process), in order to generate some persistence in the behavior, no such shocks are required in the artificial foreign exchange market described in this paper. Indeterminacy, together with the evolution of beliefs, results in the persistence in volatility of the exchange rate.

Section 7.2 describes the economic environment under consideration. Section 7.3 describes the AFEM framework and the results of simulations of two applications. Section 7.4 concludes and outlines some of the possible extensions of the basic model.

7.2 Description of the Model

The *world* economy consists of two countries. It is an overlapping generations model in which the residents of both countries are identical in terms of their preferences and lifetime endowments. The economy starts at $t = 1$ and lasts for ever. Individual agents live for two periods, and at each $t \geq 1$, $\frac{N}{2}$ new young individuals are born in each country, said to be of generation t. They are young at period t and old at period $t+1$. Each young agent of generation t is endowed with n units of labor and their own production technology for producing a single consumption good with a one-to-one labor/output transformation. When old, agents do not receive any labor endowment. No storage technology is available in the economy. Agents in both countries have the common preferences given by: $u_t[c_t^t, c_{t+1}^t] = c_t^t c_{t+1}^t$, where c_t^t is agent's consumption when young, and c_{t+1}^t is agent's consumption when old.

A government of each country issues its own unbacked currency. Supplies of both currencies, $M_{1,t}$ and $M_{2,t}$ are kept constant so that $M_{1,t} = M_1$ and

$M_{2,t} = M_2$. There are no legal restrictions on holdings of foreign currency. Thus the residents of both countries can freely hold both currencies in their portfolios. An agent of generation t solves the following maximization problem at time t:

$$\max \quad c_t^t \, c_{t+1}^t$$
$$s.t. \; c_t^t \leq n - \frac{m_{1,t}}{p_{1,t}} - \frac{m_{2,t}}{p_{2,t}}$$
$$c_{t+1}^t \leq \frac{m_{1,t}}{p_{1,t+1}} + \frac{m_{2,t}}{p_{2,t+1}}$$

where $m_{1,t}$ are the agent's nominal holdings of currency 1, $m_{2,t}$ are the agent's nominal holdings of currency 2 acquired at time t, $p_{1,t}$ is the nominal price of the good in terms of currency 1 at time t, and $p_{2,t}$ is the nominal price of the good in terms of currency 2 at time t. Agent's savings, s_t, in the first period of life, are equal to the sum of real holdings of currency 1, $m_{1,t}/p_{1,t}$ and real holdings of currency 2, $m_{2,t}/p_{2,t}$.

The exchange rate e_t between the two currencies is defined as $e_t = p_{1,t}/p_{2,t}$. When there is no uncertainty, the return on the two currencies must be equal,

$$R_{1,t} = R_{2,t} = \frac{p_{1,t}}{p_{1,t+1}} = \frac{p_{2,t}}{p_{2,t+1}}, \quad t \geq 1, \tag{7.1}$$

where $R_{1,t}$ and $R_{1.t}$ are the gross real rate of return between t and $t+1$. Rearranging (7.1), we obtain

$$\frac{p_{1,t+1}}{p_{2,t+1}} = \frac{p_{1,t}}{p_{2,t}} \quad t \geq 1. \tag{7.2}$$

From equation (7.2) it follows that the exchange rate is constant over time:

$$e_{t+1} = e_t = c, \quad t \geq 1 \tag{7.3}$$

Savings demand derived from agent's maximization problem is given by

$$s_t = \frac{m_{1,t}}{p_{1,t}} + \frac{m_{2,t}}{p_{2,t}} = \frac{n}{2}. \tag{7.4}$$

Aggregate savings that represent real world money demand are equal to the sum of young agents' savings, i.e. $S_t = Ns_t$. Since the rates of return on the two currencies are identical, the agents are actually indifferent as to which currency they hold. Because of this, equations for individual money demands are not well defined. (There is only one equation for the world real demand.) This fact results in the indeterminacy of the exchange rate.

The indeterminacy of the exchange rate proposition [7] asserts that if there is a monetary equilibrium where savings demand and money supplies are equal for an exchange rate, e, then there exists an equilibrium for any exchange rate $\hat{e} \in (0, \infty)$, $\hat{e} \neq e$. If there is an equilibrium for a price sequence, $\{p_{1,t}, p_{2,t}\}$, for the exchange rate e, we can find a sequence $\{\hat{p}_{1,t}, \hat{p}_{2,t}\}$, for the exchange rate $\hat{e}$ that results in the same sequence of real rates of return as the original price sequence and in turn in the same values of aggregate savings. The reason why this can be accomplished is the equivalence between the two currencies as savings instruments.

7.3 Description of the Artificial Foreign Exchange Market

In this section, we develop the elements of the AFEM. Agents that participate in the market make savings and portfolio decisions. They save half of the amount of the good that they produce in the first period. Given their labor endowment pattern, that is the optimal consumption decision. [2] However, they also have to make their portfolio decision, i.e. how much of their savings to invest in each currency. Let λ_t^i, $\lambda_t^i \in [0, 1]$, denote a portfolio fraction decision of agent i of generation t.

In the rational expectations equilibrium, $\lambda_t^i = \lambda$ for all i, and all t. The level of λ determines the (constant) value of the exchange rate. Note that the equilibrium value of λ cannot be deduced from the rational expectations version of the model. However, within the AFEM framework, agents have heterogeneous beliefs implying heterogeneous values of the portfolio fractions. With heterogeneous values of λ, and an assumption that $M_1 = M_2 = M$, the expressions for $p_{1,t}$ and $p_{2,t}$ are given by

$$p_{1,t} = 2M / \sum_i^N \lambda_t^i n \quad p_{2,t} = 2M / \sum_i^N (1 - \lambda_t^i) n. \tag{7.5}$$

or

$$p_{1,t} = 2M / N \bar{\lambda}_t n \quad p_{2,t} = 2M / N(1 - \bar{\lambda}_t) n \tag{7.6}$$

where $\bar{\lambda}_t$ is the average portfolio fraction. The exchange rate is then given by:

$$e_t = \frac{1 - \bar{\lambda}_t}{\bar{\lambda}_t}. \tag{7.7}$$

However, at the time when AFEM agents make their portfolio decisions, the value of e_t is not known. The agents can choose among different rules in deciding on their portfolio fraction. The choice of a rule is influenced by rule's past performance and by occasional experimentation with new rules.

[2] In the case that agents receive no labor endowment in the second period of their life, the savings decision does not depend on the rates of return on savings.

In the first environment discussed in the paper, agents are endowed with different types of moving-average rules that they use to compute exchange rate forecasts. Let $e_t^{f,i}$ be an exchange rate forecast of agent i of generation t. Then, agent i sets the value of λ_t^i equal to:

$$\lambda_t^i = \frac{1}{1 + e_t^{f,i}}. \tag{7.8}$$

Agents do not use the entire history of the exchange rates. They discard old information and employ a rolling sample. The size of the sample, T, (an even number) differs across agents and evolves over time. Agents can also choose between two types of forecasting procedures, f^1 and f^2. If f^1 is used, every sample observation is included in the computation of the moving average, and the exchange rate forecast is given by:

$$e_t^f = \frac{\sum_{k=1}^{T} e_{t-k}}{T} \tag{7.9}$$

On the other hand, if f^2 is used, only every second observation is considered in the computation of the forecast of the exchange rate, i.e.:

$$e_t^f = \frac{\sum_{k=0}^{T} e_{t-k-2}}{T/2} \tag{7.10}$$

The number of f^1 and f^2 rules also evolves over time. The actual exchange rate depends on the way individual forecasts are made and, using (8) and (7), is given by:

$$e_t = \frac{\sum_{i=1}^{N} \left(1 - \frac{1}{1 + e_t^{f,i}}\right)}{\sum_{i=1}^{N} \frac{1}{1 + e^{f,i}}}. \tag{7.11}$$

At each time period t, there are two populations of forecasting rules, one that represents the rules of the young agents (generation t) and the other that represents the rules of the old agents (generation $t-1$).[3] Only the rules of the young agents play an active role at time t. Each young agent is endowed with a binary string, of length ℓ, that has the following interpretation. The first bit of a binary string indicates whether f^1 or f^2 will be used. The bits $[2 \cdots \ell]$ encode the sample size, $T \in [1, \ldots, 64]$, i.e. the number of past observations that will be taken into account when computing the moving average of past values of the exchange rate.

---

[3] See [2] [3] for a detailed description of the implementation of the genetic algorithm in the overlapping generations environments.

The economy is initialized at the point where forecasting techniques are randomly distributed. An initial set consisting of $T_{max} = 64$ exchange rate observations is generated in the following way. For each observation, a random number between 0 and 1 is drawn from the uniform distribution. This number is interpreted as an average portfolio fraction and is used to compute the exchange rate value.

At each time period t, forecasting rules are decoded and individual forecasts are computed. Then individual portfolio fractions are calculated using these forecasts. Portfolio fractions determine the savings in terms of currency 1 and currency 2. Finally, nominal prices, exchange rate and rates of return in terms of each currency are calculated. Once the rates of return are known, second period consumption values are computed for members of generation $t-1$ and the fitness values for the forecasting rules of generation $t-1$ are calculated. A fitness of string i is given as the utility of agent i of generation $t-1$. The population of forecasting rules of generation $t-1$ is then used to obtain a population of forecasting rules for generation $t+1$.

A population of forecasting rules evolves using the genetic algorithm. Tournament selection is used as the reproduction operator. The one-point crossover takes place with probability 0.6. The probability of mutation is set to 0.033. In addition to these standard genetic operators, the election operator [1] [3] is applied as a local elitist procedure.

A *weak* form of the election operator [6] is used in the following way. After the application of the crossover operator on a pair of binary strings takes place, these two binary strings are recorded as parent 1 and parent 2. The resulting offspring strings are recorded as offspring 1 and offspring 2. Once the two offspring undergo mutation, their fitness values are calculated using the last period's rates of return. Then, the fitness of the first offspring is compared to the fitness of parent 1. If it is higher than the parent's fitness, the offspring enters as a member of the new population. However, if the parent's fitness is higher than the offspring's, the parent remains as the member of the new population. Likewise, if the fitness of offspring 2 is higher than or equal to the fitness of parent 2, the offspring 2 enters into the new population. Otherwise, parent 2 becomes a member of the new population.

Simulations of the above described evolutionary process resulted in the convergence of the economies to a single-currency equilibrium. Which of the two currencies is selected depends on a particular sequence of pseudo random numbers. Initially, the rates of return on two currencies fluctuate. There are time intervals during which $R_{1,t}$ is greater than $R_{2,t}$, and those when the direction of inequality changes sign and $R_{2,t} > R_{1,t}$. Eventually, one of the rates of return remains greater than the other long enough that it initiates a steady increase of the holdings of the currency with the higher rate of return. The final result is that agents place all of their savings in the *higher-return* currency.

The populations of forecasting techniques remain heterogeneous. Both f^1 and f^2 moving averages are represented in the populations and binary strings decode to sample lengths of different sizes. However, all of the forecasts result in the same value of λ at the end of the simulation. Figure 7.1 illustrates the behavior of the average portfolio fraction, and Fig. 7.2 the behavior of the rate of return on currency 1 in one of the simulations.

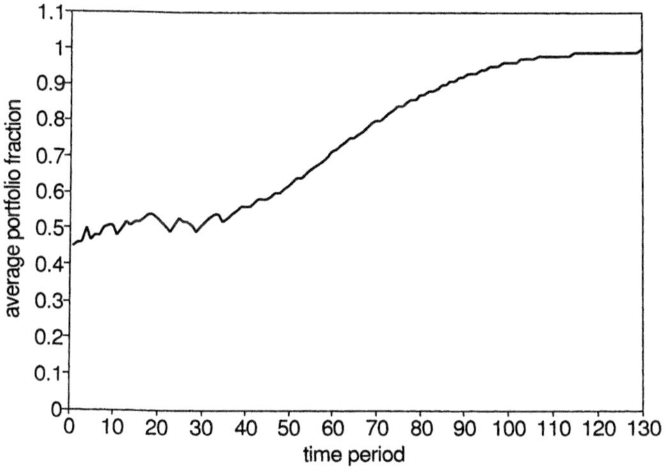

Fig. 7.1. Average portfolio fraction of f^1 and f^2 rules

After initial fluctuations, $\bar{\lambda}_t$ starts a steady increase towards the value of 1, indicating a speculative attack on currency 2. Once $\bar{\lambda}_t$ reaches the value of 1, currency 2 collapses. Examination of Fig. 7.2 reveals that after the initial fluctuations above and below 1, starting with $t = 46$, $R_{1,t}$ takes only values greater than 1. This is the interval during which $R_{1,t} > R_{2,t}$, and exactly the time when $\bar{\lambda}_t$ begins its steady increase. The end of the simulation is characterized by a slow decline of $R_{1,t}$ towards the value of 1 that is its value in the single-currency stationary equilibrium value.

Simulations of the economies in which only the size of the rolling sample evolved (and all agents used either f^1 or f^2) resulted in the same outcomes, i.e. the convergence to a single currency equilibrium. The addition of the least squares to the pool of forecasting techniques sped up the process of convergence to a single currency equilibrium. [4]

[4] Reference [9] applies the stochastic approximation algorithm to a version of this economy described in section 2. His results show convergence to an equilibrium

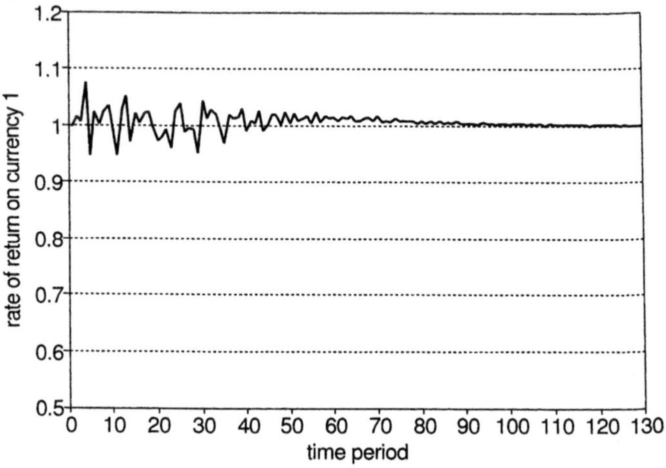

Fig. 7.2. Rate of return of f^1 and f^2 rules

While this result is interesting in light of the fact that the evolution of beliefs can result in speculative attacks, these types of speculative attacks are not observed in the actual time series. Even though the speculative attacks can occur without any apparent change in fundamentals, they end at some relatively high, but finite value of the exchange rate.

Next, we introduce another class of rules that will be represented by two overlapping populations of binary strings. With this class, a binary string encodes the value of λ_t^i. [5] Two new populations of binary strings are added to the AFEM in order to emulate the model's overlapping generations structure. These two populations that encode the values of λ represent a separate pool of rules that undergo genetic algorithm updating.

Let us denote the first class of rules that consists of f^1 and f^2 rules, the *MA* class, and the second class that consists of strings that encode values of portfolio fraction as the *P* class of rules. Both classes of rules will affect the determination of the price levels through agents' savings decisions. Thus, even though the two classes of rules that are updated separately, the evolution of each class is affected by the make-up of the populations representing the other class of rules through prices and rates of return.

with constant exchange rate. The particular level of the exchange rate selected by the adaptive algorithm depends on the initial conditions.

[5] Reference [3] showed that when this is the only class of rules used by agents, evolution results in persistent fluctuations of the exchange rate. Reference [4] showed that the time series generated in this environment exhibit chaotic behavior.

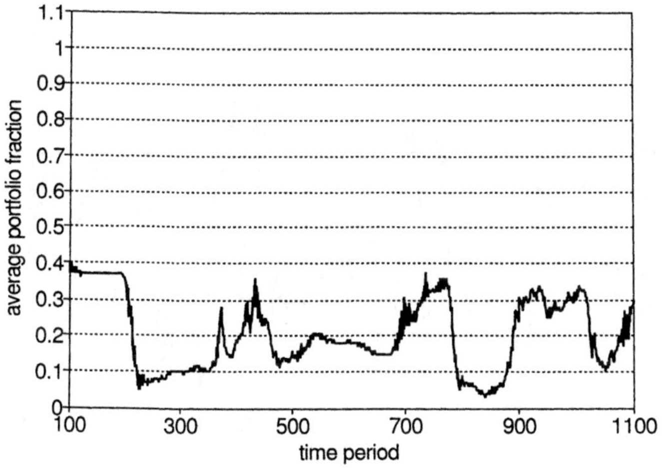

Fig. 7.3. Average portfolio fraction of MA rules

How does the addition of this class of rules affect the behavior of the economy? Figures 7.3, 7.4 and 7.5 present behavior observed in one of the simulations. Figure 7.3 shows the behavior of $\bar{\lambda}_t$ of the first class of agents, $\bar{\lambda}_t^{ma}$, and Figure 7.4 shows the behavior of $\bar{\lambda}_t$ of the second class of agents, $\bar{\lambda}_t^p$. The difference between the two is noticeable. While both exhibit wide and persistent fluctuations, the behavior of $\bar{\lambda}_t^{ma}$ is less erratic, the amplitude of fluctuations is smaller, periods of upward and downward movements are longer, and except for one instance where both $\bar{\lambda}_t^{ma}$ and $\bar{\lambda}_t^p$ take values very close to 1, $\bar{\lambda}_t^{ma}$ generally takes lower values than λ_t^p. The main impact of the addition of P class of rules on the behavior of λ_t^{ma} is that it does not converge to 1 or to 0. Instead it exhibits persistent fluctuations that do not die out over time. Fluctuations of λ_t^{ma} and λ_t^p result in continuing fluctuations of $R_{1,t}$, $R_{2,t}$ and e_t.

The co-evolution of the two classes of rules is quite interesting and is the subject of investigation. The issues being examined are: exact make-up of each of the two classes of populations, the impact of each class of rules on the behavior of the other class, the welfare implications for agents using different classes of rules, and finally the time-series properties of the simulated data.

7.4 Further Research

The paper develops a framework for studying the artificial foreign exchange market within the context of the general equilibrium monetary model with endogenous price determination. The version of the model in which agents

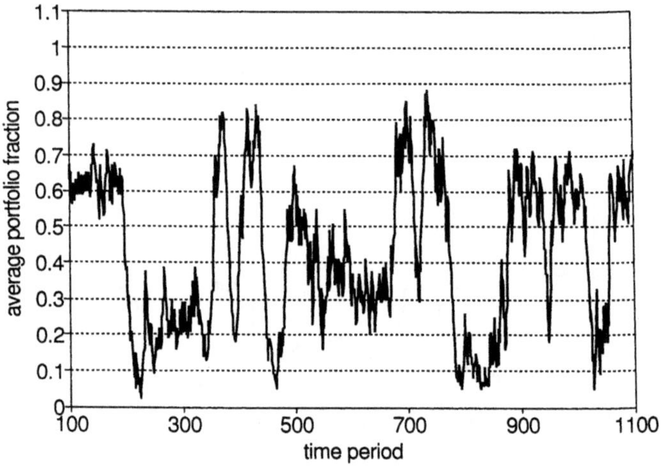

Fig. 7.4. Average portfolio fraction of P rules

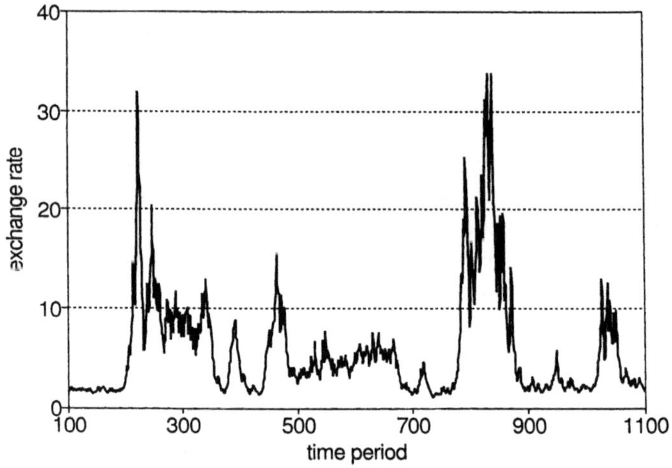

Fig. 7.5. Exchange rate

are rational does not provide a way to determine the portfolio fraction value. The reason is that agents are indifferent between the currencies that have the same rates of return in the homogenous-expectations equilibrium. The model described in this paper establishes the link between the exchange rate forecast and the portfolio decision and thus provides a way to model and examine the co-evolution of different forecasting rules.

In the model, money is the only available asset and its only role is that of the store of value. In addition, there are no restrictions on foreign currency holdings.[6] Finally, agents adopt different decision rules, and thus make heterogeneous portfolio decisions.

These features of the AFEM make it quite appropriate and conveninient for examination of the exchange rate behavior under the flexible exchange rates system. Trading in foreign exchange markets that results in observed volatility is based on differences in the expected rates of return on different currencies. Thus, the main role of money in these transactions is that of the store of value. In addition, in the world of heterogeneous beliefs, the rates of return on currencies need not be the same, and this inequality becomes the crucial driving force of the dynamics. In this respect, the AFEM captures the features of trading in real world foreign exchange markets, where rates of return on different currencies are not equalized despite a great degree of mobility and absence of restrictions on foreign currency holdings. A number of extensions of the basic framework are currently under consideration.

First, an environment that is more interesting in terms of the fundamentals will be developed, e.g. specification of different monetary and fiscal policies, definition of a stochastic process that governs the shocks to the production technology, addition of capital to the production technology, endogenous labor supply etc. Thus, the AFEM framework will allow examination of the impact of the shocks to the fundamentals and of their interaction with the dynamics that are driven by changes in agents' beliefs on the exchange rate behavior.

Second, a number of different forecasting rules will be added. We can then examine the impact of different forecasting techniques on the behavior of the exchange rate. Agents will be given an opportunity to choose among different rules and techniques, i.e. all the rules will be subjected to the evolutionary pressure. It will be interesting to examine what rules and techniques survive the selection pressure and whether the evolution results in the selection of a single rule or in the continuous extinction and reappearance of different decision rules. (The framework can also be extended to include, for example, classifier-system type of predictor rules, similar to those used in [5] and neural networks).

Third, the model presented in this paper can be used as the basis for developing a framework that can address the question of the impact of technical trading rules on foreign exchange markets. Since the era of floating exchange rates began in the early 1970s, technical analysis has been widely adopted by foreign currency traders.[7] This is partly due to the poor predictive

---

[6] Obviously, restrictions on foreign currency holdings can be added to the model.

[7] Reference [10] present the results on the issue of technical analysis by major dealers in the foreign exchange market in London.

(out-of-sample) performance of both the structural and the non-structural, time-series exchange rate models. The AFEM framework will provide an environment in which to examine the impact of different trading rules and their performance in competition with alternative forecasting techniques.

References

1. Arifovic J. (1994) Genetic Algorithm Learning and the Cobweb Model. Journal of Economic Dynamics and Control **18**, 3–28
2. Arifovic J. (1995) Genetic Algorithms and Inflationary Economies. Journal of Monetary Economics **36**, 219–243
3. Arifovic J. (1996) The Behavior of the Exchange Rate in the Genetic Algorithm and Experimental Economies. Journal of Political Economy **104**, 510–541
4. Arifovic J., Gencay R. (1998) Statistical Properties of Genetic Learning in a Model of Exchange Rate, Journal of Economic Dynamics and Control. forthcoming
5. Arthur B., LeBaron B., Palmer R., Tayler P. (1997) Asset Pricing Under Endogenous Expectations in an Artificial Stock Market. In: Arthur B., Durlauf S., Lane D. (Eds.) The Economy as an Evolving Complex System II, Addison-Wesley
6. Franke R. (1998) Behavioral Heterogeneity and Genetic Algorithm Learning in the Cobweb Model. Journal of Evolutionary Economics **8**, 383–406
7. Kareken J., Wallace N. (1981) On the Indeterminacy of Equilibrium Exchange Rates. Quarterly Journal of Economics **96**, 207–222
8. LeBaron B., Arthur W., Palmer R. (1999) Time Series Properties of an Aritifical Stock Market. Journal of Economic Dynamics and Control **23**, 1487–1516
9. Sargent T. J. (1993) Bounded Rationality in Macroeconomics. Claredon Press, Oxford
10. Taylor M. P., Allen H. (1992) The Use of Technical Analysis in the Foreign Exchange Market. Journal of International Money and Finance **11**, 304–14

8 Using an Artificial Market Approach to Analyze Exchange Rate Scenarios

Kiyoshi Izumi[1] and Kazuhiro Ueda[2]

[1] Information Science Div., ETL and
PRESTO, Japan Science & Technology Corporation.
1-1-4 Umezono, Tsukuba, Ibaraki, 305-8568, JAPAN
kiyoshi@ni.aist.go.jp
[2] Interfaculty Initiative of Information Studies, Univ. of Tokyo
3-8-1 Komaba, Meguro-ku, Tokyo, 153-8902, JAPAN

Abstract. In this study we used a new agent-based artificial market approach, to support decision-making on exchange rate policies. We first interviewed dealers and found that interaction among dealers in terms of learning had similar features to genetic operations in biology. Next, we constructed an artificial market model by using a genetic algorithm and regarding the market as a multi-agent system. Finally, using computer simulation of the model, several strategic scenarios in terms of policies to do with exchange rates were compared. As a result, it was found that intervention, and the control of interest rates, were effective measures in the stabilization of yen-dollar rates in 1998.

8.1 Introduction

Nobody had ever before experienced such a phenomenon as occurred in the Tokyo foreign exchange market over the week from 5th October 1998. In only 5 days the yen-dollar rate had dropped by about 20 yen, the same amount as the fluctuation over the whole last year. The market then fell into disorder.

Recently similar dramatic and complicated changes have been occurring in many other economic and social systems. These phenomena have forced researchers to recognize the importance of support-systems for decision-making in *realistic* economic and social situations, such as financial markets.

Decision-making in financial markets is very complicated and difficult because there are *micro-macro problems.* The micro-macro problem here is that the control of macro (market) variables cannot be reduced to the sum of decision-making by individual dealer. This is because the patterns of the dynamic behavior of macro variables are produced by interactions to do with decision-making at the micro (dealer) level.

The purpose of this study is to outline our new agent-based approach to the construction of a support system for decisions to do with foreign exchange markets.

8.2 Problems with Conventional Approaches

Conventional decision support systems have implemented static rules of decision-making from the viewpoint of individuals who do not interact with other decision-makers. Such an approach is, however confronted with the following two problems in environments in which micro-macro problems occur:

Over-fitting problems: The patterns of dynamic behavior of macro variables continuously change when micro-macro problems are present. Hence, rules for macro patterns may not work beyond sampling period, although they work well during the sampling period.

Explosions in the number of rules: When rules for macro patterns are extracted from the viewpoints of independent individuals, the number of conditions increases as samples are added. Macro patterns continuously change, so the number of rules generated by conventional approaches eventually may become enormous.

We have developed a new agent-based approach to solving these problems: an artificial market approach [1].

8.3 Framework of the Artificial Market Approach

The artificial market approach consists of three steps (Figure 8.1).

(a) **Observation in the field:** Firstly, field data were gathered by interviewing actual dealer and having them complete questionnaires, as described in section 8.4 of this paper. Then, we investigated the learning and interaction patterns of the dealers. As a result of our analysis, we were able to make hypotheses about the dealers' behavioral pattern: decision rules, learning rules, and patterns of interaction .
(b) **Construction of a multi-agent model:** Secondly, a multi-agent model was implemented on the basis of our hypotheses, as described in section 8.5. Artificial markets are virtual markets which operate on computers. They consist of a set of computer programs that act as virtual dealers, a market-clearing mechanism, and rate determination rules. The model provides the connection between the behavioral patterns of agents at the micro level and rate dynamics at the macro level.
(c) **Scenario analysis:** Finally, some good scenarios were prepared as decision-making problems on the basis of the results of simulation by using the artificial market model, as described in section 8.6. In this paper, we used the results of simulation to compare several "strategic scenarios" to do with action on exchange rate policies in 1998.

---

[1] Using this approach, we have examined some emergent phenomena in markets, such as rate bubbles [6–8]

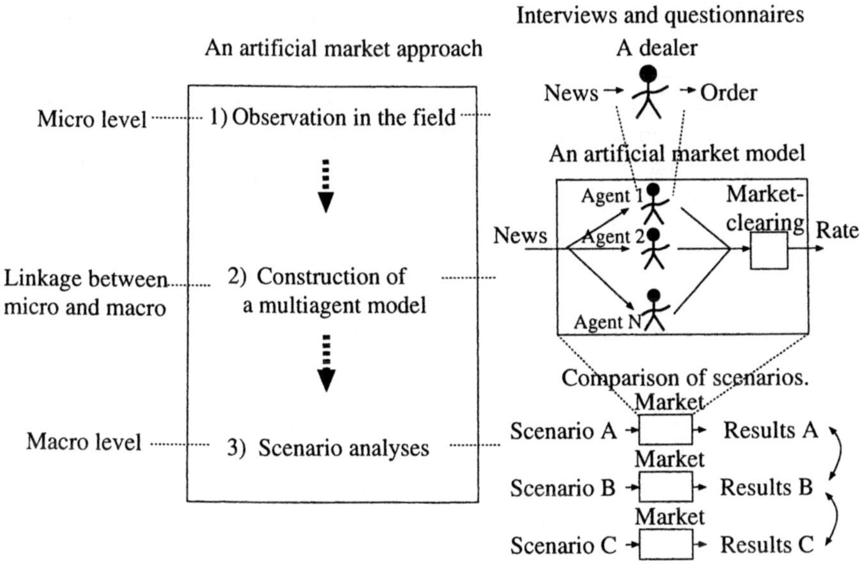

Fig. 8.1. Framework of artificial market approach.

8.4 Observation in the Field

In this section we describe our observation of actual dealers' behavior by using interviews and questionnaires. On the basis of these field data, we propose a hypothesis regarding learning by dealers. This hypothesis is also used in the construction of a multiagent model as a rule for interaction among and learning by agents.

We first observed changes in dealers' forecast rules over time by interviews and extracted several features of learning by dealers as described in section 8.4.1. The features were then verified by using data from questionnaires as described in section 8.4.2. Finally, based on the results of our fieldwork, we point out several similarities between the features of the dealers' learning process and genetic operations in biology, in section 8.4.3.

8.4.1 Interviews: Features of Learning

We interviewed two dealers who are usually engaged in yen-dollar exchange transactions on the Tokyo foreign exchange market. The first dealer (X) is a chief dealer for a bank. The second dealer (Y) is an interbank dealer for the same bank. They each had more than two years of experience on the trading desk at the time.

Interview methods The interviewees were asked to explain the dynamics of rates over almost two years, from January 1994 to November 1995 (when the interview took place). We asked each dealer to undertake the following tasks:

- Divide the whole period into several periods, according to characteristic market situations which the dealer recognized.
- Talk about factors which he had regarded as important in making rate forecasts for each period, at the time.
- Rank the factors in order of weight (importance) and give reasons for the ranking.
- Where the factors used in forecasting had changed between periods, describe the reasons for the reconsideration.

Results The division into two years and the ranking of factors are shown in Table 8.1, a and b.

From the data obtained by interviewing the two dealers, we found three basic features of the acquisition of prediction methods in the market.

Market consensus There are fashions in the interpretation of factors in markets, which are called *market consensus.* For example, the weight of the trade balance factor was not constant, although Japan always had large trade surpluses throughout these two years. The dealers said that this was because they were sensitive not to the value of economic indices but rather to the market consensus.

Communication and imitation The dealers communicated with other dealers to infer the current market consensus and thus determine which factors were regarded as important, and then replaced (some part of) their prediction method with a method that provided a better explanation of recent rate dynamics.

Learning promoted by errors in forecasts When the forecast of the interviewee had been quite different from the actual rate, he recognized the need to change his weights. At the end of Period VII in Table 8.1a, for example, Dealer X noticed that the rate had reached the level of 92 yen and the trend had changed, and suddenly changed his method of prediction.

Hypothesis From the above features, we propose the following hypothesis regarding the operation of the micro level of markets.

> *When the forecasts based on a dealer's own method of prediction differs markedly from the actual rates, each dealer replaces (at least parts of) their prediction method with other dealers' successful ones.*

Table 8.1. Results of the interviews with two dealers. The division into periods was determined by the dealers. The actual trends in rates and each dealer's forecast for each period are shown in terms of the three basic kinds of trends (downward, sideways, and upward). The factors in the forecast are listed and ranked in order of importance.

a) Dealer X

1994

	I Jan	II Feb-Jun	III Jul-Oct	IV Nov-Dec
Actual	→	↘	→	→
Forecast	→	↘	→	→
Ranking of factors	1.Value of Mark 2.Seasonal factors[*1]	1.Chart trends 2.Trade 3.Politics	1.Chart trends 2.Deviation 3.Politics	1.Seasonal factors[*1]

1995

	V Jan	VI Feb-Apr	VII May-Jul	VIII Aug-Sep	IX Oct-Dec
Actual	↗	↘	↗	↗	→
Forecast	↗	↘	↘	↗	→
Ranking of factors	1.Seasonal factors[*1]	1.Trade 2.Politics 3.Mexico 4.Chart trends		1.Deviation[*2] 2.Intervention	

b) Dealer Y

1994

	I Jan-May	II Jun	III Jul-Dec
Actual	↘	↘	→
Forecast	↘	→	→
Ranking of factors	1. Trade 1. Order[*3] 3. Chart trends	1. Rate level	1.Order[*3] 2. Chart trends

1995

	VI Jan-Feb	V Mar-Apr	VI May-Jul	VII Aug-Dec
Actual	↘	↘	→	↗
Forecast	↘	↘	→	↗
Ranking of factors	1. Politics 2. Value of Mark 2. Announcement	1. Politics 1. Order[*3] 1. Intervention	1. Chart trends 2.Order[*3]	1. Intervention 2. Politics

*1: The dealer said that rates didn't move at the beginning and end of the year.
*2: The dealer forecasted that rates would return to the previous level after large deviation.
*3: Directions and the amount of orders that the dealer received from other dealers or customers.

8.4.2 Questionnaires: Verification of Features

If the hypothesis of section 8.4.1 is correct, the frequency of successful weights in a market must be larger after a trend has changed. Thus, the following proposition holds.

Proposition
The average weight across all dealers, placed on each factor must shift towards values of successful weights.

In order to verify this proposition, we had 12 dealers respond to a questionnaire in March 1997. All of the dealers were involved in exchange transactions in banks.

Questionnaire The questionnaires were filled out just after the market trend for the value of the dollar had reversed, from an upward trend to a downward trend, in 1997. Each dealer, i, was asked the following three questions about 22 factors that might affect the determination of the yen-dollar rate.

- Write the importance of each factor, k, in the previous upward trend as one of 11 discrete values from 0 to 10: $w_i^k(t)$.
- Write the importance of each factor, k, in the current downward trend as one of 11 discrete values from 0 to 10: $w_i^k(t+1)$.
- Write the dealer's forecast before the trend changed: $\tilde{R}_i$.

The 22 factors were (1) economic activities, (2) price indexes, (3) short-term interest rates, (4) money supply, (5) trade balance, (6) employment prospects, (7) personal consumption, (8) intervention, (9) mark-dollar rates, (10) commodity markets, (11) stock prices, (12) prices of bonds, (13) short-term chart trends (under 1 week), (14) long-term chart trends (over 1 month), (15) exchange rate policy of the Band of Japan, (16) exchange rate policy of the Federal Reserve Bank, (17) trading by export and import firms, (18) trading by insurance firms, (19) trading by securities firms, (20) trading by other banks, (21) trading by foreign investors, and (22) the other factors.

Analysis The proposition implies that the average weight across all dealers placed on each factor changes towards an average across all dealers which is weighted by the accuracy of each dealer in forecasting. As mentioned in section 8.4.1, the interview data suggests that the importance of factors which can be used to provide more accurate forecast have a greater frequency after dealers have changed their opinions. Hence, if this proposition is true, the average weight across all dealers placed on each factor must change to an average which are weighted with their accuracy in forecasting.

We calculated the average weight across all dealers, $\bar{W}^k$, placed on each factor, k, in both the previous (t) and recent ($t+1$) trend.

$$\bar{W}^k(s) = \frac{1}{n}\sum_{i=1}^{n} w_i^k(s), \tag{8.1}$$

where n stands for the number of dealers, 12, and $s = \{t, t+1\}$.

Weighted averages of the importance of each factor, k, in the previous trend, t, were then calculated. The weight of a dealer i's importance is defined by using the dealer i's forecast error:

$$e_i = |\tilde{R}_i - R|, \tag{8.2}$$

where $\tilde{R}_i$ is the rate forecast by the dealer i, and R is the actual rate. The weight of dealer i's importance, f_i, is in inverse proportion to the forecast error.

$$f_i = \frac{E - e_i + 1}{\sum_{j=1}^{n}(E - e_j + 1)}, \tag{8.3}$$

where E is the maximum forecast error from among forecasts by the 12 dealers. The weight, f_i, is defined by using the difference between the maximum value of forecast error and the given dealer's forecast error. Thus, a dealer with a smaller forecast error has a greater weight value on her importance, and vice versa. That is, the weight, f_i, reflects the accuracy of dealer i's forecast. One is added to the numerator so that the importances of all dealers can make non-zero contributions to the weighted average. The denominator is necessary because the sum of weights, f_i, must be one.

The weighted average of each factor, k, is calculated as follows:

$$\bar{W}^k_{weighted}(t) = \frac{1}{n}\sum_{i=1}^{n} f_i w_i^k(t). \tag{8.4}$$

If our proposition is true, the market average after the trend has changed, $\bar{W}^k(t+1)$ must be close to the weighted average, $\bar{W}^k_{weighted}(t)$, from the market average before the trend changed, $\bar{W}^k(t)$. Thus, there is positive correlation between the two differences, the differences between $\bar{W}^k(t+1)$ and $\bar{W}^k(t)$ and the differences between $\bar{W}^k_{weighted}(t)$ and $\bar{W}^k(t)$.

$$\bar{W}^k(t+1) - \bar{W}^k(t) \propto \bar{W}^k_{weighted}(t) - \bar{W}^k(t) \tag{8.5}$$

We used the questionnaire data to test our proposition.

Results We did in fact find positive correlation between the two differences (Table 8.2). That is, successful opinions which can be used to provide more accurate forecast, appeared to spread through the market.

In summary, the hypothesis in section 8.4.1 implies that patterns of learning by actual dealers is similar to adaptation in ecosystem. In our multiagent model, the adaptation of agents in the market will be described by a genetic algorithm, based on the idea of population genetics.

Table 8.2. Correlation between differences. The number of samples is the number of factors.

Number of samples	Correlation	Significance level
22	0.284	$P < 0.1$

8.4.3 Similarities to Genetic Operations

When a dealer's prediction method is regarded as an individual in a biological framework, several similarities between the process of interaction that leads to a dealer's forecast and genetic operations in biology can be found.

Firstly, the imitative behavior is similar to the operation of selection in biology. The gene of individuals in biological populations propagate according to their fitness, meaning that the gene of fit individuals thrive and the gene of unfit individuals become extinct. Similarly, successful prediction methods spread over the market as a market consensus, but unsuccessful methods disappear.

Secondly, the accuracy of a forecast, the difference between the forecast and the actual rate, can be considered to correspond to "fitness" in a biological framework.

Finally, communication among dealers corresponds to "crossover". In biological reproduction, some of one individual's chromosomes may be exchanged for some of another individual's chromosome.

Given the similarities between the features of the interaction between dealers' forecasts and genetic operations, we used a genetic algorithm (GA) to describe learning by agents in our artificial market model. A GA is a computer algorithm that models genetic operations on the basis of population biology.

8.5 Construction of a Multi-agent Model

This section describes the construction of an artificial market: a multi-agent model of a foreign exchange-rate market. The name of the model is AGEDASI TOF (A GEnetic-algorithmic Double Auction Simulation in the TOkyo Foreign exchange market)[2].

AGEDASI TOF is an artificial market with 100 agents, as illustrated in Figure 8.2. Each agent is a virtual dealer and has dollar and yen assets. The amount of dollar and yen assets is called a *position*. The agent changes its position for the purpose of making profits. Every week of the market operation consists of five steps: (1) each agent perceives the forecast factors pertinent to the Tokyo foreign exchange market from weekly data (*perception*), then (2)

[2] AGEDASI TOF is the name of a Japanese dish, a kind of fried tofu. It's truly delicious.

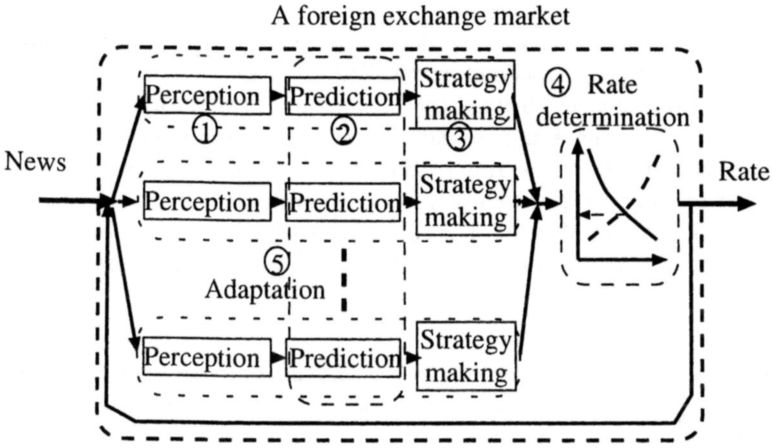

Fig. 8.2. Framework of the model.

predicts the future rate (*prediction*), and (3) determines its trading strategy (*strategy making*) every week. Then, (4) the equilibrium rate is determined from the supply and demand in the market (*rate determination*). Finally, (5) each agent improves its prediction method by learning from the other agents (*adaptation*).

8.5.1 Step 1: Perception

Each agent first interprets raw data and perceives financial and political news that is relevant to the yen-dollar rate. $x^k(t)$ is defined as the data set which is made by interpreting raw data, k, that comes in between the end of week $t-1$ and the beginning of week t. In the present study, all agents are assumed to interpret raw data in the same way. Thus the results of interpretation, the data sets $x^k(t)$ are the same for all agents.

The data set $x^k(t)$ is made up of the weekly change in 17 raw data items (Table 8.3). *External data* are defined as data on economic fundamentals, or political news (nos.1-14 in Table 8.3), because they are data on events in the real world. *Internal data* are defined as data on shot-term or long-term trends in the chart (nos.15-17 in Table 8.3), because they are calculated from rates which are generated within the market.

The values for the items range discretely from -3 to $+3$. Plus values indicate that the data change causes a depreciation in the dollar's value, according to traditional economic theories[3]. Minus values indicate appreciation. Data sets of external data items are mode up by coding news paper articles[11] and market reports[9]. For instance, a comment "the unemployment rate of

[3] We used purchasing power policy theory, the price-monetary model, and the portfolio balance model as traditional economic theories[10,3,12].

the United States has decreased greatly" is coded as "employment : −3". The datum "The chairman of Federal Reserve Bank expressed his approval of the stronger yen." is coded as "Announcement: +1". Absolute values of external data items are decided by the number of times each item appears in the newspaper articles and market reports during the week. The values are decided also by the expression of the articles, such as "greatly", "slightly", and "beyond expectations". Data sets of internal data items are made up by calculating changes of yen-dollar rates. Values of the internal data items are decided by quotients which is the changes divided by their standard deviations.

Table 8.3. Input data: US means indexes of United States. JP means indexes of Japan. (+) means that increase of data leads to a stronger dollar. (−) means that increase of data leads to a stronger yen.

Data ($x^k(t)$)	Raw data
1 Economic activities	Gross Domestic Product(US +, JP −), Industrial Production Index(US +, JP −), NAPM index(US +, JP −), Diffusion Indexes(US +, JP −).
2 Price	Consumer Price Index(US −, JP +), Producer Price Index(US −, JP +).
3 Interest rates	Official rate(US +, JP −), Fed Funds Rate(US +), Prime rate(US +).
4 Money supply	Money supply M1(US −, JP +), M2(US −, JP +), M3(US −, JP +).
5 Trade balance	The trade balance(US +, JP −), Balance of payments(US +, JP −).
6 Employment	Unemployment Rate(US −, JP +), Nonfarm Payrolls(US +).
7 Consumption	Retail sales(US +, JP −), Personal Income(US +, JP −).
8 Intervention	Buying dollar intervention(+), Selling dollar intervention(−).
9 Announcement	Announcement by a VIP about stronger dollar(+), Announcement about stronger yen(−).
10 Mark	The dollar-mark(+), and yen-mark rates(−).
11 Oil	Oil price(+).
12 Politics	Domestic political problems(US −, JP +), International conflict(+).
13 Stock	Nikkei225(JP −), Dow Jones(US +).
14 Bond	Treasury Bill(US +), Treasury Bond(US +), Government bond(JP −).
15 Short-term trend 1	Change in the last week(+).
16 Short-term trend 2	Change in short-term trend 1(+).
17 Long-term trend	Change over five weeks(+).

8.5.2 Step 2: Prediction

After perception of the above data, each agent predicts the future change of the rate. Each agent has its own weights for the 17 data items. $w_i^k(t)$ is defined as the weight assigned to datum k in agent i's prediction in week t. $w_i^k(t)$ can take any of nine discrete values $\{\pm3, \pm1, \pm0.5, \pm0.1, 0\}$. With its own importance, each agent predicts the change of the rate in the logarithm, $\Delta S(t) = S(t) - S(t-1)$, where $S(t)$ denotes the logarithm of the exchange rate in week t.

It is assumed that agent i predicts $\Delta S(t)$ as the sum of the products of $x^k(t)$ and the weight $w_i^k(t)$, over all ks. For simplicity, this sum is truncated to an integer value by a truncation function, trunc(·). Each agent makes its prediction value, $\mathbf{E}_i[\Delta S(t)]$, by calculating as follows:

$$\mathbf{E}_i[\Delta S(t)] \equiv \alpha \cdot \mathrm{trunc}\left(\sum_{i=1}^{n} w_i^k(t) x^k(t)\right), \tag{8.6}$$

where n stands for the number of data, 17, and α is a scaling coefficient. The scaling coefficient is determined by the ratio of the mean of ΔS_t to the mean of x_t^k. It is set at 0.02 for this research.

The variance in agent i's forecast is calculated as the difference between the stronger-yen factors and the weaker-yen factors as follows:[4]

$$\mathbf{Var}_i[\Delta S(t)] \equiv \frac{1}{\sqrt{|(wx_+)^2 - (wx_-)^2|}}, \tag{8.7}$$

where wx_+ denotes the sum of productions such that $w_i^k(t)x^k(t) > 0$ and wx_- the sum of productions such that $w_i^k(t)x^k(t) < 0$. wx_+ is thus the sum of the effects of stronger-yen (weaker-dollar) factors and wx_- is the sum of the effects of weaker-yen (stronger-dollar) factors. If an agent have much more factors on one side than those on the opposite side, the variance becomes smaller. If an agent have almost the same amount of stronger-dollar and weaker-dollar factors, the variance becomes larger. The variance is thus inversely proportional to the consistency of each agent's forecast. Hence, the larger the variance, the lower the confidence level of the forecast, and vice versa.

For example, suppose data for a given week are (interest: +2, trade: -1, stock: -2, long-term trend: +2), and the weights of agent i assigns to these factors are (interest: +0.1, trade: -1.0, stock: +0.1, trend: +3.0). Its forecast is then the weighted average calculated as follows,

$$\begin{aligned}\mathbf{E}_i[\Delta S(t)] &= trunc\{(+2)\times(+0.1) + (-1)\times(-1.0) + (-2)\times(+0.1) \\ &\quad +(+2)\times(+3.0)\}\times 0.02 \\ &= +7.0 \times 0.02 = +0.14.\end{aligned}$$

[4] If the terms in the denominator of equation 8.7, wx_+ and wx_- are equal, and the prediction value in equation 8.6 is zero, the optimal position in equation 8.11, calculated from equation 8.10, is zero, whatever the variance.

That is, if the logarithm of the yen-dollar rate for the previous week was $log(125\text{yen}) = 4.82$, the agent forecasts that the rate will rise to $4.82 + 0.14 = 4.96 = log(144\text{yen})$. The variance of its forecasts is calculated as follows:

$$\begin{aligned}\mathbf{Var}_i[\Delta S(t)] &= \frac{1}{\sqrt{\{2 \times (+0.1) + (-1) \times (-1.0) + 2 \times 3.0\}^2 - \{(-2) \times 0.1\}^2}} \\ &= 0.161\end{aligned}$$

8.5.3 Step 3: Strategy Making

Each agent has dollar assets and yen assets. Agent i's strategy (in terms if orders to buy or sell dollar) is determined by maximizing the expected utility, as in the standard asset pricing model, according to its prediction. Let $q_i(t)$ be the dollar assets owned by agent i at week t, and $Q_i(t)$ be the total assets (dollars and yen). Both $q_i(t)$ and $Q_i(t)$ are valued in yen. Furthermore, let $\tilde{S}_i(t) \equiv S(t-1) + \Delta S(t)$ be agent i's forecast of the logarithm of yen-dollar exchange rate at week t.

The expected return of agent i, in yen, $\tilde{P}_i(t)$, is calculated as follows.

$$\begin{aligned}\tilde{P}_i(t) &= \frac{\{\exp(\tilde{S}_i(t)) - \exp(S(t-1))\}}{\exp(S(t-1))} q_i(t) \\ &= \{\exp(\Delta S(t)) - 1\} q_i(t) \\ &\approx \Delta S(t) q_i(t).\end{aligned} \tag{8.8}$$

In AGEDASI TOF, the utility functions for all agents, $U(\tilde{P}_i(t))$, are assumed to be the same.

$$U(\tilde{P}_i(t)) \equiv -\exp(-a\tilde{P}_i(t)),$$

where $a > 0$ denotes the coefficient of risk aversion. When $\tilde{P}_i(t)$ follows the normal distribution $N(E[\tilde{P}_i(t)], Var[\tilde{P}_i(t)])$, the logarithm of the expected utility can be derived as follows[5]:

$$ln(\mathbf{E}[U(\tilde{P}_i(t))]) = \mathbf{E}[\tilde{P}_i(t)] - \frac{1}{2} a \mathbf{Var}[\tilde{P}_i(t)]. \tag{8.9}$$

Substituting equation 8.8 into equation 8.9, the logarithm of the expected utility is calculated as follows:

$$ln(\mathbf{E}[U(\tilde{P}_i(t))]) = \mathbf{E}_i[\Delta S(t)] q_i(t) - \frac{1}{2} a \mathbf{Var}_i[\Delta S(t)] (q_i(t))^2 \tag{8.10}$$

The first term in equation 8.10 is the expected return, and the second term is the risk (variance) of the position. Therefore, this equation implies that each agent tries to increase returns and reduce risks.

[5] This is done by a Taylor expansion.

Each agent is assumed to divide its whole assets between dollar and yen assets with the optimal ratio in terms of maximizing its expected utility as shown in equation 8.10. The optimal position of an agent's dollar assets $q_i^*(t)$ is as follows:

$$q_i^*(t) = \frac{1}{a}\frac{\mathbf{E}_i[\Delta S(t)]}{\mathbf{Var}_i[\Delta S(t)]}. \tag{8.11}$$

In order to make its holdings coincide with the optimal position, each agent orders the same quantity as the difference between the optimal position, $q_i^*(t)$, and the previous holdings, $q_i(t-1)$:

$$\text{quantity ordered} = \Delta q_i^*(t) \equiv q_i^*(t) - q_i(t-1). \tag{8.12}$$

If $\Delta q_t^{j*} > 0$, then the agent orders buying of dollars, that is, submits a bid. If $\Delta q_t^{j*} < 0$, then it orders selling of dollars, that is, submits a request. Each agent orders the same rate as the predicted rate, that is, buyers (sellers) are willing to buy (sell) currencies when the current rate, $S(t)$, is lower (higher) than the predicted rate.

When $\Delta q_t^{j*} > 0$,

$$\begin{cases} \text{buy dollars, } \Delta q_i^*(t) & (\text{If } S(t) \leq S(t-1) + \mathbf{E}_i[\Delta S(t)]) \\ \text{no action} & (\text{If } S(t) > S(t-1) + \mathbf{E}_i[\Delta S(t)]\ ). \end{cases} \tag{8.13}$$

When $\Delta q_t^{j*} < 0$,

$$\begin{cases} \text{no action} & (\text{If } S(t) < S(t-1) + \mathbf{E}_i[\Delta S(t)]\ ) \\ \text{sell dollars, } \Delta q_i^*(t) & (\text{If } S(t) \geq S(t-1) + \mathbf{E}_i[\Delta S(t)]\ ). \end{cases} \tag{8.14}$$

For example, the optimal amount of dollar assets for the above-mentioned agent, i, is given by:

$$q_i^* = \frac{+0.14}{0.161} = +0.87.$$

If the agent i's previous position in terms of dollar assets, $q_i(t-1)$, was -0.74, it will send an order to buy:

$$\Delta q_i^*(t) = +0.87 - (-0.74) = +1.61.$$

Agent i places an order to buy when the current rate is lower than the rate it expected, $4.82 + 0.14 = 4.96$:

$$\text{trading strategy} = \begin{cases} \text{buy dollars, } 1.61 & (\text{if } S(t) \leq 4.96) \\ \text{no action} & (\text{if } S(t) > 4.96). \end{cases}$$

8.5.4 Step 4: Rate Determination

After the submission of orders, the demand (supply) curve is made by the aggregation of orders from all agents who want to buy (sell). The demand and supply then determine the equilibrium rate, where the quantity of demand and of supply are equal (Figure 8.3). $S(t)$ is the equilibrium rate at week t.

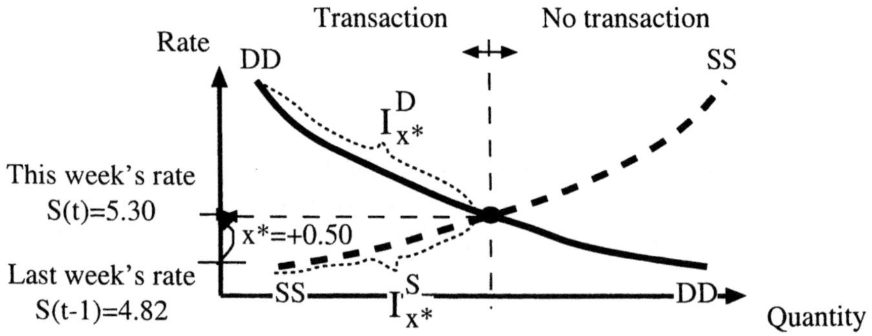

Fig. 8.3. Determination of rate.

The demand curve $\mathbf{DD}_t(x)$ is made by aggregation of all bids ($\Delta q_i^*(t) > 0$) from agents with higher order rates than x:

$$\mathbf{DD}_t(x) = \sum_{i \in I_x^D} \Delta q_i^*(t), \tag{8.15}$$

$$\left(I_x^D \equiv \{i : \Delta q_i^*(t) > 0 \text{ and } \mathbf{E}_i[\Delta S(t)] \geq x\}\right).$$

The supply curve $\mathbf{SS}_t(x)$ is made by aggregation of all requests ($\Delta q_i^*(t) < 0$) from agents with lower order rates than x:

$$\mathbf{SS}_t(x) = -\sum_{i \in I_x^S} \Delta q_i^*(t), \tag{8.16}$$

$$\left(I_x^S \equiv \{i : \Delta q_i^*(t) < 0 \text{ and } \mathbf{E}_i[\Delta S(t)] \leq x\}\right).$$

The exchange rate of the artificial market is determined as the equilibrium rate, at which the quantities demanded and supplied are equal:

$$S(t) = S(t-1) + x^*, \tag{8.17}$$

$$(\mathbf{DD}_t(x^*) = \mathbf{SS}_t(x^*)).$$

Buyers (sellers) with higher (lower) order rates can exchange currency so that their holding positions, q_i^t, coincide with their own optimal positions, q_i^*.

However, the other agents cannot exchange currency and $q_i(t)$ thus remains their previous holding positions, $q_i(t-1)$:

$$q_i(t) = \begin{cases} q_i^* & \text{if } i \in I_{x^*}^S \text{ or } I_{x^*}^D \\ q_i(t-1) & \text{otherwise.} \end{cases} \tag{8.18}$$

8.5.5 Step 5: Adaptation

After the rates have been determined, each agent improves its prediction method (combinations of weights, $w_i^k(t)$) by referring to other agents' prediction methods. Our model uses genetic algorithms (GA) to describe the interaction among agents.

When our model uses the GA in the adaptation step, each gene represents a symbol which is made by the transformation of a particular weight, $w_i^k(t)$. Each weight $w_i^k(t)$ is transformed as follows.

$$w_i^k(t) = \begin{pmatrix} +3 & +1 & +0.5 & +0.1 & 0 & -0.1 & -0.5 & -1 & -3 \\ \Downarrow & \Downarrow & \Downarrow & \Downarrow & \Downarrow & \Downarrow & \Downarrow & \Downarrow & \Downarrow \\ \mathbf{A} & \mathbf{B} & \mathbf{C} & \mathbf{D} & \mathbf{E} & \mathbf{F} & \mathbf{G} & \mathbf{H} & \mathbf{I} \end{pmatrix}$$

A chromosome represents the string of all weights for one agent:

$$\text{chromosome } \mathbf{w}_i(t) = (w_i^1(t), w_i^2(t), \cdots, w_i^{17}(t)).$$

For example, a set of weights $\mathbf{w}_i(t) = (+0.1, -3, 0, +1, \cdots, +0.5)$ becomes a chromosome **DIEB** $\cdots$ **C**. A population of chromosomes represents a set of $\mathbf{w}_i(t)$ in the foreign exchange market. Each chromosome can be regarded as an agent's belief about the exchange rate. That is, it represents which data are regarded as important causes of the changes to rates. It must be noted that beliefs can differ among agents.

In our model, the fitness of each chromosome, $F(\mathbf{w}_i(t))$, is calculated by using the difference between the predicted rate, $\mathbf{E}_i[\Delta S(t)]$, and the actual rate, $\Delta S(t) = x^*$.

$$\begin{aligned} \text{Fitness of } \mathbf{w}_i(t) = F(\mathbf{w}_i(t)) &= -|\mathbf{E}_i[\Delta S(t)] - x^*| \\ &= -|\alpha \cdot \text{trunc}\left(\sum_{i=1}^{n} w_i^k(t) x^k(t)\right) - \Delta S(t)|. \end{aligned} \tag{8.19}$$

Hence, the more precisely a chromosome predicts the rate, the greater its fitness.

Each agent changes its own belief system in order to improve its prediction by applying the three operators in Goldberg's simple GA [8]: selection, crossover, and mutation (Figure 8.4).

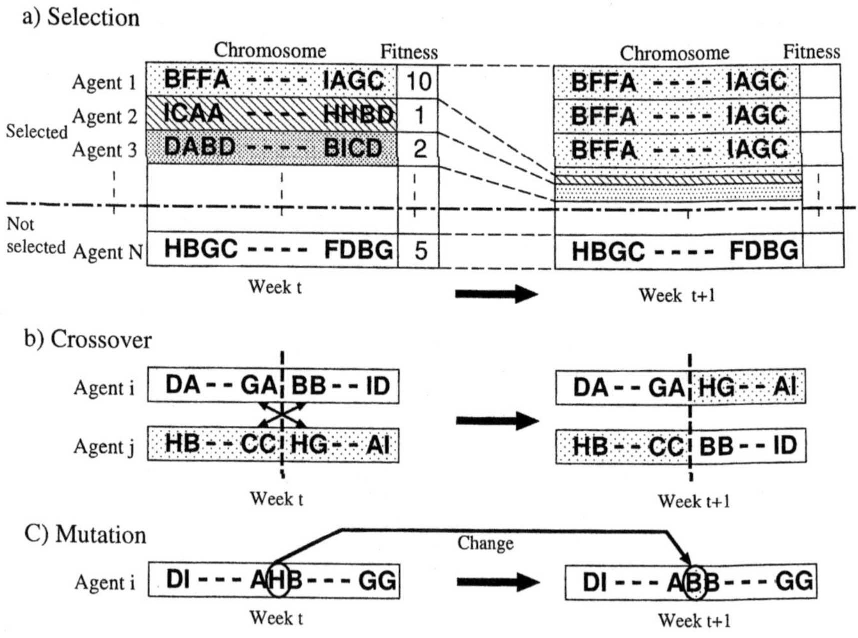

Fig. 8.4. Genetic algorithm

Selection The selection operator replaces some chromosomes with others which have greater fitness (Figure 8.4a). In a given percentage of the population, chromosomes with greater fitness are selected and copied on each generation, while the other chromosomes are maintained. The percentage of such selection operations is called a *generation gap*, G. The selection operator can be interpreted as the imitation of the belief system of another agent, which was able to predict the rate more accurately. Therefore, belief systems which predict less accurately will gradually disappear from the market. Selection is therefore regarded as the propagation of successful methods of prediction.

Crossover A pair of agents will sometimes exchange some of their weights. Here, we apply the single-point crossover (Figure 8.4b) with a prespecified crossover rate, *prcoss*). The crossover operator operates to model the agent's means of communication with other agents.

Mutation There is also a small probability of a random change in one of each agent's 17 weights $w_i^k(t)$ (Figure 8.4c). That probability is called the *mutation rate* and is denoted by *pmut*, and k is chosen with uniform probability from among $\{1, \cdots, 17\}$. This mutation operator works like an independent change in a given agent's prediction method.

After adaptation has been completed, the given week ends and the model proceeds to the next week's perception Step.

8.6 Scenario Analysis

Using our artificial market model, we tried the construction of a decision-making support system (DSS) for exchange rate policy. The goal of this DSS was to stabilize the yen-dollar rate in 1998. The strategies studied in this paper are based on the following three control variables, namely, *interest rates*, *intervention*, and *announcements*. These three variables are also included in Table 8.3, and are used by agents in *perception*. The other 14 factors in that table are considered uncontrollable.

The DSS based on AGEDASI TOF consists of two steps:

Selection of Important Factors Firstly, several "important factors" are selected by applying the procedure detailed in section 8.6.1. Then, by focusing on the dynamics of the important factors, we make some "strategic scenarios" for the control variables.

Comparison of Strategic Scenarios Secondly, computer simulations that correspond to each strategic scenario are carried out (section 8.6.2). The best scenarios are selected on the basis on results of simulation.

8.6.1 Selection of Important Factors

Firstly, we used sample data from January 1996 to December 1997 to train the agents in our artificial market. After that, according to the results of training, the "important factors" were selected from among the 14 uncontrollable factors. Finally, by focusing on the dynamics of the important factors, we made some "strategic scenarios" for the control variables.

Simulation methods In order to train the agents in our artificial market, the following procedures were repeated one hundred times: (1) an *initialization procedure* and (2) a *training procedure.*

- **Initialization-** In the initialization procedure, we randomly generated 100 agents. Each agent is represented by a vector of dimension 17 $w_i(0)$. Each component of the vector, $w_i^k(0)$, as randomly generated from the set $\{\pm 3, \pm 1, \pm 0.5, \pm 0.1, 0\}$. Each agent's initial portfolio in dollar assets and yen assets was set to zero.

$$q_i(0) = 0,\ Q_i(0) = 0.$$

- **Training-** We trained the agents in our model on real-world data according to the items listed in Table 8.3, for the period from January 1996 to December 1997. Rate determination was omitted because we used actual exchange rates. In the adaptation step, the fitness function for the GA is the cumulative absolute deviation between the agent's forecast, $\mathbf{E}_i[\Delta S(t)]$, and the actual rate, $\Delta S(t)$:

$$\text{fitness of } \mathbf{w}_i(t) = F(\mathbf{w}_i(t)) = -\sum_{\tau=0}^{t} |\mathbf{E}_i[\Delta S(\tau)] - \Delta S(\tau)|.$$

Note that we have not used equation 8.19 in this procedure.
Weekly data from January 1996 to December 1997 was applied two hundred times, so the length of the whole training period was about twenty thousand weeks. That is, GA operations ware conducted twenty thousand times over the training period.
In this study, we used the following control parameters for the GA: $\{pcross = 0.3, pmut = 0.003, G = 0.8\}$. In a similar study [6], we had carried out out-of-sample-period forecasting tests with various sets of control parameter sets on the data from 1986 to 1993, and we found that this parameter set gave rise to the smallest forecast errors. We thus used the same parameters in this study.

Results The results of training showed that the market averages of the weights assigned to the factors *Economic activities* and *Deutschmark* had the highest absolute values in the mean taken over one hundred simulation runs (Figure 8.5). That is, the agents are most sensitive to these two of the 17 factors listed in Table 8.3. Hence, we selected them as the "important factors".

Strategic scenarios Focusing on the dynamics of the important factors, we made some "strategic scenarios" regarding the control variables. The strategic scenarios are options for the solution of the decision-making problem.

We came up with the following three scenarios, when there is a continuous run of big news about on the important factors:

scenario (a): control *interest rates* in response to the news,
scenario (b): *intervene* in the markets in response to the news, and
scenario (c): make *announcements* in response to rate policies the news.

8.6.2 Comparison of Strategic Scenarios

Computer simulations were set up with a strategic scenario from (a)-(c) and actual data from the period as inputs, and the results were compared. We then selected the best scenarios according to the results of comparison.

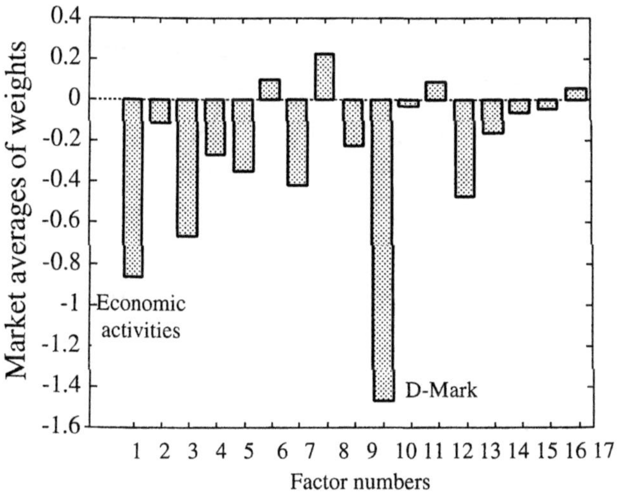

Fig. 8.5. Market averages of the weights of factors (the mean of 100 simulations).

Simulation methods Simulations for each scenario were conducted by repeating the following procedures a hundred times; (1) *initialization*, (2) *training*, and (3) *testing*. This gives us 100 simulation paths for each scenario. Since the first two procedures are the same as before (section 8.6.1), we only discuss the testing procedure here.

- **Testing Procedure -** This procedure is unlike training in that the exchange rate used is not historical data, but data that is endogenously generated from by rate determination (section 8.5.4). Hence, all internal data (nos. 15-17 in Table 8.3) are endogenously derived. The fitness function used here is the one given as equation 8.19. The rest of the procedure is the same as the training procedure.

Actual Scenario To have a benchmark for comparison, we also carried out computer simulations on the basis of real data, i.e., the three control variables were set according to their historical values.

At the end of August, the 100 simulation paths can be clearly divided into three groups. The first group consists of paths which move up to or above 136 yen. The second group consists of paths which move down to or below 116 yen. The last group consists of those paths which move between 136 yen and 116 yen. One can similarly classify the simulation paths observed at the end of December.

Combining the classifications made by the final positions for August and December, one can further categorize the 100 simulation paths into the 4 groups shown in Figure 8.6. Firstly, the simulation paths which are over 136 yen in both August and December are classified as an *upward group*. Secondly,

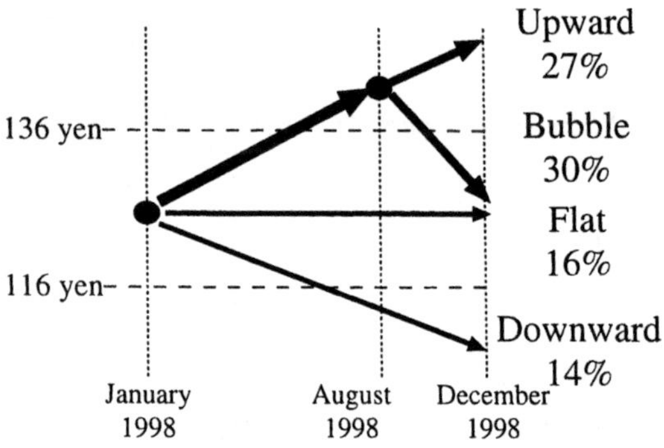

Fig. 8.6. Categorization of simulation paths.

those which are over 136 yen in August but below 116 yen in December are classified as a *bubble group*. Thirdly, those between 116 and 136 yen in both August and December are classified as a *flat group*. Those under 116 yen in both August and December are classified as a *downward group*. Almost all simulation paths belong to one of these 4 groups. The real (historical) path belongs to the bubble group, and this was also the type most commonly observed in our 100 simulations (30%, see Figure 8.6) .

Strategic Scenarios In 1998, from April to June, there was a run of good news for the U.S. economy, and a run of bad news for Japan's economy. This situation, however, changed after the depreciation of the ruble in August. Many U.S. investors suffered huge losses in the Russian currency crisis. The dollar also became relatively weak as opposed to European currencies. Consequently, from September to November, news on the economic prospects of the U.S. economy was not exciting but rather pessimistic. This news had the potential to induce severe fluctuations in the yen-dollar rate. Therefore, in order to stabilize the yen-dollar rate, we considered the three policy scenarios outline below.

Scenario (a): Set the control variable *interest rate* to +3 from April to June, and reset it to -3 from September to November.
Scenario (b): Set the control variable *intervention* to +3 from April to June, and reset it -3 from September to November.
Scenario (c): Set the control variable *announcement* to +3 from April to June, and reset it to -3 from September to November.

As shown in Figure 8.7, scenario (b) produced the highest proportion of results in the flat group. The probability of a stable yen-dollar rate became twice as great as in the actual situation. That is, if the central banks of Japan and the US had made a strong and continuous (coordinated) intervention, selling or buying dollars for yen, just after the news about the big changes

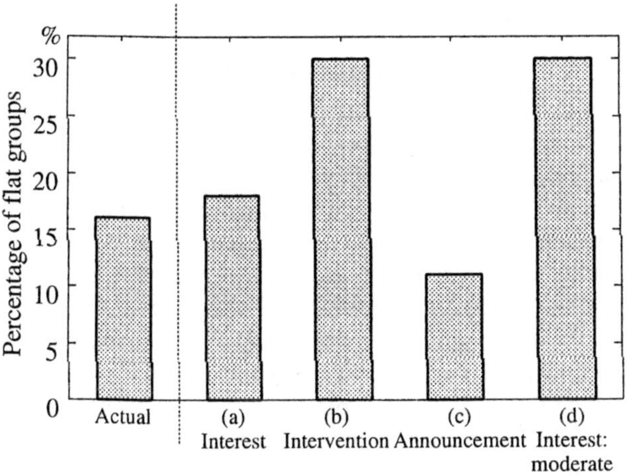

Fig. 8.7. Percentage of flat groups.

in economic activities and deutschemark rates had arrived, the probability of a stable yen-dollar rate in 1998 would have been higher. Hence, control by intervention was the best way to stabilize the rate in 1998. In scenario (a), interest rates have such a strong effect on exchange rates that the exchange rates fluctuate more strongly. Scenario (c), relying on the announcement factor, had no effect.

Considering the fact that the effect of interest rates was so strong, we made up a new scenario.

Scenario (d): *Interest rate* factors have a value of +1 from April to June and -1 from September to November.

As shown in Figure 8.7, scenario (d) produces the same probability of flat-group results as scenario (b). Therefore, moderate control of interest rates was also effective.

We also tested other scenarios in which more than two control variables were in operation, but ratios of flat-group results from these scenarios were less than those that for scenarios (b) and (d). This is because the effects of two or three factors are too strong.

Results From the above results, it was found that the following two strategies were effective in the stabilization of yen-dollar rates in 1998.

(a) Strong intervention in response to news about economic activities and the deutschmark.
(b) Moderate control of interest rates in response to news about economic activities and deutschmark.

8.7 Conclusion

In this study, we have applied an artificial market approach to decision-making about exchange rate policies. The approach consisted of fieldwork, construction of the artificial market model, and computer simulation. As a result, it was found that intervention and the control of interest rates would have been effective measures for the stabilization of yen-dollar rates in 1998.

This approach overcomes the two problems with decision-making under circumstances in which the micro-macro problems. Firstly, an artificial market model can simulate dynamic changes in important factors in the market by applying interaction among agents. Hence, our approach can trace changes in rate determination rules over time beyond sampling periods and does not over-fit simulation results to sample data. Secondly, because our approach selects several important factors in each period, we do not need to consider all potential rules. Hence the number of rules does not become enormous when data from other periods are added.

This study is the first attempt to apply an artificial society or artificial economics model to decision-making in the real world. Previous artificial society or artificial economics models [1,2,4] have focused on qualitative settings, and rarely consider the quantitative analysis of real-world situations. The results of this study show that the artificial market approach is an effective way of for analyzing real-world markets.

References

1. Arthur, W and Holland, J. H. et.al. (1997) Asset pricing under endogenous expectations in an artificial stock market. In W.B. Arthur et.al. editor, The Economy as an Evolving Complex Systems II, 15–44. Addison-Welsley Publishing.
2. Axelrod, R. (1997) The Complexity of Cooperation: Agent-Based Models of Competition and Collaboration. Princeton University Press.
3. Richard T. Baillie and Patrick C. McMahon. (1989) The foreign exchange market: theory and economic evidence. Cambridge University Press, Cambridge.
4. Epstein, J. and Axtell, R. (1996) Growing Artificial Societies: Social Science from Bottom Up. MIT Press.
5. Goldberg, D. (1989) Genetic algorithms in search, optimization, and machine learning. Addison-Wesley Publishing Company.
6. Izumi, K. and Okatsu, T. (1996) An artificial market analysis of exchange rate dynamics. In Fogel, L. J. Angeline, P. J. and Bäck, T. editors, Evolutionary Programming V, 27–36. MIT Press.
7. Izumi, K. and Ueda, K. (1999) Analysis of dealers' processing financial news based on an artificial market approach. Journal of Computational Intelligence in Finance, **7**, 23–33.
8. Izumi, K. and Ueda, K. (1998) Emergent phenomena in a foreign exchange market: Analysis based on an artificial market approach. In Adami, C., Belew, R. K., Kitano, H., and Taylor, C. E. editors, Artificial Life VI, 398–402. MIT Press.

9. Japan Center for International Finance. Weekly market report, 1992–1995.
10. MacDonald, R. (1988) Floating Exchange Rates: Theories and Evidence. Unwin Hyman, London.
11. Nihon keizai shinbun Finance outlook, Sunday supplement, 1992–1995. Nihon keizai shinbun sha.
12. Pentecost, E. (1993) Exchange Rate Dynamics: A Modern Analysis of Exchange Rate Theory and Evidence. Edward Elgar, Vermont.

9 Emulating Trade in Emissions Permits: An Application of Genetic Algorithms

Rosalyn Bell and Stephen Beare

Australian Bureau of Agricultural and Resource Economics, Australia
RBell@abare.gov.au

Abstract. Emissions permits are generally a second best option for dealing with site specific pollution. The outcome of trade in emissions permits when the economic welfare of market participants is linked spatially through production externalities is unclear. Trade will reflect the interaction of bargaining agents whose incentives vary with the relative physical location of both the buyer and seller. For the permit system to internalise the costs of pollution, information on who are the current buyers and sellers is necessary. This information corresponds to an understanding of the economic impacts of the physical externality, which in turn allows an improvement in the level and distribution of resource access or use. However, provision of such information is not characteristic of a competitive market in which rents associated with reducing the net cost of an externality are competed away. To achieve a more efficient distribution of entitlements, through the internalisation of pollution costs, the market structure must allow agents to capture these rents.

A genetic algorithm is used to emulate trading behaviour of individual agents for emission entitlements. Agents are assumed to operate independently, with each attempting to find their own optimal combination of inputs and emissions permits. The agents are linked in a simulation model through market outcomes and through a spatially dependent production externality. The process being modelled is essentially a non- cooperative evolutionary game. Agents learn that their best bidding strategy is not independent of the strategies of other market participants. In particular, the value of a permit depends on both the price and quantity of permits bought and sold by other market participants. The model is used to examine the effectiveness of emission permit schemes given a range of different market structures and trading strategies employed by market participants. The results suggest an effective emissions permit scheme may require institutional arrangements that preserve market power as opposed to atomistic competition.

9.1 Background

Tradable entitlement schemes are intended to address problems associated with poorly defined property rights governing resource access. In many cases, the externalities associated with poorly defined property rights are public. That is, the external costs and benefits are imposed on, or realised jointly, by users of a resource. These can be referred to as common entitlements, serving only to limit resource access or use. As common entitlements do not internalise the external costs and benefits to an individual user, policy makers must determine an optimal or welfare improving level of entitlements. Some examples of common entitlements would include tradable quota schemes for fisheries and emission permits to control air pollution.

The purpose of introducing trade in common entitlements is to generate and maintain an efficient distribution of entitlements.[1] In systems in which the source and impacts of externalities are site specific, the introduction of tradable entitlements may internalise at least some costs and benefits of resource use. An individual's access right to a resource at a specific site, through land ownership for example, conveys private costs and benefits that may be internalised. An example of such a system is the sequential impacts of pollution along a river where downstream users bare part of the costs of upstream pollution directly. Similarly, agricultural production decisions in one catchment may impact on the salinity of river flows used for irrigation in downstream regions ([11]).

The private benefits and costs of spatially dependent externalities are largely bilateral in nature. Hence, a common entitlement such as an emissions permit will not generally be a first best policy instrument.[2] However, the transaction costs of establishing a first best policy instrument, such as an ambient pollution permit scheme, are likely to be prohibitively high ([2] and [10]). A common entitlement scheme, such as an emissions permit, may still be a cost effective policy instrument for improving resource allocation.

Where an entitlement, such as an emissions permit, introduces property rights that partly internalise the net benefits of resource use, markets may generate information regarding a socially preferred level of resource access, as well as a more efficient distribution of entitlements. This is a potentially attractive option to policy makers in that they may only need to establish a set of trading guidelines to achieve a better resource management outcome. Individuals can retain their current level of entitlements, bare the information and transactions costs and realise any benefits of trade.

The problem in setting up institutional arrangements is that the value of a permit will depend not just on price but on the locations of other agents who buy and sell permits. Reference [9] establishes that competitive market equilibrium may not exist where there is quantity dependence and agents are only able to observe prices. An individual's marginal return from a transaction may not be equal to the market price as the marginal return will vary depending on who is buying or selling. For example, purchasing pollution entitlements from an upstream producer along a river may not be the same as purchasing entitlements from an equivalent producer downstream.

To internalise spatially dependent externalities through an emission permit, the price discovery process must eventually provide some information

[1] Reference [16] demonstrate that as the number of traders in an auction market for a common entitlement becomes large, auction prices will approach the true marginal value of the entitlement.

[2] A first best instrument would need to allow for bilateral exchanges between sources and affected sites.

on who are the current buyers and sellers. This information corresponds to an understanding of the economic impacts of the physical externality, which in turn allows an improvement in the level and distribution of resource access or use. However, provision of such information is not characteristic of a competitive market in which rents associated with reducing the net cost of an externality are competed away. To achieve a more efficient distribution of entitlements, the market structure must allow individuals to capture these rents. However, the introduction of non-atomistic trade may impose other costs. The trade-off between these costs and the benefits of the information generated through trade can be important in the design of an appropriate set of trading arrangements.

The establishment of entitlements and trading arrangements sets up a non-cooperative evolutionary game ([18]). The policy maker's goal is to put in place institutional arrangements that maximise the net pay-off from the game. In this chapter, a genetic algorithm (GA) is used to emulate such a trading game and explore the affect on the market outcome of alternative market structures. Reference [17] describes the use and properties of GAs in evolutionary games.

Individual agents are profit maximising producers linked in simulation through a spatially dependent production externality and allowed to trade in a market. Each firm uses a GA to develop a trading strategy that maximises expected net revenue from production and trade. An individual firm observes only market prices and the impact of the externality at their own location.

9.2 Model Construction and Use of GAs

9.2.1 The Basic Problem

Following on from the river example in the previous section, consider a number of identical firms that draw on a resource sequentially and the use of the resource results in a production externality that impacts on firms downstream. The physical configurations we are going to consider are shown in Fig. 9.1, the simplest case being with only three firms in a linear sequence.

It is readily demonstrated that in such an example, an emissions permit will be a second best policy instrument. If the bottom firm purchases an emissions permit from the top firm this will convey a benefit to the middle firm, as the top firm's production impacts on both. As the bottom firm does not capture the full benefits of trade with the top firm, the price that the bottom firm would be willing to pay will not be sufficient as to result in an optimal distribution of permits. In the absence of cooperation between firms, the jointness creates a public externality in trade.

As the exchange of permits between the top and middle firms or the middle and bottom firms is bilateral, there are no externalities associated with trade. However, for the middle firm, a purchase of an emissions permit from the top firm has greater value than from the bottom, as it allows the

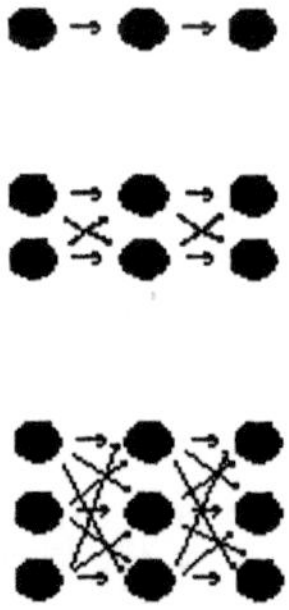

Fig. 9.1. Physical configurations of polluters

middle firm to increase its own output and reduces the impact of emissions from the upstream firm. However, in a single permit market, the market price alone will not indicate to the middle firm which firm is offering a permit. The middle firm must learn who is buying and selling by observing the affect of its trade on its returns.

The problem with incomplete or asymmetric market information is exacerbated in the second and third configurations with double and triple the number of firms, as it becomes harder for each firm to isolate the impact of its own trading strategy on its returns. Emissions from firms in one location impact on a group of firms in downstream locations, increasing the jointness of the externality as well as reducing the amount of information provided by the market. No single firm within each group is able to capture the full benefits of trade between the different locations.

Introduction of Market Power To overcome this lack of market information and the jointness of trade benefits at each location, firms in a given location may form trading blocks. That is, firms surrender their entitlements to a central body at each location. The net benefits of trade can be distributed according to the level of entitlements surrendered. However, the creation of trading blocks may create market power in one location, imposing costs on other regions and reducing the benefits of trade.

The creation of trading blocks will not address the problem of externalities in trade between locations. A second option that would address this, is for all firms to surrender their entitlements to a central trading company. Individual firms could hold shares in the trading company in proportion to the entitlement they surrendered. This trading company would simply seek to maximise its returns from the sale of permits and distribute these returns back to firms according to the proportions of surrendered entitlements.

The introduction of non-atomistic market structures may allow traders to acquire information that leads to an improvement in the distribution of permits and the level of emissions.[3] The principal advantage that a monopoly seller would introduce in a spatial model of permit trade is that permit prices may be set above those associated with trade in a competitive market in which permit traders compete away any available rents associated with a reduction in emissions.

However, this result is not guaranteed. By vesting market power in trading blocks or a central trading company, trade may still not result in an allocation of permits that is optimal and may lead to an overall decline in economic welfare. The effectiveness of the tradable permit scheme is ultimately an empirical question and the objective of developing a trade emulation algorithm is to provide a basis for making both qualitative and quantitative assessments.

9.2.2 Simulation Model Design

The trade emulation algorithm was embedded in a larger agent based simulation, implemented in EXTEND [13]. The simulation model can be used to examine a range of economic instruments for resource management with an arbitrary spatial network of n firms. Each firm is represented by an independent profit maximising algorithm. Production dependencies are represented by the network connections. Externalities that arise through a firm's production process are passed sequentially along these connections.

The production function for an individual firm i is represented by a generalised quadratic function

$$f_{it} = \mu_{it}(\alpha_{i1}x_{it} + \alpha_{i2}y_{it} + \alpha_{i3}x_{it}y_{it} - \alpha_{i4}x_{it}^2 - \alpha_{i5}y_{it}^2) \qquad \mu, \alpha_l > 0 \quad (9.1)$$

where f_{it} is firm i's output at time t, x_{it} and y_{it} are inputs to production at time t, and μ_{it} and α_{il} are production function coefficients (for $l = 1$ to 5).

An externality E_{it} incurred by firm i is assumed to arise through the use of input x in production. Specifically, for any firm i

$$E_{it} = \sum_{j \in J_i} \delta_j x_{jt} \qquad (9.2)$$

where J_i is the set of firms above firm i in the resource use network and δ_j is the proportion of firm j's usage of input x which is transferred as damage to production of firms lower in the network. The externality is assumed to have a non-linear impact on firm productivity such that

$$\mu_{it} = \frac{\mu_{0i}}{1 + \exp(-\gamma_{1i} + \gamma_{2i}E_{it})} \qquad \text{for } \gamma_{1i}, \gamma_{2i} > 0 \qquad (9.3)$$

[3] The idea that a monopoly is able to generate and use more information than a competitive market is not new. Reference [7] demonstrated that the incentive to generate information for technological advancement is at least as great under a monopoly as in a competitive industry.

Individual firms act independently to choose x, y, the quantity and reserve price of permits offered for sale and the quantity and bid price of permits to purchase, to maximise expected profit π each period from production and permit trade. That is,

$$\pi_{it} = \pi_{production} + \pi_{permits} \tag{9.4}$$

for

$$\pi_{production} = pf_{it} - cx_{it} - dy_{it} \tag{9.5}$$

where p is the market price of output, c and d are constant unit input costs. For an initial allocation of permits of w_i to firm i, if there is no central trader of permits then $\pi_{permits}$ is given by

$$\pi_{permits} = \rho_t(z_{it} - \nu_{it}) \qquad \text{for } x_{it} \leq w_i - z_{it} + \nu_{it} \tag{9.6}$$

where ρ is the traded permit price, ν_{it} is the quantity of permits purchased by firm i and z_{it} is the quantity of permits from firm i's initial allocation which are sold. If there is a central seller of permits which is maximising revenue from permit sales, then firm i's profit from permit trade is given by

$$\pi_{permits} = -\rho_t\nu_{it} + \left(\frac{w_i}{\sum_i w_i}\right)\rho_t\sum_i \nu_{it} \tag{9.7}$$

The first term of equation 9.7 is the cost of permit purchases and the second is firm i's share of revenue from permit sales by the central trader. Note that if all available permits are sold, revenue from permit sales received by firms is the same when there is a central trading company to when there is no central trading company. The values of parameters used to calibrate equations 9.1 – 9.7 are detailed in Table 9.2.

9.2.3 Emulation of Trade Using a Genetic Algorithm

To emulate trade in emissions permits, a sealed bid auction framework was utilised in which permits are distributed to the highest bidding producers first until the permit market is cleared.[4] Individual producers are assumed to operate independently with each attempting to find their own optimal combination of inputs and emission permits. The firms are linked through market outcomes and through a production externality.

Each firm engaged in trade has a set of potential trading strategies. An individual firm's trading strategy consists of four non-negative elements: a quantity offered sale, an associated reserve price on the quantity offered, a

[4] Note that in a first price sealed bid auction, there is no dominant strategy. Rather, a firm's best strategy depends on his beliefs about the bidding strategies of others. See [15] for a comparison of first price sealed bid auctions with other forms of auctioning.

quantity bid for additional permits and a bid price. To evaluate a set of trading strategies, trades are executed from the highest to successively lower bid prices until the market is cleared. The market price is the marginal bid (the lowest bid price that is accepted). Given the permit trade outcome, firms then optimise their output in sequence from the spatially highest to lowest in the network. Firm profits are determined analytically given the levels of trade and the externalities imposed by upstream firms.

To search for the optimal trading strategy for each individual firm, a genetic algorithm (GA) was utilised. The approach provides a globally robust search mechanism with which to optimise over a decision process involving uncertainty in the form of a lack of a priori knowledge, unclear feedback of information to decision makers and a time varying payoff function. Reference [12] first developed the GA approach. It has subsequently been widely employed in economics and finance research as a flexible and adaptive search algorithm in problems that have dynamic structures not easily handled with traditional analytical methods.[5] (see for example: [1] [3] [4] [5] [6] [14])

A GA performs a multi-directional search by maintaining a population of individual strategies, each with a potential solution vector for the problem. An objective function is employed to discriminate between fit and unfit solutions. The population undergoes a simulated evolution such that at each generation, the relatively fit solutions reproduce while the relatively unfit solutions die out of the population.

During a single reproductive cycle, fit strategies are selected to form a pool of candidate strategies, some of which undergo cross over and mutation in order to generate a new population. Cross-over combines the features of two parent strategies to form similar offspring by swapping corresponding segments of the parents. This is equivalent to an exchange of information between different potential solutions. Mutation introduces additional variability into tho population by arbitrarily altering a strategy by a random change. The GA is implemented using the approach described in [8], with values of the parameters used to calibrate the GA detailed in Table 9.2.

The first step in implementing the search for each firm's optimal trading strategy is to calculate a fitness value for each of the m potential trading strategies of a firm. That is, each firm has its own population of strategies and its own objective function, described by equation (4). Given there are n firms, there are m to the power n potential market outcomes. To reduce computational requirements fitness values are determined through a series of k random trials. In a single trial the trading strategies for each firm are

[5] The non-linear form of the production and damage functions, and a lack of information available to market participants regarding the precise specification of these functions and the strategies of other players means that the problem presented here is too complex to be readily solved using methods traditionally adopted by economists.

randomly ordered, generating a set of m market outcomes consisting of a quantity traded by each firm and a single market price.

A range of options exists for assigning a fitness level to each strategy, ranging from the best to the worst trial outcome for each strategy. A firm may select a strategy that performs very well against some competing strategies in an attempt to capture a greater share of the available rent associated with an overall improvement in the level of resource allocation. However, such a trading strategy can be risky as a shift in a competing firm's strategy may impose substantial trade losses. Furthermore, such a strategy may not lead to the best overall allocation of resources. A firm which selects a strategy that performs reasonably well among most competing strategies may limit the risks associated with trade but again the level of overall benefits achieved and the share of benefits captured by that firm may be lower.

In the simulations undertaken here, the trial value representing the median value was used to determine the fitness in the base case simulation, as it allowed competing strategies to converge to a reasonably stable equilibrium across the range of simulations undertaken. The impact of a range of alternative trading approaches (conservative through to aggressive) on the effectiveness of a tradable emission permit scheme was assessed for the three firm case. Fitness values were chosen at different percentiles of the sample generated from the trails, ranging from the minimum to the maximum value. For simplicity, all bidders were assumed to be equally conservative or aggressive in a given simulation.

9.2.4 Determining the Optimal Level of Resource Use with a GA

To allow comparisons of results between simulations, the maximum potential gain from trade was calculated by comparing an unregulated resource-use system with a globally optimal set of firm specific (Pigouvian) tax rates. The proportion of the gain that could be achieved through emissions trade was then compared with a corresponding second best tax instrument, a uniform emissions tax.

To determine the appropriate tax rates a central planner was introduced into the unregulated resource-use simulation. A GA was used to determine both the optimal site- specific tax rates and a uniform tax rate imposed on all firms. The fitness associated with any given tax strategy was evaluated as the sum of firm profits and tax revenue. Firms sequentially choose input levels to optimise production revenue (less tax cost), given the level of externality imposed by any upstream firms.

For firms whose emissions impose a relatively high cost on society, this single tax rate will be below the Pigouvian tax. Conversely, for those firms with relatively low cost emissions, the uniform tax rate scheme will result in lower expected individual net revenue, compared to the socially optimal outcome.

9.3 Simulation Results

The simulation results reported in Figs. 9.2 – 9.5 illustrate the relative effectiveness of a tradable emissions scheme in capturing the benefits available from reduced emissions impacts. The maximum benefits which can be derived from a change in emission levels at all sites is given by the difference between the optimal level of aggregate net revenue and aggregate net revenue when all firms act independently in production.

For a network of 3 firms (as in Fig. 9.1), around 85 per cent of these benefits are captured by firms through the introduction of a tradable emissions permit scheme. The failure of a tradeable emissions scheme to capture all of the benefits available can be attributed to the problems of externalities in trade and the jointness of trade costs and benefits.[6]

9.3.1 Impact of Basis for Selection of Trading Strategy

The effectiveness of a tradable emissions permit scheme in capturing the benefits available from reduced emissions was found to be maintained under a range of alternative trading approaches by market participants, although the variability of this outcome and the expected permit price differed significantly between approaches (Figs. 9.2 and 9.3).

Expected aggregate net revenue of all firms was greatest, and standard deviation of this revenue lowest, when firms adopt a median trading approach – that is, strategy selection based on the best outcome of a set of strategies which are expected to generate a median level return (*maxmed* in Fig. 9.2).

In contrast, selection by all individuals of a strategy that performs best against the most favourable competing strategies (*maxmax* in Figs.9.2 and 9.3) does not, in the long run, lead to the highest expected gains from trade. Market participants learn that the strategy that gave them the highest gain is unlikely to be repeated – the chosen set of best strategies of the other market participants will evolve to maximise their own individual objective functions. Trading with such strategies can be highly unstable as a small shift in one firm's strategy can result in a large change in the outcome of trade. The expected market price of an emissions permit is lower for this trading approach than for less conservative approaches because those strategies that perform best will be associated with paying, on average, a lower price for permits.

Selection by all individuals of a strategy that performs well against the least favourable strategies of other market participants (*maxmin* in Figs. 9.2 and 9.3) may mean that opportunities for higher returns are passed over. As

[6] Nevertheless, the gain achieved is higher than the 79 per cent possible from the introduction of a single uniform tax on resource use and will be preferred by firms, as some of the gains achieved under a tax scheme are tax collection revenue which may not be directly available to the taxed firms.

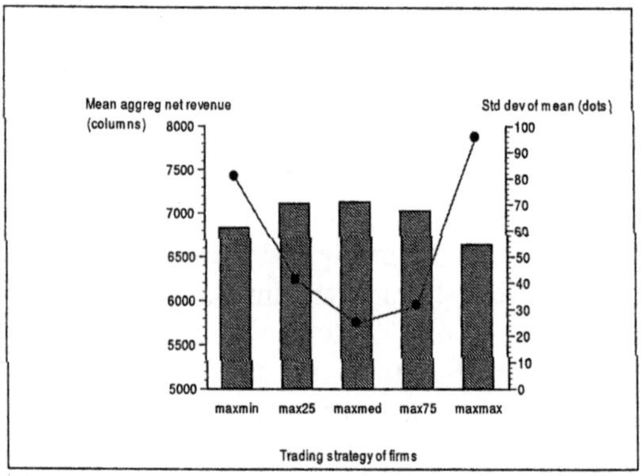

Fig. 9.2. Aggregate revenue with alternative trading strategies

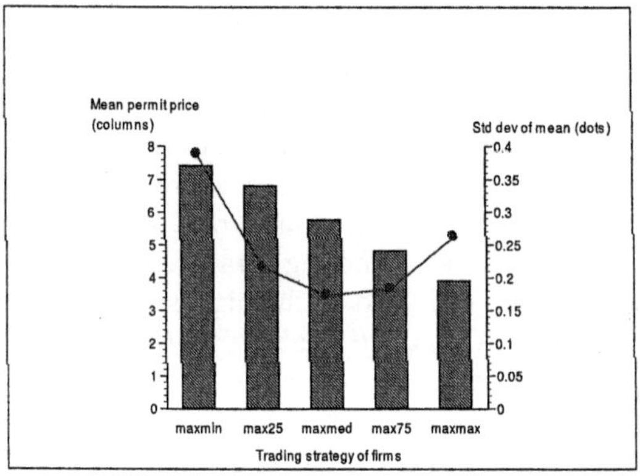

Fig. 9.3. Permit price with alternative trading strategies

in the maxmax case, a maxmin approach to trade can be highly unstable as a small shift in one firm's strategy can result in a large change in the outcome of trade. Strategies that perform well in the worst possible trading situations are associated with paying, on average, a higher price for permits than would be necessary under more favourable trading outcomes. Selection of a strategy corresponding to the 25th (*max25*) and 75th (*max75*) percentile of the sample outcomes against competing strategies, gives an outcome between the median and maxmin or the median and maxmax strategy outcomes, respectively.

9.3.2 Impact of Increased Market Size

To investigate the circumstances under which a tradable emission permit scheme can be effective as the number of participants in the market increases, simulations were conducted with double and triple the number of market participants (linked, as in Fig. 9.1). As the number of market participants was increased, the effectiveness of a simple tradable emission permit scheme declined, while the effectiveness of a single uniform tax remained largely unaffected by the number of polluters (Fig. 9.4).

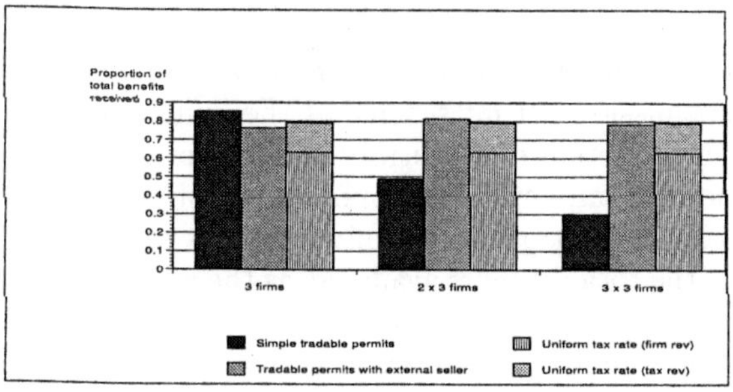

Fig. 9.4. Impact of participant numbers on the effectiveness of tradable permit schemes

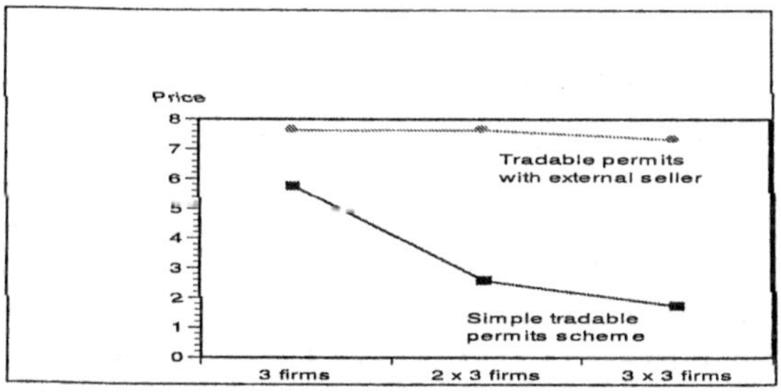

Fig. 9.5. Impact of participant numbers on permit price

The difficulty for individual firms to obtain clear information on the impacts of a trade is one of the principal reasons behind the decline in the effectiveness of trade as the number of market participants increases. The expected traded price of permits drops sharply as the number of market participants rises (Fig. 9.5), reflecting the inability of market participants to capture the benefits of reduced emissions through permit purchases. The policy maker imposing the tax, however, is assumed to be able to gather necessary information from multiple firms just as readily as from a single firm.

Table 9.1. GA parameters for alternative simulations

	Base level	**Alternate level**	**t-stat**[a]
Number of strings	20	50	0.69
Sampling frequency	400/8000	1000/8000	0.64
Mutation rate	0.0005	0.025	1.58

[a]t-stat refers to statistic for test of the difference between aggregate net revenue outcome in the 3 firm base simulation and that from an alternative simulation in which a parameter of the GA was changed. No t-stat is significantly different from zero at the 5 per cent significance level.

One practical option that could be adopted to maintain the effectiveness of a tradable emission permit scheme as the number of market participants rises, is the introduction of a central trading company to sell permits. With a central trading company as monopoly seller of permits, the proportion of benefits captured by a tradable permit scheme is maintained at around 75 to 85 per cent of the total benefits available, as the number of participants in the market is increased. This is comparable to the result achieved through the introduction of a uniform emissions tax. However, the income redistribution impacts associated with a tax may make it a less favourable option than a tradable permit scheme with a central seller of permits.

9.3.3 Robustness of Results to Change in GA Parameters

To determine the robustness of the above results to changes in the values of choice parameters for the GA, a number of additional simulations were undertaken with alternate levels for the GA parameters. The results from these simulations, reported in Table 9.1, indicate that the final fitness value does not change significantly with the number of strings in the GA population, the mutation rate, or with the proportion of the range of trading strategies available to a firm which is sampled in selection of a strategy.

While the final fitness value does not change significantly under different values of the GA parameters, the number of generations of the GA required to converge to this value does change. For example, with an increased mutation rate, the simulation result is substantially more volatile than in the base case, although the volatility does lessen after about 100 generations (Fig. 9.6).

9.4 Concluding Remarks

The simulations in this chapter demonstrate the application of GAs to the emulation of trade in emission permits. That a GA can search for a robust optimal solution when there exists uncertainty in the form of a lack of priori knowledge, unclear feedback of information to decision makers and a time varying payoff function, means that it is a particularly useful mechanism with which to emulate the processes of trade.

A tradable emission permits scheme was shown to be a relatively effective instrument to capture the benefits available to producers from reduced emissions, and the income redistribution outcome may make it preferable to the imposition of other policy options such as a uniform emissions tax.

In markets in which participants adopt extreme conservative or aggressive approaches to selecting trading strategies, tradable emission permits schemes may be less effective at capturing the benefits of reduced emissions. The variability of the final outcome and the expected price at which permits are traded are also highly dependent on the trading approach adopted by market participants. It is plausible that in a market in which agents adopt a range of trading approaches, the effectiveness of a tradable permit scheme may be not be reduced by extreme approaches.

As the number of participants in the market increases, information requirements of agents increases and the effectiveness of a tradable permit scheme declines. This decline in effectiveness can be alleviated if market power is introduced, for example, through the introduction of a central trading company that maximises the revenue from the sale of permits.

The robustness of the simulation results was examined for alternative values of the key GA parameters. The final fitness value was found to be stable, although the number of generations of the GA process required to reach the stable solution may vary under different sets of GA parameters.

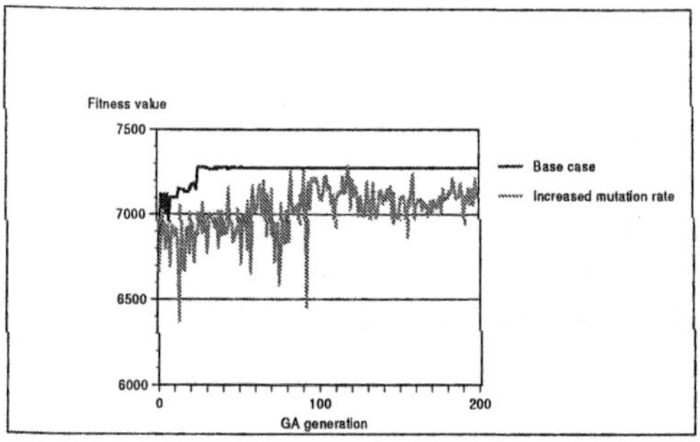

Fig. 9.6. Convergence of fitness value under alternative mutation rates

Table 9.2. Model Parameter Values

Parameter	Value
α_1	10
α_2	10
α_3	0.05
α_4	0.05
α_5	0.05
μ_0	1
γ_1	5
γ_2	0.025
δ	0.5
p	1
c	1
d	1
Population for firm optimisations (m)	20
Population for policy maker optimisations	200
Cross-Over Rate	0.06
Mutation Rate[a]	0.0025
Generations	200
Random trials (k)	30

[a] The mutation rate is expressed as the rate of mutation per bit of each population string.

9.5 Symbol Listing

π_i profit of firm i
f_i production function of firm i
x_i input to production in firm i associated with externality
y_i input to production in firm i
μ production function coefficient
α production function coefficient
E_i external cost incurred by firm i
δ_i proportion of firm i's x usage transferred as damage in network
γ social damage function coefficient
p output price
c unit cost of input x
d unit cost of input y
ρ market price paid for an emission permit
ν level of emission permit purchases
z level of emission permits from initial allocation which are sold
w Initial allocation of permits to the firm
n number of firms
k number of random trials
m number of market outcomes

Acknowledgements

The authors wish to acknowledge the invaluable assistance of Ray Hinde at ABARE with the implementation of the genetic algorithm.

References

1. Alemdar N., Ozyildirim S., (1998) A Genetic Game of Trade, Growth and Externalities. Journal of Economic Dynamics and Control **22**, 811–832
2. Atkinson S., Tietenberg T., (1987) Economic Implications of Emissions Trading Rules for Local and Regional Pollutants. Canadian Journal of Econom. **20**, 370–86
3. Beare S., Bell R., Fisher B. S. (1998) Determining the Value of Water: The Role of Risk. Infrastructure Constraints and Ownership. American Journal of Agriculture Economics **80**, December
4. Birchenhall C. (1995) Modular Technical Change in Genetic Algorithms, Computational Economics **8**, 233–53
5. Bullard J., Duffy J. (1998) A Model of Learning and Emulation with Artificial Adaptive Agents. Journal of Economic Dynamics and Control, **22**, 179–207
6. Chen S -H., Yeh C -H. (1997) Toward a Computable Approach to the Efficient Market Hypothesis: An Application of Genetic Programming. Journal of Economic Dynamics and Control **21**, 1043–1063
7. Demetz H. (1969) Information and Efficiency: Another Viewpoint. Journal of Law and Economics **11**, 1–22
8. Goldberg D. (1989) Genetic Algorithms in Search, Optimisation and Machine Learning. Addison-Wesley Publishing Company, USA
9. Green J. (1977) The Non-existence of Informational Equilibria, Review of Economic Studies **44**, 451–63
10. Hanley N., Shogren J., White B. (1997) Environmental Economics in Theory and Practice. MacMillan Press Ltd
11. Heaney A., Beare S., Bell R. (2001) Evaluating Improvements in Water Use Efficiency as a Salinity Mitigation Option in the South Australian Mallee Areas. 45th Annual Conference of the Australian Agricultural Economics Society, Adelaide, January 21–25, 2001
12. Holland J. (1997) Adaptation in Natural and Artificial Systems. University of Michigan Press.
13. Imagine That Inc. (1997) Extend User's Manual Version 4
14. Marks R. (1999) Breeding Hybrid Strategies: Optimal Behaviour for Oligopolists. Draft discussion paper.
15. Milgrom P. (1989) Auctions and Bidding: A Primer. Journal of Economic Perspectives **3**, 3–22
16. Pesendorfer W., Swinkels J. (1997) The Loser's Curse and Information Aggregation in Common Value Auctions. Econometrica **65**, 1247–1281
17. Riechmann T. (1999) Learning and Behavioral Stability: An Economic Interpretation of Genetic Algorithms. Journal of Evolutionary Economics **9**, 225–242
18. Weibull J. (1996) Evolutionary Game Theory, MIT Press

10 Cooperative Computation with Market Mechanism

Masayuki Ishinishi, Hiroshi Sato, and Akira Namatame

Dept. of Computer Science, National Defense Academy, Japan
hsato@nda.ac.jp, nama@nda.ac.jp

Abstract. In this research we investigate the algorithms for cooperative computation based on a market mechanism. We define economic agents as autonomous software entities that behave based on the market principle. They have their own computational capability and they are also designed to have their own utility function as defined by the market prices. We obtain both competitive and cooperative equilibrium solutions in an interdependent market and we discuss how a cooperative solution can be obtained through the decentralized computation of economic agents with self-interest seeking behaviors. We also provide an dynamic adaptive model for each economic agent, analyze the stability of the cooperative computation, and show that the convergence of the market depends on the internal model of each economic agent.

10.1 Introduction

There is currently a very strong interest for utilizing Internet resources based on the market principle [1] [3]. With a market-oriented computational model, we aim to describe and analyze various competitive market strategies which, if followed by a set of economic agents, will result in a decentralized computation for the efficient allocation of Internet resources, such as processor time, memory space, and communication channels. Market strategies for memory allocation and garbage collection, for instance, are based on rent payment, with economic agents paying retainer fees to those agents they wish to retain in memory, while agents that are unwanted and hence unable to pay rent are eventually evicted. Memory space, for instance, is a lower performance resource in a bountiful supply, while communication or message passing from one person to another takes time and costs money. An agent that stays in core will pay higher rent, but can provide faster service. To the degree that this is of certain value, the agent can charge more, if the increased income more than offsets the higher rent, and the agent will profit by staying in core.

In order to implement the market mechanism over Internet services, the concept of economic agents turns out to be useful. We term such an agent which behaves based on the market principle as an economic agent. Notions of self-interested behaviors are indeed the foundation of many fields [2] [4] [8]. Economic agents are selfish in the sense that they only do what they want to do and what they think is in their own best interests, as determined by their own utility which is defined as functions of the market prices. A computational market is also defined as a society of software agents with the market principle. The ability of trade and price mechanisms to combine local decisions (by diverse self-interested economic agents) into globally effective behaviors suggests that they do in fact have a value for organizing computation in large systems [9] [13]. Market strategies are outlined which, if used by agents locally, lead to decentralized resource allocation that encourages utility modification based on local knowledge. For instance, systems programming problems (such as garbage collection and processor scheduling) have traditionally been addressed in computational foundations, casting the architect in the role of an omniscient central planner. In this approach, the architect imposes a single, system-wide solution based on global aggregate statistics, precluding local choice. However, in the market-oriented approach these problems can be recast in terms of the local and decentralized decisions of many economic agents. Solutions in this framework also provide economic agents with a market price, allowing them to make profitable use of the resulting flexibility.

Another goal of this research is to understand the types of simple local interactions, based on the self-interested motivations which produce purposive and cooperative behavior as a whole. We especially address the question of how a society of economic agents with different motivations can achieve an efficient collective behavior. Economic agents necessarily do have different sets of goals, motivations, or cognitive states by virtue of their different histories, the different resources they use, the different settings they participate in, and so on. Furthermore, economic agents are driven by their own selfish motivations, and they are selfish in that they only do what they want to do and what they think is in their own best interests, as determined by their own motivations. The global behavior of economic agents is determined by the local interactions of their constituent parts. These interactions merit careful study in order to understand the macroscopic properties of the collective behavior of economic agents. We also take a close look at the question of how does a society of agents achieve a cooperative solution without losing the principle of competition. We ask the following questions about these situations: If economic agents make their locally optimal decision on the basis of imperfect information concerning other agents' decisions or goals, and incorporate expectations on how the decision will affect other agents' goals, then how will the equilibrium solution proceed? If agents make decisions on the basis of their own utility by incorporating the expectation of the other agents' behaviors, then how will their behavior be affected, and how will the type of adjustment to their behavior affect the evolution of the equilibrium situation?

10.2 A Model of Economic Agents and Definition of Equilibrium Solutions

We define a society of economic agents, $G = \{A_i, i = 1, 2, \ldots, n,\}$, and represent the utility function of each $A_i \in G$ by

$$U_i\{x_i, x(i)\} = x_i P_i\{x_i, x(i)\} \tag{10.1}$$

In equation (10.1), x_i represents the level of the activity of $A_i \in G$ and $x(i) = (x_1, \ldots, x_{i-1}, x_{i+1}, \ldots, x_n)$ represent the levels of the activities of the rest of the other agents. Hereafter, we call each activity level as a strategy, while $P_i\{x_i, x(i)\}$ represents the price scheme of agent $A_i \in G$ which is associated with the set of the activities $(x_i, x(i))$.

As a specific example, we consider the following linear price function

$$P_i\{x_i, x(i)\} = a_i - \sum_{j=1}^{n} b_{ij} x_j \tag{10.2}$$

where $a_i, b_{ij}, i, j = 1, 2, \ldots, n$, are some positive constants. We consider an interdependent market, in which each economic agent cannot simply proceed to perform its optimal decision without considering what other agents are doing. The strategy of each economic agent should be determined solely by how its decision affects other members and how the decisions of other agents affect its own utility.

We define the competitive solution as the set of strategies that optimizes each economic agent's utility function $U_i, i = 1, 2, \ldots, n$, simultaneously [4] [6]. The competitive solution is given as the set of strategies that solve the system of equations of the marginal utilities.

$$\partial U_i / \partial x_i = M_i(x_i, x(i)) = 0, \qquad i = 1, 2, \ldots, n \tag{10.3}$$

The marginal utility function of $A_i \in G$ is obtained as follows:

$$\begin{aligned} M_i(x_i, x(i)) &= P_i\{x_i, x(i)\} + x_i \partial P_i\{x_i, x(i)\} / \partial x_i \\ &= a_i - \textstyle\sum_{j=1}^{n} b_{ij} x_j - b_{ii} x_i \end{aligned} \tag{10.4}$$

A competitive solution is then given as the solution of the following system of linear equations:

$$(B + B_1)x = a \tag{10.5}$$

where B is an $n \times n$ matrix with the (i,j)th element $b_{ij}, i, j = 1, 2, \ldots, n$, which is defined as the interaction matrix of the economic agents. Term B_1 is a diagonal matrix with the i-th diagonal element being b_{ii}, and a is the column vector with the elements, $a_i, i = 1, 2, \ldots, n$. Using the property of (10.3) and (10.4), the equilibrium value of the market price at the competitive solution is obtained as follows:

$$P_i(x_1^\circ, \ldots, x_i^\circ, \ldots, x_n^\circ) = b_{ii} x_i^\circ \tag{10.6}$$

The utility associated with each agent at the competitive equilibrium is also given as

$$U_i(x_1^\circ, \ldots, x_i^\circ, \ldots, x_n^\circ) = P_i(x_1^\circ, \ldots, x_i^\circ, \ldots, x_n^\circ)x_i^\circ \\ = b_{ii}(x_i^\circ)^2 \tag{10.7}$$

From the results in (10.6) and (10.7), the price and utility of each economic agent at the equilibrium (competitive) solution do not depend on the other economic agents' activities.

We define the cooperative solution as the set of strategies that optimize the summation of the utility functions of all economic agents which is given as

$$S(x_1, \ldots, x_i, \ldots, x_n) = \sum_{i=1}^{n} U_i\{x_1, \ldots, x_i, \ldots, x_n\} \tag{10.8}$$

The cooperative solution is then obtained as the set of strategies satisfying the following equations.

$$\partial S/\partial x_i = \partial U_i/\partial x_i + \sum_{j \neq i} \partial U_j/\partial x_i = 0, \quad i = 1, 2, \ldots, n, \tag{10.9}$$

For the quadratic utility functions with the linear prices' scheme as given in (10.2), the cooperative solution is obtained as the solution of the following system of linear equations:

$$(B + B^T)x^* = a \tag{10.10}$$

where B^T is the transpose matrix of B.

We can now consider the special case in which the interaction matrix B is symmetric with the diagonal elements being the same, i.e., $b_{ii} = d$, and the off-diagonal elements are given as $b_{ij} = b, (0 < b \leq d), i, j = 1, 2, \ldots, n$. The column vector a has the same elements, i.e., $a_i = a, i = 1, 2, \ldots, n,$. The utility of each agent at the competitive solution is then obtained as follows:

$$U_i^\circ(n) = a^2 d/\{2d + b(n-1)\}^2 \tag{10.11}$$

The utility of each agent at the cooperative solution is given as

$$U_i^*(n) = a^2/4\{d + b(n-1)\} \tag{10.12}$$

The summation of the utility of each economic agent under the competitive solution is then arrived at as

$$G^\circ(n) \equiv \sum_{i=1}^{n} U_i^\circ(n) = a^2 dn/\{2d + b(n-1)\}^2 \tag{10.13}$$

The summation of the utility of each economic agent under the cooperative solution results in

$$G^*(n) \equiv \sum_{i=1}^{n} U_i^*(n) = a^2 n/4\{d + b(n-1)\} \tag{10.14}$$

We are interested here in how the utility of each economic agent may be affected if the number of economic agents in a society increases. Thus, we investigate the asymptotic property by increasing the number of agents. If we take the limits of the summations of the utility functions in (10.13) and (10.14) with the number of the economic agents, then those values converge as follows:

$$\lim_{n\to\infty} G^{\circ}(n) = 0 \tag{10.15}$$

$$\lim_{n\to\infty} G^{*}(n) = a^2/4b \tag{10.16}$$

This implies that the whole utilities of a society under the competitive solution converge to zero, meaning that the state of free competition is realized if infinitely many economic agents participate in the market, and the utility of each economic agent becomes zero. The social utility under a cooperative solution, on the other hand, converges to the constant. Figure 10.1 shows the summations of the utilities as a function of the number of agents.

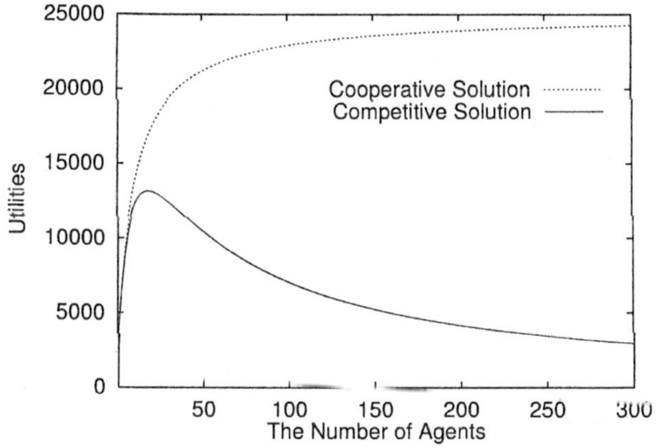

Fig. 10.1. The sums of the utilities at competitive and cooperative solutions

In the area where the number of agents is small, the sum of the utilities of the agents increases as the number of agents increases, and it becomes to decrease if the number of agents increases. Such a turning point at which the sum of utilities changes from the increasing to the decreasing is obtained as follows:

$$\partial G^{\circ}(n)/\partial n = 0 \tag{10.17}$$

By solving equation (10.17), we obtain such a turning point as the number of economic agents is given as

$$n^{*} = 2(d/b) - 1 \tag{10.18}$$

10.3 The Competitive Adaptation Using Market Prices

In this section we formulate the competitive adaptive model of each economic agent so that the competitive solution can be obtained at equilibrium for the dynamic adaptive process. The adaptive decision-making is facilitated by designing the economic agents to have a somewhat modified selfish behavior. We call this ability as mutual learning [8]. In the model of mutual learning, two types of learning may occur: the economic agents can learn to cooperate as a group, while at the same time, each economic agent also learns its own objective by adjusting its behavior. Mutual learning here requires and exchange of behaviors with the other agents. We interpret this adaptive process as mutual learning which is defined as the web of activity that emerges from competitive interactions among economic agents [12] [11].

Each agent modifies its strategy based on the current and previous performance in order to optimize its own utility. We formulate the competitive adaptation as follows: We denote the set of strategies at time period t except agent $A_i \in G$ by $x(i,t) = (x_1(t), \dots, x_{i-1}(t), x_i(t), x_{i+1}(t), \dots, x_n(t))$. Since the marginal utility of each economic agent provides information for the direction in which it can improve its utility, we describe the adaptive process of each economic agent as follows:

$$\begin{array}{l} if\ M_i\{x_i(t), x(i,t)\} > 0\ \ then\ x_i(t+1) := x_i(t) + \delta_i \\ if\ M_i\{x_i(t), x(i,t)\} < 0\ \ then\ x_i(t+1) := x_i(t) - \delta_i \end{array} \tag{10.19}$$

We implement the dynamic process in (10.19) by specifying the size of the modification as follows:

$$\begin{aligned} x_i(t+1) - x_i(t) &= (\alpha_i/b_{ii}) M_i\{x_i(t), x(i,t)\} \\ &= (\alpha_i/b_{ii})\{P_i(t) - b_{ii}x_i(t)\} \end{aligned} \tag{10.20}$$

By rewriting the relation in (10.20), we have the following equation:

$$x_i(t+1) = (\alpha_i/b_{ii})P_i(t) + (1-\alpha_i)x_i(t) \tag{10.21}$$

The dynamic action selection process must be coordinated so as to achieve globally consistent and efficient actions. The very simplest cooperation would be an exchange of facts such as current actions. This adjustment process generates a partial strategy that governs the actions of the agents. The use of directives by an agent to control another agent can be viewed as a form of incremental behavior adjustment. This adjustment process creates a partial strategy that governs the actions of the agents. The mutual learning model therefore describes how each economic agent, without knowing the others' utilities, adjusts its strategy over time and reaches an equilibrium situation. In an interdependent market consisting of many selfish agents without complete knowledge of other agents as shown in Fig. 10.2, each economic agent needs to infer the strategies, knowledge, and plans of the other agents.

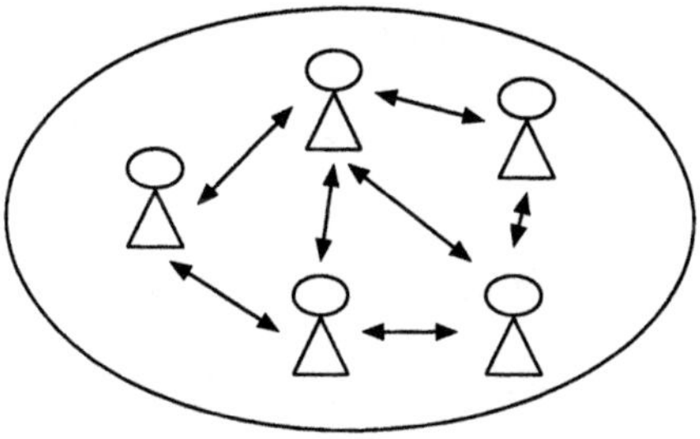

Fig. 10.2. Interdependent market

Equation (10.21) shows the decentralized computational process in which each economic agent does not need to consider the other economic agents' private knowledge. Each economic agent $A_i \in G$ only needs to care about the current market price $P_i(t)$ and its previous behavior $x_i(t)$, and will modify its behavior based on the current and previous performance in order to optimize its own utility. The concept of decentralized computation under the market mechanism is shown in Fig. 10.3. Each economic agent does not need to infer the strategies, knowledges, and plans of other agents; they only need to know the prevailing market prices.

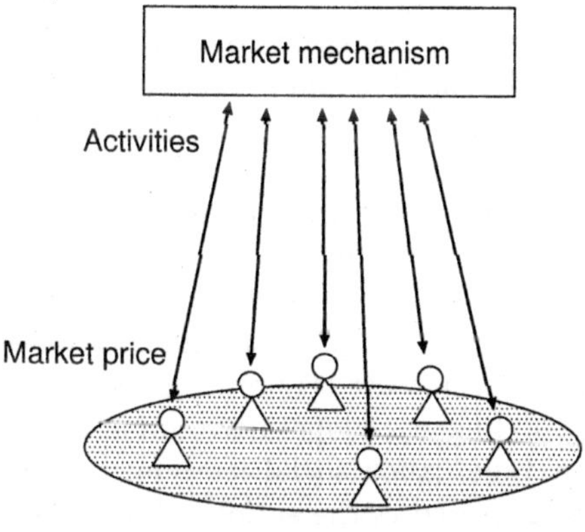

Fig. 10.3. Decentralization with the market mechanism

In equation (10.21), we define the size of α_i as the adaptive speed of each agent $A_i \in G$. We hence classify economic agents into several types, depending on their adaptive speed.

Type 1: $\alpha_i \approx 0$

In this case, the relation in (10.21) can be derived as follows:

$$x_i(t+1) \approx x_i(t) \tag{10.22}$$

Such an economic agent is insensitive to the change in the market price and insists on its own previous behavior.

Type 2: $\alpha_i \approx 1$

In this case, the relation in (10.21) can be derived as follows:

$$x_i(t+1) \approx P_i(t)/b_{ii} \tag{10.23}$$

Such an agent can be said to be very sensitive to the change in the market price.

Type 3: $\alpha_i \approx 1/2$

In this case, the relation in (10.21) can be derived as follows:

$$x_i(t+1) = P_i(t)/2b_{ii} + x_i(t)/2 \tag{10.24}$$

Such an agent is said to be neutral and cares both about the change in the market price and its previous behavior.

10.4 Social Rules that Induce Implicit Cooperation

Section10.2 showed that the conditions of individual optimality and social optimality are quite different. This implies that if each economic agent seeks its own individual optimality by adapting its behavior based on the equation (10.21), then the society as a whole faces a so-called social dilemma [5] as shown in Fig. 10.1. Our question is then stated as follows: how will the implicit cooperation proceed in a decentralized society? Here, we will consider the role of a government with the authority of designing and implementing a social rule. The government has the authority to set up the social rule, and each economic agent then pursues its profit under the social rule.

The process of building up cooperative intentions in a society of interdependent agents may be called social learning [7]. Social learning from a social perspective is grounded in the actions of many agents' activities taken together, and is not a matter of individual choice. Social learning, in this sense, is the outcome of a web of activity that emerges from the competitive interactions among agents.

We consider the following modified utility function for each economic agent by paying a tax as the social rule

$$\overline{U_i}\{x_i, x(i)\} = U_i\{x_i, x(i)\} - \lambda_i\{x(i)\}x_i \tag{10.25}$$

where $\lambda_i(x(i))$ is the tax rate on $A_i \in G$. The marginal utility function of each economic agent's modified utility is given as

$$M_i\{x_i, x(i)\} - \lambda_i\{x(i)\} - 0 \qquad i - 1, 2, \ldots, n \tag{10.26}$$

The government fixes the tax rate of each economic agent set as follows:

$$\lambda_i(x(i)) = -\sum_{j \neq i}^{n} (\partial U_j/\partial x_i)x_j \qquad i = 1, 2, \ldots, n. \tag{10.27}$$

The tax rate also indicates the influence level of the strategy of $A_i \in G$ to the utility functions of the other agents. The condition of individual optimality under the modified utility functions in (10.25) is then equivalent to the condition of social optimality as given in (10.9). With the linear market price scheme as given in (10.2), the tax rate is shown as follows:

$$\lambda_i(x(i)) = -\sum_{j \neq i}^{n} (\partial U_j/\partial x_i)x_j = \sum_{j \neq i}^{n} b_{ji}x_j \tag{10.28}$$

Economic agents are quite apparently both selfish agents and social actors. In the model of social learning, two types of learning may occur: the economic agent can learn on its own by adjusting its action, while at the same time, the government also learns to set the social rule that realizes social fairness. We define social learning as the dynamic adjustment process of economic agents and the social rule [12]. As such, the social learning model describes how each economic agent, without knowing the others' objectives, adjusts its strategy over time and reaches a cooperative equilibrium situation. The cooperative solution can be realized if each agent modifies its own utility function as defined by two terms, private utility and social utility.

The dynamic adaptation of each economic agent process under the social rule is modified as follows:

$$\begin{aligned} &if\ M_i\{x_i, x(i)\} > \lambda\{x(i)\}\ then\ x_i := x_i + \delta x_i \\ &if\ M_i\{x_i, x(i)\} < \lambda\{x(i)\}\ then\ x_i := x_i - \delta x_i \end{aligned} \tag{10.29}$$

The adjustment rate is then given as follows:

$$\delta_i = x_i(t+1) - x_i(t) = (\alpha_i/b_{ii})[M_i\{x_i(t), x(i,t)\} - \lambda_i\{x(i,t)\}] \tag{10.30}$$

where $x(i,t) = (x_1(t), \ldots, x_{i-1}(t), x_i(t), x_{i+1}(t), \ldots, x_n(t))$.

By rewriting the relation in (10.30), we have the following equation:

$$x_i(t+1) = (\alpha_i/b_{ii})P_i(t) + (1 - \alpha_i)x_i(t) - (\alpha_i/b_{ii})\lambda_i\{x(i,t)\} \tag{10.31}$$

The above adaptive model with tax rate $\lambda_i, i = 1, \ldots, n$, given in (10.26) converges to a cooperative solution. On the other hand, the adaptive model without a tax, by setting $\lambda_i = 0, i = 1, 2, \ldots, n$, converges to a competitive solution.

The social learning model also needs to specify how the government adjusts the tax rate. In this case, we provide a model as follows:

$$\lambda_i(t+1) = \beta_i[M_i\{x_i(t), x(i,t)\} - \lambda_i\{x(i,t)\}] + \lambda_i\{x(i,t)\} \tag{10.32}$$

By rewriting (10.32) and using the relation of (10.4), we have the following adaptive model:

$$\lambda_i(t+1) = \beta_i\{P_i(t) - b_{ii}x_i(t)\} + (1-\beta_i)\sum_{j\neq i}^{n} b_{ji}x_j(t) \tag{10.33}$$

If the government collects taxes from each economic agent, then the society of economic agents can realize a cooperative equilibrium without losing the principle of free competition. However, as shown in Fig. 10.4, the decentralized mechanism of realizing society's efficiency occurs by collecting the private wealth of each economic agent and turning this into government revenue.

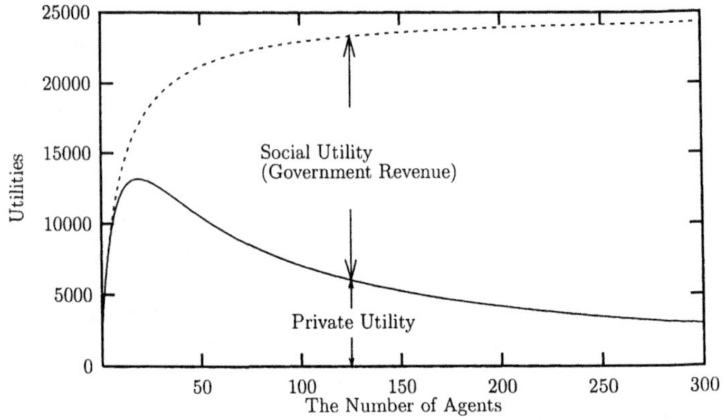

Fig. 10.4. The utilities of economic agents under the tax rule

We next consider the mechanism of redistributing the taxes to each economic agent. With this redistribution mechanism, there is no surplus on the government side. Thus, the utility of each economic agent with the rules of tax collection and distribution is given as follows:

$$\overline{U_i}(x_i, x(i)) = U_i(x_i, x(i)) - \lambda_i(x_i, x(i))\dot{x}_i + \sum_{i=1}^{n}(\lambda_i(x_i, x(i))\dot{x}_i)/n \tag{10.34}$$

10.5 Simulation Results

In this section we address the question of how a society of economic agents can achieve optimal collective behaviors as a whole. Different agents have not only different sets of goals or motivations, but they also have different interactions among agents or cognitive states by virtue of their different histories. As a result, it is shown by simulations that their collective behavior may vary with the different internal models of economic agents.

We provide the conditions of simulation as follows:
(1) Number of economic agents: 10
(2) Price scheme: $a_i = 1000$
(3) Initial strategies of each agent: $x_i(0) = 5, \quad i = 1, \ldots, 10$
(4) Adaptive speed of each agent: $\alpha_i = 0.1, \quad i = 1, \ldots, 10$
We also provide the interaction matrix B in (10.5) as follows:

We consider the interdependence structure among agents in Fig. 10.5, and their interaction matrix B is given as follows:

$$\begin{aligned} &(1)\ \partial P_i/\partial x_i = b_{ii} = 1 \quad (i = 1, \ldots, 10) \\ &(2)\ \partial P_i/\partial x_1 = b_{i1} = 0.1\ (i \neq 1) \\ &(3)\ \partial P_i/\partial x_j = b_{ij} = 0 \quad (i \neq j) \end{aligned} \tag{10.35}$$

With this interaction matrix, the influence of strategy A_1 to the market price of the other agents, $A_i, i = 2, 3, \ldots, 10,$, is significant. However, the influences of those economic agents to the market price of agent A_1 are negligible.

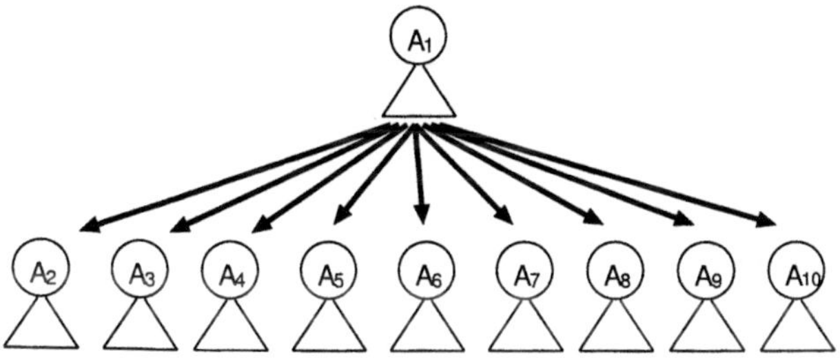

Fig. 10.5. The asymmetric interactions among economic agents

(case 1) The asymmetric and oligopolistic market without tax

Figure 10.6 shows the change in utility of each economic agent over time. Each economic agent adjusts its market behavior using the adaptive model in (10.21), and will reach a competitive equilibrium as shown in Fig. 10.1. In this case, it is observed that only one agent (A_1) can acquire the highest utility as shown in Fig. 10.6, and the so-called winner-take-all market occurs.

(case 2) The oligopolistic market with collecting taxes

The government levies a the tax for each economic agent as given in (10.28). Each economic agent then adjusts its market behavior based on the model in (10.31). Eventually, the agents reach a cooperative solution as shown in Fig. 10.1. In this case, however, the tax on agent A_1 becomes very high when compared with the other economic agents. Therefore, most of the profit of

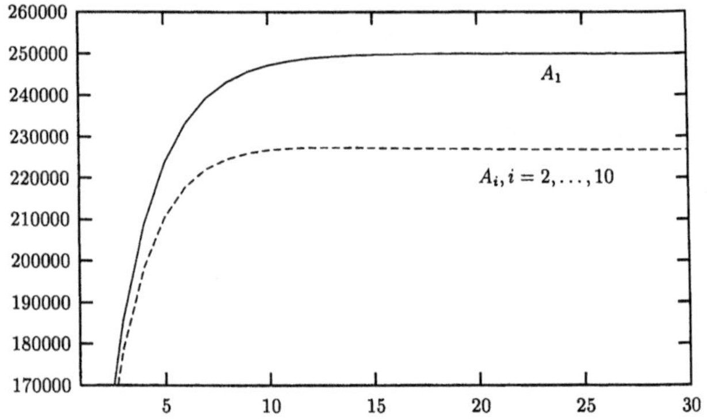

Fig. 10.6. The utility in an asymmetric market

agent A_1 is collected as taxes by the government. A reverse phenomenon henceforth occurs and the utility of agent A_1 becomes the lowest as shown in Fig. 10.7.

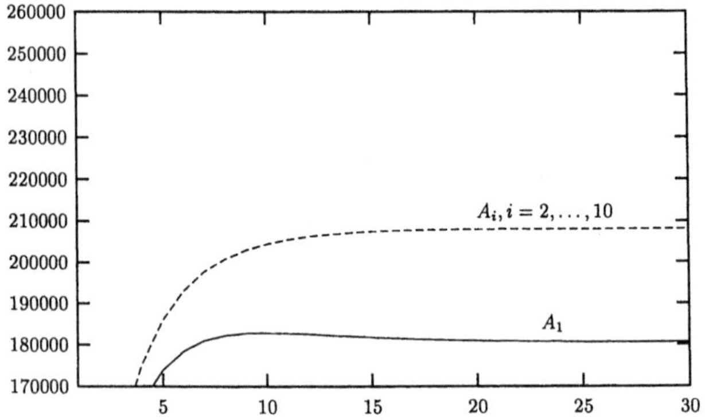

Fig. 10.7. The utility of each economic agent with taxes (case2)

(case 3) The oligopolistic market with collecting taxes and redistribution

The government in this case plays both roles of collecting taxes and subsidizing them. Each economic agent then acquires the same utility as shown in Fig. 10.8. Therefore, fairness in society is realized under such a social rule.

10.6 Conclusion

The goal of this research was to develop a model of decentralized computation through a set of economic agents that produce complex and purpo-

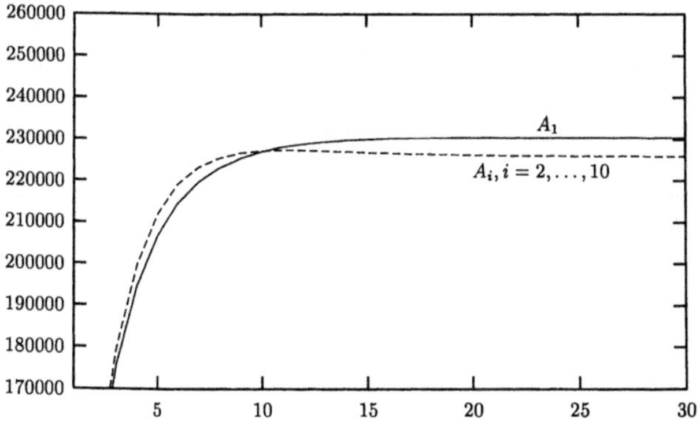

Fig. 10.8. The utility of each economic agent with taxes and subsidies (case 3)

sive group behaviors. After formulating and analyzing the interdependent decision-making problem of economic agents, we showed that an equilibrium solution can be realized through purposive local interactions based on each individual goal-seeking. Each economic agent does not need to express its objective or utility function, nor does it need to have a prior knowledge of those functions of other agents. Each economic agent adapts its own action to the actions of other economic agents, thus allowing previously unknown economic agents to be easily brought together in order to customize a group that is responsible for a specific mission. This paper also described the research of studying competitive interactions leading to a coordinated behavior. The goal of this research is to understand the types of simple local interactions which produce complex and purposive behaviors.

References

1. Adam N.Q., Yesha Y. (1997) Electronic Commerce, Springer
2. Carley K., Prietula M. (1994) Computational Organization Theory, Lawrence Erallbawn
3. Bakos Y. (1998) The Emerging Role of Electronic Marketplaces on the Internet, Communications of the ACM **41**, 35–42
4. Creps J.E. (1991) An Introduction to Modern Micro Economics, The MIT Press
5. Friedman J. (1990) Game Theory with Applications to Economics, The Oxford Univ. Press
6. Fudenberg D.E., Tirole J. (1991) Game Theory, The MIT Press
7. Gasse L. (1991) Social Conceptions of Knowledge and Action: DAI Foundations and Open Systems Semantics, Artificial Intelligence **47**, 107–135
8. Kirman A., Salmon M. (1995) Learning and Rationality in Economics, Blackwell
9. Kurose J., Simha R. (1989) A Microeconomic Approach to Optimal Resource Allocation in Distributed Computer Systems, IEEE Tran. on Computers **38**, 705–717

10. McKnight L.W., Bailey J.P. (1997) Internet Economics, The MIT Press
11. O'Hare G.M.P., Jennings N.R. (1996) Foundations of Distributed Artificial Intelligence, Wiley-Interscience
12. Shoham Y. (1993) Agent-oriented Programming, Artificial Intelligence **60**, 51–92
13. Waldspurger C. (1989) SPAWN: A Distributed Computational Economy, IEEE Tran. on Software Engineering **18**, 103–117

11 Hysteresis in an Evolutionary Labor Market with Adaptive Search

Leigh Tesfatsion

Iowa State University, Ames, IA 50011-1070, USA
tesfatsi@iastate.edu
http://www.econ.iastate.edu/tesfatsi/

Abstract. This study undertakes a systematic experimental investigation of hysteresis (path dependency) in an agent-based computational labor market framework. It is shown that capacity asymmetries between work suppliers and employers can result in two distinct hysteresis effects, network and behavioral, when work suppliers and employers interact strategically and evolve their worksite behaviors over time. These hysteresis effects result in persistent heterogeneity in earnings and employment histories across agents who have no observable structural differences. At a more global level, these hysteresis effects are shown to result in a one-to-many mapping between treatment factors and experimental outcomes. These hysteresis effects may help to explain why excess earnings heterogeneity is commonly observed in real-world labor markets.

11.1 Introduction

In the empirical labor economics literature, a labor market is said to exhibit *hysteresis* if temporary shocks appear to have persistent effects on earnings and employment histories.[1] A key concern of empirical labor economists has been the identification of possible propagation mechanisms through which hysteresis might occur.

To date, attention has largely been focused on apparent hysteresis in aggregate unemployment: namely, the protracted effects that unemployment shocks appear to have on the "natural" rate of unemployment. Reference [3]

[1] As pointed out by [8], the term hysteresis has been used in economic and econometric theory to refer to two distinct phenomena: persistence in deviations from equilibria, possibly followed by an eventual return to a previous equilibrium state; and the presence of unit/zero roots in systems of linear difference or differential equations, implying that a single temporary shock permanently changes the equilibrium path of the system. In empirical economics, however, hysteresis is used more loosely to mean that temporary shocks are observed to result in a persistent change from a previously persistent system state, even though this previously persistent system state cannot be verified to be an equilibrium and the persistent change cannot be verified to be permanent. The latter usage is followed in the current computational study.

discusses three distinct types of propagation mechanisms that have been advanced as possible explanations: lag effects arising from the difficulty of adjusting physical capital stocks; long-term labor supply effects arising from the human capital erosion resulting from unemployment; and insider-outsider effects arising from the preferential treatment of actual employees relative to potential employees in the wage bargaining process. Although [3] identifies insider-outsider effects as the most promising explanation for hysteresis in European labor markets, they also caution (pp. 270–271) about small sample problems that make this hypothesis difficult to test.

In contrast, this study focuses on a form of hysteresis routinely observed for individual work suppliers and employers in micro panel data: namely, observationally equivalent work suppliers and employers have markedly different earnings and employment histories (see, e.g., [1]). The basic question addressed in this study is whether temporary shocks in the form of idiosyncratic worksite interactions can propagate up into sustained differences in earnings and employment histories for observationally equivalent workers and employers.

Two interdependent aspects of worksite interactions are considered. Who works for whom, and with what regularity? And how do work suppliers and employers behave in these worksite interactions?

In real world labor markets, the behavioral characteristics expressed by work suppliers and employers in their worksite interactions, such as trustworthiness and diligence, depend on who is working for whom. In turn, who is working for whom depends on the behavioral characteristics that have been expressed by work suppliers and employers in their past worksite interactions. Moreover, as stressed in the efficiency-wage literature [2,15], the behavioral characteristics of work suppliers and employers can also be important determinants of worksite productivity. These behavioral characteristics thus have potentially strong effects on earnings and employment histories. Unfortunately, individual data on the behavioral characteristics of workers and employers are difficult to obtain. The potential effects of these behavioral characteristics are thus usually ignored in micro panel data studies of labor market earnings and employment; typically only observable structural attributes such as training, education, and gender are included as possible explanatory variables.

Using recently developed agent-based programming tools, however, computational labor market frameworks can be constructed in which work suppliers and employers adaptively choose and refuse their potential worksite partners and evolve their worksite behaviors over time on the basis of past worksite interactions. Consequently, the following hypothesis can now be subjected to systematic experimental investigation:

Worksite Interaction Hysteresis (WIH) Hypothesis: Temporary shocks in the form of idiosyncratic worksite interactions can result in persistently heterogeneous earnings and employment histories for work suppliers and employers with identical observable structural attributes.

This study investigates the WIH hypothesis in the context of a dynamic computational labor market framework with strategically interacting work suppliers and employers.[2] As will be clarified below, the labor market framework is a flexible computational laboratory permitting experiments with a wide variety of alternative specifications for the exogenous aspects of market structure and agent attributes. The primary purpose of this study, however, is to take a first cut at the computational study of the WIH hypothesis by specifying these exogenous aspects in relatively simple terms. Thus, as implemented for this study, the labor market framework comprises a fixed equal number of work suppliers and employers. These work suppliers and employers repeatedly participate in costly searches for worksite partners on the basis of continually updated expected utility, engage in efficiency-wage worksite interactions modelled as prisoner's dilemma games, and evolve their worksite strategies over time on the basis of the earnings secured by these strategies in past worksite interactions.

Work suppliers have identical observable structural attributes, and similarly for employers. In particular, each work supplier is assumed to have the same capacity wq, where wq is the maximum number of potential work offers that each work supplier can make. Similarly, each employer is assumed to have the same capacity eq, where eq is the maximum number of job openings that each employer can provide. Work suppliers and employers are heterogeneous with regard to their worksite strategies. However, a work supplier and employer engaged in a worksite interaction are not able to directly observe each other's strategies; they only observe the behavior and earnings outcomes flowing from the use of these strategies.

The experimental design of the study consists of the systematic variation, from high to low, of *job capacity* as given by the ratio cq/wq. Jobs are in excess supply when job capacity exceeds one, in balanced supply when job capacity is equal to one, and in tight supply when job capacity is less than one. For each tested job capacity ratio, twenty different runs are generated using twenty different pseudo-random number seed values.[3]

[2] This labor market framework was first presented in preliminary fashion in [14] as a special case of the Trade Network Game (TNG) model developed in [12,13] for studying the evolution of buyer-seller trade networks. The framework is an example of agent-based computational economics (ACE) modelling. ACE is the computational study of economies modelled as evolving decentralized systems of autonomous interacting agents. For various ACE-related resources, including surveys, readings, software, and pointers to research groups, see the ACE web site at http://www.econ.iastate.edu/tesfatsi/ace.htm.

[3] All labor market experiments reported in this study are implemented using version 105b of the Trade Network Game (TNG) source code developed by [7], which in turn is supported by SimBioSys, a general C++ class framework for

In examining the resulting run histories, particular attention is focused on the experimental determination of correlations between job capacity and the formation of persistent networks among work suppliers and employers, and between network formations and the types of persistent worksite behaviors and earnings outcomes that these networks support.

A key finding of this study is that the WIH hypothesis is strongly supported. In the presence of job capacity asymmetries, idiosyncratic worksite interactions tend to result in persistent network patterns and/or persistent behavioral patterns that support persistently heterogeneous earnings levels across employed work suppliers and across nonvacant employers. These persistent network and behavioral patterns are intermediate hysteresis effects of interest in their own right. It is therefore useful to introduce the following formal definitions:

Network Hysteresis: Temporary shocks in the form of idiosyncaratic worksite interactions result in persistently heterogenous network relationships for agents who have identical observable worksite behaviors and structural attributes.

Behavioral Hysteresis: Temporary shocks in the form of idiosyncratic worksite interactions result in persistently heterogeneous worksite behaviors for agents who have identical observable structural attributes.

As will be clarified more carefully in subsequent sections, one reason that network hysteresis arises in the labor market framework is that job search is costly. Work suppliers bear the costs of wasted time spent in submitting unsuccessful work offers to employers during the course of job search. In the presence of capacity asymmetries, these sequentially incurred job search costs can induce path-dependent networks among work suppliers and employers that support persistently heterogeneous earnings levels across employed work suppliers and across nonvacant employers even when each matched work supplier and employer pair expresses the same type of worksite behavior (e.g., mutual cooperation). Since this earnings heterogeneity arises from a structural asymmetry (e.g., tight job capacity) and not from any deficiency in the worksite strategies of the agents per se, it cannot be remedied by evolutionary selection pressures acting upon these strategies.

Behavioral hysteresis arises in the labor market framework for two reasons: differences in own worksite strategies; and differences in the strategies of worksite partners. The first reason is easy to understand. If two work suppliers have different worksite strategies, then in general they will exhibit

evolutionary simulations developed by [6]. Source code for both the TNG and SimBioSys can be downloaded as freeware at the current author's web site, along with extensive user instructions.

different worksite behaviors even if they are interacting with the same employer. The second reason is more interesting and stems from the following observation: The behavior an agent expresses in a worksite interaction is a function of the behavior that is expressed by his worksite partner. For example, a single work supplier interacting with two different employers can end up in a mutually cooperative relation with one employer and a completely hostile relation with the other, all triggered by some difference in the employers' expressed behaviors (e.g., one employer initially cooperates and the other initially defects). Thus, even if two work suppliers have identical worksite strategies and are in an identical network pattern with employers (e.g., each is working continuously for one employer), there is no guarantee they will express identical worksite behaviors unless the employers they are interacting with have identical worksite strategies.

As will be clarified below, due to the relatively greater mobility of work suppliers and to evolutionary selection pressures, work suppliers and employers tend to exhibit behavioral hysteresis in their worksite interactions only in conditions of excess job capacity. In conditions of balanced and tight job capacity, the behaviors of the agents within each agent type tend to coordinate rapidly into similar or even identical patterns. On the other hand, neither mobility nor evolutionary selection pressures can eliminate the substantial network hysteresis that tends to arise when there is tight or excess job capacity.

At a more global level, network and behavioral hysteresis result in a one-to-many mapping between treatment factors and experimental outcomes. That is, for each particular treatment, as the initial random seed value is varied across experimental runs, a small but multiple number of distinct network formations are observed to arise and persist among work suppliers and employers across runs, each supporting a distinct pattern of worksite behaviors and earnings outcomes. This finding is consistent with the many analytical two-sided labor market studies, such as [4], that establish the existence of multiple steady-state search equilibria. In the current process study, however, a histogram is obtained for each treatment showing the proportion of runs that evolve each type of network formation, which provides suggestive information regarding the size and importance of their basins of attraction.

The labor market framework is described in Sec. 11.2. In Sec. 11.3, descriptive statistics are constructed for the ex post classification of network formations, worksite behaviors, and earnings outcomes. The experimental design of the study is outlined in Sec. 11.4, and a detailed discussion of experimental findings is presented in Sec. 11.5. Concluding remarks are given in Sec. 11.6.

11.2 Labor Market Framework

The labor market framework differs in several essential respects from standard labor market models. First, it is a dynamic process model defined algorithmically in terms of the internal states and behavioral rules of work suppliers and employers rather than by the usual system of demand, supply, and equilibrium equations. The only equations that arise in the model are those used by the agents themselves to summarize observed aspects of their world and to implement their behavioral rules. Second, agents attempt to learn about the behavioral rules of other agents even as these rules are coevolving over time. Third, starting from given initial conditions, all events are contingent on agent-initiated interactions and occur in a path-dependent time line. The analogy to a culture growing in a petri dish, observed by an interested resarcher but not disturbed, is apt.

The labor market framework comprises an equal number M of work suppliers who make work offers and employers who receive work offers, where M can be any positive integer. Each work supplier can have work offers outstanding to no more than wq employers at any given time, and each employer can accept work offers from no more than eq work suppliers at any given time, where the work offer quota wq and the employer acceptance quota eq can be any positive integers.[4]

As seen in Table 11.1, work suppliers and employers are modelled as autonomous endogenously-interacting agents with internalized social norms, internally stored state information, and internal behavioral rules. Each agent, whether a work supplier or an employer, has this same general internal structure. However, work suppliers differ from employers in terms of their specific market protocols, fixed attributes, and initial endowments; and all agents can acquire different state information and evolve different worksite behavioral rules[5] over time on the basis of their past experiences. Note, in particular, that all agents have stored addresses for other agents together with internalized market protocols for communication. These features permit agents to communicate state-dependent messages to other agents at event-triggered times, a feature not present in standard economic models. As will clarified below, the work suppliers and employers depend on this communication ability to seek out and secure worksite partners on an ongoing adaptive basis.

As outlined in Table 11.2, activities in the labor market framework are divided into a sequence of *generations*. Each work supplier and employer in the initial generation is assigned a randomly generated rule governing his

[4] When wq exceeds 1, each work supplier can be interpreted as some type of information service provider (e.g., broker or consultant) that is able to supply services to at most wq employers at a time or as some type of union organization that is able to oversee work contracts with at most wq employers at a time.

[5] In principle, agents could evolve any or all of their behavioral rules, but for current study purposes only the evolution of worksite behavioral rules is considered.

Table 11.1. General form of the internal structure of an agent

```
class Agent
{
    Internalized Social Norms:
        Market protocols for communicating with other agents;
        Market protocols for job search and matching;
        Market protocols for worksite interactions;
    Internally Stored State Information:
        My attributes;
        My endowments;
        My beliefs and preferences;
        Addresses I have for myself and for other agents;
        Additional data I have about other agents.
    Internal Behavioral Rules:
        My rules for gathering and processing new information;
        My rules for determining my worksite behavior;
        My rules for updating my beliefs and preferences;
        My rules for measuring my utility (fitness) level;
        My rules for modifying my rules.
};
```

worksite behavior together with initial expected utility assessments regarding potential worksite partners. The work suppliers and employers then enter into a *trade cycle loop* during which they repeatedly search for worksite partners on the basis of their current expected utility assessments, engage in efficiency-wage worksite interactions modelled as prisoner's dilemma games, and update their expected utility assessments to take into account newly incurred job search costs and worksite payoffs. At the end of the trade cycle loop, the work suppliers and employers each separately evolve (structurally modify) their worksite behaviorial rules based on the past utility outcomes secured with these rules, and a new generation commences. The particular module specifications used in all experiments reported below will now be described in roughly the order depicted in Table 11.2.[6]

Matches between work suppliers and employers are determined using a one-sided offer auction, a modified version of the "deferred acceptance mechanism" originally studied by [5].[7] Under the terms of this auction, hereafter

[6] All experiments reported in this paper are implemented using version 105b of the Trade Network Game (TNG) source code developed in [7]. The latter study provides a detailed discussion of all module implementations. In addition, the TNG source code (with extensive comment statements and user instructions) can be downloaded as freeware from the current author's web site, permitting all module implementations to be specifically viewed in source code form.

[7] See [9] for a careful detailed discussion of [5] deferred acceptance matching mechanisms, including a discussion of the way in which the Association of American

Table 11.2. Logical flow of the labor market framework

```
int main () {

   InitiateEconomy();              // Construct initial subpopulations of
                                   //   work suppliers and employers with
                                   //   random worksite strategies.

   For (G = 1,...,GMax) {          // ENTER THE GENERATION CYCLE LOOP

                                   // GENERATION CYCLE:

      InitiateGen();               //   Configure work suppliers and employers
                                   //     with user-supplied parameter values
                                   //     (initial expected utility levels, work offer
                                   //     quotas, employer acceptance quotas,...)

      For (I = 1,...,IMax) {       //   Enter the Trade Cycle Loop

                                   //   Trade Cycle:
         MatchTraders();           //     Work suppliers and employers determine
                                   //       their worksite partners, given
                                   //       their expected utility assessments,
                                   //       and record job search and
                                   //       inactivity costs.
         Trade();                  //     Work suppliers and employers engage
                                   //       in worksite interactions and
                                   //       record their worksite payoffs.
         UpdateExp();              //     Work suppliers and employers update their
                                   //       expected utility assessments, using
                                   //       newly recorded costs and worksite
                                   //       payoffs, and begin a new trade cycle.
      }
                                   //   Environment Step:
      AssessFitness();             //     Work suppliers and employers
                                   //       assess their utility levels.

                                   //   Evolution Step:
      EvolveGen();                 //     Worksite strategies of work suppliers and
                                   //       employers are separately evolved, and
                                   //       a new generation cycle begins.
   }
   Return 0;
}
```

referred to as the *deferred choice and refusal* (DCR) mechanism, each work supplier submits work offers to a maximum of *wq* employers he ranks as most preferable on the basis of expected utility and who he judges to be tolerable in the sense that their expected utility is not negative. Similarly, each employer selects up to *eq* of his received work offers that he finds tolerable and most preferable on the basis of expected utility and he places them on a waiting list; all other work offers are refused. Work suppliers redirect refused work offers to tolerable preferred employers who have not yet refused them,

Medical Colleges since WWII has slowly evolved such an algorithm (the National Intern Matching Program) as a way of matching interns to hospitals in the United States.

if any such employers exist. Once employers stop receiving new work offers, they accept all work offers currently on their waiting lists.

A work supplier incurs a job search cost in the form of a negative *refusal payoff* R each and every time that an employer refuses one of his work offers during a trade cycle; the employer who does the refusing is not penalized.[8] A work supplier or employer who neither submits nor accepts work offers during a trade cycle receives an *inactivity payoff* 0 for the entire trade cycle. The refusal and inactivity payoffs are each assumed to be measured in utility terms.

If an employer accepts a work offer from a work supplier in any given trade cycle, the work supplier and employer are said to be *matched* for that trade cycle. Each match constitutes a mutually agreed upon contract stating that the work supplier shall supply labor services at the worksite of the employer until the beginning of the next trade cycle. These contracts are risky in that outcomes are not assured.

Specifically, each matched work supplier and employer engage in a worksite interaction modelled as a two-person prisoner's dilemma game reflecting the basic efficiency wage hypothesis that work effort levels are affected by overall working conditions (e.g., wage levels, respectful treatment, safety considerations). The work supplier can either cooperate (exert high work effort) or defect (engage in shirking). Similarly, the employer can either cooperate (provide good working conditions) or defect (provide substandard working conditions).

The range of possible worksite payoffs is assumed to be the same for each worksite interaction in each trade cycle: namely, as seen in Table 11.3, a cooperator whose worksite partner defects receives the lowest possible payoff L (sucker payoff); a defector whose worksite partner also defects receives the next lowest payoff D (mutual defection payoff); a cooperator whose worksite partner also cooperates receives a higher payoff C (mutual cooperation payoff); and a defector whose worksite partner cooperates receives the highest possible payoff H (temptation payoff). The worksite payoffs in Table 11.3 are assumed to be measured in utility terms, and to be normalized about the inactivity payoff 0 so that $L < D < 0 < C < H$. Thus, a work supplier or employer that ends up either as a sucker with payoff L or in a mutual defection relation with payoff D receives negative utility, a worse outcome than inactivity (unemployment or vacancy). These worksite payoffs are also assumed to satisfy the usual prisoner's dilemma regularity condition $(L + H)/2 < C$ guaranteeing that mutual cooperation dominates alternating cooperation and defection on average.

[8] This modelling for job search costs is equivalent to assuming: (i) each work supplier must pay a job search cost in amount -R for each work offer he makes to an employer; and (ii) each possible worksite payoff for work suppliers is increased by the amount -R, so that a work supplier is able to recoup the job search costs he incurs in making a work offer if and only if this work offer is accepted.

Table 11.3. Payoff matrix for the worksite prisoner's dilemma game

Work Supplier \ Employer	c	d
c	(C,C)	(L,H)
d	(H,L)	(D,D)

Each agent, whether a work supplier or an employer, uses a simple learning algorithm to update his expected utility assessments on the basis of new payoff information. Specifically, an agent v assigns an exogenously given initial expected utility U^o to each potential worksite partner z with whom he has not yet interacted. Each time an interaction with z takes place, v forms an updated expected utility assessment for z by summing U^o together with all payoffs received to date from interactions with z (including both worksite payoffs and refusal payoffs) and then dividing this sum by one plus the number of interactions with z.

The rule governing the worksite behavior of each agent, whether work supplier or employer, is represented as a finite-memory pure strategy for playing a prisoner's dilemma game with an arbitrary partner an indefinite number of times, hereafter referred to as a *worksite strategy*. At the commencement of each trade cycle loop, agents have no information about the worksite strategies of other agents; they can only learn about these strategies by engaging other agents in repeated worksite interactions and observing the behavioral and utility outcomes that ensue. In consequence, each agent's choice of an action in a current worksite interaction with another agent is determined entirely on the basis of his own past interactions with this other agent plus his initial expected utility assessment of the agent. Each agent thus keeps separate track of his interaction history with each potential worksite partner.

At the end of each trade cycle loop, the *utility (fitness)* of each work supplier and employer is measured by the average payoff he attained over this trade cycle loop. Average payoff is calculated as total net payoffs (negative refusal payoffs plus worksite payoffs) divided by the total number of payoffs received. The worksite strategies of workers and employers are then separately evolved by means of standardly specified genetic algorithms involving recombination, mutation, and elitism operations that are biased in favor of more fit agents.[9] This evolution is meant to reflect the formation and trans-

[9] More precisely, for each agent type (work supplier or employer), the genetic algorithm evolves a new collection of agent worksite strategies from the existing

mission of new ideas by mimicry and experimentation, not reproduction in any biological sense. That is, if a worksite strategy successfully results in high fitness for an agent of a particular type, then other agents of the same type are led to modify their own strategies to more closely resemble the successful strategy.

An important caution is in order here, however. The information that work suppliers and employers are currently permitted to have access to in the evolution step is substantial: namely, complete knowledge of the collection of strategies used by agents of their own type in the previous trade cycle loop, ranked by fitness. The evolution step is thus more appropriately interpreted as an iterative stochastic search algorithm for determining possible strategy configuration attractors rather than as a social learning mechanism per se. The resulting outcomes will be used in subsequent work as a yardstick against which to assess the performance of more realistically modelled social learning mechanisms.

11.3 Descriptive Statistics

Each of the labor market experiments reported in this study results in a one-to-many mapping between structural characteristics and outcomes. That is, when each particular experimental treatment is repeated for a range of pseudo-random number seed values, a distribution of behavioral, network, and utility outcomes is generated. Consequently, the mapping between treatment factors and outcomes must be characterized statistically.

This section explains the descriptive statistics that have been constructed to aid in the experimental determination of correlations between treatment factors and network formations, and between network formations and the types of worksite behaviors and utility outcomes that these networks support. Networks depict who is working for whom, and with what regularity. Worksite behavior refers to the specific actions undertaken by workers and employers in their worksite interactions. Finally, utility refers to the average payoff

collection of agent worksite strategies by applying the following four steps: (1) *Evaluation*, in which a fitness score is assigned to each strategy in the existing strategy collection; (2) *Recombination*, in which offspring (new ideas) are constructed by combining the genetic material (structural characteristics) of pairs of parent strategies chosen from among the most fit strategies in the existing strategy collection; (3) *Mutation*, in which additional variations (new ideas) are constructed by mutating the structural characteristics of each offspring strategy with some small probability; and (4) *Replacement*, in which the most fit (elite) strategies in the existing strategy collection are retained for the new collection of strategies and the least fit strategies in the existing strategy collection are replaced with offspring strategies. See [7] for a detailed discussion of this use of genetic algorithms in the Trade Network Game (TNG), and see [10] for a more general discussion of genetic algorithm design and use.

levels attained by work suppliers and employers as a result of job search and worksite interactions.

11.3.1 Classification of Contractual Networks by Distance

First introduced is a distance measure on persistent networks that permits the classification of these networks into alternative types. This distance measure calculates the extent to which an observed pattern of persistent agent relationships deviates from an idealized pattern that specifies relationships among agent types without consideration for the identity of individual agents within agent types. As will be seen in Sect. 11.5, this distance measure permits networks to be distinguished on the basis of the differential worksite behaviors and utility outcomes that they support. In addition, as a by-product, it provides a useful indicator of the extent to which heterogeneity in attained utility levels arises from network hysteresis.

All labor market experiments reported in this study were implemented using version 105b of the TNG source code developed in [7]. Let s denote a seed value for the pseudo-random number generator incorporated in this source code, and let E denote a *potential economy*, i.e., an economy characterized structurally by the source code together with specific values for all source code parameters[10] apart from s. The *sample economy* generated from E, given the seed value s, is denoted by (s, E).

Worksite strategies are represented as finite state machines,[11] hence the actions undertaken by any agent v in repeated worksite interactions with another agent z must eventually cycle. Consequently, these actions can be summarized in the form of a *worksite history* $H{:}P$, where the *handshake* H is a (possibly null) string of worksite actions that form a non-repeated pattern and the *persistent portion* P is a (possibly null) string of worksite actions that are cyclically repeated. For example, letting c denote cooperation and d denote defection, the worksite history *ddd:dc* indicates that agent v defected against agent z in his first three worksite interactions with z and thereafter alternated between defection and cooperation.

A work supplier w and employer e are said to exhibit a *persistent relationhip* during a given trade cycle loop T of a sample economy (s, E) if the following two conditions hold: (a) their worksite histories with each other during the course of T take the form $H_w{:}P_w$ and $H_e{:}P_e$ with nonnull P_w and P_e; and (b) accepted work offers between w and e do not permanently cease

[10] A complete annotated listing of these parameters is given in Section 11.4.

[11] A *finite state machine* is a system comprising a finite collection of internal states together with a state transition function that gives the next internal state the system will enter as a function of the current state and other current inputs to the system. For the application at hand, the latter inputs are the actions selected by a worker supplier and an employer engaged in a worksite interaction. See [7] for a more detailed discussion and illustration of how finite state machines are used to represent worksite strategies in the TNG source code.

during T either by choice (a permanent switch to strictly preferred partners) or by refusal (one agent becoming intolerable to the other because of too many defections). A persistent relationship between w and e is said to be *latched* if w works for e continuously (in each successive trade cycle), and it is said to be *recurrent* if w works for e randomly or periodically.

A possible pattern of relationships among the work suppliers and employers in the final generation of a potential economy E is referred to as a *network*, denoted generically by $K(E)$. Each network $K(E)$ is represented in the form of a directed graph in which the vertices $V(E)$ of the graph represent the work suppliers and employers, the edges of the graph (directed arrows) represent work offers directed from work suppliers to employers, and the edge weight on any edge denotes the number of accepted work offers between the work supplier and employer connected by the edge.

Let $K(s,E)$ denote the network depicting the actual pattern of relationships among the work suppliers and employers in the final generation of the sample economy (s,E). The reduced form network $K^p(s,E)$ derived from $K(s,E)$ by eliminating all edges of $K(s,E)$ that correspond to nonpersistent relationships is referred to as the *persistent network* for (s,E).

Let $V^o(E)$ denote a *base network pattern* that partially or fully specifies a potential pattern of relationships among the work suppliers and employers in the potential economy E by placing general constraints on the relationships among agent types without regard for the individual identity of agents within each type. For example, $V^o(E)$ could designate that each work supplier directs work offers to at least two employers. The collection of all networks whose edges conform to the base network pattern $V^o(E)$ is referred to as the *base network class*, denoted by $K^o(E)$.

The *distance* $D^o(s,E)$ between the persistent network $K^p(s,E)$ and the base network class $K^o(E)$ for a sample economy (s,E) is then defined to be the number of vertices (work suppliers and employers) for $K^p(s,E)$ whose edges (persistent relationships) fail to conform to the base network pattern $V^o(E)$. As will be demonstrated in Section 11.5, this distance measure provides a useful way to classify the different types of persistent networks observed to arise for a given value of E as the seed value s is varied.

11.3.2 Classification of Worksite Behaviors and Utility Outcomes

Let a sample economy (s,E) be given. A work supplier or employer in the final generation of (s,E) is referred to as an *aggressive agent* if he engages in at least one defection against another agent that has not previously defected against him. The 1×2 vector giving the percentages of work suppliers and employers in the final generation of (s,E) that are aggressive is referred to as the *aggressive profile* for (s,E). The aggressive profile measures the extent to which work suppliers and employers behave opportunistically in worksite interactions with partners who are either strangers or who so far have been consistently cooperative.

A work supplier or employer in the final generation of (s, E) is referred to as *persistently inactive* if he constitutes an isolated vertex of the persistent network $K^p(s, E)$. The 1×2 vector giving the percentages of work suppliers and employers in the final generation of (s, E) who are persistently inactive is referred to as the *p-inactive profile* for (s, E). The p-inactive profile measures the extent to which work suppliers and employers in this final generation fail to establish any persistent relationships. The p-inactive percentage for work suppliers constitutes their persistent unemployment rate, whereas the p-inactive percentage for employers constitutes their persistent vacancy rate.

A work supplier or employer in the final generation of (s, E) is referred to as a *repeat defector* if he establishes at least one persistent relationship for which the persistent portion P of his worksite history $H{:}P$ includes a defection d. Defections for work suppliers correspond to shirking episodes, and defections for employers correspond to the provision of poor working conditions. The 1×2 vector giving the percentages of work suppliers and employers in the final generation of (s, E) who are repeat defectors is referred to as the *r-defector profile* for (s, E). The r-defector profile measures the extent to which work suppliers and employers in the final generation of (s, E) engage in recurrent or continuous defections.

If, instead, a work supplier or employer in the final generation of (s, E) establishes at least one persistent relationship and his worksite history for each of his persistent relationships has the general form $H{:}c$, he is referred to as *persistently nice*. The 1×2 vector giving the percentages of work suppliers and employers in the final generation of (s, E) who are persistently nice is referred to as the *p-nice profile* for (s, E). The p-nice profile measures the extent to which work suppliers and employers in this final generation establish persistent relationships characterized by fully cooperative behavior.

By construction, each work supplier and employer in the final generation of (s, E) must either be a persistently inactive agent, a repeat defector, or a persistently nice agent.

Finally, the 1×2 vector giving the average utility (fitness) levels attained by work suppliers and employers in the final generation of (s, E) is referred to as the *utility profile* for (s, E). The utility profile measures the distribution of welfare across agent types.

11.4 Experimental Design

The labor market experiments reported in Section 11.5 are for two-sided markets comprising 12 work suppliers and 12 employers. Each work supplier has the same offer quota, wq, and each employer has the same acceptance quota, eq. Attention is focused on the effects of varying job capacity from high to low, where job capacity is measured by the ratio eq/wq. Four different settings for job capacity are tested: high excess job capacity ($eq >> wq$);

Table 11.4. Parameter values for a labor market with high excess job capacity

// PARAMETER VALUES HELD FIXED ACROSS EXPERIMENTS	
GMax = 50	// Total number of generations.
IMax = 150	// Number of trade cycles per trade cycle loop.
AgentCount = 24	// Total number of agents.
RefusalPayoff = -0.5	// Payoff R received by a refused agent.
InactivityPayoff = +0.0	// Payoff received by an inactive agent.
Sucker = -1.6	// Lowest possible worksite payoff, L.
BothDefect = -0.6	// Mutual defection worksite payoff, D.
BothCoop = +1.4	// Mutual cooperation worksite payoff, C.
Temptation = +3.4	// Highest possible worksite payoff, H.
InitExpPayoff = +1.4	// Initial expected utility level, U^o.
Elite = 67	// GA elite percentage for each agent type.
MutationRate = .005	// GA mutation rate (bit toggle probability).
FsmStates = 16	// Number of internal FSM states.
FsmMemory = 1	// FSM memory (in bits) for past move recall.
WorkSuppliers = 12	// Number of work suppliers.
Employers = 12	// Number of employers.
// PARAMETER VALUES VARIED ACROSS EXPERIMENTS	
WorkerQuota = 1	// Work offer quota wq.
EmployerQuota = 12	// Employer acceptance quota eq.

balanced job capacity ($eq = wq = 1$); tight job capacity ($eq = 1$ and $wq = 2$); and extremely tight job capacity ($eq << wq$).

The values for all remaining parameters are maintained at fixed values throughout all experiments. Table 11.4 lists these fixed parameter values along with the specific wq and eq quota values for an experiment with high excess job capacity. The parameter values in Table 11.4, together with the TNG source code, constitute a potential economy E in the sense defined in the previous section.

For each tested E, twenty sample economies (s, E) were generated using twenty arbitrarily selected seed values s for the pseudo-random number generator included in the TNG source code.[12] For each run s, the persistent network $K^p(s, E)$ was determined and graphically depicted, and the components for the four behavioral profiles (aggressive, p-inactive, r-defector, and p-nice) and the utility profile were calculated and recorded.

A base network pattern $V^o(E)$ was then specified for each tested economy E that constrained the general relationships among agent types without relying on the individual identity of agents within each agent type. This base network pattern provides the 0 point for the distance measure $D^o(\cdot, E)$ and hence is an intrinsically arbitrary normalization. However, its degree of speci-

[12] These twenty seed values are as follows: 5, 10, 15, 20, 25, 30, 45, 65, 63, 31, 11, 64, 41, 66, 13, 54, 641, 413, 425, and 212. The final fourteen values were determined by random throws of two and three die. The TNG source code used to implement the labor market framework uses pseudo-random number values in the initialization of worksite strategies, in the matching process to break ties among equally preferred worksite partners, and in genetic algorithm recombination and mutation operations applied to worksite strategies in the evolution step.

ficity governs the dispersion of the resulting distance values $D^o(s, E)$ across sample runs s and the extent to which these distance values display useful correlations with the components of the behavioral and utility profiles. In practice, then, the choice of the base network pattern for each tested economy was fine-tuned so that the resulting distance values provided a classification of networks into distinct types supporting distinct patterns of worksite behaviors and utility outcomes.

Given $V^o(E)$, the distance $D^o(s, E)$ of $K^p(s, E)$ from $K^o(E)$ was recorded for each run s, and a histogram for the distance values $D^o(s, e)$ was constructed giving the percentage of runs s corresponding to each possible distance value. Finally, as a rough stability check, the number of generations was also increased to 100 for each tested economy E and the minimum, maximum, and average values for the utility levels attained by work suppliers and employers in each of the 100 generations were graphically generated for each sample economy (s, E).

11.5 Experimental Findings

Consider a two-sided potential economy E comprising 12 work suppliers and 12 employers with a work offer quota $wq = 1$ and an employer acceptance quota $eq = 12$. These quota values imply there is high excess job capacity. Employers are forced to remain vacant unless work suppliers happen to direct work offers their way, hence employers face a substantial structural risk of vacancy. On the other hand, work suppliers face zero structural risk of having their work offers refused by employers because of limited job capacity.

The base network pattern $V^o(E)$ for this potential economy E is as follows: Each work supplier is latched to an employer, and each employer has at least one latched work supplier. A schematic diagram of this base network pattern is depicted in Fig. 11.1(a).

Descriptive statistics for the twenty sample economies (s, E) corresponding to this high excess job economy E are presented in Table 11.5(a).[13] Note that 75% of the sample economies (s, E) lie in the distance cluster 3–9. The low mean utility level 0.35 attained by employers in this distance cluster is due to three factors: a high mean p-inactivity (vacancy) rate among employers due to high excess job capacity; a high mean aggression (initial defection) rate by work suppliers; and a low mean p-nice (cooperation) rate by work suppliers that induces retaliatory defections by some employers.

The persistent networks that arise for the sample economies (s, E) in distance cluster 3–9 reveal strong network hysteresis, i.e., strong persistent

[13] In Table 11.5, for each distance cluster, the mean and standard deviation are calculated for each component of the three behavioral profiles (aggressive, p-inactive, and p-nice) and the utility profile across the sample runs lying in this distance cluster. The r-defector profiles are omitted from Table 11.5 since they can be derived from the p-inactive and p-nice profiles, see Subsection 11.3.2).

Table 11.5. Experimental findings for differential job capacities

D^o Cluster	% Runs	AGGRESSIVE		P-INACTIVE		P-NICE		UTILITY	
		w	e	w	e	w	e	w	e
3–9	75%	97% (5%)	16% (34%)	2% (3%)	40% (12%)	3% (5%)	39% (28%)	1.74 (.27)	0.35 (.14)
24	25%	2% (3%)	5% (7%)	2% (3%)	5% (7%)	98% (3%)	95% (7%)	1.39 (.02)	1.02 (.03)

(a) High Excess Job Capacity (wq=1, eq=12)

D^o Cluster	% Runs	AGGRESSIVE		P-INACTIVE		P-NICE		UTILITY	
		w	e	w	e	w	e	w	e
0–2	75%	16% (33%)	23% (39%)	1% (3%)	1% (3%)	94% (6%)	86% (26%)	1.10 (.14)	1.33 (.22)
4	10%	50% (50%)	54% (46%)	8% (8%)	8% (8%)	50% (50%)	46% (46%)	0.57 (.05)	0.86 (.57)
24	15%	0% (0%)	22% (20%)	0% (0%)	8% (0%)	89% (16%)	78% (20%)	0.24 (.08)	1.42 (.05)

(b) Balanced Job Capacity (wq=eq=1)

D^o Cluster	% Runs	AGGRESSIVE		P-INACTIVE		P-NICE		UTILITY	
		w	e	w	e	w	e	w	e
0–7	55%	2% (3%)	5% (9%)	19% (10%)	4% (7%)	81% (10%)	96% (6%)	0.30 (.05)	1.35 (.09)
13–17	15%	100% (0%)	69% (43%)	47% (14%)	19% (18%)	8% (12%)	14% (20%)	0.32 (.04)	0.76 (.13)
24	30%	100% (0%)	100% (0%)	100% (0%)	100% (0%)	0% (0%)	0% (0%)	-0.10 (0)	-0.02 (0)

(c) Tight Job Capacity (wq=2, eq=1)

D^o Cluster	% Runs	AGGRESSIVE		P-INACTIVE		P-NICE		UTILITY	
		w	e	w	e	w	e	w	e
0–6	35%	1% (3%)	1% (3%)	12% (4%)	1% (3%)	86% (7%)	96% (6%)	0.31 (.03)	1.37 (.06)
15–17	20%	10% (14%)	92% (14%)	35% (7%)	2% (4%)	17% (20%)	25% (34%)	0.35 (.17)	1.22 (.20)
24	45%	100% (0%)	100% (0%)	100% (0%)	100% (0%)	0% (0%)	0% (0%)	-0.10 (.00)	-0.01 (.00)

(d) Extremely Tight Job Capacity (wq=12, eq=1)

differences in relationship patterns for both work suppliers and employers. The typical scenario is as follows: r-defector work suppliers latch on to a proper subset of p-nice employers, with anywhere from one to four work suppliers latched to the same employer, and drive down the utility levels of these employers to small positive values. Remaining employers are left vacant with utility levels at zero (the inactivity payoff). This scenario ensures that the worksite strategies of the p-nice employers are advantaged in the evolution step relative to the worksite strategies of the employers who are left vacant. Since work suppliers and employers evolve separately, the worksite strategies of the p-nice employers tend to reproduce into the next generation, which ensures the perpetuation of a cooperative set of employers whom the work suppliers can continue to opportunistically defect against.

At least some degree of persistent heterogeneity in utility levels across employed work suppliers and across nonvacant employers is observed for each sample economy (s, E) in the distance cluster 3–9. This persistent heterogeneity in utility levels across the active agents of each agent type is primarily due to behavioral hysteresis – specifically, persistent differences in worksite behaviors between different latched work supplier and employer pairs – rather than to network hysteresis. The different distance values observed for the persistent networks arising in these sample economies are essentially a count of the number of persistently vacant employers who have degenerated into p-inactivity primarily by bad luck but also occasionally by ostracism.

Table 11.5(a) also shows that the remaining 25% of the sample economies for this E lie in a second distance cluster $D^o = 24$. The mean utility level 1.02 attained by employers in this second distance cluster is much higher than that attained in distance cluster 3–9 due to the high mean percentage of p-nice behavior exhibited by both work suppliers and employers. This mean utility level is nevertheless substantially below the mutual cooperation payoff level 1.40 due to the 5% mean p-inactivity (vacancy) rate among employers, a structural consequence of high excess job capacity that is independent of how cooperatively the employers behave in their worksite interactions. The typical pattern exhibited in this distance cluster is p-nice work suppliers randomly directing work offers among employers without latching. Note that the mean utility level 1.39 attained by work suppliers is very close to the mutual cooperation payoff level 1.40. No latching takes place in any of the sample economies in this distance cluster, and utility levels are largely homogeneous across employed work suppliers and across nonvacant employers.

Next consider the case in which the work offer quota remains at $wq = 1$ but the employer acceptance quota is reduced to $eq = 1$ so that job capacity is balanced. This change dramatically affects network formation.

Specifically, as depicted in Figure 11.1(b), the base network pattern now consists of disjoint latched pairings of one work supplier and one employer. As detailed in Table 11.5(b), 75% of the sample economies for this balanced job capacity economy E lie in distance cluster 0–2, implying that the base network pattern is by far the most predominant network formation observed. Heterogeneity in utility outcomes across employed work suppliers is essentially due to differences in job search costs incurred in the process of attaining the coordinated base network pattern. Once in this coordinated state, very little further heterogeneity in utility outcomes is observed either across employed work suppliers or across nonvacant employers.

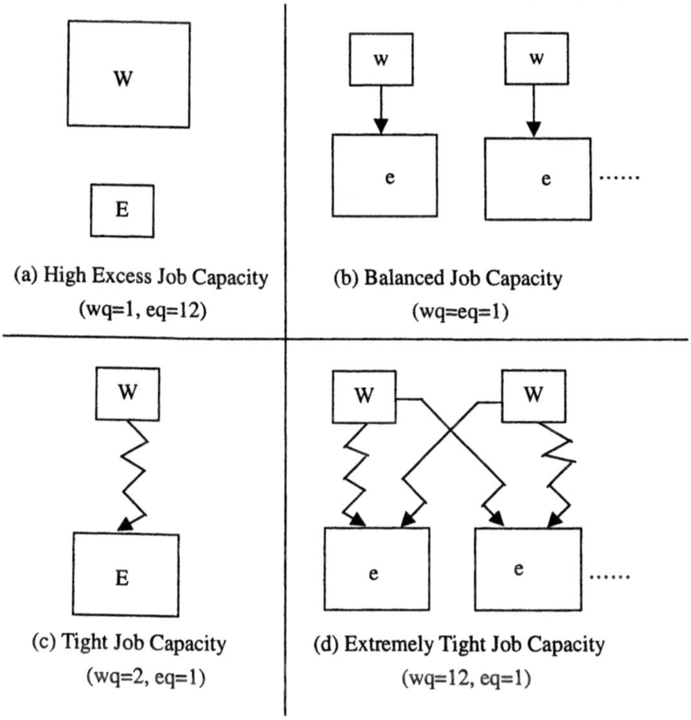

Fig. 11.1. Base network patterns for labor markets with differential job capacities. A relatively larger box for either work suppliers (W) or employers (E) under a particular job capacity treatment indicates that this agent type attains a relatively higher average utility level in the sample economies whose networks approximate the depicted base network pattern. Straight edges indicate latched (continuous) persistent relationships and zig-zag edges indicate recurrent (random or periodic) persistent relationships

More precisely, balanced job capacity favors employers over work suppliers, because work suppliers must bear the costs associated with job search. Nevertheless, the endogenous mobility of work suppliers protects them from overly opportunistic worksite behavior by employers. An employer who attempts to sustain too high a defection frequency against a work supplier will cause this work supplier to quit (redirect his future work offers elsewhere) if other employers are perceived as better earnings opportunities, or even to exit the labor force altogether. Although defections by employers occur rather frequently in the handshake portions of worksite histories, in all but one of the sample economies the employers end up expressing largely p-nice behavior rather than attempting to exploit the work suppliers' vulnerability to job search costs by engaging in repeat defections.

On the other hand, work suppliers who fail to latch when job capacity is balanced tend to accumulate large job search costs (negative refusal payoffs). Being fired by an employer for aggressive or r-defection behavior can thus be very costly for work suppliers, and most display p-nice behavior in their worksite interactions. Nevertheless, even work suppliers who succeed in established a mutually p-nice latched relationship with a single employer typically accumulate two or three negative refusal payoffs from a wide range of employers on their way to attaining this coordinated state. These job search costs, together with the aggressive (initial defection) behavior of many employers, tend to lower the mean utility level of work suppliers relative to employers.

The stability checks conducted for this balanced job capacity case reveal that many of the sample economies exhibit unsettled mean utility outcomes over generations 1 through 100 in the form of persistent drifting, bubbling, or regime shifts. The reason for this appears to be that, under conditions of balanced job capacity, networks form among work suppliers and employers in response to job search costs, yet they support largely p-nice or even *c:c* worksite behavior. Consequently, these networks tend to be nonrobust to the entrance of new, initially cooperative worksite strategies introduced in the evolution step.

As job capacity keeps tightening, work suppliers have an increasingly difficult time forming persistent relationships with employers, a finding indicated in Fig. 11.1 by the decreasing size of work supplier boxes relative to employer boxes as one moves from part (a) to part (d). This increased coordination failure is detailed in Table 11.5. Note, in particular, the growing mean percentage of work suppliers who become p-inactive (unemployed) as job capacity successively tightens.

More precisely, with tight job capacity, the typically observed experimental outcome is that each employer forms persistent relationships with a particular subset of work suppliers. These persistent relationships are recurrent in the following sense: In each trading period, the work offers received by the employer from his persistent work suppliers exceed his job capacity limits, and he accepts only a portion of these work offers by random selection. The reason for the random selection is that these persistent work suppliers tend to be largely cooperative in their interactions with their employers, so that the employers are generally indifferent regarding whose work offers to accept. Since job openings are relatively scarce, this implies that work suppliers face a risk that their work offers will be refused by employers due to capacity limitations even if they have never defected against any employer in their past worksite interactions. On the other hand, for reasons elaborated above for the case of balanced job capacity, the endogenous mobility of work suppliers still induces largely p-nice behavior among employers.

A work supplier whose work offer is refused by an employer incurs a job search cost. This causes the work supplier to lower the utility he expects to attain from any next work offer to this employer, which in turn encourages the work supplier to direct his next work offer elsewhere. A work supplier who receives too many refusals from employers eventually ceases making work offers altogether because the expected utility he assigns to each prospective employer falls below zero, the inactivity payoff level. This complete discouragement tends to occur for work suppliers in the early stages of a trade cycle loop when work suppliers are spreading their work offers among many different employers and refusal rates tend to be high.

As discouraged work suppliers leave the labor force, however, refusal rates decline and the condition of the remaining work suppliers improves. Consequently, even when job capacity is extremely tight, over half the sample economies manage to evolve to a sustainable state in which work suppliers who remain in the labor force are able to find employment at a high enough frequency to sustain their utility levels at positive levels.

Heterogeneity in utility outcomes across employed work suppliers in conditions of tight and extremely tight job capacity thus principally arises from two sources. First, work suppliers experience differential numbers of accepted work offers, resulting in differential worksite payoffs. Second, work suppliers experience differential numbers of refused work offers, resulting in differential job search costs.

These differences in worksite payoffs and job search costs result from network hysteresis. Job search costs incurred by chance determine a network of persistent relationships among work suppliers and employers in a highly path-dependent way. Some work suppliers manage to establish recurrent persistent relationships with so many employers that they are essentially guaranteed to achieve full employment in each trade cycle, whereas other "underemployed" work suppliers only manage to place a few of their potential work offers in each trade cycle. This network hysteresis supports persistent heterogeneity in utility outcomes across active work suppliers. Since this heterogeneity is fundamentally caused by a structural asymmetry (too few job openings) and not from differences in worksite strategies per se, it cannot be remedied by evolutionary selection pressures acting upon worksite strategies.

11.6 Concluding Remarks

Despite the structural simplicity of the computational labor market framework used in this study, the reported experimental findings highlight interesting network and behavioral hysteresis effects arising from idiosyncratic worksite interactions that may be important for understanding aspects of real-world labor markets. For example, it is seen that these hysteresis effects tend to support earnings outcomes that are more heterogeneous than would be predicted on the basis of the observable structural attributes of work suppliers and employers. This phenomenon is routinely observed in real-world labor markets, and is referred to as the "excess heterogeneity problem" [1]. The current study demonstrates the feasibility and potential usefulness of investigating network and behavioral hysteresis effects in an agent-based computational framework.

References

1. Abowd J. M., Kramarz F., Margolis D. N. (1999) High Wage Workers and High Wage Firms. Econometrica **67**, 251–333
2. Akerloff G., Yellen J. (1986) Efficiency Wage Models of the Labor Market. Cambridge University Press, Cambridge
3. Blanchard O. J., Summers L. H. (1990) Hysteresis and the European Unemployment Problem. In: Summers L. H. (Ed.) Understanding Unemployment. The MIT Press, Cambridge, MA, 227–285
4. Diamond P. (1982) Aggregate Demand Management in Search Equilibrium. Journal of Political Economy **90**, 881–895
5. Gale D., Shapley L. (1962) College Admissions and the Stability of Marriage. American Mathematical Monthly **69**, 9–15
6. McFadzean D. (1995) SimBioSys: A Class Framework for Evolutionary Simulations. MA Thesis, Department of Computer Science, University of Calgary, Alberta, Canada
7. McFadzean D., Tesfatsion L. (1999) A C++ Platform for the Evolution of Trade Networks. Computational Economics 14, 109–134.
8. Piscitelli L., Cross R., Grinfeld M., Lamba H. (2000) A Test for Strong Hysteresis. Computational Economics 15, 59–78.
9. Roth A., Sotomayor M. A. O. (1990) Two-Sided Matching: A Study in Game-Theoretic Modeling and Analysis. Cambridge University Press, Cambridge
10. Sargent T. J. (1993) Bounded Rationality in Macroeconomics. Clarendon Press, Oxford
11. Tesfatsion L. (1995) A Trade Network Game with Endogenous Partner Selection. ISU Economic Report No. 36, April.
12. Tesfatsion L. (1997a) A Trade Network Game with Endogenous Partner Selection. In: Amman H., Rustem B., Whinston A. (Eds.) Computational Approaches to Economic Problems. Kluwer Academic Publishers, Dordrecht, The Netherlands, 249–269
13. Tesfatsion L. (1997b) How Economists Can Get Alife. In: Arthur W. B., Durlauf S., Lane D. (Eds.) The Economy as an Evolving Complex System, II. Proceedings Volume XXVII, Santa Fe Institute Studies in the Sciences of Complexity. Addison-Wesley, Reading, MA, 533–564
14. Tesfatsion L. (1998) Preferential Partner Selection in Evolutionary Labor Markets: A Study in Agent-Based Computational Economics. In: Porto V. W., Saravanan N., Waagen D., Eiben A. E. (Eds.) Evolutionary Programming VII. Proceedings of the Seventh Annual Conference on Evolutionary Programming, Springer-Verlag, Berlin, 15–24
15. Yellon J. (1991) Efficiency-Wage Models of Unemployment. In: Mankiw N. G., Romer D. (Eds.) New Keynesian Economics, Volume 2. The MIT Press, Cambridge, MA, 113–122

12 Computable Learning, Neural Networks and Institutions *

Francesco Luna

International Monetary Fund, Washington DC
FLuna@imf.org

Abstract. After showing the positive role that an institution can play in the learning process, we motivate the intuition for why-in the context of a major change in the environment-the more rigid and strictly specialized an institution is, the longer and more complex the learning process will be of any actor subject to the by-now obsolete institution. In particular, the inductive adaptation of economic actors (firms in our setting) to a new environment (such as the one caused by transition to a market economy) is slowed by the very existence of some institution or organizational setting that had emerged in the original environment as a superior tool of induction.

12.1 Introduction

Our work is founded on a study on computable [1] inductive inference conducted by [4,5]. In particular, a neural network employed as a classifier is a tractable simplification of the original model. Furthermore, by adopting neural networks as our tool of analysis, we will be able to introduce explicitly "institutions" under various forms. The justification for the introduction of different *ad hoc* devices is certainly heuristic. Algorithmic or structural solutions adopted–and given as examples of *institutions*–will capture, time after time, relevant features that characterize the role played by an institution in the learning process. We stress data preprocessing, as well as uncertainty and complexity reduction. In other words our aim is that of suggesting a metaphor. For example, the hidden layer of a net performs most of the functions just mentioned, whereas a variation of the perceptron learning algorithm

* The author would like to thank Professors K. Velupillai, A. Leijonhufvud, and L. Punzo for their valuable comments on various partial versions of this work. Professor G. Mondello has given strong encouragement and Dr S. Bartolini endured long precious discussions. None of them is responsible for any error contained in this paper which was presented for the first time at the conference on Computing in Economics and Finance, Geneva 1996. A Human Capital and Mobility grant is gratefully acknowledged.

[1] Spear [13] is the seminal paper on learning under computability constraints. Furthermore, we want to remember Rustem's and Velupillai's [11] contribution to microeconomic theory, Zambelli's [15] attempt at systematizing the Computable Economics Research Program and the exhaustive reconsideration of economics in the light of a computability approach by Velupillai [14].

called *pocket algorithm* allows us to investigate the memory-enhancement role that certain institutions can play. We also believe that, inside the proposed framework of analysis, one may describe how institutions emerge endogenously and show, at the same time, that the form taken by them is not necessarily optimal [2] (where optimality may be in terms of simplicity, effectiveness, flexibility).

However, in the context of a major change in the environment, we try to substantiate the commonsensical forecast according to which the more rigid and strictly specialized an institution is, the longer and more complex the learning process will be of any economic actor subject to the by-now obsolete institution. We believe, of course, that the learning process will *not* be *independent* of the adaptation of the institution itself. Not accidentally, our analysis is conducted through a series of simulations of neural nets in which a particular group of neurodes (the hidden layer) cannot play an autonomous role because of the dialectical relationship that binds them to the evolutionary path of the net as a whole.

The metaphor we employ in this work is for the firm intended as organization and productive process. We like to think of a major change in the environment as the introduction of an innovation or of compulsory standards. The immediate parallel we can make is with the literature dealing with the emergence of a so called *dominant design*. Different theories (see [10] for a brief, reasoned survey) try to account for this observed phenomenon and only a subset of them rely on the eventual and universal recognition of the objective superiority of the chosen design to justify its adoption. In the case of technologies that evolve cumulatively, an early advantage may result in an industry stuck into a suboptimal productive configuration. Alternatively, there may be some externality, some synergies generated by accepting the design one's neighbor has chosen. All these arguments can be easily encompassed by the simulations performed below.

Section 12.2 sketches Gold's inductive inference model as our theoretical reference point and shows why a neural network can be considered an approximation of the original framework. In Sect.12.3 we introduce our model and justify the use of such a metaphor to represent institutions. Sections 12.4 and 12.5 present some simulation results that substantiate the positive role played by institutions in the learning process of an economic actor. On the contrary, we show–in Sects. 12.6 and 12.7–that institutions can be a nuisance

[2] Luna [8] pursues precisely this line of research. In a Cellular Automata setting, an organization emerges endogenously as a response to a complex environment. However, this form of institution is not necessarily the simplest possible one, which, on the other hand, would be obtained by perfectly rational economic actors. An Ockham's Razor approach, hence, would seem to favour the latter line of reasoning over the boundedly rational one followed here. However, at a **procedural level**, the mechanism leading to the spontaneous emergence of the organization is certainly much simpler than the one required by constrained optimization. See [12].

to the learning process, and we consider the effects of the presence of a variety of institutions for social learning in Sect. 12.8. Section 12.9 briefly concludes.

12.2 The Theoretical Reference Point

Following [5], we need to specify three elements in order to characterize a learning model.

(a) A class of "objects". The learner will receive information about some prespecified element of this set and will be required to figure out which one it is.
(b) A method of information presentation. Information concerning the hidden object to be identified is presented to the learner sequentially (let us say for $t = 1, 2, 3, \ldots$). A method of information presentation will specify what sequences are allowed. Typically, two methods are considered in the computable learning literature: the *text* and the *informant*. A text will contain only pieces of information directly related to the hidden object. On the other hand an informant will give pieces of information both in positive and negative terms.
(c) A naming relation. Also in this case there are two possibilities. The first one is a so-called *generator*, the second one is a *tester*. A generator is supposed to guess the name of the object, whereas the tester will "simply" say yes or no depending on whether the piece of information the learner is currently presented with does or does not concern the hidden object.

This framework of analysis was specified by [5] in the context of a language acquisition problem. The hidden object is in that case the grammar of some language. Pieces of information are composed of strings of characters. In the case of a text, these strings obey the implicit rules dictated by the grammar to be identified (learned). In an informant it is also made explicit whether each string presented is a positive or negative instance of the hidden grammar.

It is important to stress that a grammar is nothing but a set of rules that determine how strings of characters are to be put together. In this sense, any set of rules tending to organize some predetermined variables in order to attain some desired goal can be interpreted as a grammar. Hence, the results presented by Gold encompass a number of learning problems strictly connected to Economics.

Most of the theorems deducted in the "computable learning" literature establish boundaries that are very important since they impose some absolute limitation to a learning process (in the sense that this process is shown to be uncomputable). However, even those cases that are computable in principle may not be feasible in practice due to the extremely complex nature of the problem at hand.

We take here into consideration a particular learning model characterized by an *informant* as information presentation method and by a *tester* as naming relation. The absolute learnability results for such a model were stated by [5] and are the following:

> the learner adopting a tester as naming relation when presented pieces of information by an informant will successfully identify in the limit the hidden grammar, if such grammar is *primitive recursive* or belongs to any proper subset of the primitive recursive grammars.

What this result tells us is that not any language–hence grammar–can be learned in such setting. Furthermore, similar impossibility theorems exist for various different learning models. See [5].

As hinted before, such a theorem, however important in its own right, does not help us build a learning model which could be at least an imperfect but feasible tool of analysis.

We claim that a neural network, when employed as a classifier, is a workable example of the learning model described above. First of all, a NN is a Turing Machine (as shown by [2]) in the sense that any neural net performs some procedure so that–accepting Church's Thesis—there will exist a Turing Machine which is equivalent to it.

Second, the alphabet is the set {0,1,...,9} or equivalently {0,1}. Note that in the model considered by Gold, the learner can accept a string of any length, whereas the fixed number of input neurodes of the net may, in certain cases, put a limit to the longest input string acceptable. This, of course, may impose some extra constraints to the learning process reducing the positive learnability results, but let us stress that the uncomputability results are not challenged.

Third, the output of the net (in the considered case of a classifier) is clearly equivalent to the tester answer "yes, no".

12.3 Neural Nets and Institutions

Elsewhere, [6,7] we argued that institutions in general play an important role in the learning process of each economic agent. In particular, they may perform the following tasks:

(a) data pre-processing
(b) error reduction
(c) uncertainty and complexity reduction.
(d) memory enhancement

We want to pursue further these ideas by suggesting how "institutions" can be introduced explicitly in a model where economic actors are represented by NN. More precisely, we propose the analogy between the hidden layers of a NN and an institution. The first three characteristics can be easily recognized in a network with at least one hidden layer. As for memory-enhancement we will exploit a particular learning procedure (called the Pocket Algorithm) rather than a network physical construction. [3] Our "laboratory" for controlled experiments will be the Boolean function XOR (i.e. exclusive or). This is a function of two variables (A and B, let's say) and assumes the value TRUE if either A or B, but not both are TRUE.

As is well known, the Boolean function XOR (exclusive OR) cannot be simulated by a single-layer network; that is, the following net

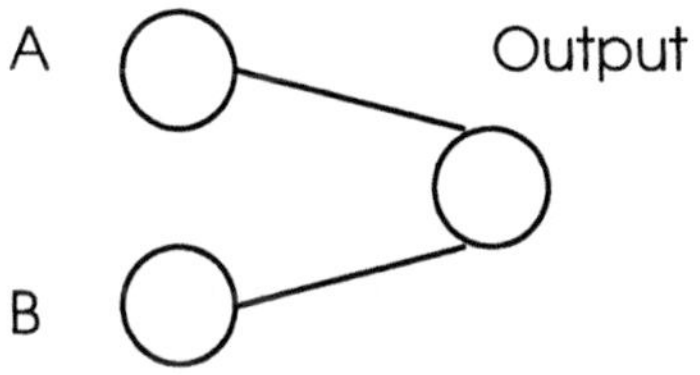

Fig. 12.1. Two-input single-layer network

composed of two input neurodes connected to a single output neurode will not be able to classify correctly the four possible instances no matter what weights and threshold are employed. The instance set is not linearly separable, whereas the discriminating power of the net described above is merely linear. Let A and B be the Boolean arguments of the XOR; the mapping performed by the net is

$$\text{net (A, B)} = w_0 + w_A A + w_B B$$

which is clearly a hyperplane. In particular, the projection of net (A, B) = 0 on the instance space determines a straight line separating the positive from the negative semispace. The classification of instances into *True* or *False* will be obtained according to whether their coordinates lie respectively in the positive or negative semispace. Graphically, the situation can be represented as in figure 12.2:

Assuming that the negative hemispace is at the right of the line depicted, a scenario equivalent to the one reproduced below is the best possible. Only (-1, -1), that is A = *False*, B = *False*, is wrongly classified as a positive instance for the simulated XOR function. By adopting a simple learning procedure such as the *perceptron learning* algorithm, the net will reach an equivalent

[3] A neural-net hidden layer performs with no doubt also a memory-enhancement function for the net as a whole. However, we have chosen to employ a particular learning algorithm to render this role more evident and to distinguish–as done below–between *psychological addiction* and *structural sclerosis*.

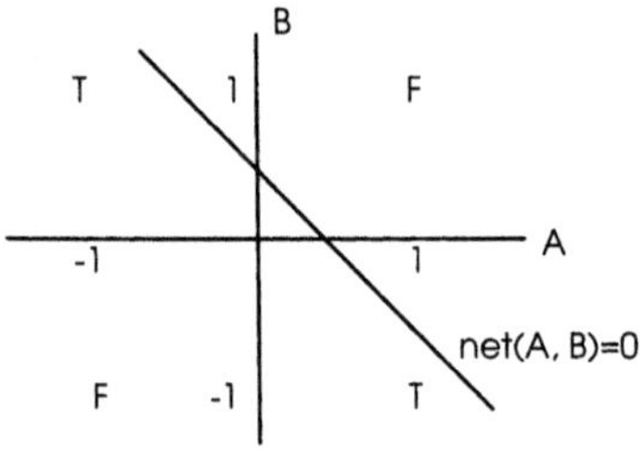

Fig. 12.2. A possible hyperplane generated by a single-layer net projected on the XOR Instance Space

configuration after a relatively small number of iterations. [4] In this way, it would classify correctly an average of 75% of all instances randomly picked from the Boolean (A, B) space. However, it can be shown that the parameters w_0, w_A, w_B will keep on changing [5] without settling for any particular value.

Clearly, the problem's complexity transcends the physical possibility of the net so that a perfect solution cannot be reached by such a structure. On the other hand, it appears obvious that if the net could somehow preserve some *memory* of its own history it would avoid the endless cycling to select a *feasible* optimum.

A variation on the theme of the perceptron algorithm called *pocket algorithm* [6] performs precisely the task of anchoring the net to a set of weights that has proved successful in the past. The net will relinquish such a configuration only when sufficient evidence has been accumulated in favor of a different set of weights.

Here is a sketched description of the algorithm:

[4] We will make this clearer below.

[5] In particular, they will cycle around a finite number of values. See [3], theorem 3.4 on page 73.

[6] Presented in [3]. The following description of the algorithm is taken from that book.

The Pocket Algorithm

0. Initialize to 0 all weights (the vector W).
1. Select a training example, *inst*, at random and the corresponding classification: *want*.
2. If W correctly classifies *inst*, i.e. $\{W \cdot inst > 0$ and $want = 1\}$ or $\{W \cdot inst < 0$ and $want = -1\}$
Then:
  2a. If the current run of correct classifications with W is longer than the run of correct classifications for the weight vector $Wpoc$ in the pocket:
    2aa. Replace the pocket weights $Wpoc$ by W, and remember the length of its correct run.
Else:
  2A. UPDATE: Modify W by adding or subtracting *inst* according to whether the correct output *want* was +1 or -1:
  $W' = W + inst * want.$
3. Go to step 1.

The simple network employing this learning algorithm will "be right" on average 75% of times. The metaphor is obvious, an institution (considered here as an acquired habit) that provides some memory resources improves the chance of success of the actors adopting such routine.

However, a complex environment can be tamed by a more sophisticated hardware, rather than by a more ingenious software. In particular, a network whose physical structure is somewhat more articulated can perfectly deal with the XOR simulation problem.

For example, the following net

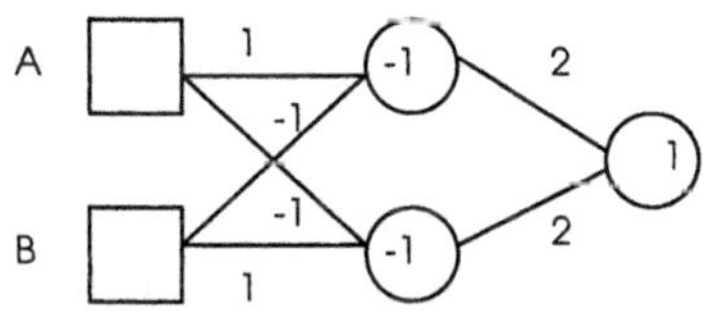

Fig. 12.3. Feedforward network with teo input neurodes and two hidden neurodes

performs the XOR function. The role played by the two hidden neurodes is that of simplifying the original problem by "disaggregating" it into two separable sub-problems. In particular, one hidden neurode will signal to the output one whether A and only A is TRUE, whereas the other hidden neurode will send an equivalent signal concerning B. Clearly, both signals cannot be on simultaneously, hence the net as a whole classifies all instances (AB, notAB, AnotB, notAnotB) correctly.

Also the following net correctly calculates the XOR function

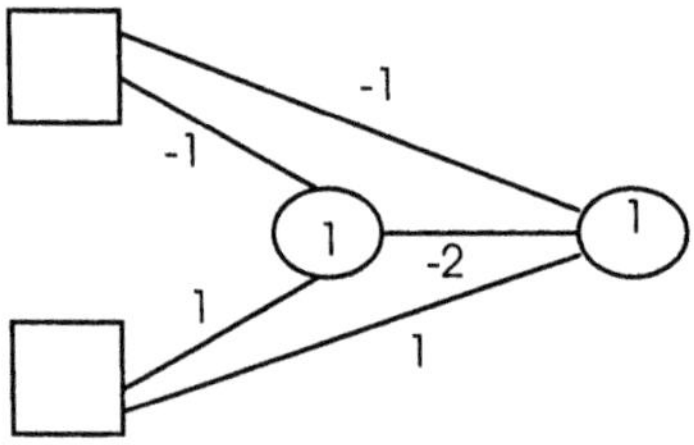

Fig. 12.4. Alternative network configuration which simulates the XOR function

In this case, the output neurode sees the original input and receives only a "suggestion" on how to combine the information.

The data pre-processing, error reduction and complexity reduction performed is already important in such simple examples. However, the analogy between neural networks and institutions goes even further. We are now interested in showing how, in certain cases, institutions may turn out to be an obstacle to a successful learning process. The risk is that of a crystallization of some *modus agendi* (represented by an institution)–which has proved effective and efficient in the past–that prevents the adaptation of an economic or social system to the emergence of a new environment. Such new environment may be the result of an exogenous and unpredictable shock or the outcome of a gradual evolution leading to an abrupt endogenous change–Catastrophe and Bifurcation Theory offer various reference points for such dynamical systems.

The shock may also be caused by a political choice such as the introduction of environmentalist production processes dictated by law. In particular, the enforcement of new stringent standards for environmental protection may cause the adoption of techniques that diverge from the state of the art either because they are based on some engineering innovation or simply because they employ well known technologies that had however been discarded as non-economical in view of the so far available exploitation of free environmental goods.

Another example at a macroeconomic level could be represented by a regime shift like the transition process to a market economy from a previously socialist system. Even in such a case, enterprises (and all economic agents) find their environment fundamentally changed: think, for example, of the effects on technology and organization imposed by the introduction of "hard budget constraints".

12.4 Memory, Confidence and Psychological Addiction

> ...it is possible to live almost without memory, and to live happily moreover (...); but it is altogether impossible to live at all without forgetting [7]

Memory is essential for a successful learning process. Under certain circumstances, even an apparently negligible memory-storage capacity can greatly enhance the learner's accomplishments. One of the roles played by institutions is precisely that of making memory resources available to the members of a society. Through the survival of the fittest, we expect superior characteristics to impose themselves in a population. Similarly, we won't be surprised if some evolutionary mechanism leads to the widespread adoption of an institution that formerly served only a minor subset of the society.

A psychological element that plays a role in the dynamics of technology and, more generally, institution adoption is *confidence*. Confidence is partially built on memory and can work as a reinforcement mechanism for the learning process. A conservative attitude based on confidence in past experience, in a logical thought construction or in a classification according to familiar categories will typically improve the quality of the learner's inductive process. Confidence will lead the learner to concentrate (perhaps unconsciously) on what is learnable either by filtering out patternless noise or by isolating the learner from patterns whose degree of complexity exceeds his own absolute capability. [8]

However, when facing an abrupt change in the environment, confidence in the "old ways" will hamper the learning process and the adaptation of individual actors to the new *status quo*. The macroscopic effect is that social learning will be prevented from being as quick as it could have been otherwise. In a somewhat colourful way we indicate as *psychological addiction* the tendency to ignore the possibility of a major change in the surroundings for no other reason than memory of the previous successes. We introduce the term in order to distinguish it–at least logically–from another sort of obstacle imposed by an institutional arrangement, which we may define *structural sclerosis*.

---

7 Friedrich Nietzsche, "On the Uses and Disadvantages of History for Life" Untimely Meditations, trans. by R. J. Hollingdale, Cambridge University Press, Cambridge UK, 1983, p. 62.

8 It may seem that the difference can be perceived only from the vantage point of an outside observer whose complexity encompasses those "very difficult" patterns. From the learner's perspective there is only patternless noise or patterns that totally trascend him and are hence utterly unexploitable. A somewhat less naive approach would be to realize that, during a genuine induction process, actual patterns are "too complex" only temporarily. Eventually, computable codes are to be disclosed.

The following simulations tend to substantiate these statements. Of course, the examples employed below are very simple; however, if interesting phenomena can be observed at this very low level of complexity, we expect to find similar features in more realistic and explicitly relevant economic settings. With an equivalent attitude, other researchers look for chaotic dynamics in nonlinear first-order difference equations. They do not really believe that those equations are a good description of reality, rather, they want to suggest how complex and rich reality can be.

We will now present two models based on the XOR problem described above. The simulations performed tend to substantiate the positive relation between memory-confidence and the learning ability of the actor. In both cases the approach followed is evolutionary so that we will consider a population of networks. Each individual deals with the same problem and observes, one after the other, the same sequence of instances. These learners, however, are not completely identical so that after a predetermined number of examples (a training session) only the fittest ones (in a sense to be made precise) will get to reproduce themselves.

12.4.1 Building up Memory

We consider here a population of 50 single-layer nets with two input neurodes and one output neurode. The initial weights and threshold for each "individual" are picked randomly from the set of integers {-5, -4,..., 5}. Each net employs the pocket algorithm as learning procedure. A gene (*msup*) composed of four bits determines the maximum memory capacity available to store the number of successes of the pocket set of weights. Hence, $msup \in \{0, 1, ..., 15\}$. However, at the beginning of each simulation, *msup* is set to 0 for every net. This implies that Wpoc is necessarily identical to W and that the actual learning procedure is simple perceptron.

A training session consists of 30 examples randomly drawn from the instance set [{A, B}, {A, notB}{notA, B}, {notA, notB}]. Each run of the simulation consists of 20 training sessions and we performed 50 such runs. After each training session the best 10 individuals are picked to produce the following "generation". The fitness index is simply the number of correct classifications that the set of pocket weights *Wpoc* scores on the sequence of examples presented in the latest training session. The 10 best individuals pass unchanged to the next generation; 39 are generated from those 10 thanks to a rudimentary cross-over procedure. In particular, two nets are randomly chosen from the selected subset and they generate two descendants. One will have the weights of the first parent and the other will assume those of the second parent. As for the memory capacity, it will be the result of a cross-over process affecting the two parents' four-bit *msup* gene. The 50th individual of the new generation will receive random integer weights and a memory capacity selected randomly ($0 < msup < 15$). This, in order roughly to represent mutation. At the end of each run we recorded the average *msup* of the population.

Given the nature of the problem (XOR simulation) we expect the "institutional arrangement" *pocket algorithm* to be widely adopted by a population. This implies, for the particular model considered here, that the population average value of *msup* should drift away from the initial value 0. Furthermore, we would like to record a tendency towards relatively high average values. The intuition behind this latter point is that a long memory will protect a net from unfortunate strings composed of the only example the net cannot classify correctly. Such strings would eventually force the net to modify the already attained optimal Wpoc. Figure 12.5 summarizes the results of this simulation.

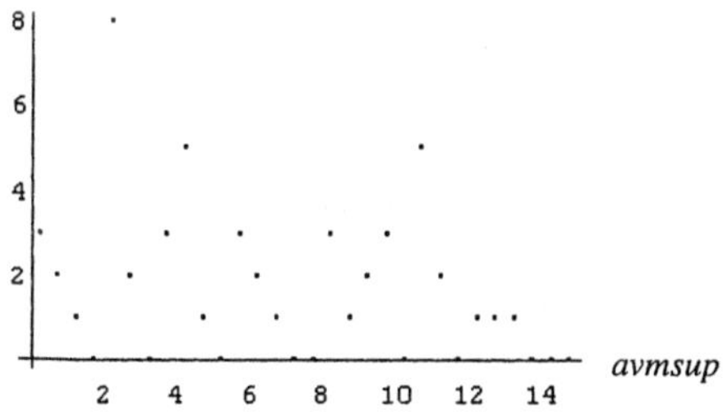

Fig. 12.5. Average *msup* distribution in 50 experiments

The horizontal axis records the average *msup*, whereas the vertical axis gives the "absolute frequency". For example, in 8 cases (out of 50 [9]) the average *msup* was between 2 and 2.5. It is evident that the tendency is towards the exploitation of memory: only 3 cases are recorded where the process seems to select a memoryless status as optimal. On the other hand, the second expected feature (the more memory the better) is not unequivocally evidenced by the simulation. We believe that the result is due in part to the relatively short number of examples in each training session (so that long "unfortunate strings" are very unlikely). This is a point we would like to investigate further when more powerful computational resources are available.

12.4.2 Memory vs Oblivion

In the second scenario we consider, individuals endowed with memory (that is, characterized by the pocket algorithm and no constraint on memory length)

[9] We realize that a much larger number of experiments would be needed to induce some regularity (if any exists) of the theoretical distribution. However, the evidence is, we believe, sufficient for our purpose.

are directly confronted with actors that cannot remember (the learning procedure is simple perceptron for these nets). XOR is still the problem this population of heterogeneous individuals is confronted with. Also in this case the aim is to show that an institution (the learning routine "pocket algorithm") by increasing the memory capacity of the learner improves his chance of success. Hence, the use of such an institution via some selection process will tend to become widespread.

There are 100 one-layer nets in the population and a random number of them follows the pocket algorithm while the rest adopts simple perceptron learning. No limitation is imposed on the storage capacity of those employing the pocket algorithm. After a training session, the best 10 individuals are selected and these will produce the new generation pretty much as it was done in the previous model. The difference is that now the gene "memory" consists of a single bit so that only two states are possible: WITH memory as opposed to WITHOUT memory. An offspring of two WITH-memory parents will necessarily follow the routine that exploits Memory (with no constraint on its length). As for the weights, it will inherit the configuration of one of the parents. However, memory related to the past successes of the inherited $Wpoc$ will be set to 0. We want in this way to capture the idea that each individual will build its own level of *confidence* in the adopted rules of behaviour. [10] Also in this series of simulations, mutation is allowed in the form of the arrival of one single new individual which, *ex ante*, has an equivalent probability of following either one of the learning procedures.

We expect that a population will eventually show universal adherence to the superior learning rule. The next picture summarizes the simulation results.

The evolution of a population [11] followed along 10 successive generations and the experiment is performed 10 times. The horizontal axis records the generation and on the vertical one we can read the proportion of the population WITH memory. Each number on the graph tells how many populations presented a particular proportion of WITH memory individuals (i.e. the proportion of actors relying on that particular institution) at a given generation. For example, on generation 6, eight populations were completely composed of WITH memory individuals, for one population about 95% of all learners employed the pocket algorithm and one population was entirely composed of memoryless actors.

[10] An interesting variation on the theme would see the offspring inherit the parent's confidence in the routines along with the routines themselves. This may better represent a "cultural bias" imposed by a society to its members. Birner [1], in his analysis of Hayek methodology and Social Philosophy, stresses that social evolution does not rely on the selection of physical characteristics, but on the imitation of successful behaviour. This would be the ratio for such an experiment.

[11] The population is composed of randomly selected individuals who, with 50% probability, adopt the pocket algorithm i.e. are WITH-memory actors.

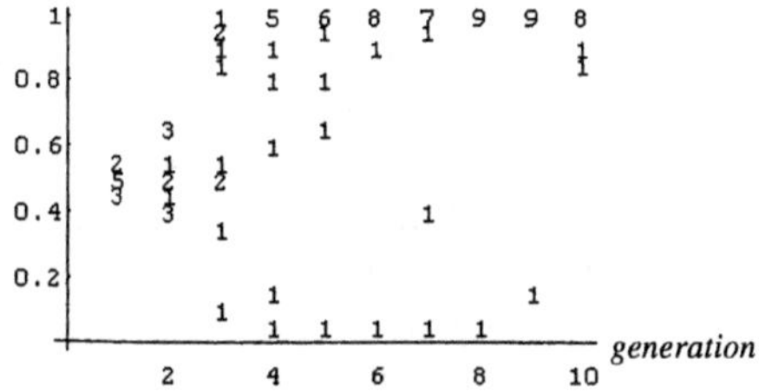

Fig. 12.6. Evolution of the distribution of WITH-memory individuals along 10 generations (10 different runs)

Clearly the expectation is confirmed. The whole population adopting the *pocket algorithm* institution is an "attractor" of the system's evolution. Even more interesting, we can see that such state is not *absorbing.* [12] This implies that even when the population is totally composed of WITH memory individuals, there is some chance a memoryless mutant will be sufficiently successful to impose, at least locally and temporarily, its rules of behavior.

Notice that were we to prevent mutation [13] in our simple stylized model, we would obtain strong path-dependent dynamics and frequent suboptimal lock-in phenomena. By looking at the picture above, it is clear that at least one population would have been forced to the "inferior" institutional configuration. An early advantage, even if caused by mere chance, is all it takes to sway the system from the *ex ante* feasible optimum.

12.5 Physical Effectiveness and Structural Sclerosis

In this section we perform an experiment very similar to the previous one; however, the institution is here represented by a more articulated physical structure of the individual learning networks. In particular, two groups of individuals are confronted: single layer nets and nets constructed as in Fig. 12.4. Obviously, the weights and thresholds employed are not those of the picture. They are chosen randomly and, by using the well known back-propagation learning algorithm, they gradually approach a configuration that correctly classify all inputs. Since the weights and thresholds can now assume values in the domain of the computable reals, we need to specify when an output

[12] We use such terms as *attractor* and *absorbing* in a rather intuitive way. We believe, however, that such notions could be rigorously referred to the underlying dynamic system

[13] We will adopt such strategy below when dealing with social learning to try to evaluate the objective obstacle imposed by a particular institutional setting to the population as a whole.

(that will $\in (-1, 1)$) is to be considered correct. We have decided that an output greater than .5 is to be taken as equivalent to 1 and an output smaller than -.5 is equivalent to -1. [14]

It is important to stress that the *distinguo* we make between an institution as a particular procedure on the one hand and as a *physical* structure on the other is quite artificial. Actually, even this second case can be reduced to the first one if we realize that a framework is nothing but a procedure of a higher hierarchical level. [15] However, we will keep on using the term physical to identify what is macroscopically structural: translated into some tangible framework (physical capital for example), or recorded into the organizational chart of an enterprise. In any case, there is a clear interdependence between the "physical" structure and the lower level routines that can be implemented thanks to such a structure.

Because of higher computational requirements, the population in these simulations is reduced to 50 individuals, whereas the slower pace of the learning algorithm has forced us to increase to 30 the cardinality of each training session. The results of the simulations are summarized in the following picture:

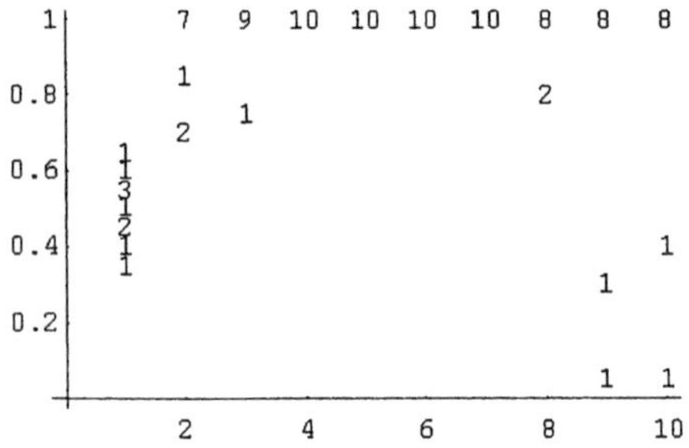

Fig. 12.7. Evolution of the proportion (vertical axis) of "figure 4" individuals in the first 10 generations(horizontal axis) of a population

With respect to the previous simulation, we observe that the adoption of the "winning" institution is much faster; however, episodes of wide and abrupt deviations are remarkable and already hint at the fragility of the institution due to its inherent rigidity.

---

[14] In other words we require a 75% level of confidence.

[15] We can refer to a Universal Turing Machine. Such an idealized computer can emulate any other TM on any particular input simply by adding–to the input–a prefix to identify the TM to be emulated. In this sense a "physical structure" (the TM) can be reduced to a meta-procedure (the prefix).

12.6 Psychological Addiction and Innovation

So far, we have considered a learning process in a stable environment. The "data generator process" has always been XOR in nature and we have substantiated the rather intuitive observation that memory, fostering confidence and a conservative attitude in the learner, improves its possibilities. An institution that plays a memory-preserving function in the context of a complex environment can, hence, be expected to expand.

We now turn to the second part of our research. We will try to show that, in case of a shock to the environment, the very presence of the institution that best dealt with the complexity of the original scenario will now hamper the learning and the adaptation process (hence the survival) of the actor relying on it.

The model employed for these simulations is similar to the one adopted in the previous exercise. There is a population composed of 100 individuals; a random proportion of them rely on the particular institution we called pocket algorithm. A generation can still be identified with a training session. The "natural" selection and construction of the following generation is precisely as before. However, after 10 generations, we introduce an innovation, a major shock to the environment.

In particular, the Boolean function to be learned–expressed in Disjunctive Normal Form–becomes the following:

$$f(A,\ B)\ =\ A \vee (\overline{A} \wedge \overline{B})$$

This function projected onto the instance set can be represented as in Fig.12.8:

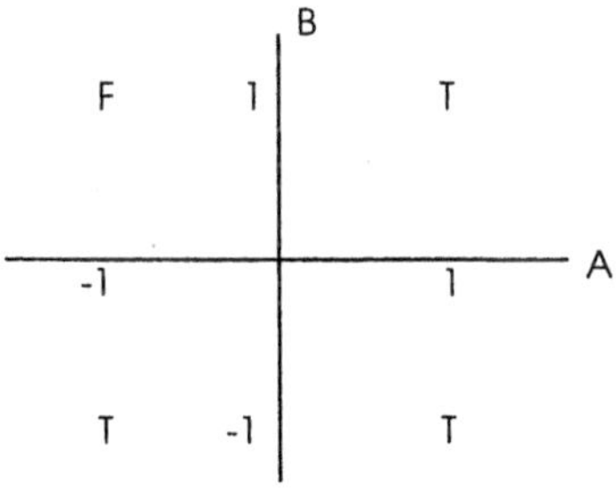

Fig. 12.8. Instance space for the Boolean function $f(A,\ B)\ =\ A \vee (\overline{A} \wedge \overline{B})$

It should be stressed that the classification problem is now linearly separable. A single-layer net is hence structurally capable of learning to simulate correctly such a function. Furthermore, we can rely on the Perceptron Convergence theorem (originally obtained by [9] to identify an upper bound for the number of steps necessary to accomplish the complete learning. In particular, given that the net configured with the following set of integer weights and threshold W* = {1, 1, -1} [16] can perfectly simulate f(A, B), the theorem states that a net initialized on W0 = {0, 0, 0}–and faced with a sequence of randomly chosen examples–will reach W* in at most (p + 1) $\|W^*\|^2$ updating steps. Where p is the number of variable inputs (i.e. A, and B). In the case of interest, hence, it will take at most (3)*3 = 9 updating steps to reach the wanted configuration.

Clearly this is an upper bound. In fact, notice that a net initialized on W0 will reach the wanted configuration after a single updating if the first example examined is *inst* = {1, 1, -1} [17] with required output *want* = {1}. However, we have chosen not to initialize a new net to the zero vector; this was done to represent heterogeneity among new comers and to increase the scope of the combinatorial optimization implicitly performed by the system. From an economic point of view, the initial random set of weights can be interpreted as the description of the natural talent and/or acquired skills of an entrepreneur entering a particular business. It is likely that these characteristics vary across individuals. We will see that this hypothesis has an interesting consequence on the simulations' results.

To render the results somewhat more macroscopic, after the innovation shock we have forced all the newcomers (one out of 100 in every generation) to be WITHOUT memory. As usual, the simulation results are summarized in the graph below. Each training session after the shock is composed of only 5 instances. In other words, we have chosen to apply the selection more frequently to model the turmoil that is likely to characterize an economic system after the introduction of a major innovation.

It is evident that 100% WITH memory individuals is no longer an attractor for the population dynamics after the shock has hit at time 11. Let us stress that

(a) most times the drifting away of the population from such a state happens in the early periods;
(b) the population is not pushed necessarily towards a 0% WITH-memory-individuals state;
(c) the very drifting is not at all a necessary event.

As for point 1, it confirms our initial hypothesis that it is the confidence in some experimented rule of behaviour that prevents swift adaptation to a new environment. However, as evidence against the old procedures accumulates and overcomes the psychological addiction to them, new routines are adopted.

[16] Where $w_0 = 1, \quad w_A = 1, \quad w_B = -1$. It is a widely used "trick" in artificial neural network to consider the threshold as the weight related to an extra variable which takes on a constant value 1.

[17] The first input is related to the threshold modification process as explained in the previous footnote.

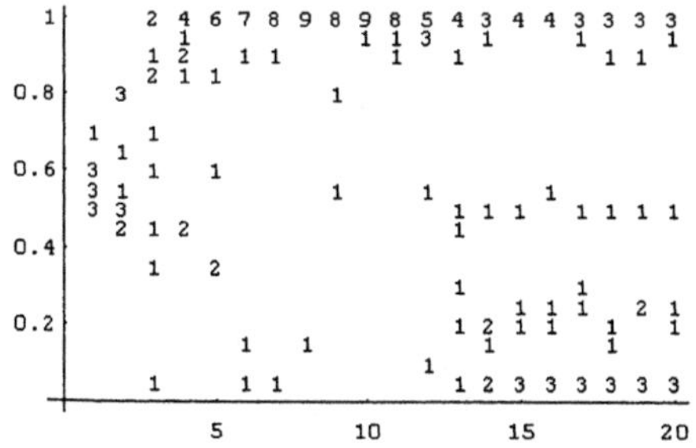

Fig. 12.9. Evolution of the distribution of WITH-memory individuals before and after a major shock to the environment

Individuals lucky enough to have survived the early "identity crisis" are now fairly equipped to compete with new comers. At this point, it is merely a matter of chance if a memoryless opponent imposes its way (through a lucky sequence of selections) to a large part of the population.

This also accounts for observation 2. Furthermore, it must be remembered that due to the particular mating process adopted, it will be the case that a WITH memory offspring may inherit from the other parent a "winning" set of weights, bypassing the identity crisis described above. Finally, and related to point 3, it can happen that no new comer will make it to the "top ten" in the early after-shock periods because of the new-arrival construction mechanism employed. [18] In this way, the learning and selection undergone by the old population will make its due course, so that later arrivals can count only on chance (rather than on an inherent inferiority of the incumbents) to impose themselves.

12.7 Structural Sclerosis and Innovation

We now proceed to replicate the previous experiment with respect to the characterization of the institution in terms of a more sophisticated physical structure. Also in this case we introduce an innovation [19] after the "superior" institution has had time to impose itself. Once more, the aim is to show that the individual relying on this previously successful structure will be disadvantaged if he/she does not abandon it in favour of a more flexible organizational or productive configuration.

---

[18] That is randomly picked weights in the integer interval {-5, 5}.

[19] As before, the innovation is characterized as the Boolean function $f(A, B) = A \vee (\overline{A} \wedge \overline{B})$.

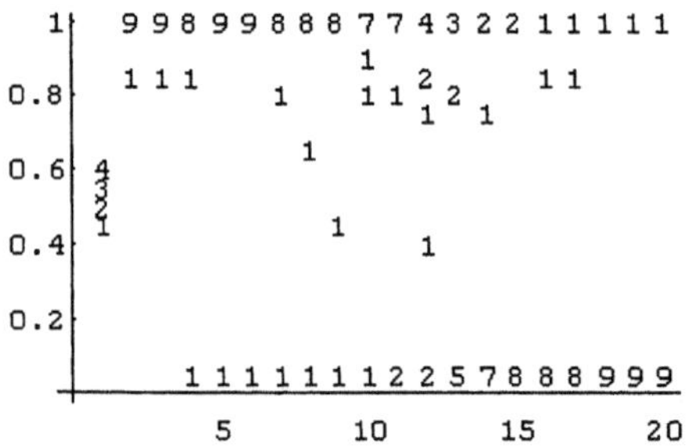

Fig. 12.10. Evolution of the proportion of figure 4 individuals before and after a major shock to the environment at t=11

The effect of the innovation as depicted in Fig. 12.10 is here more drastic than in the previous case. A structural rigidity seems more difficult to overcome than some sort of habit persistence.

12.8 Social Learning

To conclude, we ask whether, despite the selection process determining the evolution of a population, the inability to generate different institutional arrangements can hinder the social learning, making it slower than it could have been otherwise. Even in this case we tackle the issue from the perspective of psychological addiction and structural sclerosis.

In Sects. 12.6 and 12.7 we have seen that, after the shock has hit, individual members relying on once dominated institutions may be so successful as to spread among the population the adoption of those same institutions. We now posit the question whether the population as a whole gains in the process. In other words, is social learning made quicker by the simultaneous and competing adoption of different institutions? Once more the question is asked for both characterizations of the institution, that is, in terms of a procedure (pocket algorithm) and a physical structure (individuals built as in Fig. 12.4.

To try to answer the question, we will employ a slightly modified version of the previous model. In particular, dealing with an artificial reality completely under control, we can follow explicitly a counterfactual argument. After the shock, we can imagine two different populations evolving from exactly the same initial condition; mutation will be allowed in only one of the two cases. In this way, we can record the evolution of each population to evaluate precisely the effect of confidence and psychological addiction on social learning. We want to see which population as a whole learns more quickly to deal correctly with the new environment. The configurations adopted by various individuals may be different, but what interests us now is to see how long it takes for the entire population to learn $f(A, B)$.

The first picture below records the simulations' results for the institution-as-memory case. We have chosen to relate an index of confidence (at the time the shock hits) to the number of generations necessary for the population at large to adapt to the new environment. The index of confidence is the population average memory of successes obtained by the pocket weights' configuration; this is recorded on the horizontal axis. On the vertical axis we find the number of generations (i.e. training sessions) necessary to perfect the learning process. Two characters are employed to distinguish the results of the two populations. A "0" means that the population which does not count on mutation [20] has reached complete adaptation, whereas a "+" denotes the completion of the learning process by the population enriched by new arrivals. We clearly expect this latter population to be characterized by faster social learning.

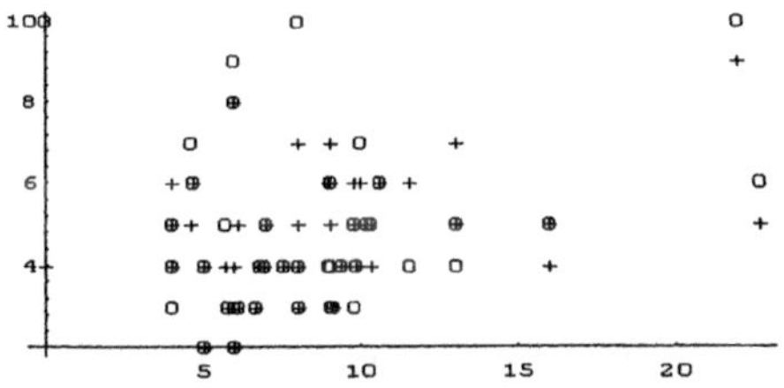

Fig. 12.11. Number of generations necessary to achieve complete social learning as a function of the confidence level (horizontal axis) in the old institution. A population with mutants is indicated by "+"; a population without memory with "0"

Once more the results do not appear to be clear cut. Actually, in a significant number of cases, it is the population deprived of new arrivals to reach complete adaptation earlier than the "more flexible" one. However, a more careful analysis shows that it is for low levels of the confidence index that "mutation free" population wins the contest. On the contrary, when confidence becomes psychological addiction, institutional flexibility, i.e. being able to quickly forget and to experiment new ways, turns out to be the winning social strategy. This confirms our hypothesis.

[20] Given that the innovation is introduced after 10 periods, the population will typically be composed entirely of WITH memory individuals. Hence, the "0" represents the completion of the process for the old institutional arrangement (going through the above mentioned crisis of identity).

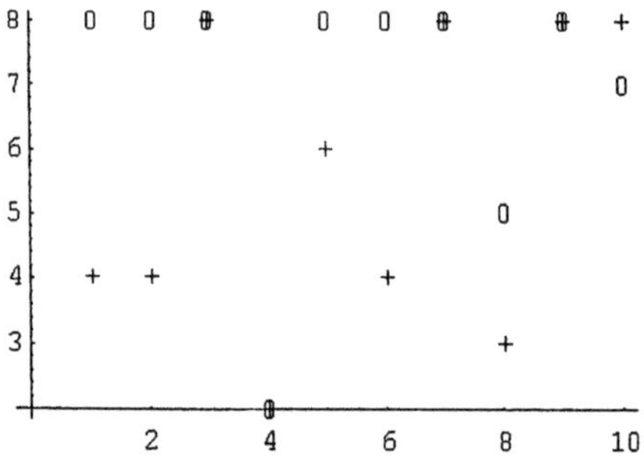

Fig. 12.12. Number of generations (vertical axis) necessary to complete social learning starting from a population of figure-4 individuals. Ten experiments (horizontal axis).Comparing the performance of "mutant" populations (+) with "mutant-free" ones (0)

The fact that for low levels of the confidence index the population relying only on the old institutional arrangement can be more successful is mainly due to the mutation process adopted. Of course a memoryless individual is introduced, but its randomly chosen weights could turn out to be far from optimal. That is so even though, in a particular training session–that could have been composed of only part of the total instance set–, the net has recorded a sufficient number of successes so that it is included among the top ten and has spread its offspring.

Our simple model already captures an interesting reality feature. There exists a trade off between, on the one hand, being new and flexible, but without any know-how that can be absorbed only with a learning by doing procedure, and, on the other hand, being experienced with wide knowledge of various aspects of the long time pursued economic activity, but crystallized on, by now, obsolete productive or organizational routines.

The next picture summarizes the results of the analogous experiment where the institution is characterized in physical terms by Fig.12.4 individuals.

In this case, the sequence of simulations is recorded on the horizontal axis while the vertical axis records the number of periods necessary to complete the "social learning" for the population deprived of new arrivals "0" and for the one where mutation is allowed "+" as in the previous picture. The conclusion is clearer in this case: only once the population deprived of new arrivals reaches a complete social learning faster than the "twin" population which is however enriched by the presence of simpler and more flexible institutional structures.

12.9 Conclusions

The framework of analysis employed in this paper is certainly simple and, at the same time, offers a relatively wide array of suggestions on how an institution may impose itself. Our simulations have shown that the evolution of an economic or social system is dialectically linked to the evolution of its institutional organization. By describing economic actors in structural terms–a well defined neural network in the present case–, rather than in terms of a notion of substantive rationality, economic theory can attain a double goal. On the one hand, it will avoid employing representative agents that know a good deal more than the economist or econometrician himself (as the economic agent often solves problems that are unsolvable according to computability theory, he/she clearly employs a notion of effective calculability that transcends Church's Thesis; such a notion does not belong to our present scientific knowledge). On the other hand, it will offer a natural way to study the endogenous emergence and the role of institutions. In a slightly modified version of the model presented here, for example, we have described how institutions emerge and find out that they are suboptimal, but feasible: a substantial step towards a more realistic description of the world.

References

1. Birner J. (1995) Models of Mind and Society: the Place of Cognitive Psychology in the Economics, Methodology and Social Philosophy of F.A. Hayek. University of Trento, mimeo
2. Fitoussi J. P., Velupillai K. (1993) Macroeconomic Perspectives. In: Barkai F., Fischer S., Liviatan N. (Eds.) Monetary Theory and Thought. Mc Millan, London
3. Gallant S. I. (1993) Neural Networks Learning and Expert Systems. MIT Press, Cambridge Massachusetts
4. Gold M. E. (1965) Limiting Recursion. The Journal of Symbolic Logic **30(1)**, 28–48
5. Gold M. E. (1967) Language Identification in the Limit. Information and Control **10**, 447–474
6. Luna F. (1993) Learning in a Computable Setting. UCLA, mimeo
7. Luna F. (1997) Learning in a Computable Setting: Applications of Gold's Inductive Model. In: H. Amman et al. (Eds.) Computational Approaches to Economic Problems. Amsterdam: Kluwer Academic Publishers
8. Luna F. (2000) The Emergence of a Firm as a Complex-Problem Solver. Taiwan Journal of Political Economy, forthcoming
9. Minsky M., Papert S. (1969) Perceptrons: An Introduction to Computational Geometry. MIT Press, Cambridge, Massachusetts
10. Nelson R. R. (1994) Economic Growth via the Coevolution of Technology and Institutions. In: Leydesdorff L., Van den Besselaar P. (Eds.) Evolutionary Economics and Chaos Theory. Pinter Publishers, London
11. Rustem B., Velupillai K. (1990) Rationality, Computability, and Control. Journal of Economic Dynamics and Control **14**, 419–432

12. Simon H. A. (1978), Rationality as Process and as Product of Thought. American Economic Association Papers and Proceedings **68**, 2, 1-16
13. Spear S. E. (1989) Learning Rational Expectations Under Computability Constraints. Econometrica **57(4)**, 889–910
14. Velupillai K. (2000) Computable Economics: The Fourth Arne Ryde Lectures. Oxford University Press, Oxford
15. Zambelli S. (1994) Computability and Economic Applications. The Research Program Computable Economics. Working Paper, Department of Economics, University of Aalborg, Denmark

13 On Two Types of GA-Learning

Nicolaas J. Vriend

Queen Mary, University of London, UK
n.vriend@qmw.ac.uk

Abstract. We distinguish two types of learning with a Genetic Algorithm. A population learning Genetic Algorithm (or pure GA), and an individual learning Genetic Algorithm (basically a GA combined with a Classifier System). The difference between these two types of GA is often neglected, but we show that for a broad class of problems this difference is essential as it may lead to widely differing performances. The underlying cause for this is a so called spite effect.

13.1 Introduction

One dimension in which we can distinguish types of Genetic Algorithms is the level at which learning is modeled. A first possibility is as a model of *population learning*, which might be called a pure Genetic Algorithm. There is a population of rules, with each rule specifying some action. The fitness of the rules is determined by all members of the population executing their specified action, and observing the thus generated feedback from the environment. On the basis of these performances, the population of rules is modified by applying some genetic operators . The second way to implement a GA is to use it as a model of *individual learning* , which is basically a Classifier System with on top of it a Genetic Algorithm.

The difference between these two types of GA is often neglected, but we show that for a broad class of problems this difference is essential. [1] This difference is due to the fact that when a GA is learning, there are actually two processes going on. On the one hand, when the rules are executed they interact with each other in the environment, generating the outcomes that determine the fitness of each rule. On the other hand, given the thus generated fitnesses of the rules, there is the learning process as such. As we will make precise below, it is the way in which these processes interact with each other that causes the essential difference between the two versions of a GA.

The phenomenon causing this essential difference between an individual and a GA is called the "spite effect". The spite effect occurs when choosing an action that hurts oneself but others even more. [2] In order to introduce the

[1] A more detailed analysis of the wider issue of the essential difference between individual and social learning and the consequence for computational analyses in a more general sense can be found in [12].

[2] The term "spiteful behavior" goes back, at least, to [3]. It has been examined in the economics (see, e.g., [9]), experimental economics (see, e.g., [2], or [4]), and evolutionary game theory literature (see, e.g., [6], [7], and [11]).

essence of the spite effect, consider the bimatrix game in Fig. 13.1 (see [6]), where T and B are the two possible strategies, and a, b, c, and d are the payoffs to the row and column player, with $a > b > c > d$. Clearly, (T, T) is the only Nash equilibrium since no player can improve by deviating from it, and this is the only combination for which this holds. Now, consider the strategy pair (B, T), leading to the payoffs (b, c). Remember that $a > b > c > d$. Hence, by deviating from the Nash equilibrium, the row player hurts herself, but she hurts the column player even more. We could also look at it from the other side. Suppose both players are currently playing strategy B, when the column player, for one reason or another, deviates to play strategy T instead, thereby improving his payoff from d to c. But the row player, simply sticking to her strategy B, would be "free riding" from the same payoff of d to a payoff c that is even higher than b. The question, then, addressed in this paper is how this spite effect influences the outcomes of an individual learning GA and a population learning GA.

	T	B
T	a, a	c, b
B	b, c	d, d

Fig. 13.1. Bimatrix game, with payoffs $a > b > c > d$

The remainder of this paper is organized as follows. In Section 13.2 we present an example illustrating this difference, which we will analyze in Section 13.3 in relation to the spite effect. Section 13.4 draws our example into a broader perspective by discussing some specific features of the example, and presents some conclusions.

13.2 An Example

Consider a standard Cournot oligopoly game. There is a number n of firms producing the same homogeneous commodity, who compete all in the same market. The only decision variable for firm i is the quantity q_i to be produced. Once production has taken place, for all firms simultaneously, the firms bring their output to the market, where the market price P is determined through the confrontation of market demand and supply. Let us assume that the inverse demand function is $P(Q) = a + bQ^c$, where $Q = \sum q_i$. Making the appropriate assumptions on the parameters a, b, and c ensures that this is a downward-sloping curve, as sketched in Fig. 13.2. Hence, the more of the commodity is supplied to the market, the lower the resulting market price P will be. We assume that the production costs are such that there are negative fixed costs K, whereas the marginal costs are k. Imagine that some firms happen to have found a well where water emerges at no cost, but each bottle costs k, and each firm gets a lump-sum subsidy from the local town council if it operates a well. Given the assumptions on costs, each firm might be willing to produce any quantity at a price greater or equal to k. But it prefers to produce the output that maximizes its profits. The parameters for the underlying economic model can be found in the appendix.

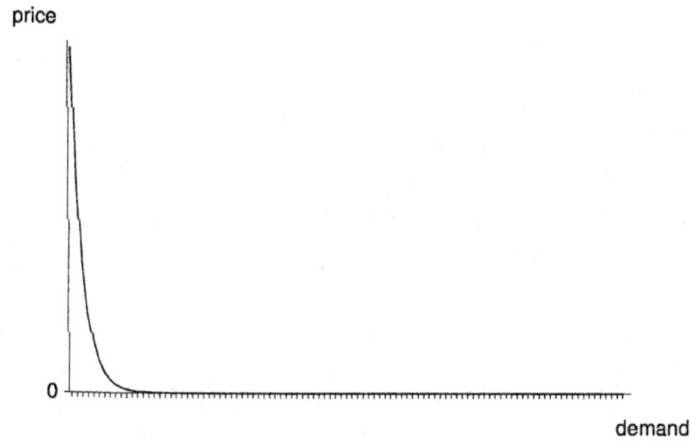

Fig. 13.2. Sketch demand function

Assume that each individual firm (there are 40 firms in our implementation) does not know what the optimal output level is, and that instead it needs to learn which output level would be good. Then, there are two basic ways to implement a GA. The first is as a model of *population learning.* Each individual firm in the population is characterized by an output rule, which is, e.g., a binary string of fixed length, specifying simply the firm's production level. In each trading day, every firm produces a quantity as determined by its output rule, the market price is determined, and the firms' profits are determined. After every 100 trading days, the population of output rules is modified by applying some reproduction, crossover, and mutation operators.[3] The underlying idea is that firms look around, and tend to imitate, and re-combine ideas of other firms that appeared to be successful. The more

[3] In each of the 100 periods between this, a firm adheres to the same output rule. This is done to match the individual learning GA (see below), and in particular its speed, as closely as possible.

successful these rules were, the more likely they are to be selected for this process of imitation and re-combination, where the measure of success is simply the profits generated by each rule. Figure 13.3 shows both the Cournot market process, and the population learning process with the GA.

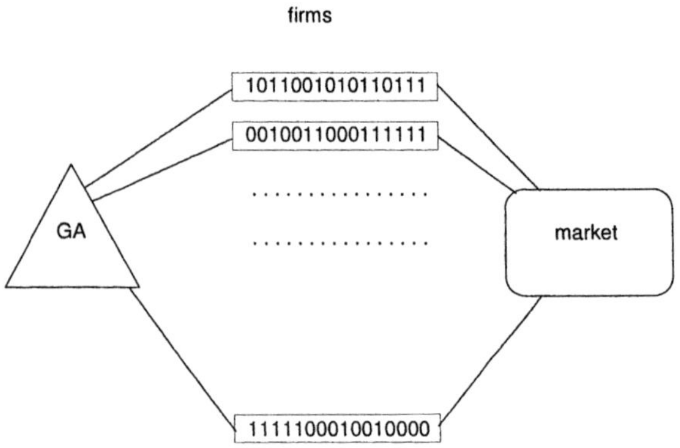

Fig. 13.3. Social learning GA

The second way to implement a GA is to use it as a model of *individual learning*. Instead of being characterized by a single output rule, each individual firm now has a set of rules in mind, where each rule is again modeled as a string, with attached to each rule a fitness measure of its strength or success, i.e., the profits generated by that rule when it was activated. Each period only one of these rules is used to determine its output level actually supplied to the market; the rules that had been more successful recently being more likely to be chosen. On top of this Classifier System, the GA, then, is used every 100 periods to modify the set of rules an individual firm has in mind in exactly the same way as it was applied to the set of rules present in the population of firms above. Hence, instead of looking how well other firms with different rules were doing, a firm now checks how well it had been doing in the past when it used these rules itself. [4] Figure 13.4 shows the underlying economic market process, and the individual learning process. [5]

Figure 13.5 presents the time series of the output levels for a representative run of each algorithm. As we see, they approach a different level. Whereas both series start around 1000, the population learning GA quickly 'converges'

[4] Hence, an alternative way to obtain the match in the speed of learning of the individual and the social learning GA would have been to endow the individual learning GA with the capability to reason about the payoff consequences for every possible output level in its set, updating all strengths every period.

[5] The parameter specification of the GA can be found in the appendix, and the pseudo-code in [12].

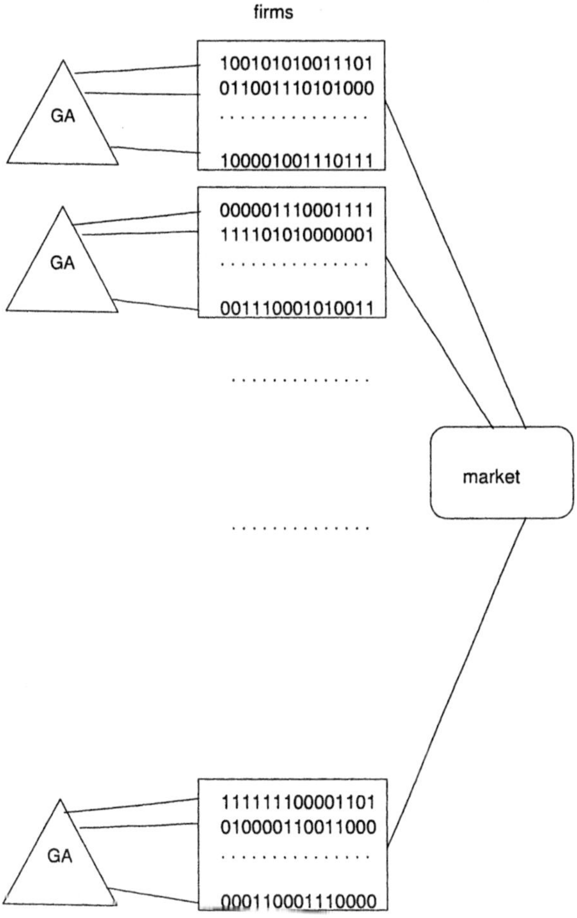

Fig. 13.4. Individual learning GA

to a level of 2000, but the individual learning GA keeps moving around a level just below 1000. [6] We want to stress that these data are generated by exactly the same identical GA for exactly the same identical underlying economic model.

[6] The 5,000 periods here presented combined with the GA rate of 100 imply that the GA has generated 50 times a new generation in each run. Each single observation in a given run is the average output level for that generation. We did all runs for at least 10,000 and some up to 250,000 periods, but this did not add new developments.

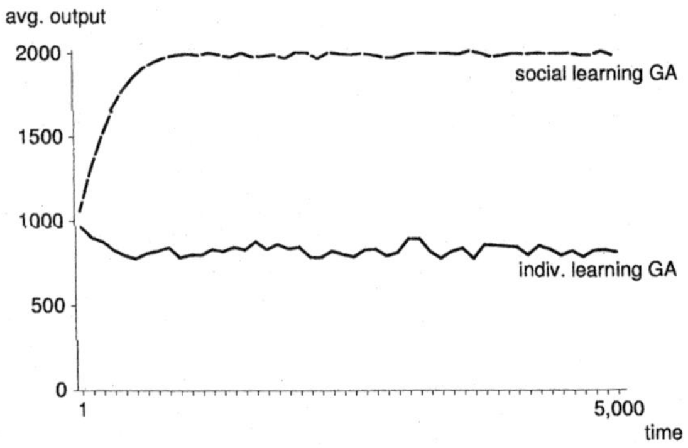

Fig. 13.5. Output levels individual learning GA and social learning GA

13.3 Analysis

We first present two equilibria of the static Cournot oligopoly game specified above for the case in which the players have complete information. The GAs do not use this information, but the equilibria serve as a theoretical benchmark that helps us understanding what is going on in the GAs. [7]

If the firms behave as price-takers in a competitive market, they simply produce up to the point where their marginal costs are equal to the market price P. Given the specification of the oligopoly model above, this implies an aggregate output level of $Q^w = 80,242.1$, and in case of symmetry, an individual Walrasian output level of $\frac{Q^w}{n} = 2,006.1$. If, instead, the firms realize that they influence the market price through their own output, still believing that their choice of q does not directly affect the output choices of the other firms, they produce up to the point where their marginal costs are equal to their marginal revenue. This leads to an aggregate Cournot-Nash equilibrium output of $Q^N = 39,928.1$, and with symmetry to an individual Cournot-Nash output of $\frac{Q^N}{n} = 998.2$.

As we see in Fig. 13.5, the GA with individual learning moves close to the Cournot-Nash output level, whereas the GA with population learning 'converges' to the competitive Walrasian output level. The explanation for this is the spite effect.

In order to give the intuition behind the spite effect in this Cournot game, let us consider a simplified version of a Cournot duopoly in which the inverse demand function is $P = a + bQ$, and in which both fixed and marginal costs

[7] The formal derivation of the two equilibria can be found in [12].

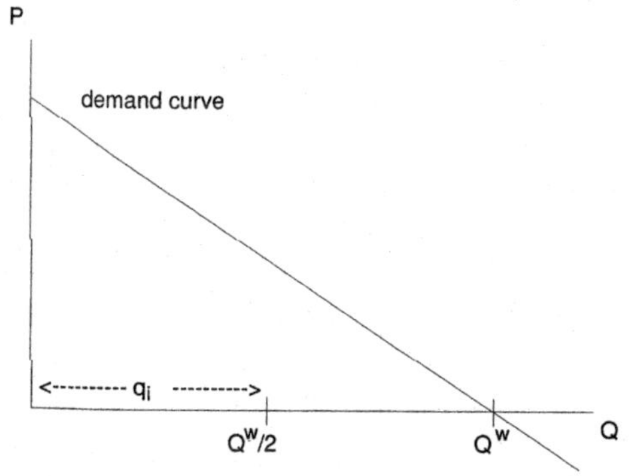

Fig. 13.6. Example simple Cournot duopoly

are zero (see [8]). The Walrasian equilibrium is then $Q^W = -a/b$, as indicated in Fig. 13.6. Suppose firm i produces its equal share of the Walrasian output: $q_i = Q^W/2$. If firm j would do the same, aggregate output is Q^W, the market price P will be zero, and both make a zero profit. What happens when firm j produces more than $Q^W/2$? The price P will become negative, and both firms will make losses. But it is firm i that makes less losses, because it has a lower output level sold at the same market price P. What happens instead if firm j produces less than $Q^W/2$? The price P will be positive, and hence this will increase firm j's profits. But again it is firm i that makes a greater profit, because it has a higher output level sold at the same market price P. In some sense, firm i is free riding on firm j's production restraint. Hence, in this Cournot duopoly the firm that produces its equal share of Q^W will always have the highest profits.

How do these payoff consequences due to the spite effect explain the difference in the results generated by the two GAs? As we saw above, the spite effect is a feature of the underlying Cournot model, and is independent from the type of learning applied. The question, then, is how this spite effect is going to influence what the firms learn. It turns out, this depends on how the firms learn.

In the population learning GA, each firm is characterized by its own production rule (see Fig. 13.3). The higher a firm's profits, the more likely is its production rule to be selected for reproduction. Due to the spite effect, whenever aggregate output is below Walras, this happen to be those firms that produce at the higher output levels. And whenever aggregate output is above Walras, the firms producing at the lowest output are most likely to be selected for reproduction. As a result, the population of firms tends to converge to the Walrasian output.[8]

---

[8] This is not due to the specifics of this simple example, but it is true with great generality in Cournot games (see [11], and [7]). In particular, it also holds when

In the individual learning GA, however, the production rules that compete with each other in the learning process do not interact with each other in the same Cournot market, because in any given period, an individual firm actually applies only one of its production rules (see Fig. 13.4). Hence, the spite effect, while still present in the market, does not affect the learning process, since the payoff generated by that rule is not influenced by the production rules that are used in other periods. Clearly, there is a spite effect on the payoffs realized by the other firms' production rules, but those do not compete with this individual firm's production rules in the individual learning process.

We would like to stress that it is these learning processes that is the crucial feature here, and not the objectives of the agents. Both the individual and the population learners only try to improve their own absolute payoffs. The only difference is that their learning is based on a different set of observations. The direct consequence is that in the social learning GA the spite effect drives down the performance level, while the performance of the individual learning GA improves over time. In other words, the dynamics of learning and the dynamics of the economic forces as such interact in a different way with each other in the two variants of the GA, and this explains the very different performance of the two GAs. [9]

13.4 Discussion

Before we draw some general conclusions, let us discuss some specific issues in order to put our example into a broader perspective. First, the spite effect we presented occurs in finite populations where the agents 'play the field'. The finite population size allows an individual agent to exercise some power,

all firms start at the Cournot-Nash equilibrium (as long as there is some noise in the system).

[9] There is one additional issue to be analyzed. As we saw above in Fig. 13.5, "convergence" with the individual learning GA is not as neat as with the population learning GA. Some numerical analysis shows that this is not a flaw of the individual learning GA, but related to the underlying economic model. The formal analysis of the Cournot model shows that there is a unique symmetric Cournot-Nash equilibrium. But in our numerical model we use a discrete version of the model, as only integer output levels are allowed. As a result, there turn out to be 1637 symmetric Cournot-Nash equilibria; for any average output level of the other firms from 1 to 1637, the best response for an individual firm is to choose exactly the same level. Hence, the outcomes of the individual learning GA are determined by the underlying economic forces, but convergence can take place at any of these Cournot-Nash equilibrium levels. As a result, the output levels actually observed in the individual learning GA depend in part on chance factors such as the initial output levels, the length of the bit string, and genetic drift. Notice that although there are multiple Cournot-Nash equilibria, they are all still distinct from the Walrasian equilibrium.

and influence the outcomes of the other agents. For example, in a Cournot model, when the population size n approaches infinity, the Cournot-Nash output level converges to the competitive Walrasian output. Finite populations are typically the case in GAs. To see why the 'playing the field' aspect is important, suppose there are many separate markets for different commodities, such that the actions in one market do not influence the outcomes in other markets, whereas firms can learn from the actions and outcomes in other markets. Since the spite effect does not cross market boundaries, if all firms in one market produce at the Cournot-Nash level, they will realize higher profits than the firms in another market producing at a Walrasian level. 'Playing the field' is typically the case in, e.g., economic models where the agents are firms competing in the same market. There are also some results concerning the spite effect with respect to, e.g., 2-person games in infinite populations with agents learning about the results in other games, but the occurrence of a spite effect becomes a more complicated matter (see [6]). Notice, however, that with an individual learning GA a spite effect can never occur.

Second, although, as we have seen, the spite effect may influence the outcomes of a coevolutionary process, one should not confuse the spite effect with the phenomenon of coevolution as such. In fact, as the bimatrix game in the introduction showed, the spite effect can occur in a static, one-period game, and is intrinsically unrelated to evolutionary considerations.

Third, the simple Cournot model we considered is not a typical search problem for a GA; not even if the demand and cost functions are unknown. The appeal of the Cournot model is not only that it is convenient for the presentation because it is a classic discussed in every microeconomics textbook, but the fact that we can derive formally two equilibria provided us also with two useful benchmarks for the analysis of the outcomes generated by the Genetic Algorithms. Hence, the Cournot model is just a vehicle to explain the point about the essential difference between individual and population learning GAs, and for any model, no matter how complicated, in which a spite effect occurs this essential difference will be relevant.

Fourth, one could consider more complicated strategies than the simple output decisions modeled here. For example, the Cournot game would allow for collusive behavior. However, as is well-known from the experimental oligopoly literature, dynamic strategies based on punishment and the building up of a reputation are difficult to play with more than two players. Moreover, considering more sophisticated dynamic strategies would merely obscure our point, and there exists already a large literature, for example, on GAs in Iterated Prisoners' Dilemma (see. e.g., [1], [5], or [10]).

Fifth, we are sure that the GAs we have used are too simple, and that much better variants are possible. However, bells and whistles are not essential for our point. The only essential aspect is the level at which the learning process is modeled, and the effect this has on the convergence level.

The general conclusion, then, is the following. We showed that the presence of the spite effect implies that there is an essential difference between an individual learning and a population learning GA. On the one hand, this means that when interpreting outcomes of a GA, one needs to check which variant is used, and one needs to check whether a spite effect driving the results might be present. On the other hand, it also has implications for the choice of the learning type of a GA. If a GA is used to model behavior in the social sciences, it seems ultimately an empirical question whether people tend to learn individually or socially. But if, instead, the GA is used to 'solve' some search problem, the presence of a spite effect implies that the performance of a population learning GA will be severely hindered.

A Appendix

Table 13.1. Parameters Cournot oligopoly model

inverse demand function	$P(Q)$	$a + b \cdot Q^c$
demand parameter	a	$-1 \cdot 10^{-}97$
demand parameter	b	$1.5 \cdot 10^{95}$
demand parameter	c	-39.999999997
fixed production costs	K	$-4.097 \cdot 10^{-94}$
marginal production costs	k	0
number of firms	n	40

Table 13.2. Parameters genetic algorithm

minimum individual output level	1
maximum individual output level	2048
encoding of bit string	standard binary
length of bit string	11
number rules individual GA	40
number rules population GA	40·1
GA-rate	100
number new rules	10
selection	tournament
prob. selection	fitness/$\sum$fitness
crossover	point
prob. crossover	0.95
prob. mutation	0.001

References

1. Axelrod R. (1987) The Evolution of Strategies in the Iterated Prisoner's Dilemma. In: Davis L. (Ed.), Genetic Algorithms and Simulated Annealing. London: Pitman
2. Fouraker L. E., Siegel S. (1963) Bargaining Behavior. New York: McGraw-Hill
3. Hamilton W. (1970) Selfish and Spiteful Behavior in an Evolutionary Model. Nature **228**, 1218–1225
4. Levine D. K. (1998) Modeling Altruism and Spitefulness in Experiments. Review of Economic Dynamics **1**
5. Miller J. H. (1996) The Coevolution of Automata in the Repeated Prisoner's Dilemma. Journal of Economic Behavior and Organization **29**, 87–112
6. Palomino F. (1995) On the Survival of Strictly Dominated Strategies in Large Populations. Institute for Economic Analysis WP 308.95. Universitat Autonoma de Barcelona
7. Rhode P., Stegeman M. (1995) Evolution through Imitation (with Applications to Duopoly). Working Paper E95-43. Virginia Polytechnic Institute & State University, Department of Economics
8. Schaffer M. E. (1988) Evolutionary Stable Strategies for a Finite Population and a Variable Contest Size. Journal of Theoretical Biology **132**, 469–478
9. Schaffer M. E. (1989) Are Profit Maximizers the Best Survivors? Journal of Economic Behavior and Organization **12**, 29–45
10. Stanley E. A., Ashlock D., Tesfatsion L. (1994) Iterated Prisoner's Dilemma with Choice and Refusal of Partners. In: Langton C. G. (Ed.) Artificial Life III. Redwood, CA: Addison-Wesley, 131–175
11. Vega-Redondo F. (1997) The Evolution of Walrasian Behavior. Econometrica **65**, 375–384
12. Vriend N. J. (2000) An Illustration of the Essential Difference between Individual and Social Learning, and its Consequences for Computational Analyses. Journal of Economic Dynamics and Control **24**, 1–19

14 Evolutionary Computation and Economic Models: Sensitivity and Unintended Consequences

David B. Fogel[1], Kumar Chellapilla[2], and Peter J. Angeline[3]

[1] Natural Selection, Inc.
dfogel@natural-selection.com
[2] University of California at San Diego, Dept. Electrical and Computer Engineering
kchellap@ece.ucsd.edu
[3] Natural Selection, Inc.
angeline@natural-selection.com

Abstract. The use of evolutionary models of complex adaptive systems is gaining attention. Such systems can generate surprising and interesting emergent behaviors. The sensitivity of these models, however, is often unknown and is rarely studied. The evidence reported here demonstrates that even small changes to simple models that adopt evolutionary dynamics can engender radically different emergent properties. This gives cause for concern when modeling complex systems, such as stock markets, where the emergent behavior depends on the collective allocation of resources of many purpose-driven agents.

14.1 Introduction

By their very nature, complex adaptive systems are difficult to analyze and their behavior is difficult to predict. These systems, which include ecologies and economies, involve a population of purpose-driven agents, each acting to obtain required resources in an environment. The conditions these agents face vary in time both as a consequence of external disturbances (e.g., weather) and internal cooperative and competitive dynamics. Moreover, such systems are often extinctive, where those agents that consistently fail to acquire necessary goods (e.g., food, shelter, monetary capital) are eliminated from the population. The essential mechanisms that govern the dynamics of complex adaptive systems are evolutionary: random variation of agents' behavior coupled with selection in light of a nonlinear, possibly chaotic, environment. By consequence, reductionist, linear piecemeal dissection of complex adaptive systems rarely provides significant insight. The behavior of each agent is almost always more than can be assembled from the "sum of its parts" and interactions with it predators and prey, its enemies and allies. The fabric of the complex systems is tightly woven and no examination of single threads of the fabric in isolation, no matter how exacting, can provide a sufficient understanding of the integrated tapestry.

Whereas the reductionism of traditional analysis fails to treat the holistic qualities of complex adaptive systems, computer simulations have been used to model low-level interactions between agents explicitly [32], [9]. Given such a model, attention is then focused on its emergent properties, patterns of observed behavior that could not be predicted easily from a linear analysis of agents' interactions. It is hoped that intricate computer simulations will provide useful tools for accurately forecasting the behavior of systems governed by the interactions of hundreds, or possibly thousands, of purposive agents acting to achieve goals in chaotic, dynamic environments [10].

In *Would-Be Worlds*, John Casti wrote "With our newfound ability to create worlds for all occasions inside the computer, we can play myriad sorts of what-if games with genuine complex systems. No longer do we have to break the system into simpler subsystems or avoid experimentation completely because the experiments are too costly, too impractical, or just plain too dangerous. ... [this] will form the basis for so-called normal science in the coming century" [10][p. xi]. With regret, this comment must be viewed as, at the very least, overly optimistic.[1] When modeling any complex system, and particularly when that system is also *adaptive* in the sense that the agents in the system are purpose-driven and evolve their behavior in light of a presumed goal, there is always a risk that the facets of the model which are omitted can have a drastic effect on the apparent behavior. This risk cannot be understated. Relying on models of complex adaptive systems as an experimental surrogate for real-world open-ended systems may have unintended negative impacts.

The degree to which models of complex adaptive systems can vary based on slight changes in the initial framework of the model is illustrated here within two simple designs: The El Farol problem and the iterated prisoner's dilemma. Both involve what may be loosely described as "economic" systems, in which players in the games must choose how to allocate their resources (i.e., make moves) in order to maximize payoffs. In each case, extending preliminary models to incorporate slightly more realistic conditions generates emergent properties that are markedly different from those observed originally. This divergence of result, which can be caused easily, should give pause for concern when contemplating a "normal" routine practice of experimenting with simulations of complex adaptive systems as a surrogate for real-world conditions.

[1] Interestingly enough, the first example in [10] is of a simulation for a National League Football game that consistently fails to predicts the outcome of the 1995 Super Bowl with respect to the Las Vegas point spread.

14.2 The El Farol Problem

14.2.1 Background

Consider an economic system that is an idealized model of agents who must predict whether or not to commit a resource in light of the likely commitment of other agents in the environment. Commitment is time dependent, iterated over a series of interactions between the agents where previous behavior can affect future decisions. To distill the essential aspects of the model, [2] suggested the following setting based on a bar, the El Farol in Santa Fe, New Mexico, which offers Irish music on Thursday nights.

Let each of N Irish music aficionados choose independently whether or not they will go to the El Farol on a certain Thursday night. Further, suppose that each attendee will enjoy the evening if no more than a certain percentage of the population N are present, otherwise the bar is overcrowded. To make considerations specific, let $N = 100$ and let the maximum number of people in the bar before becoming overcrowded be 60. Each agent interested in attending cannot collude with others to determine or estimate the density in the bar a priori; instead, the agent must predict how busy the bar will be based on previous attendance. Presume that data on the prior weeks' attendance are available to all N individuals. Based on these data, each person makes a prediction about the likely attendance at the bar on the coming Thursday night. If their prediction indicates fewer than 60 bargoers then they will choose to attend; otherwise they will stay home. The potential for paradoxical outcomes is clear: If everyone believes that the bar will be relatively vacant then they will attend, and instead it will be crowded — conversely, if everyone believes the bar will be crowded, it will be empty. Of interest are the dynamics of attendance over successive weeks.

Reference [2] offered the following procedure for determining this attendance. Each individual has k predictive models and chooses whether or not to attend the bar based on the prediction offered by their current best (or active model) measured in terms of how well it fit the available weekly attendance. The active model is dependent on the historical attendance, and in turn the attendance is dependent on each individual's active model. It is evident that the class of models used for predicting the likely attendance can have an important effect on the resulting dynamics. The specifics in [2] are not clear on which models were used (i.e., the procedure cannot be replicated), but some were suggested, including: 1) use the last week's attendance, 2) use an average of the last four weeks, 3) use the value from two weeks ago (a period two cycle detector), and so forth. Starting from a specified set of models assigned to each of the N individuals, the dynamics were completely deterministic. The results indicated a consistent tendency for the mean attendance over time to converge to 60. Curiously, a mixed strategy of forecasting above 60 with probability 0.4 and below 60 with probability 0.6, which would engender a mean attendance of 60 individuals, is a Nash equilibrium when the situation is viewed in terms of game theory [4]. This result implies that traditional game theory may be useful in explicating the expected outcomes of such complex systems.

14.2.2 Incorporating Evolutionary Learning

People, however, do not reason with a fixed set of models, deterministically iterated over time. Indeed, inductive reasoning requires the introduction of potentially novel models that generalize over observed data; restricting attention to a fixed set of rules appears inadequate. A more appropriate model of the El Farol problem would therefore include both a stochastic element, where new models were created by randomly varying existing ones, and a selective process that served to eliminate models that were relatively ineffectual. Individuals would thereby improve their predictive models in a manner akin to the scientific method and evolution [12]. The results of this variant on the method of [2] are qualitatively different and do not reflect any tendency toward stability in the limit or on the average.

Following [2], N was set equal to 100 and the bar was considered overcrowded if attendance exceeded 60. Each individual was given $k = 10$ predictive models. For simplicity, these models were autoregressive (AR) with their output made unsigned and rounded. For the ith individual, their jth predictor's ouptut was given by:

$$\hat{x}^i_j(n) = round[|a^i_j(0) + \sum_{t=1}^{l^i_j} a^i_j(t)x(n-t)|] \tag{14.1}$$

where $x(n-t)$ was the attendance on week $(n-t)$, l^i_j was the number of lag terms in the jth predictor of individual i, $a^i_j(t)$ was the coefficient for the lag t steps in the past, and $a^i_j(0)$ represented a constant bias term. Taking the absolute value and rounding the model's output ensured nonnegative integer values. Any predictions greater than 100 were set equal to 100 (predictions greater than the total population size N were not allowed). For each individual, the number of lag terms for each of its 10 models was chosen uniformly at random from the integers $\{1, \ldots, 10\}$. The corresponding lag terms (including the bias) were uniformly distributed over the continuous range $[1, 1]$.

Prior to predicting the current week's attendance, each individual evolved its set of models for 10 generations. This was somewhat arbitrary but was chosen to allow a minimal number of iterations for improving the existing models. The evolution was conducted as follows.

(a) For each individual i, one offspring was created from each of its $k = 10$ models (designated as *parents*). The number of lag terms in the offspring parent j was selected with equal probability from $\{l^i_j - 1, l^i_j, l^i_j - 1\}$. If $l^i_j = 1$, then $l^i_j - 1$ was not allowed, and similarly if $l^i_j = 10$, then $l^i_j + 1$

was not allowed (the number of lags was constrained to be between 1 and 10, with this choice also being somewhat arbitrary but sufficient to allow considerable history to affect current predictions). The number of AR terms in each offspring thus differed by at most one from its parent. The AR coefficients of the offspring were created by adding a zero mean Gaussian random variable with standard deviation 0.1 (i.e., $N(0, 0.1)$) to each corresponding coefficient of its parent. Any newly generated AR coefficients (due to an increase in the number of lag terms) were chosen by sampling from a $N(0, 0.1)$. When this step was completed, each individual had 10 parent and 10 offspring AR models.

(b) Each of the 20 models (10 parents and 10 offspring) in every individual were evaluated based on the sum of their squared errors made in predicting the attendance at the bar during the past 12 weeks. This duration was chosen as being a sufficiently long period of time to avoid a continual transient where the population of individuals would have an insufficient sample size at each step to allow for any reasonable prediction about the current week's attendance.

(c) The 10 models in each individual's collection having the lowest prediction error on the past 12 weeks of data were selected to be parent models for the next generation.

(d) If fewer than 10 generations had been conducted, the process reverted to Step 1; otherwise, each individual used their best current model (lowest error) to predict the current week's attendance. For each individual, if their prediction fell below 60 they went to the bar, otherwise they stayed home.

(e) If the maximum number of weeks was exceeded, the simulation was halted, designated as the completion of one trial; otherwise, the attendance for the week was recorded, the time incremented to the following week, and the process returned to step 1.

During the first 12 weeks, attendance at the bar was initialized with truncated samples from a Gaussian random variable with mean 60 and a standard deviation of 5. This was meant to start the system with a sufficient sample for each individual's predictors while not biasing the mean away from the previously observed average [2] and not imposing an overwhelming variability so as to make the attendance fluctuate wildly. In all, 300 independent trials were conducted, each being executed over 982 weeks (18.83 years) so as to observe the long-term dynamics of the evolutionary system.

Figure 14.1 shows the mean weekly attendance at the bar averaged across all 300 trials. The first 12 weeks exhibited a mean close to 59.5 resulting from the random initialization (the 0.5 decrement from 60 was an artifact caused by truncating samples to integer values). For roughly the next 50 weeks, the mean attendance exhibited large oscillations. This "transient" state transpired completely by about the 100th week, with weeks 101-982 displaying more consistent statistical behavior. For notational convenience, consider this

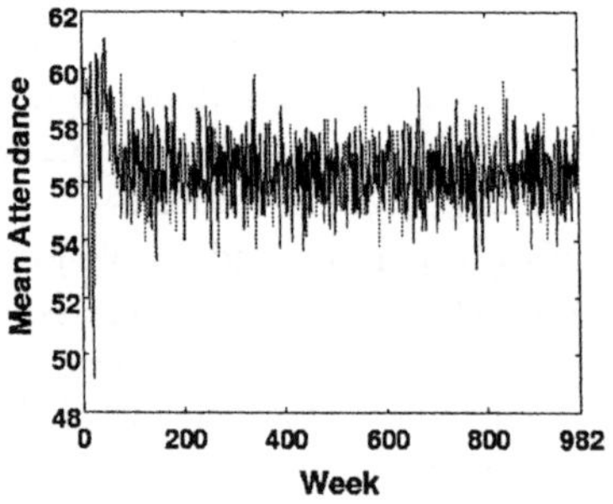

Fig. 14.1. The mean weekly attendance in the evolutionary simulation averaged across all 300 trials.

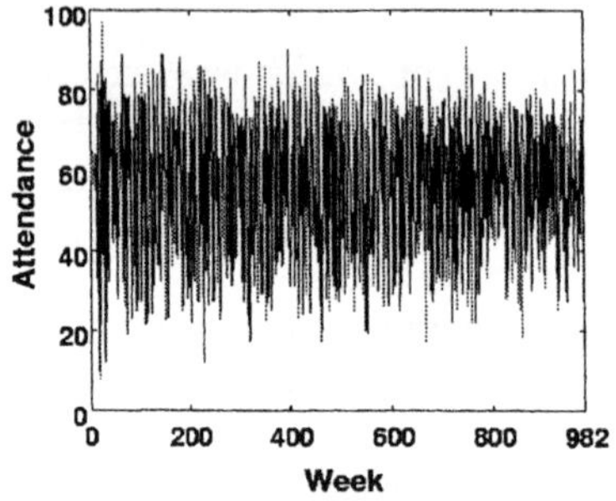

Fig. 14.2. The attendance observed in a typical trial.

period to be described as the "steady state." The mean attendance for the steady state was 56.3155 with a standard deviation of 1.0456. This is statistically significantly different ($P < 0.01$) from the previously observed mean attendance of 60 offered in [4]. Further, as a mean over 300 trials, the variability depicted in Fig. 14.1 is more than an order-of-magnitude lower than that of each single trial and the individual dynamics of each trial have been averaged out. Figure 14.2 depicts the results of a typical trial having a mean steady-state attendance of 56.3931 and a standard deviation of 17.6274. None of the 300 trials showed convergence to an equilibrium behavior around the crowding limit of 60 as observed in [2], nor were any obvious cycles or trends apparent in the weekly attendance. [2] The introduction of evolutionary learning to the system of agents had a marked impact on the observed behavior: The overall result was one of chaos and large oscillations rather than stability and equilibria. Indeed, describing the dynamics of a system with behavior as shown in Fig. 14.2 in terms of its mean does not appear useful.

Rather than seek explanations of the stochastic system's behavior in terms of stable strategies, the essential character of the weekly attendance (i.e., the system's "state") can be captured as a simple first-order random process (higher-order effects are present because of the available time window for each agent, but as shown below, these effects are not essential to describing the behavior of the system). These stochastic models have proved useful in describing the long-term behavior of many evolutionary optimization algorithms (commonly viewed as Markov chains) [28]. Such procedures are typically designed such that only the current composition of individuals in the evolving population provides a basis for determining the next-state transition probabilities, and these probabilities are invariant for a particular population regardless of time. These characteristics would also appear to hold for the agent-based system governing attendance at the bar (with the above caveats).

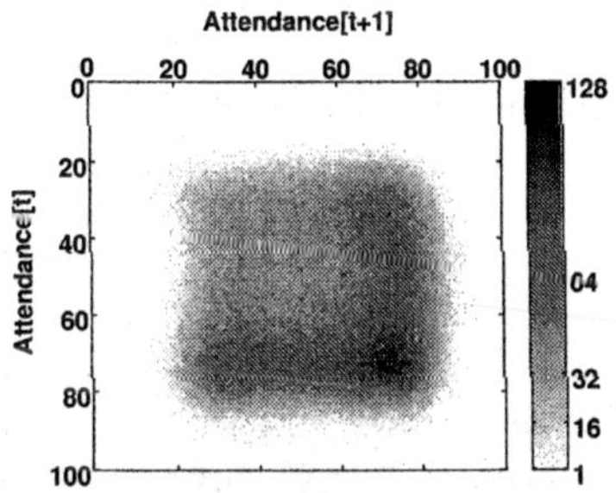

Fig. 14.3. The normalized one-step state transition matrix generated by tabulating all first-order transitions observed across all 300 trials from weeks 101 to 982. The intensity reflects the normalized frequency of occurrence.

Each of the first-order attendance transitions from week to week across all 300 trials were tabulated. These are shown in Fig. 14.3 as a normalized

[2] The attendance from weeks 101 to 982 were examined in each trial using (1) their spectral density to test for cycles, (2) regression analysis to test for a slope significantly different from zero, and (3) a computation of their fractal dimension to test for chaotic properties. None of these tests revealed any statistically significant positive findings.

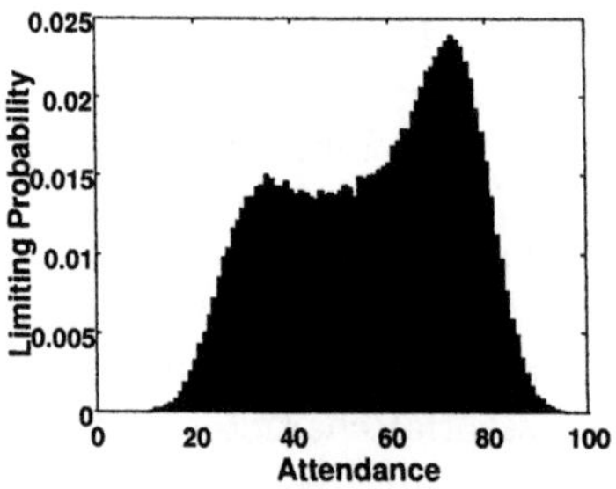

Fig. 14.4. The probability associated with each possible state [0-100] obtained by iterating the state transition matrix to its limiting distribution.

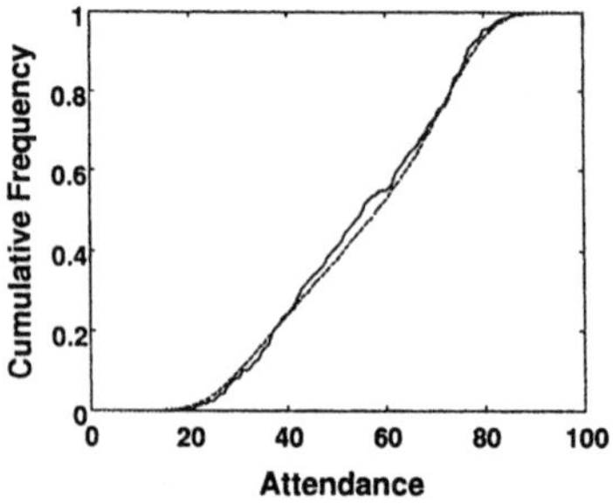

Fig. 14.5. The cumulative frequency of attendance in week 982 observed over 300 additional independent trials (solid) depicted against the cumulative distribution function obtained by summing the limiting probabilities of each state (dashed) as shown in Fig. 14.4.

state transition matrix. Under the assumption that the transition probabilities are stationary and no memory of states traversed prior to the current state is involved, the limiting probabilities of each state can be determined by raising the transition matrix to the nth power as $n \to \infty$. Beyond some value of n, the rows of the transition matrix converge to the limiting probabilities (i.e., the starting or current state is irrelevant to the long-term probabilistic behavior of the Markov chain). Figure 14.4 shows the probability mass function indicating this limiting behavior, which settled to successive differences of less than 10^{-15} after 18 iterations. To provide an independent test of the hypothesis that the behavior of the stochastic agent-based system could be captured as a Markov chain, 300 additional trials were executed and each final weekly attendance at week 982 was recorded. Figure 14.5 shows the cumulative frequency of these attendance figures, which appears to be in agreement with the cumulative distribution function obtained by summing the limiting probability masses for each state (see Fig. 14.4). A Kolmogorov-Smirnov test indicated no statistically significant difference between the preferred limiting distribution and the observed data (P > 0.3); however, one assumption for the test is that the variables in question should be continuous. Thus to provide an additional examination, the observed and expected frequencies of attendance in the ranges [0-24], [25-50], [51-75], and [76-100] were determined and a chi-square test again indicated no statistically significant different between the observed and expected frequencies ($P > 0.25$). For final corroboration, all of the weekly attendance figures for weeks 83-982 in each of the new 300 trials were tabulated. The histogram of these data appears in Fig. 14.6 and provides clear agreement with the frequencies anticipated when viewing the process as a Markov chain. Convergence to an equilibrium should not be expected from a system that is well described as a Markov chain with these limiting probabilities (cf. [2]).

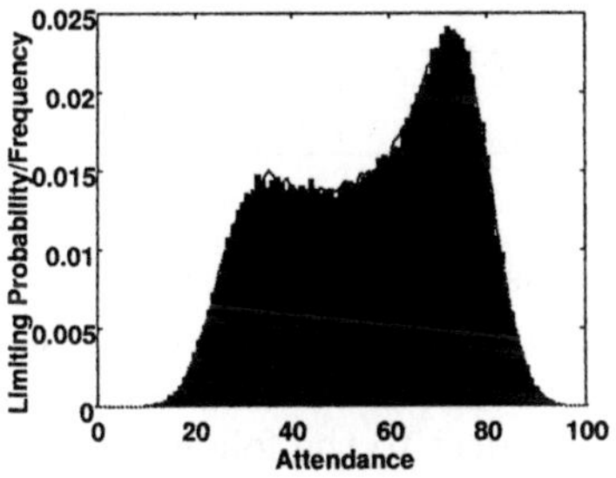

Fig. 14.6. The histogram of weekly attendance over weeks 83-982 observed over 300 additional independent trials. For comparison, the solid line represents the expected limiting probabilities generated by the prior 300 trials under the assumption that the system dynamics are captured as a Markov chain.

14.3 The Iterated Prisoner's Dilemma

14.3.1 Background

The prisoner's dilemma is an easily defined nonzero sum, noncooperative game. The term "nonzero sum" indicates that whatever benefits accrue to one player do not necessarily imply similar penalties imposed on the other player. The term noncooperative indicates that no preplay communication is permitted between the players. The prisoner's dilemma is classified as a "mixed-motive" game in which each player chooses between alternatives that are assumed to serve various motives [26].

The typical prisoner's dilemma involves two players each having two alternative actions: cooperate (C) or defect (D). Cooperation implies increasing the total gain of both players; defecting implies increasing one's own reward at the expense of the other player. The optimal policy for a player depends on the policy of the opponent [19][p. 717]. Against a player who always defects, defection is the only rational play. But it is also the only rational play against a player who always cooperates, for such a player is a fool. Only when there is some mutual trust between the players does cooperation become a reasonable move in the game.

Table 14.1. The general form of the payoff function in the prisoner's dilemma, where R (reward) is the payoff to each player for mutual cooperation, S (sucker) is the payoff for cooperating when the other player defects, T (temptation) is the payoff for defecting when the other player cooperates, and P (penalty) is the payoff for mutual defection. An entry (a, b) indicates the payoffs to players A and B, respectively.

		Player B C	D
Player	C	(R,R)	(S,T)
A	D	(T,S)	(P,P)

The general form of the game is represented in Table 14.1 (after [29]). The game is conducted on a trial-by-trial basis (a series of moves). Each player must choose to cooperate or defect on each trial. The payoff matrix that defines the game is subject to the following constraints:

$$2R > S + T$$

$$T > R > P > S.$$

The first constraint ensures that the payoff to a series of mutual cooperations is greater than a sequence of alternating plays of cooperate-defect against defect-cooperate (which would represent a more sophisticated form of cooperation [32]). The second constraint ensures that defection is a dominant action, and also that the payoffs accruing to mutual cooperators are greater than those accruing to mutual defectors.

In game-theoretic terms, the one-shot prisoner's dilemma (where each player only gets to make one move: cooperate or defect) has a single dominant strategy (Nash equilibrium) (D,D), which is Pareto dominated by (C,C). Joint defection results in a payoff, P, to each player that is smaller than the payoff R that could be gained through mutual cooperation. Moreover, defection appears to be the rational play regardless of the opponent's decision because the payoff for a defection will either be T or P (given that the opponent cooperates or defects, respectively), whereas the payoff for cooperating will be R or S. Since T > R and P > S, there is little motivation to cooperate on a single play.

Defection is also rational if the game is iterated over a series of plays under conditions in which both players' decisions are not affected by previous plays. The game degenerates into a series of independent single trials. But

Table 14.2. The specific payoff function used by Axelrod.

		Player B C	Player B D
Player A	C	(3,3)	(0,5)
	D	(5,0)	(1,1)

if the players' strategies can depend on the results of previous interactions then "always defect" is not a dominant strategy. Consider a player who will cooperate for as long as his opponent, but should his opponent defect, will himself defect forever. If the game is played for a sufficient number of iterations, it would be foolish to defect against such a player, at least in the early stages of the game. Thus cooperation can emerge as a viable strategy [21].

The iterated prisoner's dilemma (IPD) has itself emerged as a standard game for studying the conditions that lead to cooperative behavior in mixed-motive games. This is due in large measure to the seminal work of Axelrod. In 1979, Axelrod organized a prisoner's dilemma tournament and solicited strategies from game theorists who had published in the field [3]. The 14 entries were competed along with a 15th entry; on each move, cooperate or defect with equal probability. Each strategy was played against all others over a sequence of 200 moves. The specific payoff function used is shown in Table 14.2. The winner of the tournament, submitted by Rapoport, was "Tit-for-Tat":

(a) Cooperate on the first move;
(b) Otherwise, mimic whatever the other player did on the previous move.

Subsequent analysis in [5] and others indicated that this Tit-for-Tat strategy is robust because it never defects first and is never taken advantage of for more than one iteration at a time. Boyd and Lorberbaum [8] showed that Tit-for-Tat is not an evolutionarily stable strategy (in the sense of [30]). Nevertheless, in a second tournament, reported in [4], Axelrod collected 62 entries and again the winner was Tit-for-Tat.

Axelrod [5] noted that 8 of the 62 entries in the second tournament can be used to reasonably account for how well a given strategy did with the entire set. Axelrod [6] used these eight strategies as opponents for a simulated evolving population of policies by considering the set of strategies that are deterministic and use outcomes of the three previous moves to determine a current outcome. Because there were four possible outcomes for each move, there were 4^3 or 64 possible sets of three possible moves. The coding for a policy was therefore determined by a string of 64 bits, where each bit corresponded with a possible instance of the preceding three interactions, and six additional bits that defined the player's move for the initial combinations of under three iterations. Thus there were 2^{70} (about 10^{21}) possible strategies.

The simulation was conducted as a series of steps.

(a) Randomly select an initial population of 20 strategies.

(b) Execute each strategy against the eight representatives and record a weighted average payoff.
(c) Determine the number of offspring from each parent strategy in proportion to their effectiveness.
(d) Generate offspring by recombining two parents' strategies and, with a small probability, effect a mutation by randomly changing components of the strategy.
(e) Continue to iterate this process.

Recombination and mutation probabilities averaged one crossover and one-half a mutation per generation. Each game consisted of 151 moves (the average of the previous tournaments). A run consisted of 50 generations. Forty trials were conducted. From a random start, the technique created populations whose median performance was just as successful as Tit-for-Tat. In fact, the behavior of many of the strategies actually resembled Tit-for-Tat [6].

Another experiment in [6] required the evolving policies to play against each other, rather than the eight representatives. This was a much more complex environment: The opponents that each individual faced were concurrently evolving. As more effective strategies propagated throughout the population, each individual had to keep pace or face elimination through selection (this protocol of coevolution was offered as early as [7] [27] [13], see [16]). Ten trials were conducted with this format. Typically, the population evolved away from cooperation initially, but then tended toward reciprocating whatever cooperation could be found. The average score of the population increased over time as "an evolved ability to discriminate between those two will reciprocate cooperation and those who won't" was attained [6].

Several similar studies followed [6] in which alternative representations for policies were employed. One interesting representation involves the use of finite state automata (e.g., finite state machines (FSMs)) [23] [14]. Figure 14.7 shows a Mealy machine that implements a strategy for the IPD from [14]. FSM's can represent very complex Markov models (i.e., combining transitions of zero order, first order, second order, and so forth) and were used in some of the earliest efforts in evolutionary computation [11]. A typical protocol for coevolving FSM's in the IPD is as follows.

(a) Initialize a population of FSM's at random. For each state, up to a prescribed maximum number of states, for each input symbol (which represents the moves of both players in the last round of play) generate a next move of C or D and a state transition.
(b) Conduct IPD games to 151 moves with all pairs of FSM's. Record the mean payoff earned by each FSM across all rounds in every game.
(c) Apply selection to eliminate a percentage of FSM's with the lowest mean payoffs.
(d) Apply variation operators to the surviving FSM's to generate offspring for the next generation. These variation operators include:
    (a) Alter an output symbol;

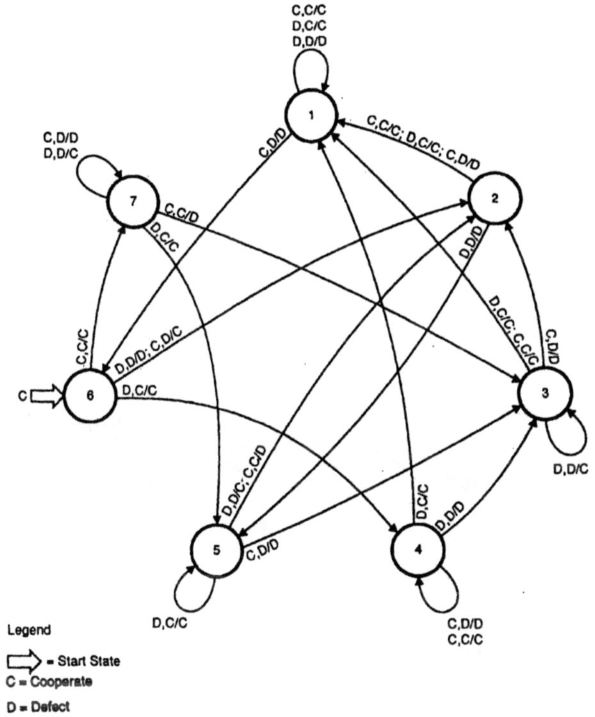

Fig. 14.7. A Mealy machine that plays the iterated prisoner's dilemma. The inputs at each state are a pair of moves that corresponds to the player's and opponent's previous moves. Each input pair has a corresponding output move (cooperate or defect) and a next-state transition.

(b) Alter a next-state transition;
(c) Alter the start state;
(d) Add a state, randomly connected;
(e) Delete a state, and randomly reassign all transitions that went to that state;
(f) Alter the initial move.

(e) Proceed to step 2 and iterate until the available time has elapsed.

Somewhat surprisingly, the typical behavior of the mean payoff of the surviving FSM's has been observed to be essentially identical to that obtained in [6] using strategies represented by lookup tables. Figure 14.8 shows a common trajectory for populations ranging from 50 to 1000 FSM's taken from [14]. The dynamics that induce an initial decline in mean payoff (resulting from more defections) followed by a rise (resulting from the emergence of mutual cooperation) appear to be fundamental. Despite the similarity of

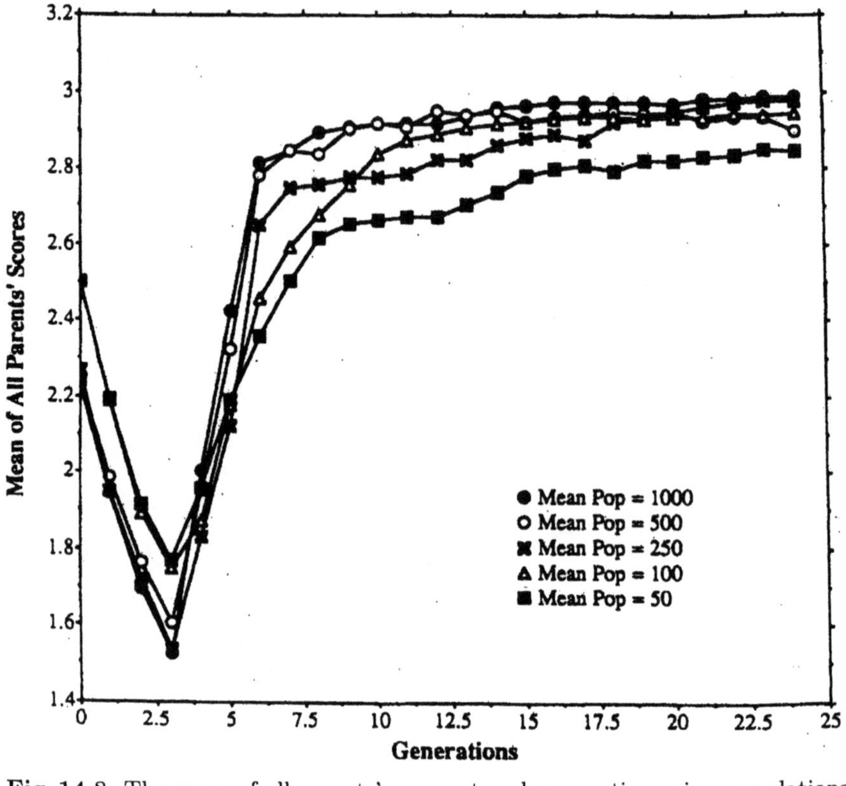

Fig. 14.8. The mean of all parents' scores at each generation using populations of size 50 to 1000. The general trends are similar regardless of population size. The initial tendency is to converge toward mutual defection, however mutual cooperation arises before the 10th generation and appears fairly stable.

results in [6] and [14], there remains a significant gap in realism between real-world prisoner's dilemmas and the idealized model offered so far. Primary among the discrepancies is the potential for real individuals to choose intermediate levels of cooperating or defecting. This severely restricts the range of possible behaviors that can be represented and does not allow intermediate activity designed to quietly or surreptitiously take advantage of a partner [33] [22]. Hence, once behaviors evolve that cannot be fully taken advantage of (those that punish defection), such strategies enjoy the full and mutual benefits of harmonious cooperation. Certainly, many other facets must also be considered, including: 1) the potential for observer error in ascertaining what the other player did on the last move (i.e., they may have cooperated but it was mistaken for defecting); 2) tagging and remembering encounters with prior opponents; and 3) the possibility of opting out of the game altogether (see [15] and [31]).

14.3.2 Evolving a Continuum of Behavior

Consider the possibility of using artificial neural networks to represent strategies in the IPD and thereby generate a continuous range of behaviors. Harrald and Fogel [18] replaced the FSM's with multilayer feedforward perceptrons (MLP's). Specifically, each player's strategy was represented by an MLP that possessed six input nodes, a prescribed number of hidden nodes, and a single output node. The first three inputs corresponded to the previous three moves of the opponent, while the second three corresponded to the previous three moves of the network itself (Fig. 14.9). The length of memory recall was chosen to provide a comparison to Axelrod [6]. The behavior on any move was described by the continuous range $[-1, 1]$, where -1 represented complete defection and 1 represented complete cooperation. All nodes in the MLP used sigmoidal filters that were scaled to yield output between -1 and 1. The output of the network was taken as its move in the current iteration.

For comparison to prior work, the payoff matrix of [6] was approximated by a planar equation of both players' moves. Specifically, the payoff to player A against player B was given by:

$$f(\alpha, \beta) = -0.75\alpha + 1.75\beta + 2.25 \tag{14.2}$$

where α and β are the moves of the players A and B, respectively. This function is shown in Fig. 14.10. The basic tenor of the one-shot prisoner's dilemma is thereby maintained: Full defection is the dominant move, and joint payoffs are maximized by mutual full cooperation.

An evolutionary algorithm was implemented as follows:

(a) A population of a given number of MLP's was initialized at random. All of the weights and biases of each network were initialized uniformly over $[-0.5, 0.5]$.

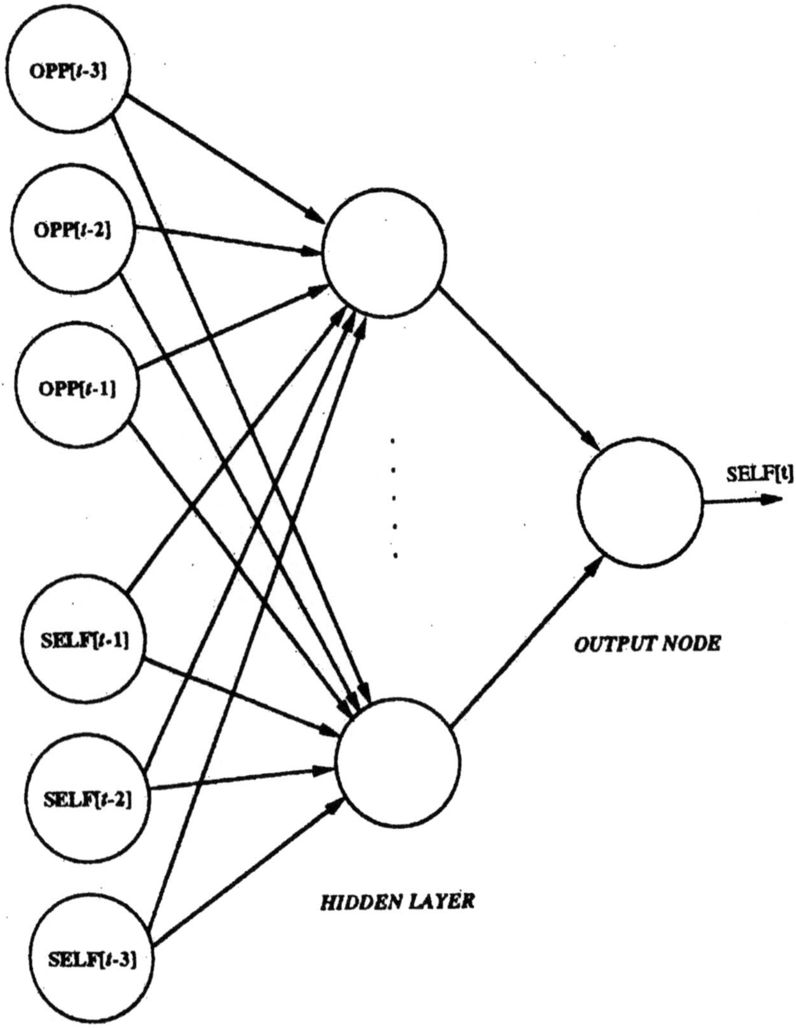

Fig. 14.9. The neural network used in [14] for playing a continuous version of the iterated prisoner's dilemma. The inputs to the network comprise the most recent three moves from each player. The output is a value between −1 and 1, where −1 corresponds to complete defection and +1 corresponds to complete cooperation.

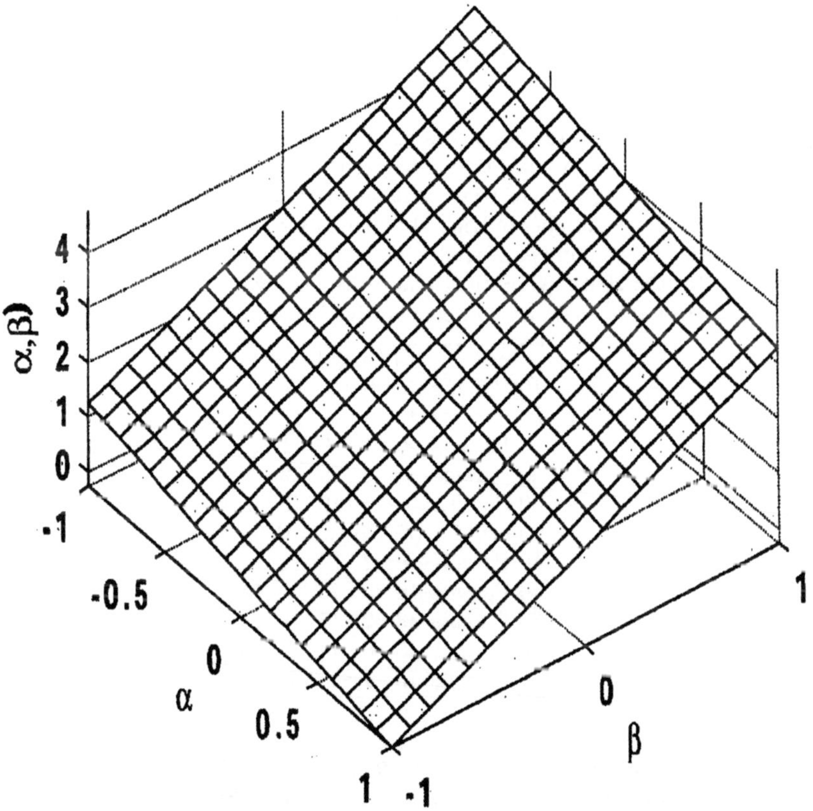

Fig. 14.10. The planar approximation to Axelrod's payoff function.

(b) A single offspring MLP was created from each parent by adding a standard Gaussian random variable to every weight and bias term.
(c) All networks played against each other in a round-robin competition (each met every other one time). Encounters lasted 151 moves and the fitness of each network was assigned according to the average payoff per move.
(d) All networks were ranked according to fitness, and the top half were selected to become parents of the next generation.
(e) If the preset maximum number of generations, in this case 500, was met, the procedure was halted; otherwise it proceeded to step 2.

Two sets of experiments were conducted with various population sizes. In the first, each MLP possessed only two hidden nodes (denoted as 6–2–1, for the six input nodes, two hidden nodes, and one output node). This architecture was selected because it has a minor amount of complexity in the hidden layer. In the second, the number of hidden nodes was increased by an order of magnitude to 20. Twenty trials were conducted in each setting with population sizes of 10, 20, 30, 40, and 50 parents.

Table 14.3. Tabulated results of the 20 trials in each setting. The columns represent: (a) the number of trials that generated cooperative behavior after the 10th generation, (b) the number of trials that demonstrated a trend toward increasing mean payoffs, (c) the number of trials that demonstrated a trend toward decreasing mean payoffs, (d) the number of trials that generated persistent universal complete defection after the 200th generation, and (e) the number of trials that appeared to consistently generate some level of cooperative behavior [14].

6–2–1	(a)	(b)	(c)	(d)	(e)
10 Parents	0	0	10	9	0
20 Parents	6	0	19	13	0
30 Parents	4	1	19	11	0
40 Parents	7	0	19	12	0
50 Parents	2	0	10	2	0
6–20–1	(a)	(b)	(c)	(d)	(e)
10 Parents	9	2	13	11	4
20 Parents	16	5	10	3	15
30 Parents	13	2	15	6	13
40 Parents	15	5	14	5	15
50 Parents	15	1	16	2	15

Table 14.3 provides the results in five behavioral categories. The assessment of apparent trends instability or even the definition of mutually cooperative behavior (mean fitness at or above 2.25) was admitted subjective, and is some cases the correct decision was not obvious (e.g., Fig. 14.11a). But in general the results showed:

(a) There was no tendency for cooperative behavior to emerge when using a 6-2-1 MLP regardless of the population size;

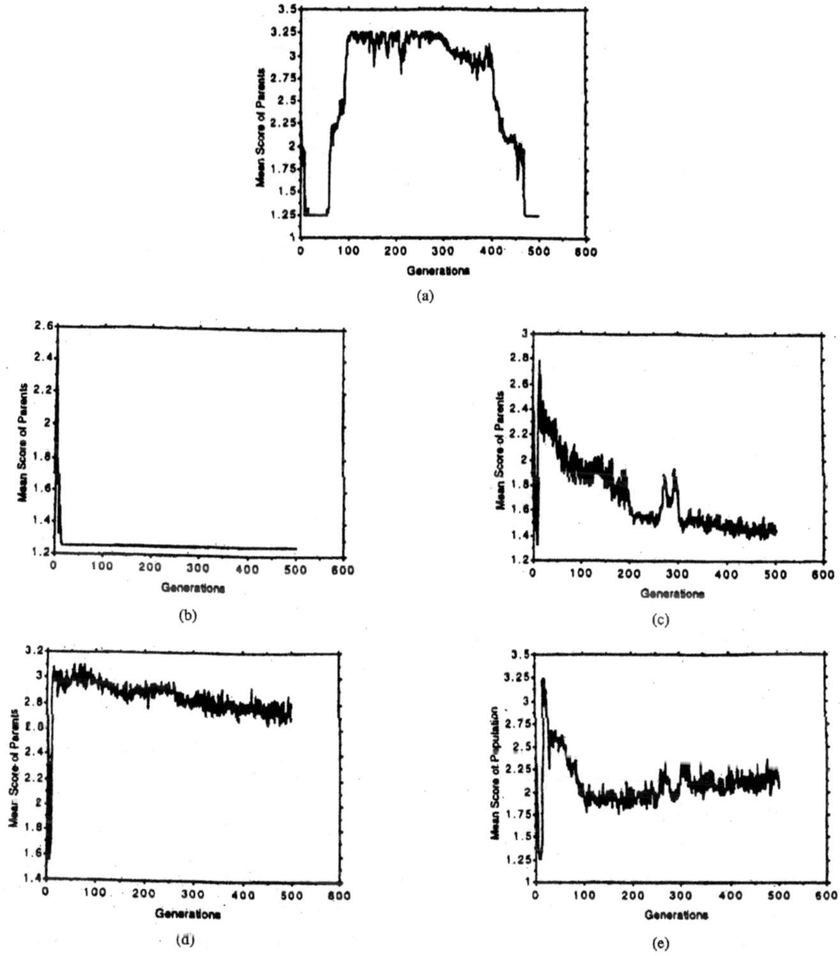

Fig. 14.11. Examples of various behaviors generated based on the conditions of the experiments: (a) oscillatory behavior (10 parents, trial ♯4 with 6–2–1 networks), (b) complete defection (30 parents, trial ♯10 with 6–20–1 networks), (c) decreasing payoffs leading to further defection (30 parents, trial ♯9 with 6–2–1 networks), (d) general cooperation with decreasing payoffs (50 parents, trial ♯4 with 6–20–1 networks), (e) an increasing trend in mean payoff (30 parents, trial ♯2 with 6–2–1 networks) from [18].

(b) Above some minimum population size, cooperation was likely when using a 6-20-1 MLP;
(c) Any cooperative behavior that did arise did not tend toward complete cooperation;
(d) Complete and generally unrecoverable defection was the likely result with the 6-2-1 MLP's but could occur even when using 6-20-1 MLP's.

Figure 14.11 presents some of the typical observed behavior patterns.

The results suggest that the evolution of mutual cooperation in light of the chosen payoff function and continuous behavior requires a minimum complexity (in terms of behavioral freedom) in the policies of the players. A single hidden layer MLP is capable of performing universal function approximation if given sufficient nodes in the hidden layer. Thus the structures used to represent player policies in the current study could be tailored to be essentially equivalent to the codings in which each move was a deterministic function of the previous three moves of the game [6]. While previous studies observed stable mutual cooperation [6], cooperative behavior was never observed with the 6-2-1 MLP's but was fairly persistent with the 6-20-1 MLP's. But the level of cooperation that was generated when using the 6-20-1 neural networks was neither complete nor steady. Rather, the mean payoff to all parents tended to peak below a value of 3.0 and decline, while complete mutual cooperation would have yielded an average payoff of 3.25.

The tendency for cooperation to evolve with sufficient complexity should be viewed with caution for at least three reasons. First, very few trials with 6–20–1 MLP's exhibited an increase in payoff as a function of the number of generations. The more usual result was a steady decline in mean payoff, away from increased cooperation. Second, cooperative behavior was not always steady. Figure 14.12 indicates the results for trial 10 with 20 parents using 6–20–1 neural networks executed over 1500 generations. The behavior appeared cooperative until just after the 1200th generation, at which point it declined rapidly to a state of complete defection. A recovery from complete defection was rare, regardless of the population size or complexity of the networks. It remains to be seen if further behavioral complexity (i.e., a greater number of hidden nodes) would result in more stable cooperation. Finally, the specific level of complexity in the MLP that must be admitted before cooperative behavior emerges is not known, and there is no a priori reason to believe that there is a smooth relationship between the propensity to generate cooperation and strategic complexity.

14.4 Discussion

The striking result of the experiments with the El Farol problem and the iterated prisoner's dilemma is that the emergent behavior of the complex adaptive system in question relies heavily on the representation for that behavior and the dynamics associated with that representation. In each case,

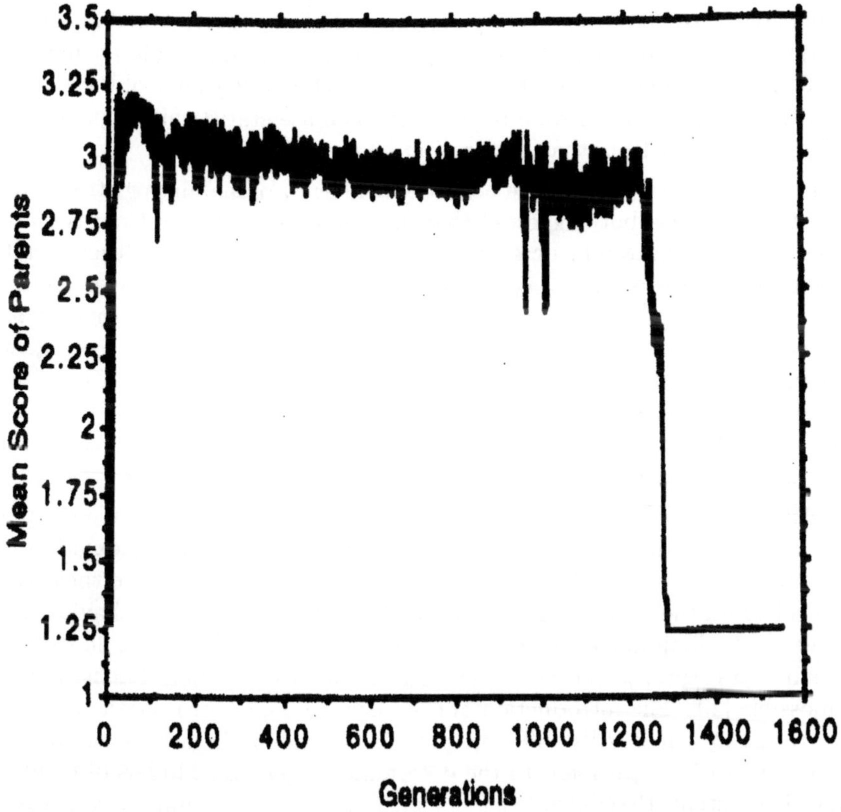

Fig. 14.12. The result of trial ♯10, with 20 parents using 6–20–1 networks iterated over 1500 generations. The population's behavior changes rapidly toward complete defection after the 1200th generation [14].

by making slight adjustments to the elements of the simulation, moving from discrete responses to continuous, or by adding in a learning process to agents, a wholly different emergent behavior was observed. Some of the early choices for working with models, particularly in the iterated prisoner's dilemma, may have been quite fortuitous, leading to "the evolution of cooperation" [5]. What would the interpretation have been if this earlier choice had instead included the option of a continuum of behaviors? Undoubtedly, things would have presented a more pessimistic outlook.

The El Farol problem studied here is only slightly more complex than that offered in [2]. It is certainly a highly idealized simulation of a market economy. Each agent in a constant-size population was only allowed linear predictive models with an AR form and a window into the past that was restricted to no more than three months. Moreover, the process for generating new models was a relatively simple mutation of existing coefficients and model structure. One could easily imagine variations that allowed agents to migrate to and from the city, employ generalized nonlinear predictive models, collaborate or collude with other agents, and so forth. Yet none of these more sophisticated procedures were required to generate statistically significantly different behavior from that obtained in [2].

Reference [2] recognized the potential deficiency of mandating strictly deterministic models but suggested that any qualitative change from the previous observations would be surprising. In retrospect, perhaps there really should be no surprise. In every case of simulating complex adaptive systems, the emergent properties are strictly dependent on the "rules" preprogrammed by the investigator. Unfortunately, the results of the interactions of agents in light of even mildly complicated rules can lead to behaviors that are "surprising." This merely reflects our own ignorance, our own inability to foresee what was predestined. This inability is heightened when faced with stochastic as opposed to deterministic models.

All models are by necessity incomplete. But there appears to be an important qualitative difference between even simple models of complex adaptive systems that rely on random variation and selection, as opposed to those that rely solely on the deterministic manipulation of fixed rules of behavior. The latter can be limited to explore only a small portion of the available space of possible strategies, while the former can be constructed such that no path is impossible [20]. The introduction of even a small degree of random variation can result in a markedly different process, stochastic in nature, with little or no qualitative agreement to the deterministic version. Models of complex adaptive systems that include agents with inductive reasoning in the face of limited information and capabilities (i.e., bounded rationality) which predict convergence to stable behavior are *ipso facto* suspect.

But should we be more suspect of models of complex adaptive systems, such as those in economics, simply by their nature? How much faith can be placed legitimately on a model of, say, a stock market, such that by tinkering with different trading rules, we might understand something about how the real-world market would react? Many such models have been constructed, one being offered by [25]. In addressing the stock market crash of October, 1987, [10] wrote: "There was no way in 1987 to safely experiment with the market mechanisms, in order to see if the rules would work without running the very real risk of inadvertently imposing trading regulations that might make the markets more unstable rather than less. With microsimulations like that of Arthur & Co., those days are now behind us. In today's high-tech world we can actually create surrogate worlds in our computers that allow us to carry out repeatable, controllable, scientific experiments on such complex systems as stock markets." The authors hope that comments such as this provide as much pause for concern as they do for optimism.

The question remains as to what role simulations of complex adaptive systems can play in the analysis of those systems. Given the limitations of accuracy and the fact that even a small deviation may lead to drastically different outcomes, quantitative conclusions (e.g., the probability of occurrence or magnitude of changes) should be considered with care. Furthermore, concluding that a certain behavior or property could not emerge from a complex system is an equally untenable conclusion because the unobserved behavior may indeed arise given only a minor modification to the model. The qualitative observations of simulated behavior can certainly be a potential source of inspiration, education, and a limited source of validation and verification. Such models can also act as an indicator of the possible range of actions that might occur, given only the minimalistic assumptions supplied in the model. Ultimately, however, a model can only provide evidence of sufficiency, in that the attributes encoded in the model are sufficient for the observed behaviors to arise in the context; necessity cannot be concluded because there may be a different set of attributes that could bring about the same qualitative effects.

Acknowledgments

The authors thank Elsevier Science Publishers and the IEEE for permission to reprint material from their previous publications [18], [17]. D. B. Fogel also thanks P.G. Harrald for prior collaborations.

References

1. Angeline P. J. (1994) An Alternate Interpretation of the Iterated Prisoner's Dilemma and the Evolution of Nonmutual Cooperation. In: Brooks R., Maes P. (Eds.) Artificial Life IV. Cambridge, MA: MIT Press, 353–358
2. Arthur W. B. (1994) Inductive Reasoning and Bounded Rationality. Amer. Econ. Assn. Papers Proc. **84**, 406–411
3. Axelrod R. (1980a) Effective Choice in the Iterated Prisoner's Dilemma. J. Conflict Resolution **24**, 3–25

4. Axelrod R. (1980b) More Effective Choice in the Prisoner's Dilemma. J. Conflict Resolution **24**, 379–403
5. Axelrod R. (1984) The Evolution of Cooperation, New York: Basic Books
6. Axelrod R. (1987) The Evolution of Strategies in the Iterated Prisoner's Dilemma. In: Davis L. (Ed.) Genetic Algorithms and Simulated Annealing. London, U.K.: Pitman, 32–41
7. Barricelli N. A. (1963) Numerical Testing of Evolution Theories. Part II: Preliminary Tests of Performance, Symbiogenesis and Terrestrial Life. Acta Biotheoretica **16**, 99–126
8. Boyd R., Lorberbaum J. P. (1987) No Pure Strategy is Evolutionarily Stable in the Repeated Prisoner's Dilemma. Nature **327**, 58–59
9. Carnahan J., Li S. -G., Constantini C., Toure Y. T., Taylor C. E. (1997) Computer Simulation of Dispersal by Anopehles Gambiae s.l. in West Africa. In: Langton C. G., Shimohara K. (Eds.) Artificial Life VI. Cambridge, MA: MIT Press, 387–394
10. Casti J. L. (1997) Would-Be Worlds, New York: John Wiley
11. Fogel L. J. (1964) On the Organization of Intellect. Ph.D. Dissertation, UCLA
12. Fogel L. J., Owens A. J., Walsh M. J. (1966) Artificial Intelligence through Simulated Evolution, New York: John Wiley
13. Fogel L. J., Burgin G. H. (1969) Competitive Goal-seeking through Evolutionary Programming. Air Force Cambridge Res. Lab., Contract AF 19(628)-5927
14. Fogel D. B. (1993) Evolving Behaviors in the Iterated Prisoner's Dilemma. Evol. Comput. **1(1)**, 77–97
15. Fogel D. B. (1995) On the Relationship Between the Duration of an Encounter and the Evolution of Cooperation in the Iterated Prisoner's Dilemma. Evol. Comput. **3**, 349–363
16. Fogel D. B. (Ed.) (1998) Evolutionary Computation: The Fossil Record, Piscataway, NJ: IEEE Press
17. Fogel D. B., Chellapilla K., Angeline P. J. (1999) Inductive Reasoning and Bounded Rationality Reconsidered. IEEE Transactions on Evolutionary Computation **3(2)**, 142–146
18. Harrald P. G., Fogel D. B. (1996) Evolving Continuous Behaviors in the Iterated Prisoner's Dilemma. BioSystems **37(1-2)**, 135–145
19. Hofstadter D. (1985) Metamagical Themas: Questing for the Essence of Mind and Pattern, New York: Basic Books
20. Hofstadter D. (1995) Fluid Concepts and Creative Analogies. New York: Basic Books, 115
21. Kreps D., Milgrom P., Roberts J., Wilson J. (1982) Rational Cooperation in the Finitely Repeated Prisoner's Dilemma. J. Econ. Theory **27**, 326–355
22. Marks R. E. (1989) Breeding Hybrid Strategies: Optimal Behavior for Oligopolists. In: Schaffer J. D. (Ed.) Proc. 3rd Int. Conf. Genetic Algorithms. San Mateo, CA: Morgan Kaufmann, 198–207
23. Mealy G. H. (1955) A Method of Synthesizing Sequential Circuits. Bell Syst. Tech. J. **34**, 1054–1079
24. Moore, E. F. (1957) Gedanken-experiments on Sequential Machines: Automata Studies. Ann. Math. Studies **34**, 129–153
25. Palmer R. G., Arthur W. B., Holland J. H., LeBaron B., Tayler P. (1994) Artificial Economic Life: A Simple Model of a Stockmarket. Physica D **75**, 264–274

26. Rapoport A. (1966) Optimal Policies for the Prisoner's Dilemma. Univ. North Carolina, Chapel Hill, Tech. Rep., no. 50
27. Reed J., Toombs R., Barricelli N. A. (1967) Simulation of Biological Evolution and Machine Learning. J. Theoretical Biology **17**, 319–342
28. Rudolph G. (1994) Convergence Analysis of Canonical Genetic Algorithms. IEEE Trans. Neural Networks **5(1)**, 96–101
29. Scodel A., Minas J. S., Ratoosh P., Lipetz M. (1959) Some Descriptive Aspects of Two-person Nonzero Sum Games. J. Conflict Resolution **3**, 114–119
30. Smith J. M. (1982) Evolution and the Theory of Games. Cambridge, U.K.: Cambridge University Press
31. Stanley E. A., Ashlock D., Tesfatsion L. (1994) Iterated Prisoner's Dilemma with Choice and Refusal of Partners. In: Langton C. G. (Ed.) Artificial Life III. Reading, MA: Addison-Wesley, 131–175
32. Taylor C., Jefferson D. (1995) Artificial Life as a Tool for Biological Inquiry. In: Langton C. (Ed.) Artificial Life: An Overview. Cambridge, MA: MIT Press, 1–13
33. To T. (1988) More Realism in the Prisoner's Dilemma. J. Conflict Resolution **32**, 402–408

Part IV

Financial Engineering

15 Tinkering with Genetic Algorithms: Forecasting and Data Mining in Finance and Economics

George G. Szpiro

Israeli Center for Academic Studies
(affiliated with the University of Manchester)
Kiriat Ono, Israel
george@netvision.net.il

Abstract. In two previous papers [13,14] genetic algorithms were presented that permit the search for dependencies among sets of data (univariate or multivariate time-series, or cross-sectional observations). These algorithms – modeled after genetic theories and Darwinian concepts, such as natural selection and survival of the fittest – permit the discovery of equations, in symbolic form, that re-create or, at least, mimic the data-generating process. This paper discusses some of the computational issues and difficulties that may arise when the genetic algorithm is applied, and suggests ways to improve the algorithm's performance.

15.1 Introduction

Genetic algorithms are search procedures that apply the notions of evolution and natural selection to computer programming. By applying the program recursively, solutions for real-life problems can be found even in relatively high-dimensional parameter space. A genetic algorithm starts out with an initial population of randomly produced solution attempts. The most successful members of the population are paired, and combine to produce offspring. In the next generation, again, the fittest individuals mate and produce offspring. Less fit individuals eventually disappear. From time to time mutations occur, that introduce renewed variety into a population that may have become too homogenous. The field was established by Holland [1975] and first applied to engineering problems by [5]. Subsequently genetic algorithms were put to good use in such diverse areas as artificial intelligence, pharmacology, chemistry, physics, acoustics, etc., and also to economics (e.g, [3,9,10,2,15,16]) and finance [1].

In this paper I utilize genetic algorithms for data mining. Their efficiency makes genetic algorithms ideally suited for the detection and identification of hidden relationships in high-dimensional search-spaces that are, in addition, characterized by the existence of multiple optima. I propose to let genetic algorithms breed mathematical equations that re-create, or at least mimic, the true data-generating formulas. The method is applicable both to time series and to cross sectional observations. In [13,14] I presented such an algorithm and tested it successfully with chaotic time-series from physics, with economic time-series, with artificial series, and also with cross-sectional data on the performance and salaries of NBA players during the 94-95 season. The method permits global predictions and forecasts, and often furnishes a deeper understanding of the dynamics of the process that underlies the data. Similar applications of the closely related method of evolutionary programming have been performed by [8], who, for example, used astronomy data as input, and received Kepler's third law on planetary motion as output. In the next section I describe the algorithm that I use to search for hidden relationships in the data. [1] In Sect. 15.3 – the main part of the paper – I address some computational issues in the application of genetic algorithms to forecasting and data-mining, and suggest ways to improve the algorithm's performance. In particular, I introduce the "residual method" and "cross-breeding", and discuss how to deal with outliers. Section 15.4 suggests some areas for future research, and Sect. 15.5 concludes the paper with a short summary.

15.2 A Primer on Genetic Algorithms

In order to apply the concepts of genetic algorithms and evolutionary programming to the breeding of mathematical formulas, I consider equations to be just strings of symbols that conform to a simple grammar: two arguments are combined by an arithmetic operator, thereby forming one of the equation's building block. The arguments can be variables, numbers, or are, themselves, self-contained blocks. These building blocks are used by the algorithm in the same way that nature uses genetic material to breed ever better formulas. The genetic algorithm that I propose consists of the following subroutines:

Initialization: The algorithm starts out with an initial population of simple, randomly created equation-strings ($j = 1, 2, \ldots, N$):

$$Equation - string_j = ((A\ Op\ B)\ \ Op\ \ (C\ Op\ D)) \tag{15.1}$$

where A, B, C, D are lagged observations of univariate or multivariate time-series, or observations of cross-sectional data, or real numbers, and Op stands for one of the four arithmetic operators. [2]

Measuring fitness: The fitness of each string is evaluated by measuring the percentage of the data's variance that is explained by the string, (i.e., by its R^2).[3]

---

[1] Section 15.2 is based, in part, on [13,14].

[2] We use "protected division", which puts an upper limit on the result if the divisor is too small.

[3] Actually, the fitness measure R^2 is modified to include a term that penalizes long strings. Thus the algorithm is discouraged from forming ever longer and more complicated equations.

Ranking: The strings are ranked in descending order of their fitness.

Choice of mate: In this routine the strings form pairs. The top-ranked strings choose first among the remaining strings. The probability of any string being chosen is proportional to its fitness. This continues until half the initial population has formed pairs. The remaining half of the population dies. Usually the strings that are ranked towards the bottom of the list tend to disappear, except for the ones that were chosen as mates by some of the fit strings.

Procreation and Crossover: Offspring are created by interchanging parts of the parent-strings. For example, if the parent-strings are of the form

$$\begin{aligned} &Parent_1: \quad (A * B)/C \\ &Parent_2: \quad (D - (E/F)) \quad , \end{aligned} \tag{15.2}$$

then Offspring1's and Offspring2's equation-strings are identical to their parents and – by crossover of the building blocks B and (E/F) – the two other offspring's equation-strings would be of the form

$$\begin{aligned} &Offspring_3: \quad (A * (E/F))/C \\ &Offspring_4: \quad (D - B) \quad . \end{aligned} \tag{15.3}$$

Mutation: A small percentage of the basic building blocks or operators are mutated, i.e., they are changed at random.

In the course of the generations the blocks of the equation-strings are combined, deleted, and re-combined, and fitter and fitter strings evolve, as evidenced by the ever higher R^2-values of the top-ranked strings. Eventually only the useful blocks survive and combine into strings, while the less useful ones disappear. Let me demonstrate how the genetic algorithm performs with the seemingly simple tent map,

$$a_t = \begin{cases} ca_{t-1} & if \quad a_{t-1} < 0.5 \\ c - ca_{t-1} & if \quad a_{t-1} > 0.5 \end{cases} \tag{15.4}$$

whose behavior, however, is notoriously difficult to predict if the underlying dynamic is not known [11]. I create a time-series of length 105, with c = 1.999 and a_0– 0.979853, and allow five lags for the algorithm.[4] After 500 to 1000 generations the algorithm bred the following strings in three instances,[5]

A0 = (A1/(((A1+(((A1+(((A1+A1)+A1)+0.3))*A1)*(((((A1+(((A1+A1)*A1)+0.1))*A1)*A1) -A1)*A1)))+0.1)+0.1))

A0 = (((A5+A5)+A1)/((A5/A1)+(((((A1*((A1*(A1*((A1*(A1*A1))+A1)))*(A5+A1)))* (((A1*(A5+(A1+A5)))*(A1+A1))*(A1+A1)))+A1)-(A1*A1))+A1)))

[4] In all the following examples we run the algorithm on series of length 105, and allow five lags. Hence the training series consists of 100 data points.

[5] To simplify notation, we will henceforth write A0 in the strings for a_t, and A1 for a_{t-1} etc.

A0 =((((((0.6*(((A1/A1)-A1)*4.1))*A1)*(((A1/A1)-A1)*4.1))+((A1/A1)-A1))*A1)+ (A1*(((A1/A1)-A1)/(((A1/A1)*-3.0)-(((A1*A1)−1.4)*5.4)))))

with R^2-values of 0.943, 0.963 and 0.978, respectively. It is quite surprising that such well-performing strings were bred, in spite of the fact that the true data-generating process was a non-arithmetic, not everywhere differentiable threshold function. The edited versions of these strings are

$$a_t = \frac{a_{t-1}}{0.2+a_{t-1}-0.3a_{t-1}^3-3.97a_{t-1}^4+0.7a_{t-1}^5+4.6a_{t-1}^6+8.0a_{t-1}^7}$$

$$a_t = \frac{a_{t-1}^2+2a_{t-1}a_{t-5}}{2a_{t-1}^2-a_{t-1}^3+a_{t-5}+8a_{t-1}^8a_{t-2}^2+12a_{t-1}^9a_{t-5}+4a_{t-1}^{10}+8a_{t-1}^{10}a_{t-5}^{11}+4a_{t-1}^{12}}$$

$$a_t = a_{t-1} + 9.086a_{t-1}^2 - 20.172a_{t-1}^3 + 10.086a_{t-1}^4 - \frac{a_{t-1} - a_{t-1}^2}{10.56 + 5.4a_{t-1}^2}$$

Figure 15.1 depicts the time series and one-period forecasts produced by the first of the above strings, on out-of-sample time-series. The fit between actual values and forecasts is amazingly good!

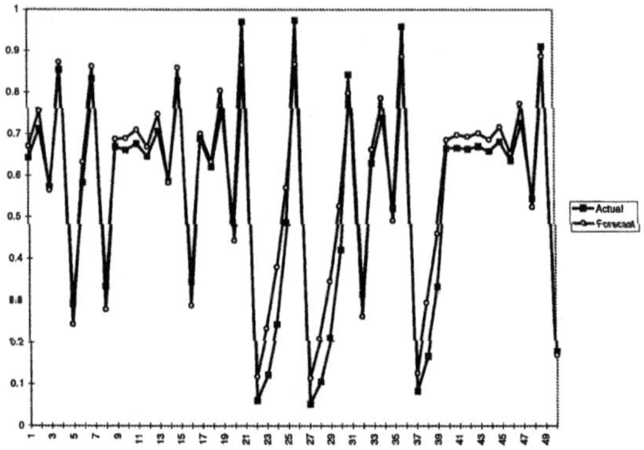

Fig. 15.1. Tent map: out-of-sample time series, and forecasts

However successful genetic algorithms may prove to be, their application to real-life problems is still more of an art than of a science. What works and what doesn't, is usually found by trial and error. Like their cousin, the neural network, genetic algorithms still consist of much fiddling and tinkering. Furthermore, successful runs of the algorithm with a high-dimensional, or otherwise complex data-set may require hundreds, if not thousands, of generations. Thus, the process of breeding successful equation-strings on a microcomputer may take many hours, and even days. At the stage at which research in genetic algorithms presently finds itself, tinkering and fiddling are still the order of the day [17], and any procedure which improves the results or speeds up the search is of importance – even if the initial gain is only marginal.

15.3 Performance Boosters

I will now suggest some methods to improve the performance of the genetic algorithm. One way to increase the speed of the algorithm is to vary the parameters.

15.3.1 Choice of Parameters

The most efficient combinations of the algorithm's parameters are usually found by trial and error. Sometimes it is obvious whether it is beneficial to increase or decrease a certain parameter, sometimes not. Let us take the number of equation-strings, and the mutation rate, for example. One wishes to have an abundant pool of basic building blocks in the initial population and ensures this with a large number of strings. But after a few generations the population generally tends to become very homogeneous and the presence of many strings implies numerous recalculations of strings with similar fitness. This just consumes computer time, and it is therefore more efficient time-wise to re-run the algorithm numerous times with fewer strings. In this case renewed variety must be introduced into the population through mutations. Hence, the researcher is advised to perform as many runs as possible, and to save computer time by using a relatively low number of strings. At the same time a relatively high mutation rate should be employed. It is also advisable to exempt the best performing strings from mutation, so that the information they carry not be lost inadvertently. Now let me propose some specific methods to enhance the results.

15.3.2 Residual Method

After the genetic algorithm bred an equation-string, the result can sometimes be enhanced, by re-running the algorithm on the residuals. This may bring about an improvement, if one component of the data generating formula dominates the other parts. During the first run of the algorithm, the "heavy" component may hide the "light" ones. Once the effect of the heavy component has been isolated, the genetic algorithm may - in the second stage - discover remaining parts of the formula. As a first example, we refer to the sunspot series[6] discussed in [13]. A first run produced the result,

$$x_t = 7.7762 + 0.8205x_{t-1}, \tag{15.6}$$

($R^2 = 0.734$), which was then improved upon in the second stage by explaining 0.468 of the remaining variance. Combining both stages, and editing, we get

$$x_t = 7.7762 + x_{t-1}\left(0.8205 + \frac{x_{t-9} - x_{t-3}}{3.9x_{t-1} + x_{t-11}}\right), \tag{15.7}$$

with an R^2-value of 0.858. It is noteworthy that on various different runs the algorithm produced very similar equation strings.

As a further example, let us create an artificial time-series of the form

$$a_t = 0.8b_{t-1} + 0.5b_{t-2}c_{t-3}d_{t-4} \tag{15.8}$$

where b_t, c_t, and d_t are random numbers, normally distributed in [-1, +1]. In one typical run, the algorithm produced an equation-string of the form

A0 =(((4.6+-9.8)*B1)/((-6.0+C2)-C2)),

within less than 50 generations, with an R^2 of 0.966. Running the genetic algorithm on the residuals, subsequently produced a string of the form,

A0 =((C3*(D4*((C1-C1)-B2)))/-3.8)

with an R^2 – which now represents the explained variance of the residuals – of 0.670. Hence, of what little variance remained unexplained after the first stage, two thirds is explained in the second stage. The edited combination of both stages is,

$$a_t = 0.8666b_{t-1} + 0.26315b_{t-2}c_{t-3}d_{t-4}, \tag{15.9}$$

with an R^2-value of 0.988. In addition to the high R^2-value, the two-stage application of the algorithm produced the correct functional form of the data-generating formula – even if the numerical parameters are somewhat off from their correct values.

Residuals need not necessarily be additive. Let us investigate an artificial series of the form,

$$a_t = b_{t-1}b_{t-2}\tilde{c}_{t-3}, \tag{15.10}$$

where b_t are random numbers, uniformly distributed in [+1.0,+10.0], and the c_t are measured with uncertainty,

$$\tilde{c}_t = c_t + \epsilon_t, \tag{15.11}$$

where c_t are uniform in [+0.5,+1.5], and the noise term lies in [-0.1,+0.1]. After about 20 generations, the algorithm bred a string of the form

A0 = ((B1*B2)-((C5/C3)-C3))($R^2 = 0.814$).

[6] As collected by the Sunspot Index Data Center. (See [4], pp. 104–105.)

Applying the algorithm to the multiplicative residuals, the algorithm bred the string

A0 = ((C3*C4)/C4)($R^2 = 0.823$) .

The edited version of the combination of both strings, i.e.,

$$a_t = b_{t-1}b_{t-2}c_{t-3} - c_{t-5} + c_{t-3}^2, \tag{15.12}$$

explains 0.944 of the data's variance.

It may be concluded, that the multi-stage method of applying the genetic algorithm to residuals can improve the results if the data-generating formula is composed of equation-parts with differing weights. The method should be attempted both with additive and with multiplicative residuals. Researcher and practitioners should be warned, however, against letting too many generations pass before embarking on the second stage. If the algorithm is allowed to linger for too long in the first stage, it may attempt – after the "heavy" part of the formula has been "identified" – to pick up components of the "light" parts and of the noise. Then, in the second stage, the algorithm would have to counterbalance these effects, at the same time as it tries to identify the "light" parts of the data-generating formula.

15.3.3 Cross Breeding

While discussing the sunspots series, at the beginning of the preceding subsection, I mentioned that on different runs, the algorithm converged to similar equation-strings, and noted that this was actually quite remarkable. This hinted at the fact, that running the algorithm several times on the same dataset, does not necessarily breed identical equation-strings. In fact, only with data-sets that are relatively free from noise, as for instance in the physical sciences, may one realistically hope for this to happen.

In economics, unexplained influences and noise pollute the data, and there may be several equations that mimic the data-generating process, with similar, or with varying degrees of success. Hence, on different runs, the algorithm may, and usually does, converge to different strings. It is advisable to re-run the algorithm multiple times for each data-set that is being considered, since on any single run, the algorithm may breed a string with a particularly low fitness. In such a case, mutations may either not suffice to pull the algorithm out of this evolutionary dead-end, or it may take in inordinate amount of time. Running the algorithm several times, increases the chances for really fit strings to evolve. It may be possible, however, to further improve on the result. Different runs of the algorithm may – in the parlance of the previous sub-section – converge to different "heavy" parts of the data-generating formula. In such a case, the average value of different "breeds" may enhance the results. Let us create the following series, for example,

$$a_t = \frac{b_{t-1}b_{t-2}b_{t-3}}{b_{t-4}} + \epsilon_t, \tag{15.13}$$

where b_t and the noise term ϵ_t are uniformly distributed in [0, +2.0], and in [-50.0, +50.0], respectively. On two occasions, the genetic algorithm bred the following strings within 20 generations,

A0 = ((B5*(-6.9+((B1/B4)+B5)))-((-7.2*B4)-(B2/B4))) ($R^2 = 0.534$)

and

A0 = ((B1/B4)/((B2/B1)-(B2-B3))) ($R^2 = 0.520$)

Using the mean of these two strings, the R^2 for the series rose to 0.546, i.e., the explained variance was increased by between two and five percent. Here we point out again, that if "cross-breeding" is to be employed, it must be done early, before the algorithm attempts to pick up noise or components of the other "heavy" part.

15.3.4 Outliers

Our genetic algorithm may get somewhat thrown off the track, if the data consist of observations whose values generally lie close together, interspersed only by some outliers. The way the algorithm works, i.e., by summing the squared errors, and then ranking the strings, errors in the outliers are heavily penalized, while good predictions for the bulk of the data contribute only little to explained variance. Hence, the algorithm does not necessarily breed equations that mimic the data-generating process for the bulk of the data. If the aim of the algorithm is to forecast outliers, this may be appropriate. If, however, the bulk of the data should be predicted, the algorithm must be adapted. A solution to the problem, is to put a cap on the maximum error when computing the fitness, which reduces the relative weight that errors in the outliers have in the computation of the fitness. As an example let us take the following equation system to generate data,[7]

$$\begin{aligned} x(t+\delta) &= x(t) - (y(t) + z(t))\,\delta\,, \\ y(t+\delta) &= y(t) + (x(t) + 0.2y(t))\,\delta\,, \\ z(t+\delta) &= z(t) + (0.2 + x(t)z(t) - 5.7z(t))\,\delta\,, \end{aligned} \tag{15.14}$$

with $\delta = 0.02$, and initial values $x_0 = -1$, $y_0 = 0$, $z_0 = 0$. The results of this equation system for integer values of t, (i.e., for every 50th step) form the time-series x_t, y_t, and z_t. The first 200 entries are discarded to let transients die out, and we apply the genetic algorithm to the next 105 observations of the z-variable. After 200 generations of a typical run, the top-ranked equation-string was,

A0 =((A1*(A1/(((A1*(((A5*A4)*A5)*A5))+(A3/(A5+A5)))+(A3+(((A1*((A1/(A4+A5))*A1))*A1)*A1)))))/(A2+(A3+((A4+(A3+((A5*A4)*A5)))*(A4/(A5+(A4*A4)))))))($R^2 = 0.986$)

[7] This differential equation defines what is known as the Rösler attractor in physics [12]. See [13] for details.

The string performs very well in terms of R^2, even on out-of-sample data, and a glance at Fig. 15.2, which depicts the actual series and the one-period forecasts, seems to confirm this. A more detailed look, however, reveals that an equation has been bred that only gets the outliers right (about 10% of the data), and, in effect, sets all other values – the bulk of the data – close to zero (Fig. 15.3).

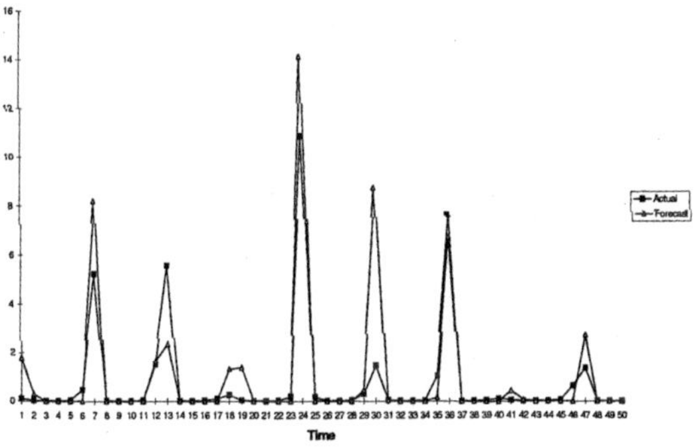

Fig. 15.2. out-of sample time-series, and forecasts

Let us now put a cap of 0.05 on the maximum that the squared error can take on. Under this condition, the following string was bred, for example, after a few hundred generations: [8]

A0 =((A5*0.1)/((A5+0.1)-((((((2.1+((-1.1*0.1)*0.1))+(((0.1+8.7)*0.1) *0.1))+A1)+0.1)*0.1)*0.1)))

Figure 15.4 depicts one period forecasts, based on the latter string, on out-of sample data. The figure shows that, while outliers are predicted incorrectly, the forecasts for the bulk of the data are now very good.

15.4 Other Problems and Suggestions for Future Research

15.4.1 Restricting the Search Space

Even though genetic algorithms are vastly more efficient than conventional search routines, it still helps to narrow down the search space, either by reducing its dimension, or by confining the values that are being searched

[8] The numerical value of R^2 for the whole data-set has no meaning in this case.

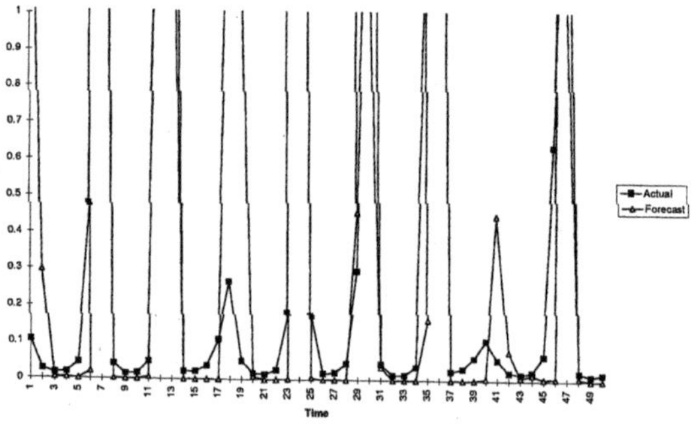

Fig. 15.3. Rössler attractor

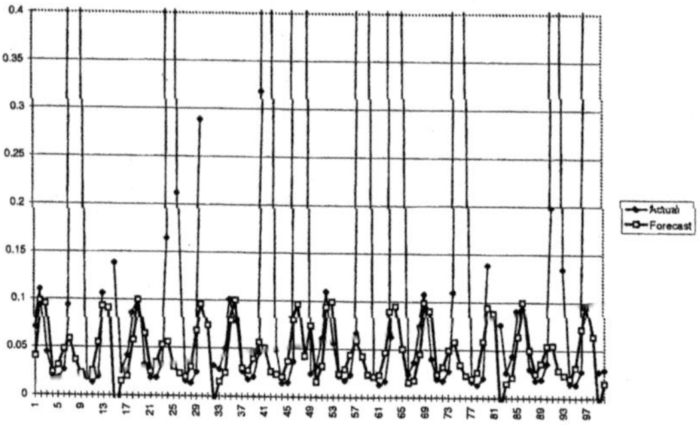

Fig. 15.4. Rösler attractor: forecasts with "capped" errors (detail)

to a more restricted range. A time-series of the well-known chaotic Hénon attractor [6],

$$x_t = 1 - 1.4x_{t-1}^2 + 0.3x_{t-2}, , \quad (15.15)$$

will serve as an example. When the number of lags was specified as 10, and the range of numbers as [-10.0,+10.0], the algorithm converged to strings with $R^2 = 0.950$ within 328 generations (median of ten runs). When the parameter space was narrowed to 6 lags and to numbers lying in [-6.0, +6.0], or to 2 lags and to numbers lying in [-2.0, +2.0] (i.e., the search space was reduced to $(0.6)^2$ and to $(0.2)^2$ of its initial size) the median number of generations that was needed to converge to strings with $R^2 = 0.950$, was 166 and 78, respectively. Looking at this from the other angle, we note however, that the required number of generations does not increase in a linear fashion with the increase in the size of the search-space. This is due to the algorithm's efficiency, which ensures that the required number of generations increases at a much slower pace than does the search space. How fast the requirements on computer time change when the dimension or the extent of the search-space change, remains an area for future research.

Sometimes an OLS regression may give a first indication of the approximate range of some of the parameter values.[9] Obviously it may also be beneficial to use OLS regression to fine-tune parameter values after the genetic algorithm has bred a string. In Subsection 3.2, for example, the result for the artificial time-series (equation 8) could be improved: after the genetic algorithm bred the equation (9), which indicated that the expressions b_{t-1} and $b_{t-2}c_{t-3}d_{t-4}$ are included in the data-generating formula, running a linear regression with these independent variables immediately gives the correct parameter-values 0.80 and 0.50.

15.4.2 Spurious Relationships

The algorithm may find an equation that simply represents spurious relationships between the variables or the lags. In order to verify that the algorithm bred meaningful expressions, it is imperative to perform out-of-sample tests. In the appendix a time-series of length 200 is given, that was produced by a pseudo random number generator. [10] The genetic algorithm with maximum lag=5 was applied to the first 105 entries and bred a string which "explained" 0.165 of the series' variance:

A0 =(((((A1-0.6)+A1)-A5)-3.8)/(((0.2/(A1-A2))/((8.4+0.6)-8.4))-8.4))

Obviously the string represents a spurious formula and nothing can be forecasted with it. This is immediately verified by applying the formula to the continuation of the data (also given in the appendix). In this context it is well to realize that the algorithm that I propose may simply breed strings that mimic the training set instead of finding the equations that underlie the data. On the other hand, the good performance on out-of-sample data in the sunspot example (Sect. 15.3.2) suggests that the equations-strings that were bred in this case do more than just mimic the data. The development of a decision criterion or of a statistical test, when to accept and when to reject a string that was bred by the algorithm, would be useful.

---

[9] At the same time one receives an initial benchmark estimate for the R^2-value that the algorithm needs to improve upon.

[10] The series may serve as a benchmark for readers who wish to test their algorithms.

15.4.3 Collinearity and Autoregression

These two phenomena pose a special problems for the algorithm. Since the values of close lags are numerically similar (in the case of autoregression), the algorithm does not penalize the equations very much when an incorrect lag is picked, and may therefore linger for many generations at a fitness level that is misleadingly high. Similar observations are made with cross-sectional data that is collinear. The usual ways of handling such problems in econometrics (for example, first differencing) may provide partial relief. Other solution must be still be found for this difficulty.

15.4.4 Comparibility of Numbers

When the numerical values of the data differ by orders of magnitude (if the observations of series A vary, say, between 0 and 0.001, and the observations of series B lie in the range 2000 to 5000) the algorithm may take an inordinate amount of time to converge to useful equations. A solution to this problem may be to employ normalized series, for example A–mean(A), or B/Mean(B).

15.5 Concluding Remarks

Genetic algorithms are general purpose search procedures that are used in a wide variety of disciplines. Their efficiency becomes particularly apparent in high-dimensional search spaces. I propose such an algorithm to seek for equations that describe the dynamic of some observed data. When presented with time-series or cross-sectional data this algorithm attempts to find the data-generating process. The algorithms is surprisingly successful but, unfortunately, also very time-consuming: it may take hours and even days until equation-strings of sufficiently high fitness have evolved. It is therefore of utmost importance, for practitioners and researchers alike, to develop techniques which speed up the process.

In this paper I discuss some methods to improve the performance of such algorithms. One such method, for example, is the re-use of residuals that remain after an initial run of the algorithm, another is the combination of the results of multiple runs. In order to deal with the problem of outliers I propose to put a maximal value on the fitness measure. Finally, I list some open problems that are suggested as areas for further research.

References

1. Allen F., Karjalainen R. (1993) Using Genetic Algorithms to Find Technical Trading Rules. Working paper. Rodney L. White Center for Financial Research, The Wharton School of the University of Pennsylvania
2. Arifovic J. (1995) Genetic Algorithms and Inflationary Economies. Journal of Monetary Economics **36**, 219–243

3. Arthur W. B. (1991) Designing Economic Agents That Act Like Human Agents: A Behavioral Approach to Bounded Rationality. American Economic Review: Papers and Proceedings, 353–360
4. Azoff E. M. (1994) Neural Networks Time Series Forecasting of Financial Markets. John Wiley, New York
5. Goldberg D. E. (1989) Genetic Algorithms in Search, Optimization and Machine Learning. Addison-Wesley, Reading, MA
6. Hénon M. (1976) A Two-dimensional Mapping with a Strange Attractor. Comm. Math. Phys. **50**, 69–77
7. Holland J. H. (1975) Adaptation in Natural and Artificial Systems. University of Michigan Press, Ann Arbor. 2nd edition 1992, MIT Press
8. Koza J. R. (1992) Genetic programming. MIT Press, Cambridge
9. Marks R. E. (1992) Breeding Optimal Strategies: Optimal Behavior for Oligopolies. Journal of Evolutionary Economics **2**, 17–38
10. Palmer R. G., Arthur W. B., Holland J. H., LeBaron B., Taylor P. (1994) Artificial Economic Life: A Simple Model of a Stockmarket. Physica D **75**, 264
11. Peitgen Heinz-Otto, Hartmut Jürgens, Dietmal Saupe (1992) Chaos and Fractals, New Frontiers of Science, Springer, Heidelberg
12. Rösler O. E. (1976). An Equation for Continuous Chaos. Physics Letters A **57**, 397–398
13. Szpiro G. G. (1997a) Forecasting Chaotic Time Series with Genetic Algorithms. Physical Review E, 2557-2568.
14. Szpiro G. G. (1997b) A Search for Hidden Relationships: Data Mining with Genetic Algorithms. Computational Economics **10 (3)**, 267–277
15. Szpiro G. G. (1997c) The Emergence of Risk Aversion. Complexity **2(4)**, 31-39.
16. Szpiro G. G. (1999) Can Computers Have Sentiments? The Case of Risk aversion and Utility for Wealth. In: Floreano D., Nicoud J-D, Mondada F. (Eds.) Advances in Artificial Life. Lecture Notes in Artificial Intelligence, Vol 1674, Springer, Heidelberg, 365–376
17. Weigend, A. (1996) Personal Communication. as quoted in the *Neue Zürcher Zeitung* (October 1996)

Appendix 1 Random Number Data Set

Training data	0.322464477	0.052910111	0.769516621
0.894388687⋆	0.771521346	0.400452254	0.195759922
0.727490384⋆	0.178755747	0.968250598	0.836776747
0.798240114⋆	0.41024004	0.022478049	0.239448676
0.152284585⋆	0.783545353	Out-of-sample data	0.096123421
0.063327661⋆	0.649381144	0.916689864	0.367669768
0.244155585	0.956871811	0.638194213	0.560753115
0.94868474	0.062085892	0.063377137	0.127924796
0.497173135	0.81558403	0.82831199	0.309430465
0.287270731	0.630649044	0.219999693	0.648009175
0.021695952	0.279647625	0.45565713	0.281787667
0.876673492	0.885965617	0.731314971	0.277066537
0.850286666	0.014120216	0.58553304	0.718874426
0.29677619	0.778923001	0.529617575	0.68857627
0.386934676	0.682676672	0.460677965	0.166630409
0.931185786	0.328324137	0.181880598	0.809892608
0.636973817	0.487538641	0.758371913	0.830017164
0.130885414	0.131379311	0.699475313	0.627290367
0.01903264	0.404736794	0.483299883	0.11235271
0.837776978	0.852682355	0.911645655	0.340281135
0.047614953	0.226111532	0.980419542	0.784658279
0.243441227	0.079728621	0.686057336	0.503265791
0.155791886	0.072586132	0.204100879	0.301017405
0.413903316	0.399945144	0.361265938	0.391517126
0.089319069	0.74688816	0.038504219	0.184317299
0.878818653	0.539510799	0.614889235	0.552588636
0.12164418	0.811865206	0.857285772	0.435631938
0.87281441	0.123368347	0.026030543	0.452624407
0.191188421	0.649721214	0.26869104	0.810125374
0.070459206	0.299097688	0.619494691	0.100152612
0.360335926	0.130647365	0.780104691	0.663872195
0.674590063	0.3015904	0.30634037	0.357545316
0.394202552	0.571234117	0.448436515	0.272594055
0.009997238	0.096055382	0.471564719	0.877717266
0.669972372	0.670286603	0.677045132	0.725554057
0.492868205	0.088734239	0.24167251	0.055342045
0.727347678	0.385997089	0.089097887	0.279894513
0.636443949	0.87141581	0.48700517	0.724169491
0.030182784	0.244034221	0.239107594	0.270900401
0.66773433	0.14235136	0.09856713	0.038800486
0.283927951	0.22665014	0.531604444	0.878609864
0.842487792	0.990939347	0.583985366	0.271119772
0.341068238	0.868931841	0.357537181	0.874560614
0.565332425	0.368359394	0.467380736	0.74591826
0.283510234	0.257627231	0.773674377	0.561809856
0.414119966	0.183794553	0.998937211	0.618811779
0.923042744	0.313509059	0.124609267	0.515773087
0.370197711	0.795856041	0.523563108	0.366490627
0.603620698	0.709254101	0.163966219	0.262819994
0.134954922	0.869514095	0.317478122	0.019453365
⋆ used as lags	0.841529191	0.634060294	

16 Forecasting Ability But No Profitability: An Empirical Evaluation of Genetic Algorithm-Optimised Technical Trading Rules

Robert Pereira

Investment Solutions & Quantitative Analytics
Merrill Lynch Investment Managers
Melbourne, Australia 3000
rocketrob1969@hotmail.com

Abstract. This paper evaluates the performance of several popular technical trading rules applied to the Australian share market. The optimal trading rule parameter values over the in-sample period of 4/1/82 to 31/12/89 are found using a genetic algorithm. These optimal rules are then evaluated in terms of their forecasting ability and economic profitability during the out-of-sample period from 2/1/90 to the 31/12/97. The results indicate that the optimal rules outperform the benchmark given by a risk-adjusted buy and hold strategy. The rules display some evidence of forecasting ability and profitability over the entire test period. But an examination of the results for the sub-periods indicates that the excess returns decline over time and are negative during the last couple of years. Also, once an adjustment for non-synchronous trading bias is made, the rules display very little, if any, evidence of profitability.

16.1 Introduction

Forecasting the future direction of share market prices is an important, but difficult exercise. Both technical and fundamental analysis have been used for this purpose, with varying success. Initial studies of technical analysis by [2] and [16] were unable to find evidence of profitability and thus concluded that technical analysis is not useful. More recently, there has been a renewed interest in this topic; see [7,10,3].

Technical analysis uses only historical data, usually consisting of only past prices but sometimes also includes volume, to determine future movements in financial asset prices. This method of forecasting is commonly used by foreign exchange dealers, who are mostly interested in the short term movements of currencies; see the survey of the London foreign exchange market by [29]. However, technical analysis is also used to forecast share prices. A survey by [11] reveals that investment analysts consider technical analysis as an important tool for forecasting the returns to different classes of assets. This widespread use of technical analysis in financial markets is surprising to most academics, since this behavior is irrational given the implications of the efficient market and random walk hypotheses for investment and speculation.

Under an efficient market it is expected that prices follow a random walk and thus past prices cannot be used successfully to forecast future prices. Therefore, the most appropriate investment strategy is the buy and hold strategy which consists of holding the market portfolio. It is not expected that any other strategy can consistently beat or outperform the market.

Although criticized by economists, most notably [25], technical analysis has received an increasing amount of attention by academics. Numerous studies examining technical trading rules applied to various shares and share market indices, have uncovered evidence of predictive ability and profitability; see [28,8,6,19,27]. There are also studies which have found some evidence of predictive ability but no profitability once reasonable adjustments are made for risk and trading costs; see [12,20,7,10,3].

The majority of these studies have examined trading rules where both the rules and their parameter values were chosen arbitrarily. However this approach leaves these studies open to the criticisms of data-snooping and the possibility of a survivorship bias; see [23] and [9] respectively. By choosing trading rules based on an optimisation procedure utilising in-sample data and testing the performance of these rules out-of-sample, this bias can be avoided or at least reduced. This approach is taken by [26] and [3], by employing a genetic programming approach to discover optimal technical trading rules for the foreign exchange market and US share market respectively.

In this paper the forecasting ability and economic profitability of some popular technical trading rules applied to the Australian share market are investigated using a standard genetic algorithm optimisation procedure. [1] The approach adopted in this study differs from the genetic programming approach for two reasons. First, since the objective of this study is not to discover new trading rules but rather to examine popular, commonly used trading rules.[2] Second, there is a potential problem associated with the use of genetic programming, since this artificial intelligence technique was only recently developed by [21]. Therefore it is unrealistic to evaluate the performance of trading rules discovered by the genetic programming approach prior to the date of the development of this technique.

The next section of the paper describes the technical trading rules examined in this study. Section 16.3 develops the genetic algorithm methodology used in trading rule optimisation. Section 16.4 explains the performance measures that are used to evaluate trading rule forecasting ability and economic profitability. Section 16.5 considers an empirical investigation of the perfor-

[1] To the author's knowledge there is only one other study considering the performance of technical trading rules applied to the Australian share market. The performance of the filter rule applied to various individual Australian shares was examined by [4], who was unable to find any significant evidence of profitability.

[2] Focusing exclusively on the popular and commonly used rules, does introduce the possibility of a survivorship bias. But since it is difficult, if not impossible, to include the entire universe of all technical trading rules, there is always a danger of survivorship bias in any performance study.

mance of the genetic algorithm-optimised technical trading rules applied to the Australian share market. Finally, Section 16.6 provides some conclusions and directions for possible future research.

16.2 Technical Trading Rules

Trading rules are used by financial market traders to assist them in determining their investment or speculative decisions. These rules can be based on either technical or fundamental analysis. This study considers only rules based on technical indicators. A technical indicator is a mathematical formula that transforms historical data on price and/or volume into a single number. These indicators can be combined with price, volume or each other to form trading rules. Some of the more popular technical indicators used by traders include: channels, filters, momentum, moving averages and relative strength indices. Reference [1] provides an excellent description of the different technical indicators used in trading.

16.2.1 Determination of the Investment Position

Trading rules return either a buy or sell signal which together with a particular trading strategy determines the trading position that should be taken in a security or market. The trading strategy considered in this study is based on a simple market timing strategy, consisting of investing total funds in either the share market or a risk free security. If share market prices are expected to increase on the basis of a buy signal from a technical trading rule, then the risk free security is sold and shares are bought. However, if the rule returns a sell signal, it is expected that share market prices will fall in the near future. As a result, shares are sold and the proceeds from the sale invested in the risk free security.[3]

16.2.2 Different Types of Rules

Two general types of technical trading rules are considered - rules based on either moving averages or order statistics.

Moving average rules Moving averages are used to identify trends in prices. A moving average (MA) is simply an average of current and past prices over a specified period of time. An MA of length θ is calculated as

[3] This strategy excludes the possibility of short selling, which in general is difficult to conduct in the Australian share market due to certain legal restrictions. By using a stock index futures contract, the market portfolio can be sold short to establish a negative position in order to profit from a fall in prices. This is not considered here, but left for possible future work.

$$MA_t(\theta) = \frac{1}{\theta}\sum_{i=0}^{\theta-1} P_{t-i} \tag{16.1}$$

where

$$\forall\theta \in \{1, 2, 3, ...\}.$$

By smoothing out the short-term fluctuations or noise in the price series, the MA is able to capture the underlying trend in the price series over a particular period of time. An MA can be used to formulate a simple trend-following rule also referred to as a momentum strategy.

A simple MA rule can be constructed by comparing price to its trend, as represented by the MA. If the price rises above the MA, then the security is bought and held until the price falls below the MA at which time the security is sold. This simple rule can be modified to create the filtered MA rule and the double MA rule. A filtered MA rule is similar to the simple MA rule, except it includes a filter which accounts for the percentage increase or decrease of the price relative to its MA. The purpose of this filter is an attempt to reduce the number of false buy and sell signals, which are issued by a simple MA rule when price movement is nondirectional. This rule operates by returning a buy signal if the price rises by X percent above the MA and then returning a sell signal only when the price falls by X percent below the MA at which time the security is sold. In contrast to the previous two rules, a double MA rule compares two MAs of different lengths. With this rule if the shorter length MA rises above the longer length MA from below then the security is bought and held until the shorter MA falls below the longer MA at which time the security is sold.

A more general MA rule can be specified by considering two moving averages and a filter.[4] This Generalised MA (GMA) rule can be represented by the binary indicator function

$$S(\Theta)_t = MA_t(\theta_1) - \left(1 + (1 - 2S_{t-1})\frac{\theta_3}{10^4}\right) MA_t(\theta_2) \begin{cases} > 0, 1 \\ \leq 0, 0 \end{cases} \tag{16.2}$$

where

$$\forall\theta_1, \theta_2 \in \{1, 2, 3, 4, ...\}, \theta_1 < \theta_2$$
$$\forall\theta_3 \in \{0, 1, 2, ...\}$$

and the MA indicator is defined by Equation 16.1. This function returns either a one or zero, corresponding to a buy or sell signal respectively, which indicates the trading position that should be taken at time t. The lengths of the short and long MAs are given by parameters θ_1 and θ_2, which represent

[4] Obviously, this general rule could be extended to include more MAs, filters and other technical indicators.

the number of days used to calculate the MAs. The parameter θ_3 represents the filter parameter in terms of basis points; where one hundred basis points equivalent to one percent.

The three different MA rules discussed above are nested within the GMA rule. These rules can be derived individually by imposing certain restrictions on Equation 16.2:

1. *Simple MA*: $\theta_1 = 1$, $\theta_2 > 1$ and $\theta_3 = 0$

$$S(\Theta)_t = P_t - MA_t(\theta_2) \begin{cases} > 0, 1 \\ \leq 0, 0 \end{cases}$$

2. *Filtered MA*: $\theta_1 = 1$, $\theta_2 > 1$ and $\theta_3 > 0$

$$S(\Theta)_t = P_t - \left(1 + (1 - 2S_{t-1})\frac{\theta_3}{10^4}\right) MA_t(\theta_2) \begin{cases} > 0, 1 \\ \leq 0, 0 \end{cases}$$

3. *Double MA*: $1 < \theta_1 < \theta_2$ and $\theta_3 = 0$

$$S(\Theta)_t = MA_t(\theta_1) - MA_t(\theta_2) \begin{cases} > 0, 1 \\ \leq 0, 0. \end{cases}$$

Rules based on order statistics Technical trading rules can also be based on order statistics, such as the maximum and minimum prices over a specified period of time. The filter and channel rules are two examples which use local maximum and minimum prices. Reference [8] refer to this rule as the trading range break-out rule. The maximum price $P_t^{\max}(\phi)$ and the minimum price $P_t^{\min}(\phi)$ at time t given a historical price series consisting of ϕ observations are

$$P_t^{\max}(\phi) = Max\left[P_{t-1}, ..., P_{t-\phi}\right] \quad (16.3)$$
$$P_t^{\min}(\phi) = Min\left[P_{t-1}, ..., P_{t-\phi}\right] \quad (16.4)$$

where

$$\forall \phi \subset \{1, 2, 3, ...\}.$$

The channel rule is founded on the idea of support and resistance levels which are related to the market forces of demand and supply. The support level is achieved at a price where buying power dominates selling pressure, effectively placing a floor on the level of prices. The resistance level, which is the opposite of support, is defined as the price where selling pressure exceeds buying power forcing down the price and effectively creating an upper level or ceiling in prices. With the channel rule the resistance (or support level) is defined using the maximum (or minimum) price over the most recent historical period of prices consisting of ϕ observations as defined by Equations 16.3 and 16.4 respectively. This rule returns a buy (or sell) signal when price breaks through its current resistance (or support) level from below (or above) to above (or below) this level.

The filter rule is based on the idea that when price rises above (or drops below) a certain level, it will continue to rise (or fall) for some period of time. The filter rule operates by returning a buy signal when price increases by X percent above a previous low and a sell signal once the price falls by X percent below a previous high. The original filter rule of Alexander (1964) defines the previous low (or high) implicitly using the minimum (or maximum) price from a historical series commencing on the date of the most recent transaction.

The filter rule can be generalised by explicitly choosing the amount of data to use in order to determine the previous low or high. This can be done by introducing a parameter ϕ which specifies a fixed length for the historical price series used to calculate the maximum or minimum price, similar to the channel rule. Furthermore the original channel rule as outlined above, can also be generalised by introducing a filter parameter. Similar to the filtered MA rule, this rule will only return a buy (or sell) signal if the price exceeds the maximum (or minimum) price by X percent.

Since both the channel and filter rules use order statistics, these two rules can be nested within a single decision rule. This Generalised Order Statistic (GOS) rule (Generalised Order Statistic rule (GOS rule)) is represented by the indicator function

$$S(\Phi)_t = P_t - \left(1 + (1 - 2S_{t-1})\frac{\phi_2}{10^4}\right)\left(P_t^{\max}(\phi_1)\right)^a \left(P_t^{\min}(\phi_1)\right)^b \begin{cases} > 0, & 1 \\ \leq 0, & 0 \end{cases} \tag{16.5}$$

where

$$a = \phi_3 S_{t-1} + (1 - \phi_3)(1 - S_{t-1})$$
$$b = \phi_3(1 - S_{t-1}) + (1 - \phi_3)S_{t-1}$$

$$\forall\, \phi_1, \phi_2 \in \{1, 2, 3, ...\}.$$

The parameter ϕ_1 represents the length of the historical price series used in determining either the maximum or minimum price, ϕ_2 is the filter parameter given in basis points and ϕ_3 is a binary parameter defined as

$$\phi_3 = \begin{cases} 1, & \text{Filter rule} \\ 0, & \text{Channel rule} \end{cases} \tag{16.6}$$

where the value one and zero represent the filter and channel rules respectively. The rule can be generalised further by allowing the parameter ϕ_1 to be derived either implicitly (as is the case with the channel rule) or explicitly (as is the case with the filter rule).

The filter and channel rules outlined above, can be derived from the GOS rule by imposing certain restrictions on Equation 16.5:

1. *Filter rule*; ϕ_1 = number of days since last transaction, $\phi_2 > 0$ and $\phi_3 = 1$

$$S_t = P_t - \left(1 + (1 - 2S_{t-1})\frac{\phi_2}{10^4}\right)\left(P_t^{\max}(\phi_1)\right)^{S_{t-1}}\left(P_t^{\min}(\phi_1)\right)^{(1-S_{t-1})}\begin{cases} > 0, & 1 \\ \leq 0, & 0 \end{cases}$$

2. *Channel rule*; $\phi_1 > 0$, $\phi_2 = 0$ and $\phi_3 = 0$

$$S_t = P_t - \left(P_t^{\max}(\phi_1)\right)^{(1-S_{t-1})}\left(P_t^{\min}(\phi_1)\right)^{S_{t-1}}\begin{cases} > 0, & 1 \\ \leq 0, & 0 \end{cases}$$

16.3 Genetic Algorithm Methodology

16.3.1 Optimisation

The choice of technical trading rule parameter values has a profound impact on the profitability of these rules. In order to maximise trading rule profitability, parameter values must be chosen optimally. In this optimisation problem, it is important to be aware of two issues. First, there are a large number of possible parameter values. Second, the profit surface is characterised by multiple optima; see [26] and [3]. Genetic algorithms are a very efficient and effective approach to this type of problem.

Efficiency refers to the computational speed of the optimisation technique. Through a recombination procedure known as crossover and by maintaining a population of candidate solutions, the genetic algorithm is able to search quickly through the profitable areas of the solution space. Effectiveness refers to the global optimisation properties of the algorithm. Unlike other search or optimisation techniques based on gradient measures, a genetic algorithm avoids the possibility of being anchored at local optima due to its ability to introduce random shocks into the search process through mutations. Since a genetic algorithm is an appropriate global optimisation method, it can be used to search for the optimal parameter values for the GMA and GOS trading rules given by Equations 16.2 and 16.5 respectively.[5]

Genetic algorithms were originally developed by [18]. They are a class of adaptive search and optimisation techniques based on an evolutionary process. By representing potential or candidate solutions to a problem using vectors consisting of binary digits or bits, mathematical operations known as crossover and mutation, can be performed. These operations are analogous to the genetic recombinations of the chromosomes in living organisms. By performing these operations, generations of new candidates can be created and evolved over time through an iterative procedure. However, there do exist restrictions on the process of crossover so as to ensure that better performing

[5] Another global optimisation technique is simulated annealing. However, this study focuses exclusively on a standard genetic algorithm.

candidates are evolved over time. Similar to the theory of natural selection or survival of the fittest, the better performing candidates have a better than average probability of surviving and reproducing relative to the lower performing candidates which eventually get eliminated from the population. The performance of each candidate can be assessed using a suitable objective function. A selection process based on performance is applied to determine which of the candidates should participate in crossover, and thereby pass on their favourable traits to future generations. It is through this process of "survival of the fittest" that better solutions are developed over time. This evolutionary process continues until the best (or better) performing individual(s), consisting of hopefully the optimal or near optimal solutions, dominate the population.[6]

16.3.2 Problem Representation

Potential solutions to the problem of optimisation of the parameters of the GMA rule defined in Equation 16.2 can be represented by the vector

$$\mathbf{y}_1 = [\theta_1, \theta_2, \theta_3]. \tag{16.7}$$

For the GOS rule given by Equation 16.5, candidates can be represented by the vector

$$\mathbf{y}_2 = [\phi_1, \phi_2, \phi_3, \phi_4] \tag{16.8}$$

where ϕ_3 is defined above given by Equation 16.6, while ϕ_4 is a dummy variable defined by

$$\phi_4 = \begin{cases} 1, & \phi_1 \text{ is determined implicitly} \\ 0, & \phi_1 \text{ is determined explicitly.} \end{cases} \tag{16.9}$$

In order to use a genetic algorithm to search for the optimal parameter values for the rules considered above, potential solutions to this optimisation problem are represented using vectors of binary digits. Binary representation is necessary in the standard genetic algorithm for the application of the recombination operations. These vectors also known as strings, are linear combinations of zeros and ones, for example [0 1 0 0 1]. A binary representation $\mathbf{x} = [x_1, x_2, x_3, ..., x_n]$ is based on the binary number system which has a corresponding equivalent decimal value given by $\sum_{i=1}^{n}(2^{n-i})x_i$. For example, the decimal equivalent of the vector $[0\ 1\ 0\ 0\ 1] = (2^4 \times 0) + (2^3 \times 1) + (2^2 \times 0) + (2^1 \times 0) + (2^0 \times 1) = 8 + 1 = 9$.

[6] [17] provides a detailed description of the mathematical operations involved, the programming and applications of genetic algorithms.

Binary representation of the GMA rule The periodicity of the two MAs have a range defined by $1 < \theta_1 \leq L_1$ and $2 < \theta_2 \leq L_1$, where L_1 represents the maximum length of the moving average. The filter parameter has a range given by $0 \leq \theta_3 \leq L_2$, where L_2 represents the maximum filter value. For this study $L_1 = 250$ days and $L_2 = 100$ basis points.[7]

In order to satisfy the limiting values given above, the binary representations for θ_1 and θ_2 are each given by a vector consisting of eight elements. For the filter parameter (θ_3) a seven bit vector is required. Therefore, the binary representation for the GMA rule can be defined by a row vector consisting of twenty three elements stated as

$$\mathbf{x}_1 = [\mathbf{x}_{11}, \mathbf{x}_{12}, \mathbf{x}_{13}] \tag{16.10}$$

where

$\mathbf{x}_{11}$ = subvector consisting of eight elements (binary representation of θ_1)
$\mathbf{x}_{12}$ = subvector consisting of eight elements (binary representation of θ_2)
$\mathbf{x}_{13}$ = subvector consisting of seven elements (binary representation of θ_3).

Binary representation of the GOS rule The parameter on the channel rule ϕ_1 represents the number of the most recent historical observations used to calculate either the maximum or minimum price. This parameter is restricted to the values $1 \leq \phi_1 \leq 250$. Therefore, the binary representation is given by a vector consisting of eight elements. The range for ϕ_2 is given by $0 < \phi_2 \leq 1000$ basis points. Thus a vector consisting of ten elements is used and the decimal equivalent values are restricted to the desired range. Therefore, the binary representations for the order statistics based rule can be defined by a row vector consisting of twenty elements stated as

$$\mathbf{x}_2 = [\mathbf{x}_{21}, \mathbf{x}_{22}, \mathbf{x}_{23}, \mathbf{x}_{24}] \tag{16.11}$$

where

$\mathbf{x}_{21}$ = subvector consisting of eight elements (binary representation of ϕ_1)
$\mathbf{x}_{22}$ = subvector consisting of ten elements (binary representation of ϕ_2)
$\mathbf{x}_{23}$ = subvector consisting of one element (binary representation of ϕ_3)
$\mathbf{x}_{24}$ = subvector consisting of one element (binary representation of ϕ_4).

16.3.3 Objective Function

The ultimate goal of the genetic algorithm is to find the combination of binary digits for the two vectors $\mathbf{x}_1$ and $\mathbf{x}_2$, representing the parameter values

[7] These limiting values are consistent with what is used in practice. Also, results from a preliminary investigation, indicated that higher parameter values generally produced losses.

given by $\mathbf{y}_1$ and $\mathbf{y}_2$, which maximises an appropriate objective function. Each candidate's performance can be assessed in terms of this objective function, which can take numerous forms depending upon specific investor preferences. Given that individuals are generally risk averse, performance should be defined in terms of both risk and return. The Sharpe ratio is an example of a measure of risk-adjusted returns. The Sharpe ratio is given by

$$SR = \frac{\bar{r}}{\sigma\sqrt{Y}} \tag{16.12}$$

where $\bar{r}$ is the average annualised trading rule return, σ is the standard deviation of daily trading rule returns, while Y is equal to the number of trading days per year. This formulation is actually a modified version of the original Sharpe ratio which uses average excess returns, defined as the difference between average market return and the risk-free rate.

Trading rules as defined by the indicator functions given in Equations 16.2 and 16.5, return either a buy or sell signal. These signals can be used to divide the total number of trading days (N), into days either "in" the market (earning the market rate of return r_{mt}) or "out" of the market (earning the risk-free rate of return r_{ft}). Thus the trading rule return over the entire period of 0 to N can be calculated as

$$r_{tr} = \sum_{t=1}^{N} S_{t-1} r_{m,t} + \sum_{t=1}^{N} (1 - S_{t-1})\, r_{f,t} - T(tc) \tag{16.13}$$

where

$$r_{m,t} = \ln\left(\frac{P_t}{P_{t-1}}\right)$$

which includes the summation of the daily market returns for days "in" the market and the daily returns on the risk-free security for days "out" of the market. An adjustment for transaction costs is given by the last term on the right hand side of Equation 16.13 which consists of the product of the cost per transaction (tc) and the number of transactions (T). Transaction costs of 0.2 percent per trade are considered for the in-sample optimisation of the trading rules.

16.3.4 Operations

Selection, crossover and mutation are the three important mathematical operations in any genetic algorithm. It is through these operations that an initial population of randomly generated solutions to a problem can be evolved, through successive generations, into a final population consisting of a potentially optimal solution. The search process which ensues is highly efficient and effective because of these operations.

Selection involves the determination of the candidates for participation in crossover. The genitor selection method, a ranking-based procedure developed by [30], is used in the genetic algorithm employed in this study. This approach involves ranking all candidates according to performance and then replacing the worst performing candidates by copies of the better performing candidates. In the genetic algorithm developed in this paper a copy of the best candidate replaces the worst candidate.

The method by which promising (better performing) candidates are combined, is through a process of binary recombination known as crossover. This ensures that the search process is not random but consciously directed into promising regions of the solution space. As with selection there are a number of variations, however single point crossover is the most commonly used version and the one adopted in this study.

To illustrate the process of crossover, assume that two vectors $A = [1\ 0\ 1\ 0\ 0]$ and $B = [0\ 1\ 0\ 1\ 0]$ are chosen at random and that the position of partitioning is randomly chosen to be between the second and third elements of each vector. Vectors A and B can be represented as $\left[1\ 0 \vdots 1\ 0\ 0\right] = [A_1\ A_2]$ and $\left[0\ 1 \vdots 0\ 1\ 0\right] = [B_1\ B_2]$ respectively, in terms of their subvectors. Recombination occurs by switching subvector A_2 with B_2 and then unpartioning both vectors A and B, producing two new candidates $C = [1\ 0\ 0\ 1\ 0]$ and $D = [0\ 1\ 1\ 0\ 0]$.

In contrast to crossover, mutation involves the introduction of random shocks into the population, by slightly altering the binary representation of candidates. This increases the diversity in the population and unlike crossover, randomly re-directs the search procedure into new areas of the solution space which may or may not be beneficial. This action underpins the genetic algorithms ability to find novel inconspicuous solutions and avoid being anchored at local optimum solutions. Mathematically, this operation is represented by switching a binary digit from a one to a zero or vice versa. However, the probability of this occurrence is normally very low, so as to not unnecessarily disrupt the search process.

This operation can be illustrated by an example. Assume that the third element in vector $C = [1\ 0\ 0\ 1\ 0]$ undergoes mutation. The outcome of this operation changes the binary representation of vector C slightly, producing a new candidate represented by $E = [1\ 0\ 1\ 1\ 0]$.

16.3.5 Procedure

The genetic algorithm procedure can be summarised by the following steps:

(a) Create an initial population of candidates randomly.
(b) Evaluate the performance of each candidate.
(c) Select the candidates for recombination.

(d) Perform crossover and mutation.
(e) Evaluate the performance of the new candidates.
(f) Return to step 3, unless a termination criterion is satisfied.

The last step in the genetic algorithm involves checking a well-defined termination criterion. If this criterion is not satisfied, the genetic algorithm returns to the selection, crossover and mutation operations to develop further generations until this criterion is met, at which time the process of the creation of new generations is terminated. The termination criterion adopted, is satisfied when either one of the following conditions is met:

(a) the population converges to a unique individual,
(b) a predetermined maximum number of generations is reached,
(c) there has been no improvement in the population for a certain number of generations.

This latter condition ensures that the genetic algorithm cannot continue indefinitely.

16.3.6 Parameter Settings

The genetic algorithm has six parameters settings $\{b, p, c, m, G_1^{\max}, G_2^{\max}\}$, defined as:

b = number of elements in each vector,
p = number of vectors or candidates in the population,
c = probability associated with the occurrence of crossover,
m = probability associated with the occurrence of mutation,
$G_1^{\max}$ = maximum number of generations allowed,
$G_2^{\max}$ = maximum number of iterations without improvement.

These parameters can effect both the efficiency and effectiveness of the genetic algorithm search. Small (large) values for $b, p, m, G_1^{\max}, G_2^{\max}$ and a large (small) value of c result in rapid (slow) convergence; see [17] or [5] for an explanation of these results. The greater the rate of convergence the lower the computational time that is required to reach the solution. However, if the rate of convergence is too rapid the solution space is not adequately searched, potentially missing the optimal solution (or better solutions to the one found).

Table 16.1 displays the parameter values that are used by the genetic algorithm for each of the trading rules. The choice of b values was discussed above in Section 16.3.1, while the choice of values for p, c, m, $G_1^{\max}$ and $G_2^{\max}$ was guided by previous studies (see [5], Chapter 7) and experimentation with different values.

Table 16.1. Genetic algorithm parameters

Rule	b	p	c	m	$G_1^{\max}$	$G_2^{\max}$
GMA	23	150	0.6	0.005	250	150
GOS	20	150	0.6	0.005	250	150

16.4 Performance Evaluation

16.4.1 Economic Profitability

The true profitability of technical trading rules is hard to measure given the difficulties in properly accounting for the risks and costs associated with trading. Trading costs include not only transaction costs and taxes, but also hidden costs involved in the collection and analysis of information. Transaction costs of 0.1 percent per trade are used to investigate trading rule performance. Since according to [28], large institutional investors are able to achieve one-way transaction costs in the range of 0.1 to 0.2 percent. However, given that different individuals face different levels of transaction costs, the break-even transaction cost is also reported in the results section; see [6]. This is the level of transaction costs which offsets trading rule revenue with costs, leading to zero trading profits.

To evaluate trading rule profitability, it is necessary to compare trading rule returns to an appropriate benchmark. Since the trading rules considered in this paper restrict short selling, they do not always lead to a position being held in the market and therefore are less risky than a passive buy and hold benchmark strategy, which always holds a long position in the market. Therefore, the appropriate benchmark is constructed by taking a weighted average of the return from being long in the market and the return from holding no position in the market and thus earning the risk free rate of return. The return on this risk-adjusted buy and hold strategy can be written as

$$r_{bh} = \alpha \sum_{t=1}^{N} r_{f,t} + (1-\alpha) \sum_{t=1}^{N} r_{m,t} - 2(tc) \tag{16.14}$$

where α is the proportion of trading days that the rule is out of the market. This return represents the expected return from investing in both the risk-free asset and the market according to the weights α and $(1-\alpha)$ respectively. There is also an adjustment for transaction costs incurred due to purchasing the market portfolio on the first trading day and selling it on the last day of trading.

Therefore, trading rule performance relative to the benchmark can be measured by excess returns

$$XR = r - r_{bh} \tag{16.15}$$

where r represents the total return for a particular trading rule calculated from Equation 16.13 and r_{bh} is the return from the appropriate benchmark strategy given by Equation 16.14. Since investors and traders also care about the risk incurred in deriving these returns, a Sharpe ratio based on excess returns can be calculated using Equation 16.12, where $\bar{r}$ represents annualised excess returns given by Equation 16.15 and σ is the standard deviation of the daily excess returns.

16.4.2 Predictive Ability

To investigate the statistical significance of the forecasting power of the buy and sell signals, traditional t tests can be employed to examine whether the trading rules issue buy (or sell) signals on days when the return on the market is on average higher (or lower) than the unconditional mean return for the market.

The t-statistic used to test the predictive ability of the buy signals is

$$t_{buy} = \frac{\bar{r}_{buy} - \bar{r}_m}{\sigma\sqrt{\frac{1}{N_{buy}} + \frac{1}{N}}} \tag{16.16}$$

where $\bar{r}_{buy}$represents the average daily return following a buy signal and N_{buy} is the number of days that the trading rule returns a buy signal. The null and alternative hypotheses can be stated as

$$H_0 : \bar{r}_{buy} \leq \bar{r}_m$$
$$H_1 : \bar{r}_{buy} > \bar{r}_m.$$

Similarly, a t-statistic can be developed to test the predictive ability of the sell signals. To test whether the difference between the mean return on the market following a buy signal and the mean return on the market following a sell signal is statistically significant, a t-test can be specified as

$$t_{buy-sell} = \frac{\bar{r}_{buy} - \bar{r}_{sell}}{\sigma\sqrt{\frac{1}{N_{buy}} + \frac{1}{N_{sell}}}} \tag{16.17}$$

where the null and alternative hypotheses are

$$H_0 : \bar{r}_{buy} - \bar{r}_{sell} \leq 0$$
$$H_1 : \bar{r}_{buy} - \bar{r}_{sell} > 0.$$

Another test of whether the rules have market timing or forecasting ability is based on an approach suggested by [13]. This test is based on the following regression

$$r_{m,t} - r_{f,t} = \alpha + \beta S_t + \varepsilon_t \tag{16.18}$$

where $r_{m,t}$ and $r_{f,t}$ are the return at time t for the risky asset or market portfolio and the risk-free security respectively, ε_t is a standard error term and S_t is the trading rule signal. To test whether a particular rule has market timing ability the regression given in Equation 16.18 is estimated using OLS and the following hypothesis test is conducted

$$\begin{aligned} H_0 &: \beta = 0, \text{ no market timing ability} \\ H_1 &: \beta > 0, \text{ positive market timing ability.} \end{aligned}$$

16.4.3 Statistical Significance

The bootstrap method proposed by [15] has been applied in finance for a wide variety of purposes; see [24]. Reference [22] use this method for the purpose of testing the significance of trading rule profitability, while [8] use trading rules on bootstrapped data as a test for model specification. In this study, a bootstrap approach similar to [22] is used to test the significance of both the predictive ability and the profitability of technical trading rules.

To use the bootstrap method a data generating process (DGP) for market prices or returns must be specified a priori. The DGP assumed for prices in this study is the simple random walk with drift

$$\begin{aligned} \ln P_{t+1} &= \mu + \ln P_t + \varepsilon_t \\ \varepsilon_t &\sim IID\ N(0, \sigma^2) \end{aligned} \tag{16.19}$$

where μ represents the drift in the series, $\ln P$ is the natural logarithm of the price and ε is the stochastic component of the DGP. Since continuously compounded returns are defined as the log first difference of prices, then the above DGP given in Equation 16.19 implies an IID normal process with a mean of zero for returns.

The bootstrap method can be used to generate many different return series by sampling with replacement from the original return series. The bootstrap samples created are pseudo return series that retain all the distributional properties of the original series, but are purged of any serial dependence. Each bootstrap sample also has the property that the DGP of prices is a random walk with drift. From each bootstrap a corresponding price series can be extracted, which can then be used to test the significance of the predictive ability or profitability of a particular rule. This is done by applying the rule to each of the pseudo price series and calculating the empirical distribution of the trading rule profits or the statistic of interest. P-values can then be calculated from this distribution.

To test the significance of the trading rule excess returns the following hypothesis can be stated

$$H_0: \quad XR \leq \bar{XR}^*$$
$$H_1: \quad XR > \bar{XR}^*.$$

Under the null hypothesis, the trading rule excess return (XR) calculated from the original series is less than or equal to the average trading rule return for the pseudo data samples ($\bar{XR}^*$). The p-values from the bootstrap procedure are then used to determine whether the trading rule excess returns are significantly greater than the average trading rule return given that the true DGP is a random walk with drift. In a similar way, the market returns following buy and sell signals and the Sharpe ratio can also be bootstrapped to test the significance of the predictive ability and profitability of the trading rule studied.

The bootstrap procedure involves the following steps:

(a) Create Z bootstrap samples, each consisting of N observations, by sampling with replacement from the original return series.
(b) Calculate the corresponding price series for each bootstrap sample given that the price next period is $P_{t+1} = exp(r_{t+1})P_t$.
(c) Apply the trading rule to each of the Z pseudo price series.
(d) Calculate the performance statistic of interest for each of the pseudo price series.
(e) Determine the P-value by calculating the number of times the statistic from the pseudo series exceed the statistic from the original price series.

16.5 An Empirical Application

16.5.1 Data

The data consists of the daily closing All Ordinaries Accumulation index and the daily 90 day Reserve Bank of Australia bill dealer rate. The data is collected over the period 4/1/1982 to 31/12/97, consisting of 4065 observations.[8] For the purpose of avoiding the possibility of data-snooping, the total period is split into an in-sample optimisation period from 4/1/1982 to 31/12/89 and an out-of-sample test period from 2/1/1990 to 31/12/97.

Table 16.2 provides summary statistics for the continuously compounded daily returns (r_m) on the All Ordinaries index.[9] The return series has characteristics common to most financial time series and the results are broadly consistent with previous studies. The autocorrelation coefficients are also

[8] The data was obtained from the Equinet Pty Ltd data base.

[9] The continuously compounded daily returns are calculated as the natural logarithm of the first difference of the orignal price series.

Table 16.2. Summary statistics for daily returns

	1982-97	1982-89	1990-97
Sample size	4064	2028	2035
Mean	0.0533	0.0676	0.0381
Std. dev.	1.0214	1.1956	0.8107
Skewness	-6.0385	-7.4628	-0.3231
Kurtosis	163.7916	173.2427	8.9921
Maximum	6.2228	5.5994	6.2228
Minimum	-28.7495	-28.7495	-7.4286
ρ_1	0.1114	0.1214	0.1222
ρ_2	-0.0400	-0.0413	-0.0413
ρ_3	0.0696	0.1119	0.1111
ρ_4	0.0869	0.1300	0.1295
ρ_5	0.0493	0.0751	0.0752
ABP(10)	22.7308	19.1460	19.2501
ABP(30)	53.8039	46.1164	46.3069

reported for the first five lags. These coefficients show evidence of highly significant low-order positive autocorrelation.[10] Significant first-order serial correlation in share indices is a well known stylized fact due to the inclusion of thinly-traded small shares in share market indices. Therefore, this result is not surprising given that the All Ordinaries Accumulation index is comprised of over 300 shares, which includes a significant amount of small shares. There also appears to be significant higher order serial correlation as indicated by the heteroscedasticity-adjusted Box-Pierce Q statistic (ABP). These results seem to be consistent across the in-sample optimisation and out-of-sample test periods.

16.5.2 Trading Rule Parameter Values

A genetic algorithm was programmed and then used to search for the optimal parameter values using the All Ordinaries Accumulation index data during the in-sample optimisation period.[11] The GA-optimal parameter values for the trading rules found during the in-sample period based on transaction costs of 10 basis points are reported in Table 16.3. The returns and the Sharpe ratios are high, even compared to the buy and hold return of 18.79 percent per annum and the corresponding Sharpe ratio of 0.86 percent per unit of standard deviation. The best GMA rule can be described as a 14 day MA rule with a 64 basis point filter, while the best GOS rule can be described as a 9 day channel rule with a 21 basis point filter.

[10] The 95% confidence interval is ± 0.0314, which is calculated using the formula $\pm\frac{2}{\sqrt{n}}$, where n is the number of observations.

[11] All programs are written in GAUSS, version 3.2, which are available from the author upon request.

Table 16.3. Trading rule parameter values

Rule	Parameter values	$\bar{r}$	SR	No.	$\bar{r}$	SR	Best after	No. of iters	Time (mins)
Best over ten trials					Average over 10 trials				
GMA	(1,14,64)	36.1	3.0	8	35.8	3.0	119	229.7	203.9
GOS	(9,21,0,0)	36.0	3.1	10	36.0	3.1	123	245.3	207.8

The Sharpe ratio (SR) is calculated as the ratio of annualised returns ($\bar{r}$) to standard deviation. The number of times the best rule was found in 10 trials (No.) is given in the fifth column. The average number of iterations completed until the best rule was found (Best after) is reported in column 8. The average number of iterations completed for one trial (No. of iters) is reported in the second last column.

In order to investigate the important properties of effectiveness and efficiency, the genetic algorithm is run over ten trials for each rule. The effectiveness of the genetic algorithm's ability to search for the optimal parameter values is investigated by observing how many times the best rule is found over ten trials; given in the fifth column of Table 16.3. It appears that the genetic algorithm is reasonably efficient, since in 90 percent of the runs the genetic algorithm has found the same best rule.

Another measure of the effectiveness of the genetic algorithm is to find good rules in terms of the performance criteria used, not necessarily the best or optimal rule. This is evaluated by considering the average of the annualised returns and Sharpe ratios over the ten trials for each rule. The results indicate that the genetic algorithm may not have perfect accuracy, but on average finds rules very close to the best rule.

The efficiency of the genetic algorithm as a search or optimisation technique is measured by considering the time it takes to find good rules. On average the genetic algorithm took over three and half hours to run, whereas an exhaustive grid-search procedure would have taken many hours, if not days.[12] Obviously, a more efficient genetic algorithm could be developed, but this is not pursued in this study.

16.5.3 Performance Evaluation

It should not be surprising to observe high in-sample performance for the genetic algorithm-optimised trading rules. Rather, it is more interesting and important to examine how these rules perform out-of-sample.

Economic profitability The out-of-sample performance statistics are reported in Table 16.4. In terms of the annualised excess returns (XR) and the corresponding Sharpe ratio (SR), both rules are able to outperform the

[12] All genetic algorithm runs were conducted on a Pentium 233 MHz desktop PC.

Table 16.4. Performance statistics for GA-optimised share market rules

Rule	$\bar{r}$	XR	SR	tc^*	$\bar{T}$	T_w/T	$\bar{r}_w$	$\bar{r}_L$	$Max\ D$
Panel A: Full Sample (1990-97)									
GMA	10.82	2.45	0.37	53	7.36	44.07	1.72	-0.88	-14.10
GOS	11.16	2.91	0.44	62	6.86	45.45	1.92	-0.98	-10.97
Panel B: Sub-period results for the GMA rule									
1990-91	12.48	5.91	0.82	99	7.53	40.00	1.87	-0.97	-6.11
1992-93	16.53	3.59	0.62	69	7.44	46.67	2.20	-0.81	-7.64
1994-95	8.02	2.83	0.45	58	7.50	46.67	1.27	-0.78	-5.40
1996-97	5.20	-3.37	-0.46	0	8.53	41.18	1.20	-0.82	-14.31
Panel C: Sub-period results for the GOS rule									
1990-91	11.27	4.64	0.65	87	7.03	35.71	2.58	-1.24	-9.35
1992-93	16.36	3.73	0.64	79	6.45	53.85	2.09	-0.87	-9.10
1994-95	8.35	3.16	0.51	64	7.00	42.86	1.50	-0.71	-5.67
1996-97	7.01	-1.39	-0.19	3	8.53	47.06	1.31	-0.94	-11.02

The Sharpe ratio (SR) is the ratio of annualised excess returns (XR) to standard deviation. The break-even level of transaction cost is given by tc^*. Trading frequency $\bar{T}$ is measured by the average number of trades per year. T_w/T represents the proportion of trades that yield positive excess returns. The average excess return on winning and losing trades is given by $\bar{r}_w$ and $\bar{r}_L$ respectively. The maximum drawdown $Max\ D$ represents the largest drop in cumulative excess returns.

appropriate benchmarks after allowing for transaction costs of 10 basis points per trade. These results remain positive, as long as transaction costs are below 0.53 and 0.62 percent per trade as indicated by the break-even costs (tc^*). The rules trade roughly seven times per year and produce positive excess returns for approximately 45 percent of the trades. An indication of their riskiness is given by the maximum drawdown ($Max\ D$), which measures the largest drop in the cumulative excess return series. In terms of this measure of risk, both rules are much less risky than the buy and hold, which has a maximum drawdown of -69 percent.

The robustness of the results is investigated across different non-overlapping sub-periods. Four 2 year sub-periods are investigated during the out-of-sample period from 1990 to 1997. A sub-period analysis of the performance results, show that this good performance deteriorates over time. In the last couple of years neither rule is able to outperform the benchmark.

Predictive ability To examine the forecasting ability of the rules, the signals are investigated both individually and together. The results for the predictive ability of the trading rules are reported in Table 16.5. Both rules display some evidence of significant predictive ability as indicated by the t-

Table 16.5. Predictive ability - share market rules

Rule	N_{buy}	$\bar{r}_{buy}$	σ_{buy}	t_{buy}	N_{sell}	$\bar{r}_{sell}$	σ_{sell}	t_{sell}	$t_{buy-sell}$	t^*
Panel A: Full Sample (1990-97)										
GMA	1159	0.065	0.729	0.922	876	0.003	0.906	-1.033	1.699	1.669
GOS	1097	0.069	0.731	1.037	938	0.002	0.894	-1.077	1.834	1.830
Panel B: Sub-period results for the GMA rule										
1990-91	284	0.065	0.793	0.796	222	-0.048	0.951	-0.883	1.458	1.455
1992-93	303	0.100	0.700	0.636	209	0.020	0.739	-0.799	1.245	1.248
1994-95	271	0.052	0.745	0.551	237	-0.016	0.804	-0.588	0.987	0.972
1996-97	298	0.033	0.680	-0.201	208	0.062	1.103	0.218	-0.364	-0.335
Panel C: Sub-period results for the GOS rule										
1990-91	271	0.058	0.813	0.658	235	-0.033	0.924	-0.693	1.171	1.199
1992-93	292	0.101	0.696	0.642	220	0.023	0.742	-0.758	1.214	1.223
1994-95	255	0.056	0.748	0.610	253	-0.016	0.798	-0.600	1.049	1.037
1996-97	275	0.047	0.665	0.031	231	0.043	1.079	-0.030	0.052	0.050

All the statistics reported in this table are defined and discussed in Section 16.4.

statistics in the second last column of Table 16.5. This result is confirmed by the final column in the table which reports the t-statistic based on the [13] market timing test. However, individually the buy and sell signals do not seem to have any significant predictive ability. In addition to this overall significant predictive ability, all the rules issue buy (or sell) signals when the excess returns on the market are on average less (or more) volatile as indicated by the volatility of returns following buy (σ_{buy}) and sell (σ_{sell}) signals respectively.

A sub-period analysis of these results indicates that the difference between the average return following buy signals and the average return following sell is not significant in all periods. This is also true for the market timing test. However, the ability of the rules to buy when volatility in the market is low and sell when volatility is high, appears to be robust across different time periods.

Statistical significance The bootstrap approach outlined in Section 16.4 is used to study the statistical significance of trading rule profitability and predictive ability in the out-of-sample test period. The simulated p-values for the various measures of performance are given in Table 16.6. The results for the entire out-of-sample period provide evidence that the performance based upon the original series are statistically different to the performance from a random walk with drift. However this statistical significance deteriorates over time as can be seen from the sub-period results. In general, these results confirm those reported in Tables 16.4 and 16.5.

Table 16.6. Bootstrap p-values

Rule	XR	SR	$\bar{r}_{buy}$	σ_{buy}	$\bar{r}_{sell}$	σ_{sell}	$\bar{r}_{buy-sell}$	$\sigma_{buy-sell}$
Panel A: Full Sample (1990-97)								
GMA	0.000	0.000	0.000	1.000	0.000	1.000	0.000	0.998
GOS	0.016	0.018	0.018	0.996	0.018	0.960	0.044	0.946
Panel B: Sub-period results for the GMA rule								
1990-91	0.076	0.082	0.078	0.976	0.026	0.894	0.086	0.844
1992-93	0.104	0.110	0.130	0.746	0.238	0.892	0.112	0.842
1994-95	0.124	0.126	0.110	0.842	0.220	0.838	0.148	0.716
1996-97	0.592	0.570	0.596	0.986	0.046	0.390	0.582	0.118
Panel C: Sub-period results for the GOS rule								
1990-91	0.106	0.110	0.106	0.926	0.104	0.872	0.122	0.758
1992-93	0.094	0.098	0.108	0.806	0.208	0.892	0.114	0.836
1994-95	0.130	0.134	0.110	0.798	0.254	0.828	0.154	0.728
1996-97	0.464	0.462	0.458	0.984	0.048	0.542	0.480	0.108

The measures reported in this table are described in Tables 16.4 and 16.5

16.5.4 Return Measurement Bias

It is important to evaluate the sensitivity of the results to the significant persistence in returns, which are reported in Table 16.2. Since the existence of thinly traded shares in the index can introduce a non-synchronous trading bias or return measurement error. Therefore, these returns might not be exploitable in practice. To investigate this issue, the performance of the trading rules is simulated based on trades occurring with a delay of one day. This should remove any first order autocorrelation bias due to non-synchronous trading.

As can be seen from Table 16.7, the rules are still profitable over the out-of-sample test period, although there has been a substantial reduction in performance. Furthermore, there appears to be weak, if any, evidence of predictive ability. However, both rules still retain the property of being in the market when return variability is low and out of the market when return variability is high.

The break-even transaction costs have been reduced to approximately 0.25 percent per trade. This probably lower than the costs faced by most financial institutions. Since stamp duty and taxes are also incurred on all trades, which have been ignored in this evaluation of trading rule performance.[13] Also during volatile periods liquidity costs, as reflected by the bid-ask spread, could increase substantially. Even for large shares this increase could be in the or-

[13] Stamp duty costs per trade in Australia are currently 0.15 percent. Taxation costs vary across institutions and different individuals.

Table 16.7. Return measurement sensitivity–share market rules

							Bootstrap p-values			
Rule	XR	SR	tc^*	$\bar{T}$	$MaxD$	t^*	$\bar{r}_{buy}$	σ_{buy}	$\bar{r}_{sell}$	σ_{sell}
GMA	0.57	0.09	28	7.36	-21.36	0.879	0.156	1.000	0.002	0.810
GOS	0.39	0.06	25	8.11	-23.91	0.866	0.158	1.000	0.002	0.790

The performance statistics contained in columns two to seven are described in the notes to Table 16.4. The last four columns contain bootstrap simulated p-values for the statistics described in Section 16.4.

der of 0.5 to 1 percent. Thus, there does not appear to be sufficient evidence to conclude that the trading rules are economically profitable.

16.6 Conclusion

This paper has outlined how a genetic algorithm can be used to optimise technical trading rules and considered an application of this methodology to the Australian share market. The results indicate that there exists some evidence of overall market timing ability, but individual buy and sell signal have only, at best marginal forecasting power for the next days returns. Surprisingly, the rules appear to be able to distinguish between periods of low and high volatility. This is an interesting issue which was not investigated in this study, but is left for future research.

Both the GMA and GOS rules were able to outperform the benchmark strategy over the out-of-sample test period, taking into account both trading costs and risks. However, a sub-period analysis of the results indicates that the performance of both rules deteriorates over time. This performance is substantially reduced once the trading rule returns are adjusted for non-synchronous or thin trading.

In conclusion, there appears to be some evidence of forecasting ability, but probably little or no evidence of profitability once a reasonable level of trading costs has been considered. Since the break-even costs for these trading rules do not appear to be high enough to exceed realistic trading costs.

References

1. Achelis S. B. (1995) Technical Analysis from A to Z. Probus Publishing, Chicago.
2. Alexander S. S. (1964) Price Movements in Speculative Markets: Trends or Random Walks, No. 2. In: Cootner P. (Ed.) The Random Character of Stock Prices. MIT Press, Cambridge, 338–372
3. Allen F., Karjalainen R. (1999) Using Genetic Algorithms to Find Technical Trading Rules. Journal Financial Economics **51**, 245–271

4. Ball R. (1978) Filter Rules: Interpretation of Market Efficiency, Experimental Problems and Australian Evidence. Accounting Education **18**, 1–17
5. Bauer R. J. Jr. (1994) Genetic Algorithms and Investment Strategies. Wiley Finance Editions, John Wiley and Sons, New York
6. Bessembinder H., Chan K. (1995) The Profitability of Technical Trading Rules in the Asian Stock Markets. Pacific Basin Finance Journal **3**, 257–284
7. Bessembinder H., Chan K. (1998) Market Efficiency and the Returns to Technical Analysis. Financial Management **27**, 5–17
8. Brock W., Lakonishok J., LeBaron B. (1992) Simple Technical Trading Rules and the Stochastic Properties of Stock Returns. Journal of Finance **47**, 1731–1764
9. Brown S., Goetzmann W., Ross S. (1995) Survival. Journal of Finance **50**, 853–873
10. Brown S., Goetzmann W., Kumar A. (1998) The Dow Theory: William Peter Hamilton's Track Record Reconsidered. Working Paper, Stern School of Business, New York University
11. Carter R. B., Van Auken H. E. (1990) Security Analysis and Portfolio Management: A Survey and Analysis. Journal of Portfolio Management, Spring, 81–85
12. Corrado C. J., Lee S. H. (1992) Filter Rule Tests of the Economic Significance of Serial Dependence in Daily Stock Returns. Journal of Financial Research **15**, 369–387
13. Cumby R. E., Modest D. M. (1987) Testing for Market Timing Ability: A Framework for Forecast Evaluation. Journal of Financial Economics **19**, 169–189
14. Dorsey R. E., Mayer W. J. (1995) Genetic Algorithms for Estimation Problems with Multiple Optima, Nondifferentiability, and Other Irregular Features. Journal of Business and Economic Statistics **13**, 53–66
15. Efron B. (1979) Bootstrap Methods: Another Look at the Jackknife. Annals of Statistics **7**, 1–26
16. Fama E., Blume M. (1966) Filter Rules and Stock Market Trading. Journal of Business **39**, 226–241
17. Goldberg D. E. (1989) Genetic Algorithms in Search, Optimisation, and Machine Learning. Addison-Wesley, Reading
18. Holland J. H. (1975) Adaptation in Natural and Artificial Systems. University of Michigan Press, Ann Arbor
19. Huang Y -S. (1995) The Trading Performance of Filter Rules On the Taiwan Stock Exchange. Applied Financial Economics **5**, 391–395
20. Hudson R., Dempsey M., Keasey K. (1996) A Note on the Weak Form of Efficiency of Capital Markets: The Application of Simple Technical Trading Rules to the U.K. Stock Markets– 1935-1994. Journal of Banking and Finance **20**, 1121–1132
21. Koza J. R. (1992) Genetic Programming: On the Programming of Computers By the Means of Natural Selection. MIT Press, Cambridge
22. Levich R. M., Thomas L. R. (1993) The Significance of Technical Trading-Rule Profits in the Foreign Exchange Market: A Bootstrap Approach. Journal of International Money and Finance **12**, 451–474
23. Lo A. W., MacKinley A. G. (1990) Data Snooping Biases in Tests of Financial Asset Pricing Models. The Review of Financial Studies **3**, 431–467

24. Maddala G. S., Li H. (1996) Bootstrap Based Tests in Financial Models. In: Maddala G. S., Rao C. R. (Eds.) Handbook of Statistics, V XIV. Elsevier Science, 463–488
25. Malkiel B. (1995) A Random Walk Down Wall Street, 6th edn. W. W. Norton, New York
26. Neely C. J., Weller P., Dittmar R. (1997) Is Technical Analysis in the Foreign Exchange Market Profitable? A Genetic Programming Approach. Journal of Financial Quantitative Analysis **32**, 405–426
27. Raj M., Thurston D. (1996) Effectiveness of Simple Technical Trading Rules in the Hong Kong Futures Market. Applied Economic Letters **3**, 33–36
28. Sweeney R. J. (1988) Some New Filter Rule Tests: Methods and Results. Journal of Financial and Quantitative Analysis **23**, 285–301
29. Taylor M. P., Allen H. (1992) The Use of Technical Analysis in the Foreign Exchange Market. Journal of Money and Finance **11**, 304–314
30. Whitley D. (1989) The GENITOR Algorithm and Selection Pressure: Why Rank-Based Allocation of Reproductive Trials is Best. In: Schaffer D.J. (Ed.) Proceedings of the Third International Conference on Genetic Algorithms, Morgan Kaufmann, San Mateo, 116–121

17 Evolutionary Induction of Trading Models

Siddhartha Bhattacharyya and Kumar Mehta

Information and Decision Sciences, College of Business Administration,
University of Illinois at Chicago
sidb@uic.edu
kmehta1@uic.edu

Abstract. Financial markets data present a challenging opportunity for the learning of complex patterns not readily discernable. This paper investigates the use of genetic algorithms for the mining of financial time-series for patterns aimed at the provision of trading decision models. A simple yet flexible representation for trading rules is proposed, and issues pertaining to fitness evaluation examined. Two key issues in fitness evaluation, the design of a suitable fitness function reflecting desired trading characteristics and choice of appropriate training duration, are discussed and empirically examined. Two basic measures are also proposed for characterizing rules obtained with alternate fitness criteria.

17.1 Introduction

Financial markets data present a challenging opportunity for the learning of complex patterns not otherwise readily discernable. The existence of such discernible patterns in financial time-series data and their use in formulating profitable trading models has long been debated. While the investment industry makes routine use of technical analysis and decision rules, academic opinion has for long been almost unanimous on the view that such rules do not exist. Recent research, however, indicates that traditional academic conclusions may be premature. Reference [7] note that simple trading strategies hold significant prediction potential for the Dow Jones Index; [17] replicated and extended these results for the foreign exchange markets. Facilitated by computing resources and availability of high-frequency data [12] in recent years, a number of studies have undertaken non-linear and complex systems approaches (see [6] for a review) and observed fractal patterns in high-frequency market data [22]. Simulations using multi-agent systems also point out shortcomings to the traditional academic viewpoint [2]. Following on such findings, studies have sought the use of artificial intelligence (AI) learning techniques for data mining in financial markets data and learning of decision rules for trading in financial markets.

Powered by greater computing resources, recent years have seen an increasing use of neural and evolutionary computing techniques in financial decision aiding. Such techniques are used for data mining in financial time-series data to learn complex patterns not discernable otherwise. It has been noted that the use of learning techniques is changing the way investment decisions are made; the broader potential impact of the use of such techniques on financial markets operation and their regulation underscores the need for a systematic investigation [6]. While the use of neural networks for analysis of financial markets data has received wide attention [29], interest in genetic algorithm based approaches is more recent, with a series of studies noting its advantages [1] [3] [4] [24] [25] [27].

This paper reports on our ongoing research effort examining issues in utilizing genetic search in the data mining of financial time series. Genetic search here is aimed at discovering complex patterns informative for formulating decision rules for trading. Such decision rules propose trading recommendations for financial assets based on their past price history as given by the time series of the asset under consideration.

A trading model seeks to capture market movement patterns to provide trading recommendations in the form of a signal: a +1 indicating a "buy" signal, -1 a "sell" signal. Price history information is usually summarized in the form of variables called indicators. Reference [28] provide a detailed account of such trading models in the foreign exchange markets. The representational structure specified determines the patterns of indicator combinations searchable, and the manner of rules that can thereby be obtained. While the literature reports different flat string representations for machine learning with GAs [11] [10], a number of studies in the context of financial markets are based on the tree-structured genetic programming (GP) [16] representation. The use of the regular GP representation and operators can however pose problems arising from the closure property restriction [4]. We examine issues pertaining to the representation of trading decision rules, and propose a simple flat-string representation used in this study. The learning of investment strategies using GAs can, in general, include consideration of fundamental factors like price-equity ratio, market capitalization, etc., or be based on technical analysis that concentrates on learning patterns in the time series of price history data. This study, in keeping with its data-mining focus, takes the latter approach.

Being driven by a survival of the fittest policy, credit assignment to individual candidate solutions forms a critical aspect of the genetic search process. The fitness function encodes stated search objectives, and much of the power and appeal of GAs arises from the flexibility accorded in the formulation of fitness functions. While trading models are learnt with the general objective of maximizing profits or returns, the exact form of the fitness function can impact the nature of models learnt. Alternate specifications of learning objectives can lead to different search paths, and will yield solutions with diverse characteristics. Here, we briefly review different fitness functions reported in the literature, and examine the nature of decision models they are likely to yield. Drawing attention to performance desirables in trading models, alternate fitness functions are specified (see [5] for a detailed account of fitness function related issues in financial time-series data) and empirically examined using data on the Standard and Poor's composite (S&P500) index. The S&P index provides a well understood benchmark data series that has been used in numerous prior studies pertaining to financial markets data analysis.

A second crucial issue related to fitness evaluation in time-series data is that of training period duration. The training horizon used impacts the nature of rules obtained and their predictability over time. Longer training horizons are generally sought, in order to discern sustained patterns with robust training data performance that extends well into the predictive period. With changing market characteristics in dynamic environments, however, patterns that persist over time may be unavailable, and shorter-term patterns may hold higher predictive ability, albeit with shorter predictive periods. Such potentially useful shorter-term patterns may be lost when the training duration covers much longer periods. Too short a training duration can, of course, be susceptible to over-fitting – learned rules may fail to capture the essential principles underlying price movements, but instead may model the specific movements in the training data. We report on empirical findings regarding the use of different training horizons.

The next section discusses the representation of trading models in genetic search, and describes the rule structure used. The following section elaborates on fitness evaluation and describes the proposed fitness functions. Experimental results are then provided, and the final section highlights key findings and notes issues for future research.

17.2 Representation of Trading Models

In using genetic search for machine learning of rules, different representational forms have been proposed [10]. Key amongst these are the Michigan approach to classifier systems and the Pitt approach. In the former, each population member represents a partial solution to the learning task, and a collection of rules from a population provide a complete solution; specialized schemes for credit apportionment like the bucket brigade are required. The Pitt approach is closer to the traditional GA, with each population member representing a complete solution. Many applications are for classification, with rules typically containing a conjunctive or disjunctive description of concepts.

For induction of trading models from financial markets data using GAs, a candidate solution in the population is taken to represent a complete trading decision rule. These are typically of the form: IF condition THEN action, where condition details some pattern in the financial time series data, and action prescribes an out-of-market (sell) or in-market (buy) signal. Induced patterns in the condition are usually expressed in terms of indicators that provide different summaries of the time-series data; the moving average and volatility of the price series over different time ranges form commonly used indicators. An example of such a rule is:

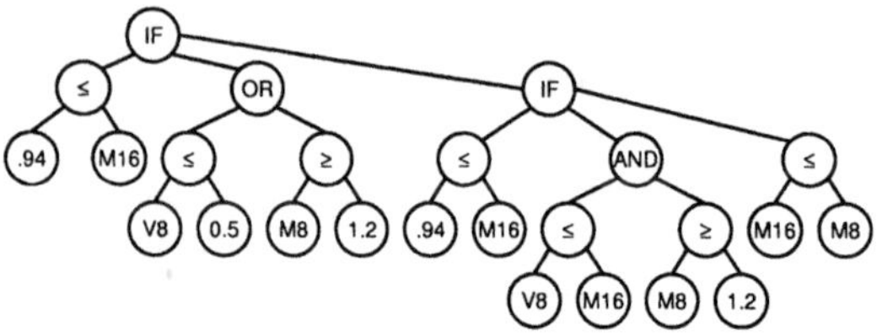

Fig. 17.1. Example of a trading model in GP [4]

IF (moving-average(4) >= moving-average(10))
AND (volatility(10) <= volatility(30))
THEN Buy.

Complex indicators seeking to capture various market behaviors can also be used [23].

The tree structured GP representation admits more complex combinations of terms and has been used in specifying trading models [1] [4] [25]. An example of a model in GP is provided in Fig. 17.1. GP trees, however, often tend to get overly complex, forming redundant structures of repeated and similar subtrees [1] [4]. Such solutions, comprised of complex combinations of functions and indicators, are often difficult to understand and interpret. Basic interpretability of models and behavior of model components, besides being generally desirable, is also key to furthering our understanding of trading model capabilities, usage, and guidelines for their design. It forms a particularly important criterion here since the models obtained through automated learning are subject to manual analysis and tailoring (for example to test for sensitivity to certain market conditions or client preferences) prior to implementation. Further, a large portion of population members in earlier GP studies was noted to constitute ill-defined and thus useless solutions, resulting in wasteful search - these arise from the use of regular GP operators that do not follow domain related semantics.

The crossover and mutation operators in regular GP are guided by the closure property [16], which requires that all components and representational constructs be handled uniformly, thereby allowing arbitrary sub-tree expressions to be recombined. Unrestricted operators, however, can yield solutions that are invalid and difficult to interpret–for instance, a comparison between a price and a volatility term, a logical ANDing of a comparison (Boolean return) and a price term, or an IF function specifying just a moving-average price term in the condition. Data-typing related mechanisms have been proposed in the literature to help circumvent the closure property in GP search([21] [15]). In the context of trading models for foreign exchange rates, [4] propose a set of semantic restrictions to guide the use of different representational constructs and GP search operators to yield semantically valid models.

A trading decision rule r at any time t, specifies

$r_t = X_{ht} \rightarrow S$

where X_{ht} gives the price history up to time t; and $S \in \{0, 1\}$ specifying a trading signal (the 0 and 1 being interpreted to imply either an "out of market" or "in the market" position, as desired). In this study, rather than use a complex GP tree structured representation, a simpler and more traditional GA based representation is used for specifying trading models. The Standard & Poor's Composite Index (S&P500) is taken as the time series data in this study. In addition to using past high, low and closing prices in the rules, a range of indicators may be used for specifying trading models. While some suggest use of simple moving average rules [7] [9], others have considered more complex indicators such as momentum, exponential moving averages, etc. [23]. The learning of trading rules, in general, involves learning good indicators, as well as combinations of these in defining good rules. Rules then seek appropriate combinations of these for identifying patterns in the time-series data. The indicators used in this study are categorized into three types in order to ensure their meaningful combination in rules:

- High, Low and Closing prices at specific times t_i (point measure)
- Moving averages of the prices for multiple time window Δt_i (trend measure)
- Variance of high and low prices for multiple time windows, Δt_i (volatility measure)

A trading rule takes the form:

IF <Condition>THEN <Signal>,

where, the condition is a logical combination of terms, each term being of the form

<indicatordays-ago><arithmetic operator><indicatordays-ago>.

A term thus specifies a comparison between two indicators, each evaluated at a certain point in (past) time. Both conjunction (AND) and disjunction (OR) operators can be used in combining terms in the condition part of a rule. A trading model then follows the following representational form: ([ ] denotes optional and { } denotes required constructs)

rule:
    condition: logical-expression
    signal
logical-expression:
    [logical-unary-operator] term_T [{logical-binary-operator} logical expression]
term_T:
    {indicator_T days-ago}{arithmetic-operator} {indicator_T days-ago}
logical–binary-operator: one of
    AND, OR
logical-unary-operator:
    NOT

comparison-operator: one of
<=, >=
signal: one of
buy(1), sell(0)
indicator_T: one of
high, low, close
moving-average {days-ago}
volatility {days-ago}

Since each type of indicator (price, moving-average, volatility) represents a different measure, only indicators of the same type are allowed within a term. A term may thus compare values of two moving-average indicators, or of two volatility indicators, or of two price values (high, low, or close prices).

To help evaluate differences in nature of rules learnt under different fitness functions, two rudimentary measures are defined to characterize learnt rules: the complexity of the rule, and the specificity of the rule. Complexity is defined as the number of possible combination of terms under which a rule generates a TRUE as a signal. Complexity would thus characterize the possible different circumstances under which a rule is likely to generate a "sell" signal (in the loss minimization approach). This is defined for different operators as follows:

- The comparison operator has a complexity of 1.
- The complexity of an OR operator is the sum of complexities of the concerned terms.
- The complexity of an AND operator is the product of the complexities of the concerned terms.

Specificity of a rule is defined as the minimum number of terms required to be satisfied for a given rule to fire. While complexity measure the possible different number of scenarios under which a rule is likely to signal "sell", it does not measure the number of terms which characterize each of these scenarios. We characterize the minimum number of terms that define a scenario as the specificity of a given rule. Specificity for different operators is defined as follows:

- The specificity of a comparison operator is 1
- The specificity of an OR operator is the minimum of the specificities of the concerned terms.

The specificity of an AND operator is the sum of the specificities of the concerned terms.

For the experiments reported, a maximum of three terms of each indicator-type is allowed. The maximum possible complexity of a rule is hence 8, and the minimum possible complexity is 1.

17.3 Fitness Function

The objectives encoded in the fitness function play a crucial role in determining the nature of rules learned. The fitness function should thus embody performance desirables of trading rule sought. While maximization of profits generated from use of the rule is a generally sought performance criterion, closer analysis reveals that nuances in fitness function formulation are key to learning rules exhibiting desired characteristics. This section examines certain key aspects to fitness evaluation, and proposes a fitness function design for trading model induction.

The total realized returns earned by a managed portfolio (adjusted for the riskiness of the portfolio compared to index returns such as S&P500) is typically used as a performance measure for a portfolio manager in an investment firm. Drawing an analogy between performance of a trading model and that of an investment manager, the continuously compounded total returns has been suggested as an "obvious" choice of fitness criterion for evaluating trading rules obtained by evolutionary learning methods [1]. Certain key differences, however, between the final evaluation of a portfolio's performance and the decision process followed by the investment manager should be noted. Portfolio managers evaluate their decisions and update their trading decisions on a case by case basis, and not based on an aggregate of decisions taken over the entire trading period. Furthermore, the measure of performance of a portfolio manager is an overall performance measure, which does not evaluate the process and means adopted by the decision maker for optimizing his/her performance. The manager may choose to maximize or minimize one or more measures that ultimately lead to maximization of the overall performance. However, framing the fitness function to maximize the overall objective does not guarantee maximization or minimization of the underlying measures that might be of importance to the manager.

The performance of a trading decision model is determinod by evaluating it on the time-series data (the price history of the asset under consideration) over a certain period. The in-market and out-of-market position decisions as given by the model are computed over the period of evaluation on the data, and provide the basic measures for performance assessment. Each *trade* provides a unit of measurement, and is defined by a "buy" decision at time t_b and a "sell" decision at some later time t_s. The merit of a single trade is usually measured in terms of returns realized through the trade, and is calculated as

$$r_s = \frac{P_{t_s}\ (1-c)\ -\ P_{t_b}\,(1+c)}{P_{t_b}\,(1+c)} = \frac{P_{t_s}\,(1-c)}{P_{t_b}\,(1+c)} - 1 \qquad (17.1)$$

where r_s = realized return from an individual decision, P_{t_s} = selling price of the asset, c = transaction cost as percentage of selling price, and P_{t_b} = buying price of the asset.

Table 17.1. Illustrative example (N denotes trade number)

	Returns from Individual Trades						
Rule	$N = 1$	$N = 2$	$N = 3$	$N = 4$	$N = 5$	Aggregate Return	Compounded Return
1	1.00	1.00	1.00	1.00	1.00	5.00	1.05101
2	1.50	0.50	1.00	1.00	1.00	5.00	1.05098
3	1.25	1.25	1.25	1.25		5.00	1.05095
4	2.00	1.75	0.25	1.00		5.00	1.05085
5	2.50	0.50	1.00	1.00		5.00	1.05083

17.3.1 Compounded Returns

For a series of N trades over time period T, the merit of undertaken decisions is determined by the *total realized return.* Since the total realized return is based on reinvestment of returns from individual trades, it can be expressed as a continuously compounded series:

$$r_{c_N} = \prod_{i=1}^{i=N} (1 + r_{s_i}) - 1 \tag{17.2}$$

where r_{s_i} = realized return from the ith trade.

Such continuously compounded returns (equation 17.2) have been used for evaluating trading rules in foreign exchange markets ([17]). Since continuous compounding implies reinvestment of returns realized from individual decisions, considering two rules that earn the same amount of aggregate returns (summation of simple returns), the rule with a higher number of trades (and thus lower returns per trade) would evaluate to receive a greater fitness. The example in Table 17.1 illustrates this effect of using the compounded total return as a fitness measure. Consider five different rules, each providing equal total aggregate return of 5 units. Based on compounded returns, however, note that the rule with lowest mean return per trade (arising from it having the maximum number of trades) and lowest variance in mean returns per trade evaluates as best. While the bias towards rules exhibiting a lower variance in mean returns per trade may be desirable, the bias for rules generating higher number of trades poses a serious problem. Rules that trade very frequently have a greater tendency to adhere to what might essentially be noise in the data. Furthermore, use of realized returns evaluates a rule based on the ratio of selling price to buying price, and thus ignoring the possible opportunity losses during the holding period (more detailed explanation provided later).

Another industry measure of portfolio performance is the amount by which the portfolio's returns exceed that of the market (after adjusting for the riskiness of the portfolio). Reference [1] use this measure of continuously compounded excess returns (equation 17.3) as a fitness criterion during training:

$$\Delta r_{cN} = r_{cN} - r_{cBH} \tag{17.3}$$

where
Δr_{cN} = continuously compounded returns in excess of the market returns,
r_{cN} = continuously compounded returns realized from the trading rule, and
r_{cBH} = continuously compounded returns from a buy and hold strategy (buy on day one and sell on the last day of the period considered).

Note that, in evaluating multiple trading models, the second part of equation (17.3) remains constant irrespective of the signals and returns generated by the different trading rules; only the first term of the above equation is a variable, which the rules seek to maximize in order for excess returns to be maximized. Thus, using equation 17.3 as a fitness criterion is identical to the use of continuously compounded returns as specified in equation 17.2, raising the same concerns.

17.3.2 Realized vs. Non-Realized Returns

Considering a simple aggregate of realized returns as a fitness function (in-lieu of compounded realized returns), its maximization entails maximizing the realized returns from each trade (the first term), and minimizing the number of trades N

$$r_{sN} = \sum_{i=1}^{i=N} \left[\frac{P_{s_i}(1-c)}{P_{b_i}(1+c)} \right] - N \tag{17.4}$$

Here, maintaining prolonged positions with infrequent trading will tend to increase the realized returns during the training period. Use of this aggregation of simple realized returns as a fitness criterion will thus tend to favor rules that trade infrequently and maintain prolonged market positions. A second drawback arises from the fact that the function in equation 17.4 does not account for any non-realized (opportunity) losses incurred by the rule while maintaining a particular market position. The use of non-realized returns for performance assessment, as in [27], helps alleviate these biases. While a realized return is obtained only when a trading rule specifies taking an out-of-market position from an ensuing in-market position, non-realized returns are calculated at regular intervals of time irrective of trades conducted.

17.3.3 Transaction Costs and Risk-Free Rate

Transaction costs, present another concern in evaluating trading models. Note that a transaction cost is incurred only when an actual trade is conducted. The inclusion of a transaction cost in the fitness function, besides being realistic, also serves as a check against rules that switch positions very frequently leading to an overfit to small fluctuations in the market data. Early experimentation, however, revealed that if this barrier to entry and exit from the market is too high, obtained rules tend to rarely adopt out-of-market positions. This is due to a rule obtaining zero returns when being out-of-market. To counter this effect, we allow a rule to earn a risk-free rate during the days when it specifies out-of-market positions.

Note that the trading rule structure used, in general, can include consideration of terms and indicators from the risk-free rate series. In the absence of an explicit consideration of risk-free rate series terms in the rules, however, use of risk-free rate values can introduce unaccountable biases – though the risk-free rate can vary independently of the primary asset price series, the rules here do not have a means to account for high versus lower risk free rate periods. Rules here can be biased towards adopting out-of-market positions solely to take advantage of high prevailing risk-free rates; the same rules, however, have no way of predicting high rate periods and thus report poor performance on out-of-sample (test) data. To counter possible bias towards overfitting during high risk-free rate period(s), a constant risk free rate value equal to the daily risk-free rates averaged over the training period is used.

17.3.4 Fitness Function Used

The above discussion suggests the following design for the fitness function:

$$r_{s_t} = \sum_{t=1}^{t=T} \left[\frac{M_t P_t (1 - \theta_t c)}{P_{t-1}} \right] - \sum_{t=1}^{t=T} [(1 - M_t)\ \bar{r}_{f_t}] \tag{17.5}$$

where

r_{s_t}= total non-realized return over a period T,

M_t = market position signal, TRUE (=1) implying in-market position, out-of-market otherwise

θ_t = trade signal, assuming a value of 1 whenever a held market position is reversed, indicating that a trade (buy or sell) has been conducted,

T = total number of periods in training period, and

$\bar{r}_{f_t}$ = average of risk-free rate over the training period T.

Since all rules in the population are evaluated on the same training data, T is a constant over different candidate rules; maximization of first term in the above expression thus does not favor rules that trade infrequently. Consideration of non-realized returns in the first term of the equation also penalizes rules for any non-realized losses incurred while holding particular market positions, thus rewarding rules for improved accuracy in prediction, and further removing the bias towards rules that tend to maintain prolonged market positions. The second term in the abover expression is from the inclusion of the average risk-free rate.

17.3.5 What to Predict: Buy or Sell?

Noting that a rule's condition can capture and specify patterns for either taking in-market or out-of market positions, two alternate interpretations are useful. A rule's condition evaluating to TRUE is typically considered as a signal to buy, with a FALSE indicating a signal to sell. This approach, termed *profit-maximizing approach* specifies taking a default out-of-market position, with being in the market only when the rule's condition evaluates as TRUE. Obtaining returns in excess of the market (a buy and hold strategy) using this approach requires that a rule learn useful patterns that specify periods for being in the market. An alternative interpretation, termed the loss minimizing approach, is to consider a rule's condition evaluating to TRUE as a signal to step out of the market, with the default position being to stay in-market.

Under conditions where the price data series exhibits a general increasing trend over time and a simple buy-and-hold strategy provides greater than zero returns, identifying conditions for staying out of market (and staying in market at other times) can be easier. Since the rules here can focus on the fewer periods when it should stay out of market. For example: identification of a precise common pattern identifying even a small number of days when the market goes down allows a rule to earn returns in excess of the market. Rules learnt using the *loss minimization approach* can also be expected to yield more precise (and possibly robust) rules too.

17.3.6 Adequacy of Training Period Duration

As noted, the designed fitness function takes into account possible biases that might occur due to inappropriate coding of the desired objective. However, one key question remains – what constitutes an adequate training period duration T ? With too short a training period, the rules obtained may lack robustness. On the other hand, with too long a training period, the rules may be unable to capture more recent profitable patterns. To investigate the potential impact of training period duration on performance, four different lengths of training time spreading from 2.5 year to 10 years (intervals of 2.5 years) is used with the fitness function of Equation 17.5 and using the loss minimization approach.

17.4 Experimental Study

17.4.1 Data

The daily high, low and closing prices for the Standard & Poor's Composite Index (S&P500) from Jan 1, 1983 to Mar 31, 1997 is taken as the time series data for this study. The 90-day treasury bill rates for the corresponding period were used as the existing risk-free (RF) rate. Data spanning the 10-year period from Jan 1st, 1983 to Dec 31st, 1992 is used for training and the remaining data is used for testing the trading models obtained. No data on dividends declared by the firms is used. Note that the training data includes the market crash of October 19th, 1987. Initial experiments revealed that fitness values of rules tended to be dominated by a few correct decisions around this crash date, with poor decisions at other times. To eliminate this bias, the data following the stock market crash was corrected by the crash amount, effectively eliminating the market crash from the used data.

17.4.2 Indicators

As mentioned above, three types of indicators are used in specifying trading rules for this study. - high, low and closing prices providing point measures, moving averages of prices providing trend measures, and variances of prices giving volatility measures. These are specified using different time horizons, for example, closing-price taken 3 days ago, or moving-average over the past 50 days. In the interest of simplicity and given our primary interest in examining the impact of alternate fitness criteria and training horizons, a pre-specified set of indicators is used. For the trend and volatility measures, time windows Δt_i of 5, 10, 20, 50, 100 days are used.

17.4.3 Experimental Results

Two sets of experiments were run to evaluate the impact of (1) alternate fitness function specifications and (2) different training durations. As mentioned, two fitness criteria are considered, corresponding to the profit-maximizing and loss-minimizing approaches. Nine independent runs of the GA were conducted for the criterion, each with a different random number seed. In keeping with the measures used in fitness evaluation during training, performance on the test period considers the aggregation of non-realized returns; that is, no reinvestment of obtained profits is considered (compounded returns take such reinvestment into account). It should be noted that a compounding of returns will yield higher values than those reported here; such compounded returns is also more realistic when comparing against returns from other portfolios. Results reported below show the excess returns (over that returned by a buy-and-hold strategy) provided by the obtained trading rules. The first set of experiments correspond to a 10 year training period.

Figure 17.2 shows the average cumulative returns in excess of a buy-and-hold (BH) strategy for the trading rules obtained (averages across all rules at quarterly points). Interestingly, for all rules obtained using both the fitness criteria, the cumulative excess returns begin to show a decline after two years into the test period. This indicates a possible change in the underlying data patterns from that captured by the rules from the training data; the rules here are no longer equally applicable, signaling a need for re-training. For subsequent results reported, the test period is thus considered from Jan 1, 1993 to Dec 31, 1994.

Table 17.2. Trading model performance characteristics

	Training Data		Test Data	
	Profit-max	Loss-Min	Profit-max	Loss-Min
Average performance	6.9	7.2	4.41	4.44
Std. Dev (performance)	1.3	1.3	2.1	1.7
Avg. trades per year	36.9	33.5	44.7	38
Std. Dev(Avg. trades/year)	16.15	3.9	15.3	3.8

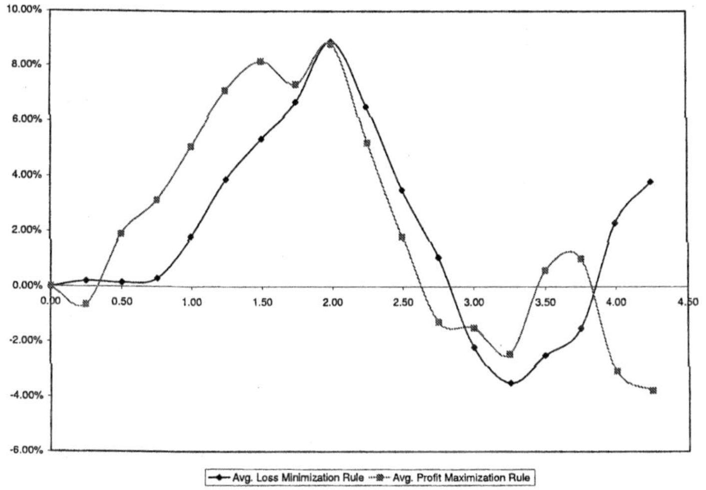

Fig. 17.2. Average trading rule performance

Table 17.2 shows the annualized mean excess returns and number of trades for the profit-max and loss-min rules learnt. Though both fitness approaches exhibit similar excess returns in the test period, the loss-minimizing rules are found to trade less frequently. Figure 17.3 shows a scatter plot of the cumulative excess returns versus the average number of trades per year. An improvement in rule performance is noticed as the number of trades made by the respective rules increase; beyond a certain extent, however, an increasing number of trades is associated with declining performance. The plot also reveals a more compact clustering of rules learnt with the loss minimization criterion ; this indicates that rules learnt using this fitness criterion are stable across multiple independent GA runs with different random seeds. To further examine differences in rule characteristics, Figure 17.4 plots the specificities and complexities of rules obtained with the two fitness approaches. Here, a clear distinction in nature of rules is observed, with the loss-minimizing rules being more specific than those obtained with the profit maximizing criterion.

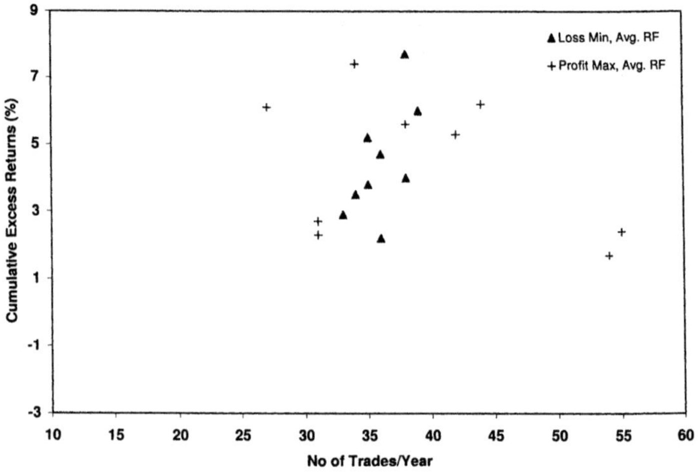

Fig. 17.3. Trading rule behavior

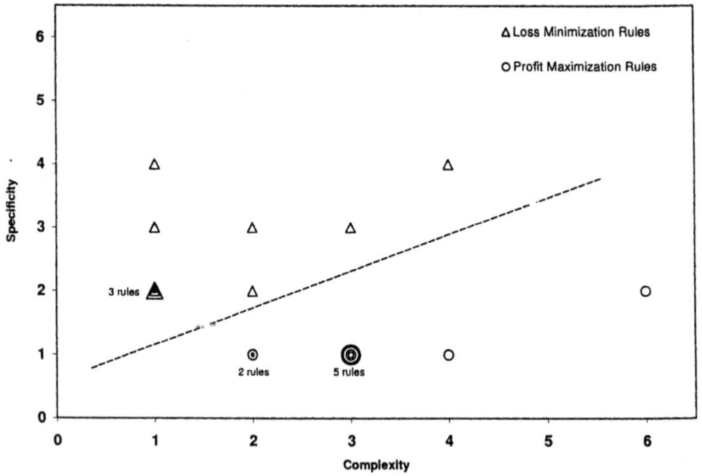

Fig. 17.4. Nature of rules obtained

When no risk free rate is included in the fitness function, no significant excess returns over the test period were observed ([5] give detailed findings). The number of trades executed by the rules was also noticed to be lower without consideration on a risk-free rate in the fitness evaluation - in keeping with expectations, rules here tend to hold prolonged in-market positions. Note further that with provision of a risk free rate during out-of-market periods, the riskiness of learnt trading rules is necessarily lower than that of the buy-and-hold strategy. Obtained rules show significantly lower variance in the daily non-realized returns when the fitness criterion uses the average risk-free value.

Exploring possible causes for the noticed decline in performance beyond the second year of the test period, the average number of trades per year was found to be significantly higher in the third and fourth year of the test period. The average number of trades per year for the first two years of the test period, however, remains similar to that for the training period. This is indicative of some change in the underlying data series, with the data patterns embodied in the rules and learnt from the training data no longer being present beyond the first two years of the test period. Furthermore, the increase in trading frequency for rules learnt with average risk-free value is less than that for the rules learnt without a risk-free rate; sharper declines in performance here are associated with the higher increase in trading frequencies of the rules learnt without a risk-free rate component in the fitness function.

A second set of experiments was directed at investigating the effects of different training period durations. A total of four different training durations were considered: T=10-year, T=7.5-year, T=5-year and T=2.5-year periods; maintaining the same test period as above, rules were obtained using these different T preceeding years for training. Again, nine independent GA runs were conducted, and results report the average annual excess returns.

Figure 17.5 shows the the means of performance values over the different GA-runs; it plots the training period performance, and in order to to highlight observed findings, shows the test period performance for the two-year test period as well as for the one-year test period–rules were again found to show a sharp decline in performance in the third and fourth years into the test period for all training durations. As can be expected, rules obtained using shorter training durations show higher training performance, but with poorer performance in the test period–the 2.5-year trained rules exhibit no significant positive excess returns. Rules learnt over the 7.5-year training horizon are seen to have the lowest shrinkage in performance from the training to the test period, while also yielding highest excess returns in the test period (both in the first year and first two years). For the 10-year training period, the GA obtains rules that show higher performance in the training period than the 7.5 year trained rules - but with poorer performance in the first year of prediction; perfromance for the two-year prediction period is however close for both the 7.5-year and 10-year rules. On an average, rules from the 7.5-year

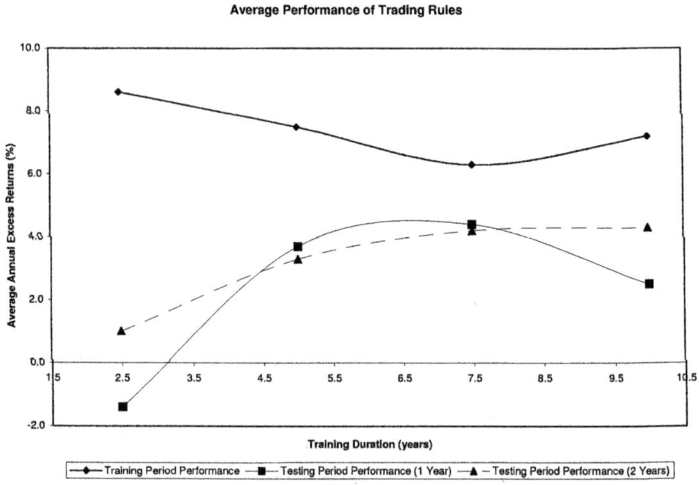

Fig. 17.5. Average performance with different training durations

training period earn 1.9% and 0.1% higher excess returns for the first year and the first two years of testing respectively than the 10-year trained rules. Note also that considering a one year prediction period, the 5-year trained rules are better than the 10-year trained rules. Further, though the 10-year rules show significantly lower performance than the 2.5 year rules for the training period, they earn on an average 3.8% and 3.4% higher returns for the first and first two years of the test period respectively.

Figure 17.6 plots the mean shrinkage, measured as ((training-performance– test-performance)/ training-perfromance) - considering both the 1-year and 2-year test periods- for rules obtained with different training durations. While longer training durations are generally seen to indicate more robust rules, note that the 10-year rules show increased shrinkage for the immediate year of prediction. Figure 17.7 shows the average number of trades per year for rules resulting from different training durations. Lower trading frequency is noticed to correspond to more robust performance (lesser shrinkage). Higher training performance for the shorter training duration rules is also seen to arise through higher trading frequency - these rules show poor predictability.

17.5 Discussion

Genetic algorithms provide a useful tool for data mining in financial time series data, and this study has considered their utilzation for trading model induction. While the representation used determines the manner of patterns

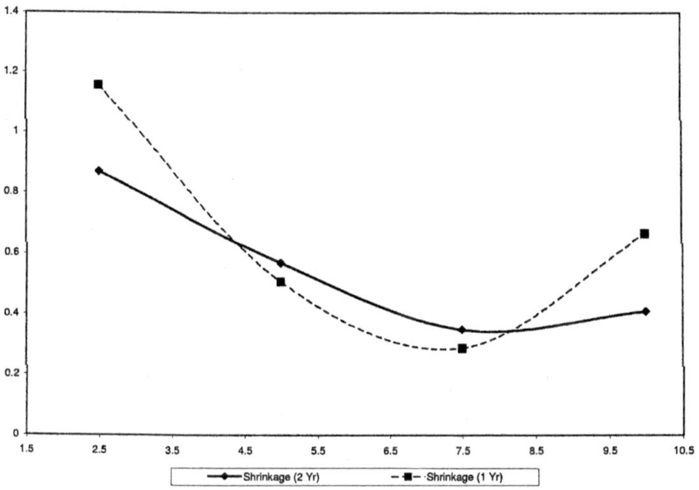

Fig. 17.6. Shrinkage in perfromance with different training durations

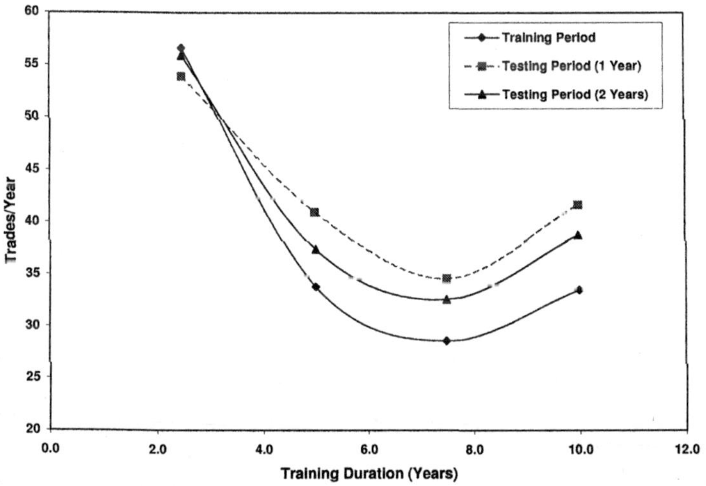

Fig. 17.7. Average annual trades with different training durations

and rules obtainable, fitness evaluation is key to achieving desired performance objectives. Different formulations of the fitness function can yield rules with varying behavior. While generally used measures for assessing the performance of trading models seem, at first, natural candidates for use in fitness evaluation for genetic search, closer analysis reveals that nuances in fitness function formulation are key to learning rules exhibiting desired characteristics. Since the search proceeds from a population of randomly generated solutions, an initial rule that makes a few arbitrary good decisions can quickly get to dominate the search. It is thus essential that the fitness function be free of potentially harmful biases.

Though continuously compounded returns are usually taken as measure of ultimate trading model performance, when used for fitness evaluation during genetic search, it embodies a problematic bias towards solutions that trade more frequently; such rules, consequently, tend to pick up noise in the data, leading to overfitting. The simple use of realized returns in fitness evaluation is also noted to be flawed in that it does not penalize potential solutions for any non-realized opportunity losses incurred during periods of their maintaining a single position. Fitness measures based on an aggregation of simple (non-compounded) non-realized returns are thus proposed. Transaction costs are found desirable for inclusion in the fitness function. The use of a risk-free rate component during out-of market positions is also suggested. Since obtained patterns from the data can pertain to conditions favorable for taking either in-market or out-of-market positions, two alternate approaches for learning are adopted: in the profit-maximization approach, rules evaluating to TRUE specify taking an in-market position; in the loss-minimization approach, a rule evaluating to TRUE specifies taking an out of-market position. While these different approaches are found to yield similar perfromance, the solutions can have dissimilar characteristics. Two simple measures for the complexity and specificity of rules show differences in the nature of rules obtained with the alternate fitness criteria. Rules learnt with the profit maximization approach - the patterns here capture conditions for being in-market - are found to be of greater complexity, but with lower specificity.

All the models obtained were found to perform poorly after the second year into the test period. Results show that a lack of uniformity in trading frequency across different time horizons can be indicative of some change in the underlying data characteristics. While the average number of annual trades for the first two years of the test period was similar to that observed during training, the decline in perfromance after the second year of the test period corresponds to a sudden increase in trading frequency . The patterns embodied in the rules are here no longer profitable, indicating a need for re-training, possibly on data from more recent price history.

While inductive learning in general seeks as large a training data set as can be obtained, a similar rationale need not necessarily apply for the case of financial time series data. With changes in market characteristics and in the regulatory environment, data patterns can undergo fundamental changes over time. Data over too long a length of time considered for training can confound the learning process with multiple, possibly overlapping or even conflicting patterns being present in the data. An overly narrow time span, however, may not include sufficient data points to present fundamental patterns in price movements - obtained rules, here, will have learnt individual price movements rather than any general relevant pattern, with subsequent overfit and shrinkage in performance on unseen test data. The extent into the past price history that the training data spans, can thus, be crucial for effective learning. The effect of training on data spanning different time horizons was examined in a second set of experiments using four decreasing training durations. As expected, very short training durations were found to yield rules with shorter predictive horizons, with larger training durations also being susceptible to lower future applicability. Intermediate training durations were observed to give models exhibiting least shrinkage and robust performance.

Given that rules learnt under different evaluation objectives can embody patterns with diverse characteristics, trading decisions in combined consideration of multiple rules can yield enhanced performance. Such decisions, arising from a some consensus mechanism amongst multiple rules, can also provide added robustness and reliability. Schemes for such combination of learnt patterns or rules and performance implications are under investigations, and some preliminary results are reported in [18].

A simple rule structure with only certain basic indicator types were considered in this study. The proposed fitness functions and obtained results, however, are applicable with more complex rule structures, including those explored using genetic programming as in [1] [4] [24]. The use of more sophisticated indicators and pattern templates for capturing various market behaviors as can be reflected in the price data series, can provide interesting insights and need further investigation. The design of specialized genetic search operators for enhanced learning capability in dynamic time series data is another avenue for future research.

While the focus of the current study has been on discerning patterns from financial time series to be used for trading decisions, computational schemes can also prove useful for examining various aspects of market behavior [2] [6]. The focus here is on obtaining an understanding of diverse market phenomenon, with interpretability of learnt patterns of key concern. Genetic learning, with its ability to capture complex non-linear patterns in an understandable representational format, are preferable over neural networks. The flexibility accorded in formulation of search objectives as fitness-functions provides an added advantage. As emphasized in this study, however, the design of fitness functions and appropriate use of training durations needs careful attention to detail.

References

1. Allen F., Karajalainen R. (1999) Using Genetic Algorithms to Find Technical Trading Rules. Journal of Financial Economics 51(2), 245–71.

2. Arthur W. B., Holland J. H., LeBaron B., Palmer R., Tayler P. (1997) Asset Pricing Under Endogenous Expectations in an Artificial Stock Market. In Arthur W. B., Durlauf S. and Lane D. (Eds.), The Economy as an Evolving Complex System II, Addison-Wesley, pp. 15–44.
3. Bauer R. J. Jr. (1994) Genetic Algorithms and Investment Strategies. John Wiley and Sons, NY
4. Bhattacharyya S., Pictet O. V., Rumbach G. (1998) Representational Semantics for Genetic Programming Based Learning in High-Frequency Financial Data. In: Koza J. R. (Ed.) Proceedings of the Third Annual Genetic Programming Conference. Wisconsin, MD, Morgan Kaufmann, 11–16
5. Bhattacharyya S., Mehta K. (1999) Genetic Learning in Financial Time Series Data: Focus on Fitness Measures. In Review. 1999
6. Brock W. A. (1996) Asset Pricing Behavior in Complex Environments. In: Arthur W. B., Lane D., Durlauf S. (Eds.) The Economy as an Evolving Complex System II
7. Brock W., Lakonishok J., LeBaron B. (1992) Simple Technical Trading Rules and the Stochastic Properties of Stock Returns. Journal of Finance **47(5)**, 1731–1764
8. Cootner P. H. (1962) Stock Prices: Random vs. Systematic Changes. Industrial Management Review **3(2)**, 24–45
9. Dacorogna M. M., Muller U. A., Jost C., Pictet O. V., Olsen R. B., Ward J. R. (1995) Heterogeneous Real-Time Trading Strategies in the Foreign Exchange Market. The European Journal of Finance **1**, 383–403
10. DeJong K., Spears W. M., Gordon D. F. (1993) Using Genetic Algorithms for Concept Learning. Machine Learning **13**, 161–188
11. Goldberg D. E. (1989) Genetic Algotrithms in Search, Optimization and Machine Learning. Addison-Wesley, Reading, MA
12. Goodhart C., O'Hara M. (1995) High-Frequency Data in Financial Markets: Issues and Applications. London School of Economics and Johnson Graduate School of Management, Cornell University
13. Guillaume D. M., Dacorogna M. M., Davé R. D., Müller U. A., Olsen R. B., Pictet O. V. (1997) From the Bird's Eye to the Microscope: A Survey of New Stylized Facts of the Intra-daily Foreign Exchange Markets. Finance and Stochastics **1**, 95–129
14. Harrald P. G., Kamstra M. (1997) Evolving Artificial Neural Networks to Combine Financial Forecasts. IEEE Transactions on Evolutionary Computation **1(1)**, 40–52
15. Janikow C. Z., DeWeese S. (1998) Processing Constraints in Genetic Programming with CGP2.1. In: Koza J. R. et al. (Eds.) Proceedings of the Third Annual Conferenceon Genetic Programming. Wisconsin, Madison,Morgan Kaufmann, 173–180
16. Koza J.R. (1992) Genetic Programming: On the Programming of Computers by Means of Natural Selection, MIT Press
17. LeBaron B. (1991) Technical trading rules and regime shifts in foreign exchange, Santa Fe Institute Working Paper 91-10-044
18. Mehta K., Bhattacharyya S.. Combining Rules Learnt Using Genetic Algorithms for Financial Forecasting. In Angeline P. (Eds.) Proceedings of the 1999 Congress on Evolutionary Computation (CEC-99), Washington D.C, IEEE, 1245-1252

19. Michalewicz Z. (1994) Genetic Algorithms + Data Structures = Evolution Programs, 2nd Edition, 1994, Springer-Verlag
20. Mitchell M. (1996) An Introduction to Genetic Algorithms, 1996, MIT Press
21. Montana D. J. (1995) Strongly Typed Genetic Programming. Evolutionary Computation, 3(2), 199–230
22. Müller U. A., Dacorogna M. M., Davé R. D., Pictet O. V., Olsen R. B., Ward J. R. (1993) Fractals and Intrinsic Time–A Challenge to Econometricians. Invited presentation at: the XXXIXth International AEA Conference on Real Time Econometrics, 14–15 Oct 1993 in Luxembourg, and at: the 4th International Parallel Problem Solving from Nature - Applications in Statistics and Economics Workshop, 22–26 Nov 1993 in Ascona (Switzerland)
23. Müller U. A. (1989) Indicators for Trading Systems. Technical Report UAM.1989-08-10, Olsen and Associates, Zurich
24. Neely C., Weller P., Ditmar R. (1997) Is Technical Analysis in the Foreign Exchange Market Profitable? A Genetic Programming Approach. Journal of Financial and Quantitative Analysis **32(4)**, 405–427
25. Oussaidene M., Chopard B., Pictet O. V., Tomassini M. (1996) Parallel Gentic Programming and Its Application for Trading Model Induction. Parallel COmputing (Netherlands), 23(8), 1997, 1183–1198
26. Pantazopoulos K. N., Tsoukalas L. H., Bourbakis N. G., Brun M. J., Houstis E. N. (1998) Financial Prediction and trading Strategies Using Neurofuzzy Approaches. IEEE Transactions on Systems, Man and Cybernetics-Part B: Cybernetics **28(4)**, 520–531
27. Pictet O. V., Docorogna M. M., Chopard B., Shirru M. O. R., Tomassini M. (1995) Using Genetic Algorithms for Robust Optimization in Financial Applications. In: Parallel Problem Solving from Nature–Applications in Statistics and Economics Workshop, Germany
28. Pictet O. V., Dacorogna M. M., Müller U. A., Olsen R. B., Ward J. R. (1992) Real-time Trading Models for Foreign Exchange Rates. Neural Network World **2(6)**, 713–744
29. Weigend A. S., Abu-Mostafa Y. S., Refenes A. P. N. (1996) Decision Technologies for Financial Engineering: Proceedings of the Fourth International Conference on Neural Networks in the Capital Markets (NNCM-96). World Scientific, Singapore

18 Optimizing Technical Trading Strategies with Split Search Genetic Algorithms

Raymond Tsang[1] and Paul Lajbcygier[2]

School of Business Systems, Monash University
Raymond.Tsang@infotech.monash.edu.au
Paul.Lajbcygier@infoteh.monash.edu.au
http://www.bs.monash.edu.au/webpage/public/pStaffDetails.cf

Abstract. Genetic algorithms can be used to search a space and find a near optimal solution to a problem. Standard genetic algorithms are composed of three operators: reproduction, crossover and mutation. Mutation is the occasional random alteration of the value of a bit-string. The role of the mutation operator is to introduce some randomness into the search. Many researchers erroneously consider mutation unimportant when compared to reproduction and crossover. They argue that the mutation rate is so low that it may as well be non-existent. However, mutation can prevent the search from ending in a local optima. A variant of the standard genetic algorithm, that splits the population into two and applies a high mutation rate to one of the sub-populations, is proposed and tested in this study. The idea is that the high mutation rate will permit the sub-population to 'jump' out of a local minima. It is found that the approach optimizes well on a test-bed of functions. The same approach is then applied to a practical optimization problem in finance.

18.1 Introduction

Genetic algorithms (GAs) are a type of heuristic search algorithm commonly used for the optimization of complex problems. Over the years, the development of genetic algorithms has paid relatively little attention to the "mutation" operator when compared with other genetic operators. Research has shown that the mutation operator actually plays a very significant role in the performance of genetic algorithms. As a result, a novel genetic algorithm, named the Split Search Genetic Algorithm (SSGA) was designed to fully utilize the mutation operator. The algorithm is based on combining the characteristics of the mutation operator with distributed genetic algorithms.

The ability of the algorithm was first tested against a canonical genetic algorithm in a test bed of functions to assess its ability as a general optimization tool. To assess the SSGA's performance in more applied problems, it was also tested in the optimization of technical trading strategies - trading a portfolio of eight future commodities. The result from the optimization shows that it provides a more profitable strategy than a naive buy-and-hold and the canonical genetic algorithm. The application of genetic algorithms in the area of futures trading also highlighted several important issues which are discussed, in particular, the issue of robustness in a noisy environment, and the question of determining the appropriate fitness objective.

18.2 GAs and Mutation

Since the publication of Kenneth De Jong's paper *An Analysis of the Behavior of a Class of Genentic Adaptive Systems* and Holland's *Adaptation in Natural and Artificial Systems* in 1975, a great deal of research has been devoted to the development of faster and more powerful genetic algorithms. These works range from finding optimal parameter settings for the algorithm [17], to improving its selection and crossover schemes [26], or the design of various "add-ons" to the algorithm to improve its general performance [36,28]. Simple, or canonical, genetic algorithms are composed of three operators: reproduction, crossover and mutation. in this study, we focus on mutation. Mutation is the occasional random alteration of the value of a bit string. A review of the currently available literature on genetic algorithms reveal that only a small portion, estimated to be around 0.5%, is focused on the mutation operator [2].

This relative lack of interest in mutation is due to the common view of the mutation operator. As [36] quoted from John Koza, the author of "Genetic Programming":

> Mutation is a sideshow in nature and a sideshow in genetic algorithms. I usually keep it turned off.

However, there are some studies which do not support this opinion. For example, [20] performed experiments to identify the importance of mutation and the crossover operators in the optimization of a complex problem. And rather than a "side show", their results indicated that mutation significantly affected the performance of genetic algorithms. Kempen stated that:

> ... (the) results indicate that mutation may not considered to be a background operator, but plays an essential role during the search.

Research by [29] also supports the idea that mutation is by no means a "sideshow". They found that the crossover operator has a greater impact on the population's fitness at the earlier stages of the optimization process but can be stuck in a local optima without the introduction of new information by operators such as mutation. Earlier work by [14] and [5] also proposed that the mutation operator play an equally important role as the other genetic operators in the optimization process. With these works demonstrating the importance of the mutation operator, we have decided to place further focus on it.

18.2.1 A Closer Look at the Mutation Operator

While the process of mutation is very simple, where one gene of a binary chromosome is switched from 0 to 1 or 1 to 0, the process is responsible for the restoration or the creation of lost or unexplored genetic material in the population. Many researches have found that this characteristic helps prevent the algorithm from premature convergence [15,18]. Although some researches have also shown that genetic algorithms in a mutation free environment seem to produce better results, Fogel [15] believes that:

> Rather than indicate the superiority of crossover mutation... [the results] may simply indicate the general utility of taking big steps away from poor solutions instead of small steps.

Furthermore, because of the usually small amount of information in optimization problems, whether a multi-dimensional move (when crossover is applied) is more efficient than a single-dimensional move (when mutation is applied) will differ according to the application [5]. But as a general rule, mutation is applied at a relatively low rate in most cases, for example, [30] used 0.005 mutation in his experiments, [12] tested values between 0.001 to 0.02, and [26] other researchers have used 0.01 as the mutation probability.

At lower mutation rates, the process of convergence will be less likely to be disrupted, but it also means the algorithm is more likely to fall victim to local optima because the search space is not as thoroughly explored. A high mutation rate, on the other hand, does not necessarily imply improved results, as found by [31]:

> Large values of [mutation] transform the GA into a purely random search algorithm, while some mutation is required to prevent the premature convergence of the GA to suboptimal solutions.

Hence, the ideal system is one which converges efficiently with minimal disruption but also maintains a level of diversity where the search space is effectively searched. For example, work by [1] has shown that using a Distributed Genetic Algorithm (DGA), or what is referred to as a "subpopulation-type parallelised GA" , coupled with a protection scheme for strong individuals, can help to improve the performance of the GA.

Conceptually, DGAs are the same as PGAs, where a population is divided into separate sub-populations. But rather than needing to be ran on parallel processors, DGAs can run on a single computer processor. Studies performed by [33,25] have shown that the DGAs are much more effective than traditional GAs when using the same population size.

The motivation of the SSGA in this research is quite similar to [1], where we try to protect good genetic material, but instead of using an "add-on" scheme, the SSGA does this through the mutation operator.

18.3 The SSGA Explained

In most genetic algorithms, the mutation rate must be high enough to introduce an adequate amount of new genetic material throughout the optimization process but must not be too high to cause too much disruption so as to delay the convergence of the population [31].

To achieve this system, this research proposes combining the concepts from Distributed Genetic Algorithms (DGAs) with the use of two different mutation rates. The SSGA splits the population into two groups and applys different mutation rates to the two groups. The group with the higher mutation rate is called the **experimental group**, while the group with the lower mutation rate is called the **control group**.

During the optimization process, each group separately perform reproduction and mutation operations (See Figs. 18.1 and 18.2). To ensure the two populations work together, at the end of each generation, the best and worst individual of the two groups be identified and the best individual is replicated and used to replace the other group's worst individual. The term "invasion" has been coined to describe this process (Fig. 18.3).

The invasion strategy permits two complementary outcomes:

(a) **Diversity of population** is ensured by the experimental group's high mutation rate, which causes new genetic materials to be introduced all the time.

(b) **Preservation of strong individuals** is ensured by the control group's low mutation rate. Individuals with high fitness values are placed in the control group where the mutation rate is low. As a result, the cross-over operator will have a much longer time to manipulate the genetic materials within the control group, with relatively little (or even without) influence of mutation.

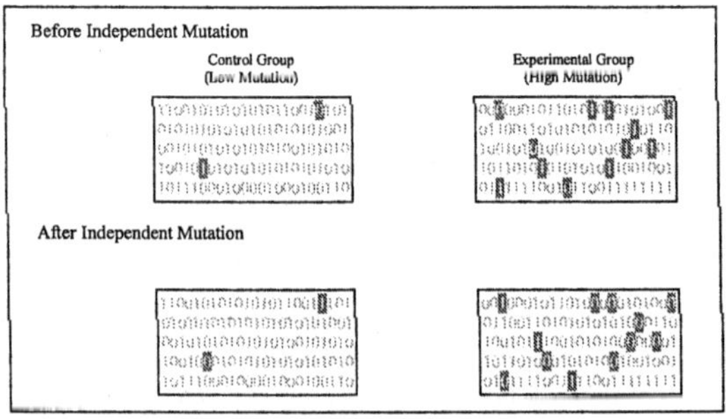

Fig. 18.1. Control groups have low mutation while experimental group has high mutation

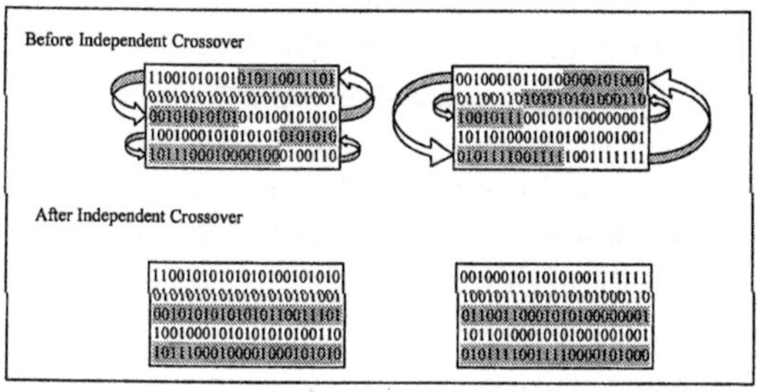

Fig. 18.2. Each group performs crossover independently.

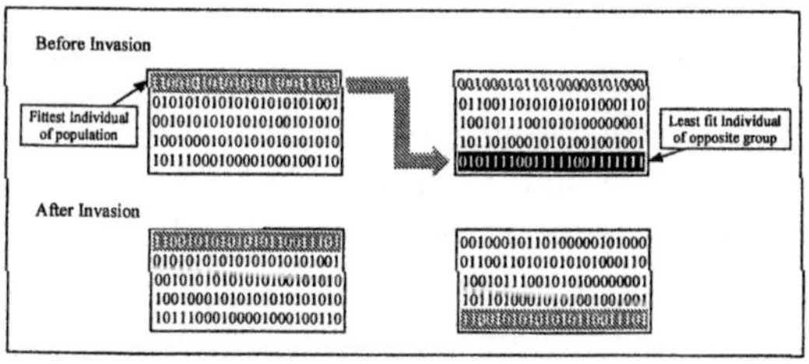

Fig. 18.3. The fittest individual (gray colored) of the entire population is copied, and used to replace the least fit individual (black colored) of the opposite group.

In essence, the experimental group with its high mutation rate allows the system a much more randomized search, while the control group, on the other hand, concentrates on a more localized search for the optimal.

Figure 18.4 is the pseudo code for the SSGA.

```
Initilization
Create random population
Do while Convergence = False
        Evaluation –    Calculate and record fitness for all individuals
        Invasion –      The best individual of the population will replace the worst individual of
                        opposite group
        Parent Selection and Reproduction – Randomly cross-over individuals within group
        Mutation –      Randomly mutate individual from each group
Loop
```

Fig. 18.4. Pseudo code for the SSGA.

However, by dividing the population into two smaller groups, the available genetic material for manipulation by the genetic operators is reduced, which in turn compromises their searching abilities. The extent and the effect of this on the GA's performance is unclear. So to understand this more clearly, experimentation with different control group size on a common test bed was conducted. From these tests, two goals were achieved. First, performance of the SSGA was assessed and the optimal control group size was determined. Secondly, the SSGA can also be rated against a Standard Genetic Algorithm (STDGA) in terms of its performance.

18.4 Preliminary Testing

The test environment for the preliminary tests was ten mathematical functions, representing various types of problems that are commonly faced in general optimization. Five of these functions were the ones used in De Jong's 1975 thesis. Used by many researchers as the benchmark test bed, these functions represent "common difficulties in optimization problems in an isolated manner" [36].

A further set of five functions (see Fig. 18.6–Fig. 18.10 below) from the First International Contest on Evolutionary Optimisation [24] was selected to complete the test bed of ten functions. These "modern equivalent" of the De Jong functions present some extremely challenging problems for GAs. While some are similar to the De Jong functions, the ICEO functions in general have different number of dimensions with added complexity.

$$f_6(\bar{x}) = \sum_{i=1}^{5}(x_i - 1)^2 \quad x_i \in [-5.000000, 5.000000], i = 1, \cdots, 5 \qquad \text{(ICEO 1)}$$

$$f_7(\bar{x}) = \frac{1}{4000}\sum_{i=1}^{5}(x_1 - 100)^2 - \prod_{i=1}^{5}\cos(\frac{x_1 - 100}{\sqrt{i}}) \quad x_i \in [-600.0000, 600.0000], i = 1, \cdots, 5 \qquad \text{(ICEO 2)}$$

$$f_8(\bar{x}) = -\sum_{i=1}^{30}[1/\sum_{j=1}^{5}(x - A(i,j)^2 + C_i)] \quad x_i \in [0, 10.0000], i = 1, \cdots, 30, j = 1, \cdots, 5 \qquad \text{(ICEO 3)}$$

$$f_9(\bar{x}) = -\sum_{i=1}^{5}\sin(x_i)\cdot\sin^{20}(\frac{i\cdot x_i^2}{\pi}) \quad x_i \in [0, 3.142], i = 1, \cdots, 5 \qquad \text{(ICEO 4)}$$

$$f_{10}(\bar{x}) = -\sum_{i=1}^{5}c_i(\exp(\frac{-1}{\pi}\sum_{j=1}^{5}(x - A(i,j))^2))\cdot\cos(\pi\cdot\sum_{j=1}^{5}(x - A(i,j))^2) \quad x_i \in [0, 10.0000], i = 1, \cdots, 5, j = 1, \cdots, 5 \qquad \text{(ICEO 5)}$$

The preliminary testing involved testing 9 versions of SSGAs which were tested over the entire test bed. These versions of SSGAs are as follows:

10-90SSGAs	where	10% Control Group : 90% Experimental Group
20-80SSGAs	where	20% Control Group : 80% Experimental Group
30-70SSGAs	where	30% Control Group : 70% Experimental Group
40-60SSGAs	where	40% Control Group : 60% Experimental Group
50-50SSGAs	where	50% Control Group : 50% Experimental Group
60-40SSGAs	where	60% Control Group : 40% Experimental Group
70-30SSGAs	where	70% Control Group : 30% Experimental Group
80-20SSGAs	where	80% Control Group : 20% Experimental Group
90-10SSGAs	where	90% Control Group : 10% Experimental Group

By testing different sizes of control group we could derive some relationships between the performance of the SSGAs and the group sizes. Furthermore, four different population sizes were tested 10, 25, 50, and 100. These sizes were derived from previous benchmark studies [35]. The following table (Table 18.1) summaries the 36 different SSGAs configurations that were tested:

Table 18.1. Table of different group size configurations tested. Each version of the SSGA consist of two groups, a control group and an experimental group. The group sizes differ depends on the version of SSGA. A version 40-60SSGA means a control group size of 40% and an experimental group size of 60%. For a population size of 25, this would mean 10 individuals will belong to the control group while the remaining 15 will belong to the experimental group.

	Population Size 10		Population Size 25		Population Size 50		Population Size 100	
	Control Group Size	Experiment Group Size	Control Group Size	Experiment Group Size	Control Group Size	Experiment Group Size	Control Group Size	Experiment Group Size
10-90SSGA	1	9	3	23	5	45	10	90
20-80SSGA	2	8	5	20	10	40	20	80
30-70SSGA	3	7	8	18	15	35	30	70
40-60SSGA	4	6	10	15	20	30	40	60
50-50SSGA	5	5	13	13	25	25	50	50
60-40SSGA	6	4	15	10	30	20	60	40
70-30SSGA	7	3	18	8	35	15	70	30
80-20SSGA	8	2	20	5	40	10	80	20
90-10SSGA	9	1	23	3	45	5	90	10

To smooth out random effects, each configuration was tested 10 times.

Hence, the total number of test trials performed for the 9 SSGAs and 1 STDGA is:

10 function tests X10 trials X4 population sizes X (9 SSGA + 1 STDGA) = 400 trials.

The parameter settings of the GAs were selected from previous benchmark studies performed using the same test bed [35]. Following the recommendations made by the benchmark test a crossover rate of 95% is used. For the SSGA, the mutation rate for the control group will be 0, to give a mutation free environment. The experimental group, on the other hand will have a mutation rate of 8%. For the STDGA, the mutation rate was set at 5%, which was the optimum mutation rate. All of these mutation rates were derived as a result of benchmark testing from [35].

For both algorithms, termination of the search will occur when the best solution hasn't been improved for the last 10,000 fitness evaluations. To ensure the system will stop, the maximum number of fitness evaluations is 1,000,000.

18.5 Preliminary Results

To assess the algorithms, their performances were classified into two types: effectiveness, and efficiency,

(a) **Effectiveness** is the quality of the result. Effectiveness can be measured by the optimum fitness value that the algorithm could find in testing.
(b) **Efficiency** is the amount of resources that the algorithm uses before the termination criterion is met. To measure this, the number of function evaluations was used.

To ensure the results are comparable across all the tests, a common index is used to rank the solutions between 1 to 0. This ratio, called the Relative Deviation Index (RDI), and was first used in [21]. It is an index used to grade solutions that the algorithm produces. For minimization problems, the RDI is as follows:

$$RDI = \frac{(T_W - T_a)}{(T_W - T_B)}$$

where:
T_a = the solution for a given parameter set
T_B = the parameter set which gave the best solution
T_W = the parameter set which gave the worst solution
Hence, under the RDI ratio, the closer the solution (T_a) is to the best solution (T_B), the higher it would be rated. The optimum solution is given an RDI value of 1 (when $T_a = T_B$), whereas the worse solution will get 0 (when $T_a = T_W$). Furthermore, by multiplying the RDI efficiency ratio (number of function evaluations) and RDI effectiveness (fitness value) ratio, an overall performance indicator called the **Performance Ratio** (PR) can be derived. PR is a final "definitive" value that can be used to judge an algorithm in terms of both its efficiency and effectiveness. This ratio is useful for identifying overall performance because it penalizes an algorithm's performance if it achieves a high RDI in efficiency but not effectiveness, or vise versa.

To perform the comparison, the algorithm, which produced the highest median (RDI fitness value (RDI-FIT) for effectiveness, or RDI function evaluation (RDI-FE) for efficiency), is identified first. The sample of results from this algorithm is then tested against the other nine algorithms', using the

Wilcoxon Rank test. This non-parametric tests the null hypothesis (H_0) of no difference between two samples, and therefore, can determine if the algorithm's superior performance was statistically significantly better than the others. Comparisons were made in terms of RDI-FE, RDI-FIT and PR, to assess the most efficient, effective and overall performance respectively.

Each Wilcoxon test produces a p-value which is an indication of the probability of error from the rejection of the H_0. These p-values are published in tabular form (C10 from the leftmost column to C90 and then STDGA in the rightmost column) and presented in Table 18.2. Each row of the table represents one of the ten tests from the test bed, while the 10 columns represents the p-values of from testing the optimum algorithm (marked with OPT and colored in dark gray) against the particular version of SSGA or the STDGA.

In this research we say that the H_0 is rejected and said to be statistically significant when the p-value is equal to or less than 0.1 (i.e. 90% certain that the difference between the two samples is significant). These cells were colored in light gray on the p-value tables. So for example, in the first row of Table 18.2, C30 achieved the highest RDI-FE, and its performance was only significantly better than C20, C60 and C70. Compared to the remaining algorithms, including STDGA, its performance was only marginally better. (i.e. not statistically significant)

Table 18.2. P-values for the rejection of the null hypothesis that there are no differences between the optimum algorithm and the other nine algorithms in terms of number of fitness evaluations used to reach convergence (RDI-FE).

FE	C10	C20	C30	C40	C50	C60	C70	C80	C90	STDGA
DeJong 1	0.7254	0.0758	OPT	0.1102	0.7218	0.0005	0.0223	0.1516	0.2217	0.6790
DeJong 2	OPT	0.0480	0.0377	0.0015	0.0000	0.0002	0.0438	0.0120	0.0087	0.5036
DeJong 3	0.2522	0.2145	0.1384	0.0453	0.1436	0.3789	0.2235	OPT	0.6685	0.8852
DeJong 4	0.8512	0.6407	0.9463	OPT	0.0833	0.0149	0.2290	0.0438	0.0023	0.5933
DeJong 5	0.0859	0.5222	OPT	0.8625	0.6755	0.2309	0.2423	0.0766	0.7039	0.7728
ICEO 1	0.0000	0.0000	0.0003	0.0005	0.0052	0.0502	0.2309	0.5285	OPT	0.0000
ICEO 2	0.2309	0.2092	0.5933	0.7146	OPT	0.2726	0.3408	0.0931	0.2562	0.7987
ICEO 3	0.7435	0.2623	0.0044	0.0922	0.0502	0.0050	0.0013	0.0015	0.0404	OPT
ICEO 4	0.6826	0.1906	0.0009	0.0237	0.0008	0.0858	0.0000	0.0001	0.0000	OPT
ICEO 5	0.0606	0.0428	0.2309	0.0323	0.0066	OPT	0.0008	0.0359	0.0066	0.2109

Results from Table 18.2 shows that the SSGA is not statistically different from STDGA in terms of efficiency, although the SSGA seems to require fewer fitness evaluations than STDGA, except for ICEO 3 and ICEO 4. For these two problem functions, the STDGA was able to perform significantly better than most SSGAS. Hence, in a very small way, the SSGA may be less efficient than the STDGA.

In terms of effectiveness however, the SSGA is shown to be statistically significantly better than STDGA (see Table 18.3). The results indicate that

Table 18.3. P-values for the rejection of the null hypothesis that there are no difference between the optimum algorithm and the other nine algorithms in terms of the fitness value reached (RDI-FIT).

FIT	C10	C20	C30	C40	C50	C60	C70	C80	C90	STDGA
DeJong 1	0.0000	0.0000	0.0000	0.0000	0.0000	0.0043	0.0134	0.3971	OPT	0.0000
DeJong 2	0.0003	0.1056	OPT	0.7467	0.2249	0.8849	0.1301	0.7719	0.2630	0.0044
DeJong 3	0.0000	0.0000	0.0134	0.0325	0.0325	0.7851	0.7269	OPT	0.6983	0.0000
DeJong 4	0.0004	0.0068	0.0128	0.0012	0.2022	0.1987	0.0781	OPT	0.4300	0.0000
DeJong 5	OPT	0.3690	0.7467	0.9671	0.1786	0.3969	0.2186	0.0182	0.0069	0.0000
ICEO 1	0.0000	0.0000	0.0003	0.0005	0.0052	0.0502	0.2309	0.5285	OPT	0.0000
ICEO 2	0.9578	0.9846	0.7003	0.9923	0.4386	0.3581	0.7691	0.7691	OPT	0.0108
ICEO 3	0.2602	OPT	0.7924	0.4019	0.0859	0.8746	0.4875	0.4407	0.2938	0.0003
ICEO 4	0.0000	0.0102	0.1236	0.2092	0.0543	0.0682	OPT	0.6545	0.6441	0.0000
ICEO 5	0.0104	0.0013	0.0005	0.0002	0.0044	0.0051	0.0573	0.6195	OPT	0.0000

Table 18.4. P-values for the rejection of the null hypothesis that there are no difference between the optimum algorithm and the other nine algorithms in terms of its performance ratio.

PR	C10	C20	C30	C40	C50	C60	C70	C80	C90	STDGA
DeJong 1	0.0033	0.1351	0.3306	0.3652	0.5952	0.0510	0.1277	0.3760	OPT	0.0001
DeJong 2	OPT	0.0476	0.0404	0.0014	0.0000	0.0001	0.0136	0.0016	0.0003	0.5243
DeJong 3	0.0002	0.0057	0.0711	0.0758	0.2070	0.5624	0.3859	0.9048	OPT	0.0000
DeJong 4	0.1410	0.3258	0.9200	0.4238	0.6932	0.4996	0.6842	OPT	0.9200	0.0359
DeJong 5	0.0850	0.5006	OPT	0.8738	0.5933	0.2145	0.1644	0.0619	0.2502	0.6720
ICEO 1	0.0045	0.1901	OPT	0.3027	0.4637	0.2399	0.0104	0.0001	0.0000	0.0000
ICEO 2	0.5222	0.8436	0.8436	0.8211	0.9195	0.6407	0.8587	0.7110	OPT	0.1397
ICEO 3	0.6153	OPT	0.6842	0.2807	0.0372	0.1253	0.0546	0.0012	0.0035	0.0071
ICEO 4	0.0774	OPT	0.2807	0.1934	0.0019	0.4350	0.1429	0.0021	0.0008	0.0000
ICEO 5	0.4696	0.0093	0.1302	0.0395	0.0476	0.5496	0.1029	0.5369	OPT	0.0022

Table 18.5. P-values for the rejection of the null hypothesis that there are no difference between the optimum algorithm and the other nine algorithms in terms of its RDI-FE, RDI-FIT and PR values.

	C10	C20	C30	C40	C50	C60	C70	C80	C90	STDGA
RDI-FE	OPT	0.5562	0.8325	0.4574	0.0921	0.0867	0.0184	0.0406	0.0494	0.8386
RDI-FIT	0.0000	0.0028	0.0236	0.0354	0.0978	0.2567	0.5412	OPT	0.9618	0.0000
PR	0.0521	0.3138	OPT	0.2803	0.1592	0.3956	0.2228	0.0725	0.0596	0.0000

those with larger control group sizes seem to perform better for most algorithms (although DeJong 2, DeJong 5 and ICEO 3, prefer a smaller control group size).

So ultimately, it is important to determine whether the improvement in effectiveness is worth the compromise in efficiency. To answer this, we turn to the PR value.

According to the results from the PR values (in Table 18.4), the SSGA produced better results than the STDGA. The optimum algorithms in all of the test functions were SSGAs with either a very low control group or a very high one. Based on the 4000 trials that were undertaken, the C30 SSGA, with 30% in its control group and 70% in the experimental group, was found to produce an overall more optimal result than other configurations (see Table 18.5). Compared with a standard genetic algorithm (STDGA), the C30 SSGA was also found to produce better results when the over all performance was considered (see Table 18.5). This is especially true when the PR ratio is assessed, because the C30 was statistically significantly better than STDGA.

In terms of population size, the tests revealed that the best population size was 25. [1]

The preliminary testing has thus shown that the SSGA can generate better results than STDGAs in function optimization. The testing of its abilities in the real life problem of optimizing technical trading strategy for trading in futures will now be performed.

18.6 Financial Application: Technical Trading Strategies

Trading strategies are basically a systematic approach for trading in a market. They are made up of primarily two components: first, a component that generates trading signals, telling the trader to buy, sell or hold based on some sort of analysis; and second, a component which performs the trading based on the signals generated by the first component.

Technical trading strategies (TTS), are a type of trading strategy which use technical analysis to perform price analysis. The output of technical analysis is typically a "conclusion" as to the trend of the historical prices (i.e. rising trend, falling trend, or non-trending). From this, a trading signal can be produced to recommend a position to take by the trader.

Research over the years has shown that technical trading strategies are able to generate economically and statistically significant amount of profit [13] [7]. Various research improving the profitability of TTS has also received a lot of attention in recent years. Some of these studies include work by [7,3,32,34] [28,27,23,10].

[1] This was performed by grouping the results based on the different population sizes. The results are not presented here.

However, in terms of optimizing TTS using GAs, only a hand-full of work exists, these include: [3,6,9,11,27,23]. Furthermore, all of these studies focused on either using TTS in the stock markets or in the trading of foreign exchange, rather than on derivatives such as future contracts which is undertaken in this study.

18.6.1 Trading of Futures

Future contracts, are commitments to buy or sell a specific commodity of designated quality at a specific price and at a specified date in the future [16]. The main purpose of a future contract is to minimize the risk exposed due to uncertainty about the future price of a commodity by deciding the price today.

Future contracts are traded in standardized exchanges. Traders who wish to trade in these exchanges must open a trading account with a certain amount of money. This account is used by the traders to pay or receive profit or losses that are incurred daily. The settlement (or the realization) of daily profit and loss by the trader is called the *Mark-to-Market* system. Throughout the trading of futures, the trader has to keep a minimum amount of money in the account. This amount is a function of the amount of risk that the trader is exposed to by trading in the futures. Currently, most major futures exchanges of the world uses a system developed by the Chicago Mercantile Exchange (CME) called the Standard Portfolio Analysis of Risk (SPAN) to calculate this risk and set the margin.

18.6.2 Implementing TTS in Futures Trading

To simulate trading futures using TTS, data on future commodities were collected. In their raw form, data on futures prices cannot be used because each futures contract has a different maturity date. To form a historical series that can be used for trading, a technique called "splicing" was used to join together prices from different contracts. [2]

In total, ten years of daily high and low, closing and opening historical data was collected from eight commodities between the period of 1 Oct 1988 to 31 Sept 1989. These prices are quoted in the form of an index, which can only move up or down in fixed units called **tick size** which is different depending on the commodity. Tick size for a US Treasury Bond future, for example, is 1/32 points, and each of these point represents the value of US$31.25, this value is commonly known as the **tick value**. So if the index say, moved from 100.0000 to 100.09375, then in terms of ticks, it has moved by 3 ticks (that is, $0.09375 \div 1/32 = 3$). If a buy position was entered into

[2] Unfortunately, data for margins were not available for testing. As a result, the mark-to-margin system could not be included in the trading simulation.

when the price of 100.0000, and sold when the price was 100.09375, then the monetary profit from this "trade" is US\$93.75 (that is, $31.25 \times 3 = 93.75$).

The following are the specifications for the eight futures that were used by this research:

Table 18.6. The tables are summaries of the specifications of the eight commodities used in this research.

(KC) Coffee 'C' Futures		(CT) Cotton No.2 Futures	
Exchange	CSCE	Exchange	NYCE
Contract Unit	37,500 lbs	Contract Unit	50,000 lbs (approx 100 bales)
Quotation	US cents per pound	Quotation	US cents per pound
Tick Size	0.05 cents	Tick Size	0.01 cents
Tick Value	USD 18.75	Tick Value	USD 5
(CL) Light, Sweet Crude Oil Futures		**(GC) Gold Futures**	
Exchange	NYME	Exchange	COMEX
Contract Unit	1,000 barrels (42,000 gallons)	Contract Unit	100 troy ounces
Quotation	USD per barel	Quotation	USD per troy ounce
Tick Size	USD 0.01	Tick Size	USD 0.10
Tick Value	USD 10	Tick Value	USD 10
(DM) Deutschemark Futures		**(JY) Japanese Yen Futures**	
Exchange	CME	Exchange	CME
Contract Unit	DEM 125,000	Contract Unit	JPY 12,500,000
Quotation	USd per Deuschemark	Quotation	USD per JPY
Tick Size	USD 0.0001	Tick Size	USD 0.000001
Tick Value	USD 12.50	Tick Value	USD 12.50
(US) US Treasury Bond Futures		**(SP) S & P 500 Stock Index Futures**	
Exchange	CBOT	Exchange	CME
Contract Unit	USD 100,000 face value	Contract Unit	Index * USD 500.00
Quotation	Points	Quotation	Index Points
Tick Size	1/32 point	Tick Size	0.1 points
Tick Value	USD 31.25	Tick Value	USD 25.00

To fully utilize the data, the ten years of data were split into yearly groups, each year is then used for both optimization and testing. The following table can illustrate the way this is achieved:

Table 18.7. The table illustrates how the 10 years of data was split into 9 pairs of "optimization and test" data pair.

Oct 1988 to Sept 1989	Oct 1989 to Sept 1990	Oct 1990 to Sept 1991	Oct 1991 to Sept 1992	Oct 1992 to Sept 1993	Oct 1993 to Sept 1994	Oct 1994 to Sept 1995	Oct 1995 to Sept 1996	Oct 1996 to Sept 1997	Oct 1997 to Sept 1998
Optimize	Optimize	Optimize	Optimize	Optimize	Optimize	Optimize	Optimize	Optimize	
	Test	Test	Test	Test	Test	Test	Test	Test	Test

So, for example, the data set of Oct 1988 to Sept 1989 is used for optimization, the solution derived from this period is then applied in the period Oct 1989 to Sept 1990 for testing. Once this is finished, the time period, Oct 1989 to Sept 1990, is then re-used for optimization and again, the derived solution is then tested in the Oct 1990 to Sept 1991 period. So in this way, 9 pairs of "optimize and test" time periods are created from the 10 years of data.

18.6.3 Technical Trading Rule

For this research, the filter rule [13], and the filtered moving average [22] were chosen for optimization. These rules are very simple in design and are commonly used by traders [19].

Perhaps one of the most commonly used technical trading rules is the filter rule (FR). The one used in this research is the same one used by [13].

The filter rule is basically a way of detecting changes in the price trend. As the prices fluctuate through out a trend, a filter of a certain size is applied to the lowest low and the highest high so far. If the day's closing price close against the low (high) beyond the filter's area, then a buy (sell) signal is generated, otherwise an exit signal is generated, recommending the trader to exit the market.

A moving average filter rule (MA), is basically a linearly weighted moving average, coupled with the use of filters. First, an x day moving average is generated, a y per cent "filter" is then created above and below this moving average to form a "band". If the current day's price close above (below) the band, a buy (sell) signal is generated. If the price happens to be in between the upper and lower filter, an exit signal is generated. This type of filter is also known as a "percentage envelope" or "volatility band".

The signals generated daily from the filter rule and moving average filter rules are fed into a trading strategy for trading.

18.6.4 TTS Design

Based on the signals from the trading signals generated today (S_t) and the previous day's (S_{t-1}), an action is determined, using the following decision table.

Table 18.8. S_{t-1} is the previous day's signal, St is today's signal. A long position is the purchase of a future that gives the buyer the right to purchase the commodity in the future. A short position is the purchase of a future that gives the buyer the right to sell the commodity in the future.

S_{t-1}	S_t	Action
Hold	Hold	No action
Hold	Buy	Open a long position
Hold	Sell	Open a short position
Buy	Hold	Close a long position
Buy	Buy	No Action
Buy	Sell	Close a long position, simultaneously open a short position (a reverse)
Sell	Hold	Close a short position
Sell	Buy	Close a short position, simultaneously open a long position (a reverse)
Sell	Sell	No Action

On the next trading day, the action is performed and the contract is assumed to be executed at the average of the opening and closing price of that day. This contract is then held until one of the following criteria is satisfied:

(a) When a signal is given by the technical trading rule to exit or reverse the position (See Table 18.8).
(b) When a signal to exit trade is given because stop-loss level has been breached.

The stop-loss mechanism is the same as the one suggested by [4] to help minimize "draw downs". It operates by recording the cumulative loss incurred since the beginning of a trade. When this exceeds the level at which the amount of loss is no longer tolerable (the stop-loss level), the algorithm is forced to abandon the current trade on the next available price, that is, the next day's average. The analysis of the market is restarted after a "cooling down" period. In the case where the trade signals from the trading rules conflict with the stop-loss level's signal, the stop-loss signal is given priority.

Graphically, the entire process can be drawn on a time line, this is shown in Fig. 18.5.

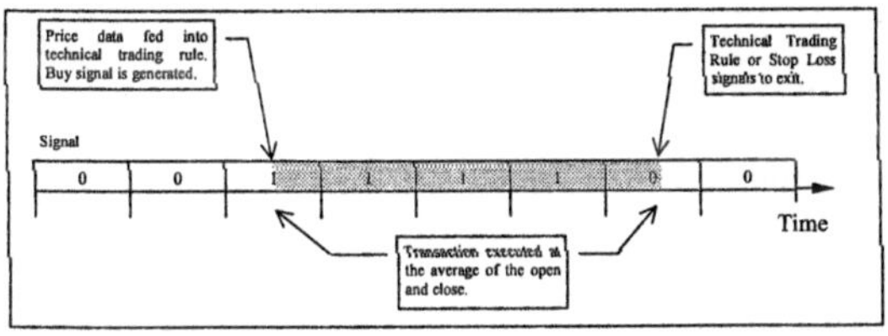

Fig. 18.5. The time line illustrates how the signals produced by the technical trading strategy will cause the simulator to enter or exit the market. Note that even though the signal is given at the start of the day, the position is not opened until half way through the day, that is, the price is assumed to be between the high and low of the day. Note that the stop-loss mechanism has not been included.

At the end of each trading year, the average daily return is calculated and is divided against the standard deviation of the daily return to calculate the overall profitability (fitness) of the TTS as a unit of variability of the return, this measure is commonly known as the Sharpe ratio. By optimizing the TTS's Sharpe ratio, it is hoped that the return of the TTS will be maximized while the variability of the return will be minimized [28,8].

Furthermore, to improve robustness of the TTS in out-of-sample trading, this research simultaneously optimized the trading of all eight commodities, so that each TTS must perform well across all eight commodities. It was hoped that by doing so, and with the use of the Sharpe ratio, the solution reached would be more robust in the relatively volatile environment of the derivatives market.

So in summary, the following are the list of variables that were optimized in the simulation for the TTS:

Filter Rule:

- Filter Size (0.001 to 1.000) – this is the size of the filter in terms of percentages

Moving Average Filter Rule:

- Moving Average Period (1 to 20) – this is the length of the moving average in terms of days. So a 5 day moving average, for example, would mean the average was calculated from the previous 5 days of prices.
- Filter Size (0.001 to 1.000) – this is the size of the filter that is placed above and below the moving average

For both the trading rules, the following two variables were also optimized:

- Stop loss level (0.001 to 1.000) – this is the size of the stop-loss level. A size of 0.0 would mean that no loss is tolerated, so a size of say 0.5 would mean half the trade can be lost before the trade is abandoned.
- Cooling Period/Wait Period (1 to 20) – this is the amount of time the trading system waits before technical analysis is restarted, after the stop loss signal has been triggered off.

The following figure outlines the entire trading strategy.

```
For each day
        For X market
                Determine if X market is tradable on current day
                If yes, then
                    Determine today's signal
                    Signal may indicate to enter, exit or reverse, depending on current position.
                    If already in a trade, assess whether stop loss is breached.
                    Trade based on the signals.
                End if
        Next X market
        Calculate Daily Return
Next day
Calculate Performance (Sharpe Ratio)
```

Fig. 18.6. The figure represents the pseudo code used to simulate the trading of futures

18.6.5 Parameters of the SSGA and STDGA

The parameter settings of the GAs were be the same as those used in stage one testing. The crossover rate for both the STDGA and the SSGA was 0.95.

The mutation rate for the STDGA was 0.05 while the SSGA's control and experimental group mutation rates was and 0.08 respectively. The control group sizes will be 30%, while the experimental group will have 70%. Population size for both algorithms will be 25.

Termination of search will occur when no improvement in the optimum value occurs for 5,000 fitness evaluations. Again, as a precaution, the maximum number of fitness evaluation is set at 1,000,000.

18.6.6 Results from the Optimization of TTS

Results are compared against a naive buy and hold (BH) strategy. (Only one contract was bought/sold at any one time.) From Table 18.9, the average daily return of the optimized moving average (MA) was much lower than the BH's. The 5% trimmed mean, as well as the median, shows that the moving average was a much poorer performer than the BH and the strategy would have made a $8,742 loss over the 9 year period, as compared with the possible $47,218 profit that is possible from the BH strategy (Table 18.9). The TTS only recommended an average of 39 days of being in market out of the possible 247. These differences in performance between the MA and the BH were statistically significant, as shown by the p-value from Table 18.10.

The Filter Rule (FR) however, was able to produce a result that was higher than the BH strategy in terms of the average total amount of profit of $79,166. It had the same median as the BH strategy and its mean and trimmed mean was marginally lower than the BH. Unfortunately, the Mann Whitney U test indicated that the difference was not statistically significant. The TTS recommended an average of 82 days of being in market.

Table 18.9. Table of the mean, 5% trimmed mean and median of average commodity daily profit the two TTS in the out-of-sample period.

	Out-of-Sample Average Daily Return				
	Ave. No. of days in Mkt per yr†	Mean	5% Trimmed Mean	Median	Profit/Loss
BH	247	5.46	7.24	10.00	47,218.13
MA	39	-25.72	-23.02	-25.00	-8,742.24
FR	82	4.62	6.17	10.00	79,166.82

Note: †There were only an average of 247 trading days per year.

So it seems that the TTS is marginally better than a BH strategy where the difference between the two is not greatly significant. In terms of the better TTS, the results indicate that the FR is much better than the MA. An analysis of the total number of the trading behavior of the FR shows that the average amount of win and the % of winning trades are much better than MA, although this is compromised by a higher average amount of loss.

Table 18.10. Table of the mean, 5% trimmed mean and median of average profit/loss per trate for the two TTS in the out-sample period and the p values from statistical test of no difference between the two TTS and BH. Note that no mean, trimmed mean or median statistic exists for the BH strategy because it assumes a single buy position.

Out-of-Sample Average Trade Return						
	Ave. No. of Trades over 9 years	Mean	5% Trimmed Mean	Median	Sum	p value
BH	1	-	-	-	47,218.13	-
MA	396.25	-22.06	-201.47	-234.38	-8,742.24	0.000
FR	501	158.02	-70.93	-183.75	79,166.82	0.611

Table 18.11. The table presents the average winning and loosing amounts per trade from the out of sample trading of the four TTS.

Out-of-Sample Trade Return					
	Ave. No. of Winning	Ave. Amount of Win	Ave. No. of Losses	Ave. Amount of Loss	% of Winning Trade
MA	123	1,538.6243	273	-724.5857	0.3104
FR	191	2,269.5800	310	-1,145.7316	0.3817

The reason for this marginal performance of the FR rule can be better explained by the following histogram of the distribution of the trades that were made.

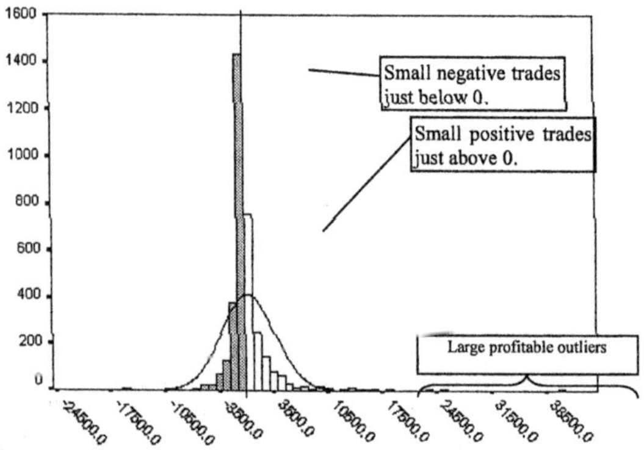

Fig. 18.7. Histogram of trade distribution for FR. The y axis is the frequency while the x axis is the amount of profit or loss for the trades.

The histogram shows that although the number of winning trades were much fewer than the losing trades, the small number of large profitable outliers have made the TTS profitable. This is primarily due to the use of the Sharpe Ratio for optimization. The TTS is trying to produce a high Sharpe ratio by capturing the occasional large profitable trades from spurious price movements to get higher returns, while making a large number of small losses to keep the standard deviation low.

To reduce this effect, a new version of FR, called FR2, was tested. There are two major changes to the FR TTS: First, a take-profit mechanism similar to the stop-loss level is used to prevent the TTS from seeking large spurious returns by limiting the amount of profit that the TTS can obtain in any one trade. In this prototype version, the take-profit level has been arbitrary set at twice the level of the stop-loss level. Second, the number of winning trades was used as the objective function rather than Sharpe ratio maximization. In this way, the TTS will optimize to improve its return through making a greater number of positive returning trades, where the amount of these profits will be "controlled" by the take-profit level to discourage it from optimizing for spurious profit.

18.6.7 Results of FR2

Comparing FR to FR2, the number of trades has been reduced from 501 to 145 but the sum of its profit over the out-of-sample period was higher than the FR, a gain of $21,465.68 (see Table 18.12). Also the mean daily return of FR2 was higher than FR and BH (see Table 18.13).

In terms of percentage of winning trade, FR2's was 42% compared to FR's 38% (See Table 18.14). Furthermore, the ratio of profit to loss in dollars has been increased from 1.9808 to 2.3745. Combining these two factors means the FR2 is winning more frequently and is picking trades that are more profitable. The median in Table 18.15 also supports this hypothesis: it rises from $840.0002 to $1,837.4980, a gain of $997.4978, while the loss in the median fell from -$605.0005 to -$1,201.876, an increased loss of $596.8755.

However, a Mann Whitney U test of the H_0 that there is no difference between FR and FR2 cannot be rejected with a p-value of 0.664. Hence, the improvement of profitability in terms of daily return (with a p-value of 0.914) and per trade by FR2 is only marginal and not statistically significant (See Table 18.16).

When compared with BH, the p-value of 0.749 indicates that FR2's improvement was not significant, although for the in-sample period, the p-value was significant at 0.079, see Table 18.16.

Again, if we look at a histogram of the distribution of the trades, we can see that the reason why the null hypothesis cannot be rejected is because its trading behavior is still very similar to FR, even though it produced much more profitable results (Fig. 18.8). However, note the portion of small

Table 18.12. Table of the mean, 5% trimmed mean and median of average profit/loss per trade for FR and FR2 in the out-of-sample period. Note that no mean, trimmed mean or median statistic exists for the BH strategy because it assumes a single buy position.

	Out-of-Sample Average Trade Return				
	No. of Trades	Mean	5% Trimmed Mean	Median	Sum
BH	1	-	-	-	47,218.13
FR	501	158.02	-70.93	-183.75	79,166.82
FR2	145	694.02	70.4777	-249.37	103,408.56

Table 18.13. Table of the mean, 5% trimmed mean and median of average commodity daily profit for FR and FR2.

	OutSample Average Daily Return			
	Ave. No. days in Mkt per year	Mean	5% Trimmed Mean	Median
BH	247	5.46	7.24	10.00
FR	82	4.62	6.17	10.00
FR2	40	12.2516	4.0889	10.00

Table 18.14. The table presents the average winning and loosing amounts per trade from the out-of-sample trading of FR and FR2.

	Out-of-Sample Trade Return					
	No. of Winning	Ave. Amount of Win	No. of Lossing	Ave. Amount of Loss	Ave. ratio of Profit:Loss	% of Winning Trade
FR	191	2,269.5800	310	-1,145.7316	1.9809	0.3817
FR2	121	4038.5794	169	-1700.6100	2.3745	0.4172

Table 18.15. The table presents of the median of winning and loosing trades in the out-of-sample trading for the FR and FR2.

	Out-of-Sample Trade Return			
	No. of Winning	Median of Winning Trade	No. of Loosing	Median of Loosing Trade
FR	191	840.0002	310	-605.0005
FR2	121	1,837.4980	169	-1,201.876

Table 18.16. Table of p-values from statistical test of no difference between FR2 and BH.

	p-value of H_0FR2 vs BH	
	p-value	Higher Mean
In Sample	0.079	FR2
Out Sample	0.749	FR2

negative trades have reduced significantly while the portion of positive trades have increased.

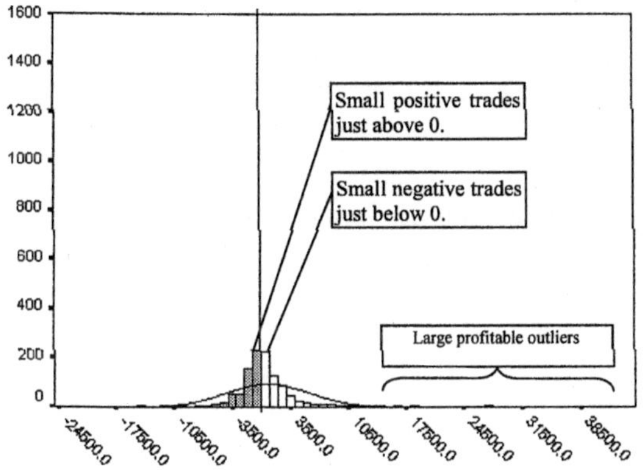

Fig. 18.8. Histogram of trade distribution for FR2.

18.6.8 Comparing SSGA against STDGA in TTS Optimization

The final part of the analysis is to look at whether the SSGA can produce more profitable results than the STDGA. To answer this, we look at the graph of cumulative profits for FR and FR2 TTSs in in-sample optimization and out-of-sample trading by SSGA and STDGA (Figs. 18.9 – 18.11).

The in-sample graphs for FR and FR2 shows the SSGA's cumulative profit is consistently higher than the STDGA's. As for the out-of-sample performance, the SSGA was again, shown to be the more profitable optimizer. However, for FR2, the graph shows that the STDGA was actually producing much more profitable results than SSGA, but during the years of 1996 to 1997, FR2 became much more profitable, this is primarily due to the capture of large profitable trades during those years.

Statistical tests of no difference between the SSGA and the STDGA performance indicates that their performance were not statistically significantly different in FR or FR2 (See Table 18.17). However, note that the p-values for FR2 indicates that SSGA was almost found to be significantly better than STDGA with p-values of 0.212 and 0.148 for in and out sample performance.

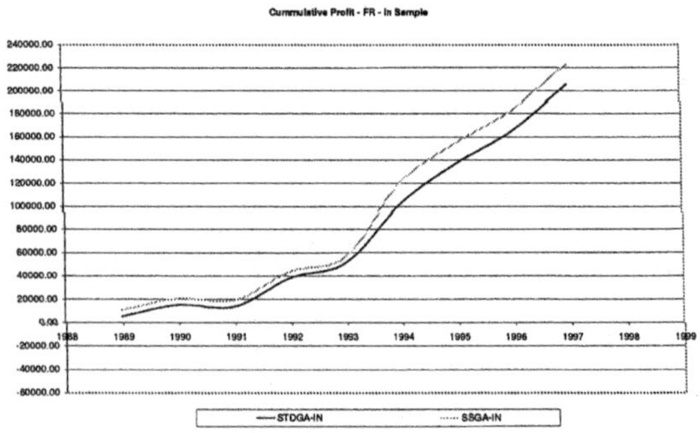

Fig. 18.9. Graph of the cumulative profit of in-of-sample testing from trading using FR

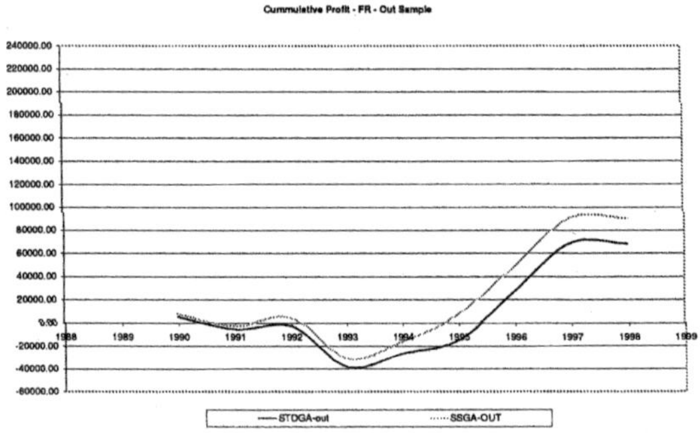

Fig. 18.10. Graph of the cumulative profit of out-of-sample testing from trading using FR.

Table 18.17. Table of p-values from the testing of whether there are statistical differences between SSGA and STDGA performance in FR and FR2.

	P-values from the Mann Whitney U Test between SSGA and STDGA				
FR	p-value	Higher Mean	FR2	p-value	Higher Mean
Std – In	0.943	SSGA	In	0.212	SSGA
Std – Out	0.947	SSGA	Out	0.148	SSGA

18.7 Conclusion

Genetic algorithms have been used for solving complex optimization problems ever since their introduction by Holland in 1975 (1992). Over the years, relatively little attention has been paid to the mutation operator due to its relatively low profile in the genetic algorithm literature. Of the handful of

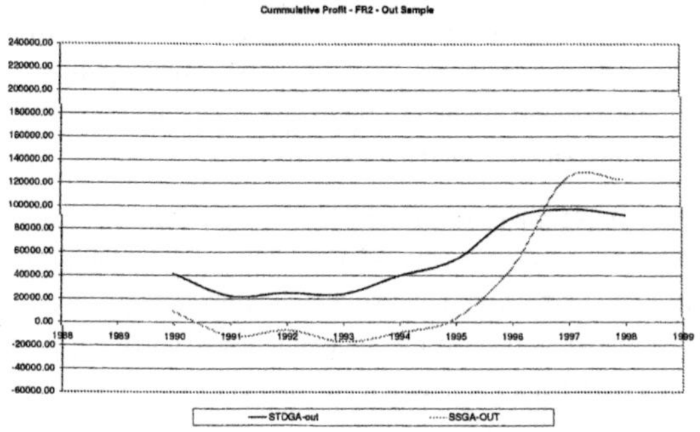

Fig. 18.11. Graph of the cumulative profit of out-of-sample testing from trading using FR.

publications that do focus on the mutation operator, most are concerned with altering the nature of the mutation operator rather than with novel methods of applying the mutation.

In this paper, the Split Search Genetic Algorithm, or SSGA for short was developed by using concepts from DGAs and used the two extreme behaviors of the mutation operator to improve performance of the standard genetic algorithms.

The novel SSGA was benchmarked using a test bed of function problems and revealed that it performed significantly better than the STDGA. The algorithm was then further tested in the optimization of technical trading strategies (TTS) in the trading of future commodities.

The optimization results in out-of-sample testing indicated that a filter rule is more profitable than either an MA TTS or the passive BH strategy. However, the difference in performance was not statistically significant. Analysis of the TTS's trading distribution shows that it is getting many small losses and trying to catch spurious large profits due to the use of the Sharpe ratio, which favors this behavior, as the objective value. As a result, a novel TTS that focuses on improving the number of winning trades, rather than on maximizing profitability, was developed and tested. It showed that the new TTS, called FR2, was more profitable, while performing fewer trades. It also showed a reduction in the number of small losses.

In all cases of TTS optimization, the SSGA showed that it was more profitable in both in-sample optimization and out-of-sample trading. Although the result of the SSGA's application in futures trading were not impressive, the research was able to reveal several important issues in the use of optimization tools for the optimization of TTS, all of which presents some interesting directions for future work.

Firstly, the mutation rate in this research was set according to previous benchmark studies. An interesting extension to the SSGA would be an investigation into the use of varying mutation rates, or mutation rates that change during the optimization process: for example, the use of an exponentially reducing mutation rate, or a self-adapting mutation rate which changes according to the performance of the population.

Another area of interest is in the SSGA's parallelism: for example, instead of having only two groups, future research may consider the use of 3, 4, or N number of groups. In that case, the research may incorporate a " community" model that has been proven to have interesting effects on genetic algorithms' performance [17]. Alternatively, work could also be performed in the area of designing SSGAs with group sizes that change throughout the optimization process.

Other potential research areas may also be considered from the issues that were raised from the application and testing of SSGA in TTS optimization.

One of the drawback of this research is the lack of margin data for the trading of futures. In particular, because the historical data for the margins that were set by the exchanges for the futures traded was unavailable, it was difficult to determine the appropriate measures for the daily return. For future research in optimizing future trading, a more realistic simulation could be achieved if the mark-to-margin system could be incorporated into the future trading simulation. In this case, rather than using the passive buy and hold strategy, a more appropriate benchmark may also be developed for measuring the performance of TTS.

In terms of the TTS, future studies may choose to study the TTS in more detail. This research has shown that a novel type of TTS, the FR2, was found to have better performance when compared with other TTS that were tested. Its was successful because it optimized for the number of profitable trades, rather than trying to maximize the profit, and also to ensure that the TTS will ultimately make a profit, take-profit and stop-loss mechanisms were implemented. The improvement in profitability as a result of implementing these mechanisms showed that "trading techniques", are equally, if not, more important, than the technical trading rules that were used. And a possible future direction of research may involve more focus in optimizing these trading techniques. More importantly, this may include work in optimizing trade exit and entry techniques. Alternatively, one may also choose to look at the strategy for choosing the optimum portfolio mix, or how to optimize resource (money) allocation.

Finally, future research could also be performed in optimizing robustness in the solution to further improve the out-of-sample performance so a less volatile profit curve could be produced. This is especially important in the case of TTS optimization as the environment of the simulation changes and robust solutions are required to ensure performance is maintained in the out-of-sample period. A starting point in this area maybe the implementation of robustness mechanisms, like those suggested by [28] into the SSGA.

References

1. Adachi N., Yoshida Y. (1995) Accelerating Genetic Algorithms: Protected Chromosomes and Parallel Processing. International Conference on Genetic Algorithms in Engineering Systems: Innovations and Applications, 76–81
2. Alander J. T. (1999) Evolution of Genetic Algorithms – A brief bibliographical review [Online]: ftp.uwasa.fi/cs/report99-2/main-a4.ps.z
3. Allen F., Karajalainen R. (1999) Using Genetic Algorithms to Find Technical Trading Rules. Journal of Financial Economics 51(2), 245–71.
4. Atiya A (1996) An Analysis of Stops and Profit Objectives in Trading Systems. Proceedings of the Third International Conference on Neural Networks in the Capital Markets, London, World Scientific
5. Back T. (1991) Self-Adaptation in Genetic Algorithms. Proceedings of the First European Conference on Artificial Life. Paris, France, MIT Press
6. Bauer R. J. (1994) Genetic Algorithms and Investment Strategies: John Wiley & Sons
7. Brock W., Lakonishok J., and LeBaron B. (1992) Simple Technical Trading Rules and the Stochastic Properties of Stock Returns. Journal of Finance **47(5)**, 1731–1764
8. Choey M. and Weigend A. (1997) Nonlinear Trading Models Through Sharpe Ratio Maximization. International Journal of Neural systems **8(4)**, 417–431
9. Colin A. (1994) Genetic Algorithms for Financial Modeling. In: G. Deboeck (Eds.) Trading on the Edge : Neural, Genetic, and Fuzzy Systems for Chaotic Financial Markets. Wiley, New York
10. Conrad J., Kaul G. (1998) An Anatomy of Trading Strategies. The Review of Financial Studies **11(3)**, 489–519
11. Deboeck G.J. (1994) Trading on the Edge. Toronto: John Wiley & Sons, Inc
12. DeJong K.A. (1975) An Analysis of the Behavior of a Class of Genetic Adaptive System. Computer an Communication Sciences, Michigan University of Michigan, 256
13. Fama E. F., Blume M. E. (1966) Filter Rules and Stock-Market Trading. Journal of Business **39**, 226–241
14. Fogarty T. C. (1989) Varying the Probability of Mutation in Genetic Algorithms. Proceedigns of the Thrid International Conference on Genetic Algorithms, Arlington, Morgan Kaufmann Publisher
15. Fogel D. B. (1995) Evolutionary Computation: Toward a New Philosophy of Machine Intelligence. New York: Institute of Electical and Electronics Engineers Press
16. Gardner M. J., Mills D. L. (1994) Managing Financial Institutions. Fort Worth: The Dryden Press, Harcourt Brace College Publishers
17. Goldberg D.E. (1989) Genetic Algorithm in Search, Optimization and Machine Learning. Addison-Wesley Publishing Company Inc.
18. Holland J. H. (1993) Adaptation in Natural and Artificial Systems. London: Massachusetts Institute of Technology

19. Kaiser T. (1998) [Personal Communication], Endeavour Hills, Victoria, Australia.
20. Kampen A. H. C. V., Buydens L. M. C. (1996) The Effectiveness of Recombination. Proceedings of the 2nd Nordic Workshop on Genetic Algorihtms and their Applications (2NWGA), Vaasa (Finland), University of Vaasa
21. Kim J., Kim Y. (1996) Simulated Annealing and Genetic Algorithms for Scheduling Products with Multi-level Product Structure. Computers Operations Research **23**, 857–868
22. Murphy J. J. (1986) Technical Analysis of the Furtures Markets: A Comprehensive Guide to Trading Methods and Applications. New York: Prentice-Hall
23. Neely C., Weller P., Dittmar R. (1997) Is Technical Analysis in the Foreign Exchange Market Profitable? A Genetic Programming Approach. Journal of Finaicial Quantitative Analysis **32(4)**, 405–426
24. OTC (1996) (Optimization-Technology-Center) NEOS Guide Optimization Tree [Online]: http://www.mcs.anl.gov/home/oct/Guide/optWeb
25. Pavel, Ivan S., Jan R. (1995) Multilevel Distributed Genetic Algorithms. International Conference on Genetic Algorithms in Engineering Systems: Innovations and Applications. University of Sheffield, UK, IEEE Press.
26. Park J -M., Park J -G., Lee C -H., Han M -S (1993) Robust and Efficient Genetic Crossover Operator: Homologous Recombination, International Joint Conference on Neural Networks, 2975–2978
27. Pereira R. (1996) Selecting Parameters for Technical Trading Rules Using a Genetic Algorithm. Journal of Applied Finance and Investment **1(3)**, 27–34
28. Pictet O., Dacorogna M., Dave R., Chopard B., Schirru R., et al. (1995) Genetic Algorithms with Collective Sharing for Robust Optimizsation in Financial Applications. Neural Network World **5(4)**, 573–587
29. Sendhoff B., Kreutz M., and Seelen W. v. (1997) Causality and the Analysis of Local Search in Evolutionary Algorithms. Bochum Ruhr-Universitat Bochum
30. Sinclair M. (1993) Comparison of the Perforamnce of Modern Heuristics for Combinatorial Optimization on Real Data. Computers and Operations Research **20(7)**, 687–695
31. Srinivas M., Patnaik L. M. (1994) Adaptive Probabilities of Crossover and Mutation in Genetic Algorithms. IEEE Transactions on Systems, Man and Cybernetics **24(4)**, 656–666
32. Sweeney J. (1996) Campaign Trading: Tactics and Strategies to Exploit the Markets. New York: John Wiley & Sons, Inc
33. Tanese R. (1989) Distributed Genetic Algorithms. Proceedings of the Third International Conference on Genetic Algorithms, Arlington, Morgan Kaufmann Publishers
34. Taylor S. J. (1994) Trading Futures Using a Channel Rule: A Study of the Predictive Power of Technical Analysis with Currency Examples. The Journal of Futures Markets **14(2)**, 215–235
35. Tsang R. (1997) Genetic Algorithms and Simulate Annealing, A Baseline Comparison. School of Business Systems, Melbourne, Monash University, 113
36. Yuret D. (1994) From Genetic Algorithms to Efficient Optimization. Department of Electrical Engineering and Computer Science, Massachusetts, Massachusetts Institute of Technology

19 GP Forecasts of Stock Prices for Profitable Trading

Mahmoud Kaboudan

Management Science & Information Systems, Penn State Lehigh Valley
mak7@psu.edu

Abstract. This chapter documents how GP forecasting of stock prices used to execute a single-day-trading-strategy (or SDTS) improves trading returns. The strategy mandates holding no positions overnight to minimize risk and daily trading decisions are based on forecasts of daily high and low stock prices. For comparison, two methods produce the price forecasts. Genetically evolved models produce one. The other is a naive forecast where today's actual price is used as tomorrow's forecast. Trading decisions tested on a small sample of four stocks over a period of twenty days produced higher returns for decisions based on the GP price forecasts.

19.1 Introduction

In this chapter GP is used to evolve equations that emulate the dynamics of daily stock price-movements and use those equations to produce forecasts that increase profitability from adopting a single-day-trading-strategy (or SDTS). SDTS uses forecasts of daily high and low stock prices to formulate trading decisions. According to this strategy, trading takes place only if a set of conditions is satisfied. These conditions are described here as trading rules and guidelines the investor should follow to achieve profitability. By design, SDTS takes into consideration the possibility that stock prices may move aggressively in response to news that may occur after the market closes on one day and before it opens the following day. Such news may affect high and low prices of a specific stock, or may affect the entire market including prices of that particular stock. Hence use of this strategy is more effective in the absence of news that affects daily prices of a particular stock. Buying at a price close to the daily low and selling the same day at a price close to the high, irrespective of the order of trading, seems to generate respectable trading profits over time. Buying then selling (i.e., taking a long position) or selling then buying (i.e., taking a short position) the same day for a profit seems to have some advantage that can be attributed to avoiding any overnight risk that may offset already experienced gains. The main ingredient for such trading strategy is a forecast of tomorrow's high and low prices of a stock. Reliable long-term forecasts of stock prices are simply nonexistent, and according to the financial literature, the best forecast for tomorrow's price is today's. The random walk model [15] and many other studies that followed provide sufficient evidence that stock prices follow such well-known behavior. Although there are other studies that document deviations from random walk (see [27] for a complete review), perhaps forecasts generated using GP methodology may be more compelling evidence of existing deviations. Alternatively, if GP forecasts of daily highs and lows does in fact produce better profits than a naive forecast where the best prediction of tomorrow's highs and lows are today's prices, then GP poses a serious challenge to this particular financial theory. It is important to realize that such challenge is imposed mainly by advancements in computational sciences. After all, disciplines such as neural networks, genetic algorithms, evolutionary programming, and genetic programming came into existence decades after the random walk theory was formulated. The new methods have superior predictive and forecasting abilities that make the possibility of producing acceptable stock-price forecasts a logical development.

This study is not first in applying GP to financial markets. However, it is an early study that documents the possibility of making high returns from trading using GP forecasts and SDTS. To achieve more profitable trading, this study aims to clearly explain this strategy then use GP to furnish the main ingredient needed to execute it successfully. Therefore, this Chapter contributes to the areas of finance and GP applications. Financial analysts should find a unique simple trading strategy that is rather capable. In GP applications, it provides a foundation useful in forecasting time-series with attention given to evolving models that predict and successfully forecast dynamically complex data. The application of genetic programming to financial markets has been rather limited with fewer than fifty papers found in the literature. Among them are studies by [1,2,4–12,14,19,28,29,32]. Since high monetary financial rewards for reasonably accurate forecasts of financial data and advancements in genetic computations now furnish relatively superior forecasts, one would expect far more GP applications by financial analysts in the future.

In a way, this paper investigates what has been debated in the financial literature for decades – namely market efficiency. Reference [23] was the first to suggest that stock returns follow a "random walk". He implied that successive price changes are random or independent of each other. Alternatively, $(P_t - P_{t-1}) = u_t$, where u_t is a random variable drawn independently with $E(u_t) = 0$, and where $E(\cdot)$ is the expected value or mean. Evidence of market efficiency during the 1960s and 1970s flooded the financial literature until the 1980s when anomalies were documented. See [16] and [26] for example. Evidence supporting or refuting market efficiency is based on analysis of price returns. Returns are defined as the percentage change in price over time or $R_t = (P_t - P_{t-1})/P_{t-1}$. Intuitively, traders are most interested in the price level (P_t), and one should be evaluating forecasts of P_t instead. Evaluating returns is important if these can be useful in computing accurate and statistically defendable price forecasts. But returns have been shown to be either random or very highly complex. Studies that document non-random behavior of returns and evaluate their chaotic behavior ([3,18,20,30]) foster the notion of unpredictable returns. Whether random or non-random with complex dynamics, the question of whether genetic programming can predict them remains enigmatic. Accordingly, genetically evolved models (or GEMs) predict and forecast price levels (and not returns) here.

Assuming that the reader is familiar with the basic concepts of GP, this Chapter will proceed as follows: In the next Section, SDTS is explained. Before using GP to evolve stock-price models and applying SDTS, a Section introduces the data and shows that it is GP-predictable using a test based on an index that measures predictability of time series. GP prediction and forecasts of daily high and low prices of four stocks are in a Section following data introduction. SDTS utilizes those forecasts and reports trading profits in the next Section. Some remarks conclude the Chapter.

19.2 SDTS

A single-day-trading-strategy is one where a trader buys a stock at a price close to the daily low and sells at a price close to the daily high, regardless of the order trading occurs. Buying long then selling to close the position or selling short then buying to cover the position during the same day makes intuitive sense. First, financial markets tend to react to news as it occurs. Short term breaking news introduces short run uncertainties about market conditions. Those uncertainties affect the outcome of any investment strategy. Second, because market conditions change (for whatever reasons) from fairly stable to very volatile, and companies that seem to be doing well sometimes face difficulties, buying and holding a stock for an extended period results in net returns that, although positive, are less than optimal. Consider the following investment scenarios. Assume first that an investor buys a stock, holds it for a few months or even years, and experiences positive returns on the investment. Given that stock prices do not continuously rise, assume that the price rose first, then company or economy problems forced a price-decline only for such problems to disappear a few months later and the price restored lost gains. Under this very common scenario, an investor misses the opportunity of increasing returns by trading that same stock several times during such period of volatility. Of course it is also conceivable that negative returns may materialize after holding a stock for an extended period of time. Under this second scenario, an investor could have abated long term losses by closing positions daily (as SDTS suggests). What makes such strategy more appealing is fact that long term stock-price forecasts remain a fantasy. If long term forecasts of prices were possible, financial markets' efficiency and positive returns would disappear.

There are five rules to follow when using SDTS. Following them should minimize trading risk and a trader should use discretion to determine when to close a position quickly once the forecast proves worthless for the day. Here they are:

Rule 1: *Decisions to trade or not to trade during a specific day is contingent upon the "forecasted spread".* If the forecasted spread (or difference between the forecasted daily high and low prices) is not sufficiently large, there is no money to be made even if the forecasts are perfect. In other words, the first rule to follow demands calculating the forecasted spread before the market opens and an order is placed. If the forecasted spread falls below a specific threshold difference for a particular stock, the investor should not trade the stock during that day. That difference should be sufficiently large to justify the risk taken. An investor then makes a decision once a forecast is ready and before the market opens.

Rule 2: *The investor should avoid trading if the stock price opens significantly above or below the forecasted high or low, respectively.* This may occur when unexpected news breaks after the forecast is completed. (Breaking news during the day becomes a fortunate or unfortunate event if one is already invested. It introduces uncontrollable risk that renders any forecast useless.)

Rule 3: *The investor should trade a large volume of shares (=1,000 shares) with a discount brokerage firm.* Single-day trading by design takes advantage of small changes in the stock price. Thus, trading a small quantity will not be sufficient to offset the risk taken and transaction fees. The discussion below assumes trading quantities of 1,000 shares. (Lowering this suggested quantity is still profitable but up to some minimum level that can be easily computed by the investor.) Trading with a discount brokerage helps benefit from relatively small price changes without sacrificing profits in commission.

Rule 4: *The investor should close every position by the end of the trading day.* Since trading is based on a one-day forecast, every position must be closed daily even if sometimes experiencing loss. This helps avoid greater gambling losses in subsequent days for which no forecast is available to base a decision on.

Rule 5: *The investor should trade only heavily traded stocks.* Given that one's trading decisions are based on forecasts, it is only logical to expect that the probability that forecasted prices will occur is very low if the selected stock trades a light volume. Therefore, to increase the probability of trading at the forecasted prices, one should trade stocks that are well known and have heavy trading volume. Although subjective, it is perhaps safe to assume that heavily traded stocks average a minimum of one millions shares exchanging hands daily. This restriction is also imposed to reduce the trading risk involved. SDTS mandates closing positions daily, and for less frequently traded stocks, closing may force accepting an undesirable price.

Table 19.1. SDTS guidelines

Spread	1.25 < 1.50	1.5 < 1.75	1.75 < 2.00	2.00
If $\lvert FL - OP\rvert < \frac{3}{8}$ Or if $\lvert AL - FL\rvert < \frac{3}{8}$	Long	Long	Long	Long
If $\frac{3}{8} \le \lvert FL - OP\rvert < \frac{1}{2}$ Or $\frac{3}{8} \le \lvert AL - FL\rvert < \frac{1}{2}$	No Trade	Long	Long	Long
If $\frac{1}{2} \le \lvert FL - OP\rvert < \frac{5}{8}$ Or $\frac{1}{2} \le \lvert AL - FL\rvert < \frac{5}{8}$	No Trade	No Trade	Long	Long
If $\frac{5}{8} \le \lvert FL - OP\rvert < \frac{3}{4}$ Or $\frac{5}{8} \le \lvert AL - FL\rvert < \frac{3}{4}$	No Trade	No Trade	No Trade	Long
If $\frac{3}{4} \le \lvert FL - OP\rvert$ Or $\frac{3}{4} \le \lvert AL - FL\rvert$	No Trade	No Trade	No Trade	No Trade
If $\lvert OP - AH\rvert < \frac{3}{8}$ Or $\lvert FH - AH\rvert < \frac{3}{8}$	Short	Short	Short	Short
If $\frac{3}{8} \le \lvert OP - AH\rvert < \frac{1}{2}$ Or $\frac{3}{8} \le \lvert FH - AH\rvert < \frac{1}{2}$	No Trade	Short	Short	Short
If $\frac{1}{2} \le \lvert OP - AH\rvert < \frac{5}{8}$ Or $\frac{1}{2} \le \lvert FH - AH\rvert < \frac{5}{8}$	No Trade	No Trade	Short	Short
If $\frac{5}{8} \le \lvert OP - AH\rvert < \frac{3}{4}$ Or $\frac{5}{8} \le \lvert FH - AH\rvert < \frac{3}{4}$	No Trade	No Trade	No Trade	Short
If $\frac{3}{4} \le \lvert OP - AH\rvert$ Or $\frac{3}{4} \le \lvert FH - AH\rvert$	No Trade	No Trade	No Trade	No Trade

Guidelines to follow for making trading decisions are in Table 19.1. The first row lists different spread conditions to follow. Determined intuitively and experimentally, a minimum spread of \$1.25 must first occur before any other condition is evaluated. This subjectively defined threshold level is selected to provide a sufficiently large buffer to absorb forecasting errors. The defined spread increases with forecast errors as shown. A long, short, or no trade decision is identified from the intersection of one of the four spread intervals defined in the first row with any of the conditions in the first column in the Table. There, FL is forecasted low price, FH is forecasted high price, OP is actual opening price, AL is actual low price that occurred thus far in the trading day, AH is actual high price that occurred thus far in the trading day, and $|.|$ is for absolute value of those differences. The listed restrictions are also determined intuitively, and because they are subjective, investor's intuition is necessary when making decisions. The Table then has ten groups of mutually exclusive conditions. Each group has two mutually inclusive conditions within it. An example explains how to use these guidelines. Assume that the forecasted spread for a particular stock is predicted to be \$1.625 prior to commencement of trading. This spread >\$1.25 and falls within the interval defined in the third column of the Table (1.5<1.75). One then takes a long position only if (a) $(0 \le |FL - OP| < 3/8)$, (b) $|AL - FL| < 3/8)$, (c) $(3/8 \le |FL - OP| < 1/2)$, or (d) $(3/8 \le |AL - FL| < 1/2)$ is satisfied after trading starts. More simply, these conditions state that one should take a long position if the absolute difference between the forecasted low and opening price or if the absolute difference between the actual low for the elapsed trading period and the forecasted low is less than \$1/2. Alternatively, one takes a short position if the absolute difference between the forecasted high and opening price or if the absolute difference between the actual high for the elapsed trading period and the forecasted high is less than \$1/2. No trading should take place otherwise.

Like any stock trading strategy, SDTS is risky. Following the rules and guidelines helps minimize risk. Here is how SDTS is executed to minimize losses. If one decides to trade and an order is placed, as soon as it is filled, one places the order to close the position at a price close to the forecasted high or low depending on whether the position taken was long or short. A trader then observes price movements closely until the position is closed. If the price does not move in the expected direction, that position must be closed even if sometimes at loss.

19.3 The Data

Only four stocks are selected to apply SDTS to here. Two of the stocks trade on the NYSE: Chase Manhattan Corp (or CMB) and General Motors Corp (or GM), and two on the NASDAQ: Dell Computer Corp (or DELL) and MCI-Worldcom (or WCOM). The data used covers the period starting February 8, 1999 and ends August 5, 1999 providing a total of 126 trading days per stock. (All data was obtained from Dryfus Brokerage Services at http://www.edreyfus.com/.) The first six days are lost degrees of freedom to accommodate lags in the evolved models. Each model is evolved using 100 data points or sequence of daily prices. The remaining twenty days are used to evaluate the resulting one-day-ahead forecast GEMs produce. (Forecasting a large number of stocks and analyzing trading decisions for a much longer period is an investigation the author is currently completing.) Using the data on the four stocks, GP's task is to search for the best equation to forecast daily high and the best equation to forecast daily low prices for each of the twenty days per stock. Early investigations showed that forecasting more than one-day ahead generates worthless predictions. This is one reason why a different equation is evolved for each price series during each of the twenty trading days. A second reason is less compelling but perhaps deserves the attention provided in forthcoming work. This second reason has to do with the dynamics of the stock market. According to preliminary investigation results, GP seems to produce different specifications that include different explanatory variables for the same stock during different days of the week. This may be the result of the markets seasonal dynamics, although it is yet too early to deliver statistical confidence in such conclusion. Whether GP can in fact evolve equations that predict stock prices or not is important to investigate before actually investing the time and effort in producing prediction models. [21] proposed a test that measures GP's predictive ability. The idea of the test is rather simple. Given a time series, one evolves a large number of fast autoregressive GP models. (The study suggests completing more than thirty runs to provide a statistically acceptable sample of results.) Producing such models is made fast by limiting population size to 500 and number of generations to forty per run. This means that GP evolves only 20,000 equations for each one of the 30 runs. The sequence of data in that series is then randomly shuffled using [13]'s bootstrap method. This dismembers any orderly dynamical pattern in the series. One then evolves the same number of fast autoregressive GP models using the same configuration as for the original series. If the series before shuffling contained unpredictable patterns, the ratio of the fitness measure of sum of squared errors (or SSE) before to SSE after running must approach unity. If that series contained predictable patterns, the ratio must be less than unity. The test statistic is the average ratio of the (before and after shuffling) SSEs subtracted from one. Formally, this test statistic is:

$$\eta = max\left\{0, \frac{1}{n}\sum_{i=1}^{n}\left(1 - \frac{SSE_Y}{SSE_S}\right)_i\right\} \quad (19.1)$$

where n is the number of runs completed without the run getting trapped in a local minimum solution, Y is the original series before shuffling, and S is the shuffled sequence of that series. This ratio produces an index with a minimum of zero since it is conceivable that the mean of the SSE ratios be greater than one. This may happen if shuffling actually introduced some pattern into an originally random data sequence. It may also happen if GP is not successful in depicting any pattern in the original data but is somehow successful in finding pattern in the shuffled data.

To apply this test to stock prices, fast autoregressive GP models were evolved for the original data and for its shuffled counterpart. To produce the shuffled data, rather than shuffling prices, periodic changes in prices were computed first. Only these changes were shuffled using the bootstrap method, then added back again. This provides 'scrambled' price series with random changes to determine whether returns contain GP-predictable information. The results of this experiment suggest that at the 5% level of significance, prices are at least 0.620.2 predictable. When a GP evolved equations to predict returns and their shuffled counterpart, returns were found to be at most 0.280.03 predictable at the 5% significance level. These results only suggest that returns are less predictable than prices when only lagged dependent variables are used to explain variation in the dependent variable. Since GP's predictive ability relies upon the appropriate selection of explanatory variables, this test then only demonstrates that autoregressive models do not have all variables that explain variations in returns. However, given that interest is in producing price forecasts and given the higher chance for success this test demonstrated when predicting prices versus returns, the next Section contains price models using GP.

19.4 GEMs and Their Price Forecasts

GPQuick computer code written in C++ [31] produces all GP runs in this study. (It is available from http://www.lv.psu.edu/mak7/gems/Links.htm# Soft~klipp/research/ssprob.) GPQuick replicates an original GP program in LISP by [24]. Both are computer programs designed to randomly evolve regression models and determine the one that replicates the series' data generating processes best. This Section contains a brief description of how GPQuick is used to generate models.

The program randomly selects and combines explanatory variables and operators to generate a pre-specified number of equations known as a population. It keeps the fittest 10% (equations with the lowest SSE statistics) of that first population and includes them in the next generation of equations. The other 90% are replaced by new equations. The new ones are produced using a combination of crossover and mutation processes. In crossover, one section of an equation is randomly exchanged with another section from a different equation. In mutation one section of an equation is replaced by a different set of variable(s) and operator(s). New regression models continue to evolve until a pre-specified program termination condition is reached. The program then saves the fittest equation and its statistics in an output file. In this study, evolved models serve two purposes: (a) to evaluate series' predictability, and (b) to forecast with. Inputs to execute the program differ accordingly. The program demands providing the following inputs needed to evolve the regression models:

(a) It must be given individual files containing input data. The dependent and each of the independent variables should be in separate files. To test predictability, the independent variables are lagged-dependent ones. To evolve the fittest model, other variables are used.
(b) Whether evolving models to evaluate predictability or to forecast with, the program must be given operators to select from. The operators used throughout are: +, -, *, %, exp, sqrt, ln, sin, and cos, where %, exp, sqrt, and ln are protected operators. These protections follow standards commonly used in all GP software. They provide solutions to computational-problems computers may encounter during the search for an equation. Such solutions are intentionally formulated to yield less fit equations than those that the search already identified without protections do. (They prevent program halting during execution.) The operator % is a protected division that returns a value=1 if the denominator in a division = 0. The operator exp is a protected exponential which returns e^{10} if e is raised to a power more than ten. The operator sqrt is a protected square root operation that returns the square root of the absolute value of any argument. Finally the operator ln calculates the natural logarithm of the absolute value of nonzero arguments and returns zero otherwise.
(c) The program is also given instructions to use a random number generator that produces random constants when evolving equations.
(d) It must be given population size . Population size is 2,000 equations in runs for testing and 5,000 in runs for forecasting. Smaller population sizes produce faster runs but less accurate results. Since accuracy is needed when actually searching for forecast models, it is only logical that the population size be larger.
(e) The program must be given a maximum length of arguments in an equation. This maximum is set equal to fifty.
(f) It must be given a condition to terminate the program. The program is set to terminate if one of the equations produces SSE < 0.00001 or if the number of equations evolved reaches 100,000.

These inputs were selected according to suggestions made in a different study by [22]. They tested different configurations of population size, number of generations, and data sample size while comparing performances of three GP programs written in C++. The above parameters were selected to produce reliable results within reasonable time as determined by trial and error exercises performed by [22].

One of the important decisions to make when attempting to find a best fit equation (using GP or any other method) is the selection of the explanatory variables. To forecast stock prices using GP, one has to think of the most pertinent variables to include as terminals. After experimenting with many variables, 25 variables were identified as best. They are all daily data taken directly from the market. One to five lags of each of the following variables make up the list: closing price, opening price, high price, low price, and traded volume (or CP, OP, HP, LP, and VOL, respectively). These variables are given to the program as terminals to select from when evolving an equation to forecast HP or LP. After evolving many equations, it was evident that they differ from day to day for the same stock and among stocks. Given the large number of equations evolved for just four stocks over twenty trading days, only two selected best equations evolved for each stock's high and low prices are reported in Appendix A. Table 19.2 presents the R^2 fitness measure of the best-fit reported sample of equations. It is evident that GP's ability to evolve an equation that replicates the dynamics of price behavior varies among stocks and not markets (NYSE and NASDAQ). Given the small sample of stocks evaluated here, no attempt to explain the reason for such variation will be made since there is little to gain. There is little to gain because equations more successful in predicting the training set may not necessarily forecast better than slightly worse performing ones. (See [17] who found that a model that predicts well does not necessarily generate better forecasts.) It is then more important to evaluate the forecasts.

Table 19.2. R^2 performance measure for selected equations

Stock	Date	Highs	Date	Lows
CMB	7/9/99	0.90	7/20/99	0.92
CMB	8/2/99	0.90	7/28/99	0.91
DELL	7/21/99	0.93	7/14/99	0.90
DELL	8/4/99	0.92	7/26/99	0.91
GM	7/16/99	0.97	7/13/99	0.96
GM	7/29/99	0.98	8/3/99	0.97
WCOM	7/12/99	0.87	7/23/99	0.79
WCOM	7/27/99	0.83	8/5/99	0.77

One-day-ahead forecasts for high and low prices of the selected twenty days per stock differed in their levels of accuracy. A summary of the forecast results is Table 19.3 and plots of them are in Figs. 19.1a&b – 19.4a&b. Ta-

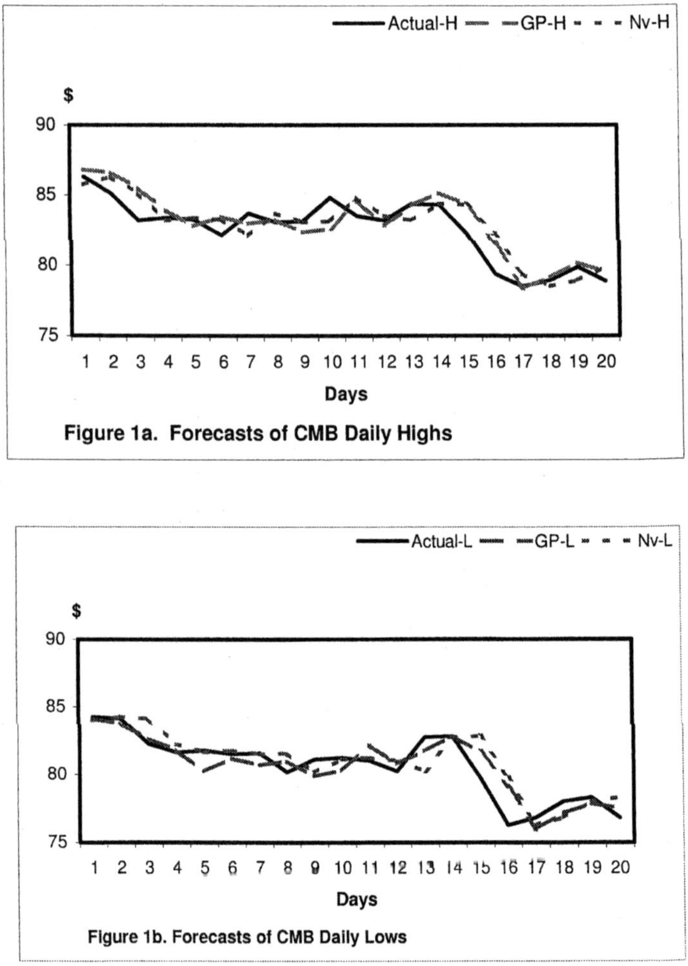

Fig. 19.1. Forecasts of CMB daily highs and lows

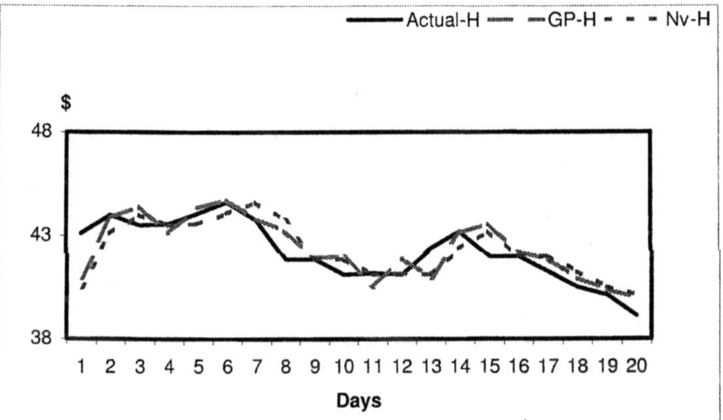

Figure 2a. Forecasts of DELL Daily Highs

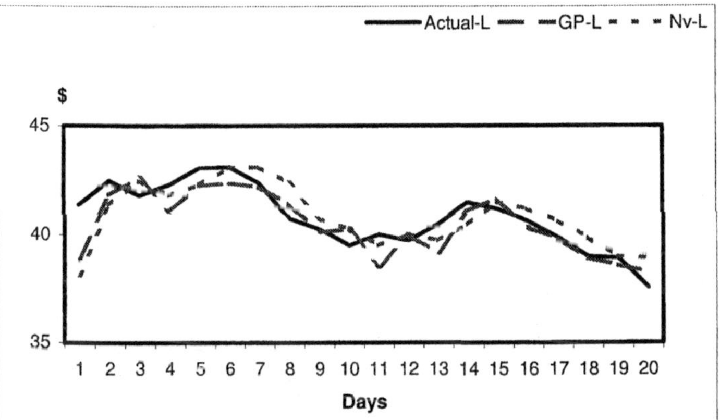

Figure 2b. Forecasts of DELL Daily Lows

Fig. 19.2. Forecasts of DELL daily highs and lows

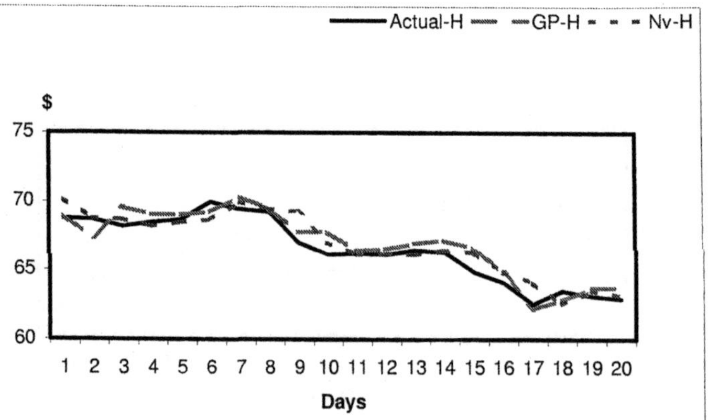

Figure 3a. Forecasts of GM Daily Highs

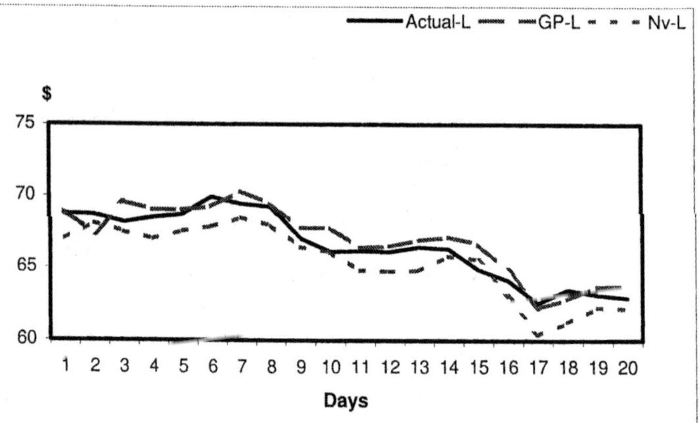

Figure 3b. Forecasts of GM Daily Lows

Fig. 19.3. Forecasts of GM daily highs and lows

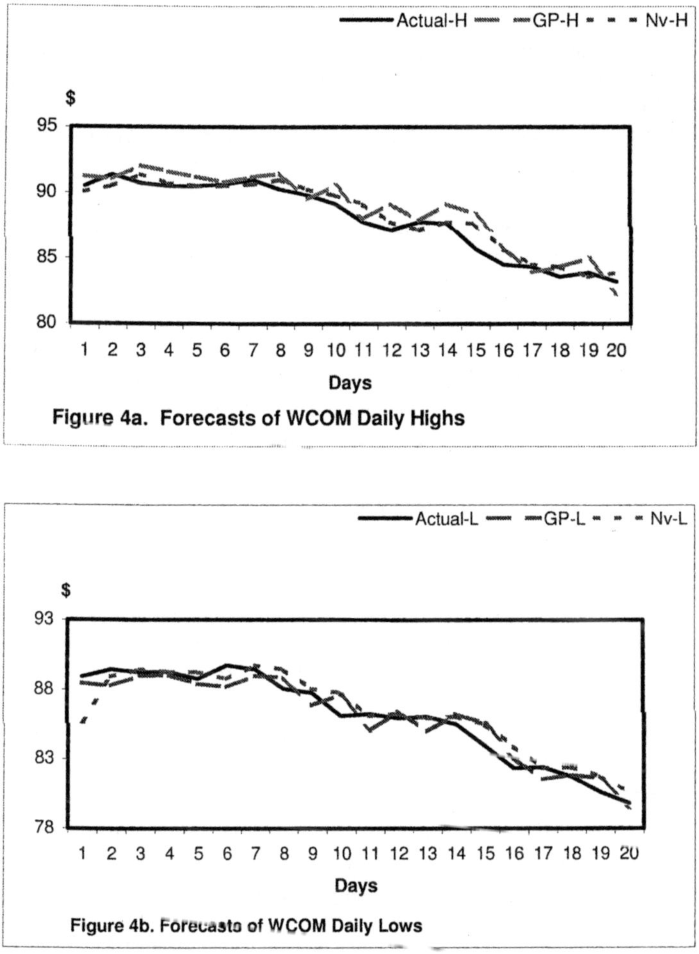

Fig. 19.4. Forecasts of WCOM daily highs and lows

Table 19.3. MSE, NMSE, and returns over twenty-day forecast period

	Daily Highs		Daily Lows		%
	MSE	NMSE	MSE	NMSE	Returns
CMB:					
GP forecast error	1.42	0.28	1.17	0.21	34.02
NM forecast error	1.55	0.30	1.97	0.36	27.16
DELL:					
GP forecast error	0.78	0.35	0.81	0.37	67.27
NM forecast error	0.96	0.44	1.13	0.51	14.87
GM:					
GP forecast error	0.76	0.14	0.76	0.14	21.80
NM forecast error	0.81	0.14	1.70	0.30	12.43
WCOM:					
GP forecast error	1.32	0.16	0.83	0.08	33.63
NM forecast error	0.59	0.07	1.34	0.13	31.17

ble 19.3 contains forecast mean square errors (MSE) and forecast normalized MSE (NMSE) for each stock. They are reported twice for daily highs and twice for daily lows: once to measure GP's forecast performance and once to measure the performance of a naive model (or NM) forecast. NMSE is defined as

$$NMSE = \frac{MSE}{s_y^2} \tag{19.2}$$

where s_y^2 is the observed respective price sample forecasted. The resulting computation approaches zero when a perfect prediction is generated and approaches one when the resulting model is doing no better than predicting the mean value of the data set on average. These measures document the superiority of GP's forecast over a random walk or NM forecast in all instances except in forecasting WCOM's daily high prices. It is interesting to note that stocks possessing the highest and lowest R^2 estimates in Table 19.2 have the lowest MSE and NMSE. Figures 19.1a&b– 19.4a&b confirm that GP produces a forecast different than and, for the most, more superior than that of a naive forecast. Further, and as indicated in Table 19.3, better forecasts do not guarantee higher investment returns as discussed in the next Section.

19.5 Trading Profits

This Section documents the benefits an investor may experience when GP is used to forecast stock prices and use them to implement SDTS. The last column in Table 19.3 presents the percentage returns on investments using such strategy. For all four stocks and over the twenty-day period, returns from trading based on the GP forecasts were higher but with distinct differences. Trading DELL generates the highest returns for trades based on GP forecasts. Table 19.2 shows that models evolved to depict price dynamics of DELL are equally acceptable as models evolved for CMB and less acceptable than those for GM. Yet GM's returns are the worst among those three, and CMB's returns using the GP forecast are not much better than the returns using NM. Returns on trading WCOM were rather impressive using NM. Before commenting on these results Tables 19.4– 19.7 present SDTS' execution first. Each Table contains actual prices (Actual-H for the high and Actual-L for the low) and GP forecasts (GP-H and GP-L, respectively) of daily prices of each stock in its upper half. It also reports the spread between the forecasted daily high and low to use in decision-making according to the rules and guidelines outlined above and summarized in Table 19.1. The column reporting spread is followed by the actual daily closing price (CP) and opening price (OP). These are added to help explain some of the trading decisions made since they indicate the direction of change in a stocks price over the day's trading hours. The column labeled 'trade' shows the decision made. A decision to take a long position is identified by 'Long', to take a short position is identified by 'Short', and to refrain from trading is identified by 'NT'. The buying and selling prices in dollars follow. Profit is calculated by assuming that an investor will buy a quantity of 1,000 shares and will pay only 50% of the amount, assuming that trades will occur only on margin. Trading on margin is logical since an investor will not pay interest by borrowing the other 50% from the brokerage house for trades executed and closed the same day. Trading cost is assumed zero since many brokerage firms offer trades for commission of less than $10. Finally, the last column reports the percentage returns made on each trade. The second half of each Table replicates the same information but for trading decisions and returns using the naive forecast for comparison with returns using GP's forecast.

Table 19.4. CMB price-forecasts and trading decisions

Date	Actual-H	GP-H	Actual-L	GP-L	Spread	CP	OP	Trade	Buy @	Sell @	$ Profit	Return
7/9/99	86.313	86.805	84.250	84.107	2.698	85.188	85.875	Short	84.375	86.250	1875	4.35
7/12/99	85.125	86.636	84.125	83.743	2.893	84.125	84.688	Long	84.250	85.000	750	1.78
7/13/99	83.188	85.520	82.250	82.628	2.892	82.813	83.125	NT				
7/14/99	83.375	83.947	81.625	81.779	2.168	81.750	82.938	Short	81.875	83.250	1375	3.30
7/15/99	83.250	82.790	81.750	80.250	2.540	82.125	81.750	Long	81.875	82.750	875	2.14
7/16/99	82.125	83.450	81.500	81.185	2.265	81.813	82.125	NT				
7/19/99	83.688	82.987	81.563	80.677	2.309	82.125	82.125	Short	81.625	82.750	1125	2.72
7/20/99	83.063	83.219	80.188	80.967	2.252	80.875	82.313	Short	81.000	82.875	1875	4.52
7/21/99	83.125	82.319	81.125	79.875	2.444	81.313	81.875	NT				
7/22/99	84.813	82.575	81.250	80.301	2.274	83.313	81.250	NT				
7/23/99	83.500	84.624	81.000	82.166	2.457	81.563	82.813	NT				
7/26/99	83.188	82.830	80.250	80.794	2.036	82.875	80.438	Long	80.875	82.750	1875	4.64
7/27/99	84.375	84.257	82.750	81.734	2.523	83.750	83.625	Short	82.875	84.125	1250	2.97
7/28/99	84.313	85.155	82.813	82.819	2.337	82.938	83.500	NT				
7/29/99	82.250	84.347	79.750	81.707	2.641	80.250	81.750	Long	81.875	82.125	250	0.61
7/30/99	79.375	81.581	76.250	79.103	2.477	77.063	79.000	NT				
8/2/99	78.500	78.361	76.813	75.977	2.384	78.250	76.938	NT				
8/3/99	78.938	79.118	78.000	77.127	1.990	78.813	78.250	NT				
8/4/99	79.875	80.162	78.313	77.851	2.311	78.438	79.063	Short	78.250	79.750	1500	3.76
8/5/99	78.875	79.498	76.813	77.486	2.012	78.625	78.438	Long	77.500	78.750	1250	3.23
								Total returns using GP's forecast				34.02

Date	Actual-H	GP-H	Actual-L	GP-L	Spread	CP	OP	Trade	Buy @	Sell @	$ Profit	Return
7/9/99	86.313	85.750	84.250	84.000	1.750	85.188	85.875	Short	84.375	85.625	1250	2.92
7/12/99	85.125	86.313	84.125	84.250	2.063	84.125	84.688	Long	84.375	85.000	625	1.48
7/13/99	83.188	85.125	82.250	84.125	1.000	82.813	83.125	NT				
7/14/99	83.375	83.188	81.625	82.250	0.938	81.750	82.938	NT				
7/15/99	83.250	83.375	81.750	81.625	1.750	82.125	81.750	Long	81.875	83.125	1250	3.05
7/16/99	82.125	83.250	81.500	81.750	1.500	81.813	82.125	NT				
7/19/99	83.688	82.125	81.563	81.500	0.625	82.125	82.125	NT				
7/20/99	83.063	83.688	80.188	81.563	2.125	80.875	82.313	Short	81.625	82.875	1250	3.02
7/21/99	83.125	83.063	81.125	80.188	2.875	81.313	81.875	NT				
7/22/99	84.813	83.125	81.250	81.125	2.000	83.313	81.250	Long	81.375	83.000	1625	3.99
7/23/99	83.500	84.813	81.000	81.250	3.563	81.563	82.813	NT				
7/26/99	83.188	83.500	80.250	81.000	2.500	82.875	80.438	Long	81.125	83.000	1875	4.62
7/27/99	84.375	83.188	82.750	80.250	2.938	83.750	83.625	Short	82.875	83.125	250	0.60
7/28/99	84.313	84.375	82.813	82.750	1.625	82.938	83.500	Short	82.875	84.250	1375	3.26
7/29/99	82.250	84.313	79.750	82.813	1.500	80.250	81.750	NT				
7/30/99	79.375	82.250	76.250	79.750	2.500	77.063	79.000	NT				
8/2/99	78.500	79.375	76.813	76.250	3.125	78.250	76.938	Long	77.000	78.375	1375	3.57
8/3/99	78.938	78.500	78.000	76.813	1.687	78.813	78.250	Short	78.125	78.125	0	0.00
8/4/99	79.875	78.938	78.313	78.000	0.938	78.438	79.063	NT				
8/5/99	78.875	79.875	76.813	78.313	1.562	78.625	78.438	Long	78.500	78.750	250	0.64
								Total returns using NM's forecast				27.16

Although the sample of stocks considered here is rather small and the forecast period is short, the following comments can be made about the results achieved using SDTS and the implications of using GP-forecasting:

(a) SDTS in and by itself appears to be a strategy that if properly executed would generate returns that exceed those of the average S&P 500. S&P returns average about 20% annually. Returns from trading any of the four stocks are much more using either forecast. The higher return on investment (or ROI) may be attributed to reduced risk associated with absence of overnight holding of a stock.

(b) Enhancing returns by using GP-forecasts is possible from trading specific stocks and surely not all. Gains from using GP forecasts to improve returns from trading WCOM for example were statistically nonexistent. Conversely, gains from using them to improve returns from trading DELL were undeniable. This means that a trader who intends to use SDTS and GP forecasting should identify those stocks that can yield such high returns first, then trade only those stocks.

Table 19.5. DELL price-forecasts and trading decisions

Date	Actual-H	GP-H	Actual-L	GP-L	Spread	CP	OP	Trade	Buy @	Sell @	$ Profit	Return
7/9/99	43.125	40.936	41.375	38.915	2.021	42.813	39.906	NT				
7/12/99	44.000	43.865	42.500	41.830	2.034	43.625	43.500	Short	42.625	43.750	1125	5.14
7/13/99	43.500	44.415	41.813	42.617	1.798	42.250	42.500	Long	42.625	43.375	750	3.52
7/14/99	43.563	43.129	42.313	41.088	2.041	43.500	42.625	Short	42.750	43.000	250	1.16
7/15/99	44.063	44.336	43.063	42.268	2.068	43.938	43.563	Short	43.125	43.875	750	3.42
7/16/99	44.625	44.729	43.125	42.376	2.353	43.250	43.969	Short	43.250	44.500	1250	5.62
7/19/99	43.750	43.817	42.375	42.211	1.606	42.438	43.500	Short	42.500	43.625	1125	5.16
7/20/99	41.875	43.253	40.750	41.440	1.813	41.125	41.875	Long	41.500	41.750	250	1.20
7/21/99	41.875	41.946	40.250	40.125	1.821	41.250	41.375	Short	40.375	41.750	1375	6.59
7/22/99	41.125	41.987	39.500	40.285	1.701	39.625	40.813	NT				
7/23/99	41.188	40.512	40.031	38.534	1.978	41.000	40.063	Short	40.125	40.500	375	1.85
7/26/99	41.125	41.908	39.750	40.000	1.907	40.000	40.313	Long	40.125	41.000	875	4.36
7/27/99	42.375	40.959	40.500	39.167	1.792	42.125	40.625	NT				
7/28/99	43.188	43.116	41.500	41.093	2.023	42.750	41.938	Long	41.625	43.000	1375	6.61
7/29/99	42.000	43.570	41.188	41.669	1.900	41.250	41.563	Long	41.750	41.875	125	0.60
7/30/99	42.000	42.204	40.625	40.332	1.873	40.875	41.500	Short	40.750	41.875	1125	5.37
8/2/99	41.250	41.872	39.875	39.830	2.042	39.938	40.375	Short	40.000	41.125	1125	5.47
8/3/99	40.500	40.936	39.000	38.927	2.009	39.438	40.375	Short	39.125	40.375	1250	6.19
8/4/99	40.125	40.358	38.938	38.588	1.771	39.125	39.313	Short	39.000	40.000	1000	5.00
8/5/99	39.125	39.981	37.563	38.291	1.691	38.750	39.000	NT				
								Total returns using GP's forecast				67.27

Date	Actual-H	GP-H	Actual-L	GP-L	Spread	CP	OP	Trade	Buy @	Sell @	$ Profit	Return
7/9/99	43.125	40.500	41.375	38.125	2.375	42.813	39.906	Short	41.500	40.375	-1125	-5.57
7/12/99	44.000	43.125	42.500	41.375	1.750	43.625	43.500	Short	42.625	43.000	375	1.74
7/13/99	43.500	44.000	41.813	42.500	1.500	42.250	42.500	Long	42.625	43.375	750	3.52
7/14/99	43.563	43.500	42.313	41.813	1.687	43.500	42.625	NT				
7/15/99	44.063	43.563	43.063	42.313	1.250	43.938	43.563	Short	43.125	43.500	375	1.72
7/16/99	44.625	44.063	43.125	43.063	1.000	43.250	43.969	NT				
7/19/99	43.750	44.625	42.375	43.125	1.500	42.438	43.500	Long	43.250	43.625	375	1.73
7/20/99	41.875	43.750	40.750	42.375	1.375	41.125	41.875	NT				
7/21/99	41.875	41.875	40.250	40.750	1.125	41.250	41.375	NT				
7/22/99	41.125	41.875	39.500	40.250	1.625	39.625	40.813	NT				
7/23/99	41.188	41.125	40.031	39.500	1.625	41.000	40.063	NT				
7/26/99	41.125	41.188	39.750	40.031	1.157	40.000	40.313	NT				
7/27/99	42.375	41.125	40.500	39.750	1.375	42.125	40.625	NT				
7/28/99	43.188	42.375	41.500	40.500	1.875	42.750	41.938	Short	41.625	42.250	625	2.96
7/29/99	42.000	43.188	41.188	41.500	1.688	41.250	41.563	Long	41.625	41.875	250	1.20
7/30/99	42.000	42.000	40.625	41.188	0.812	40.875	41.500	NT				
8/2/99	41.250	42.000	39.875	40.625	1.375	39.938	40.375	Long	40.500	41.125	625	3.09
8/3/99	40.500	41.250	39.000	39.875	1.375	39.438	40.375	NT				
8/4/99	40.125	40.500	38.938	39.000	1.500	39.125	39.313	Long	39.125	40.000	875	4.47
8/5/99	39.125	40.125	37.563	38.938	1.187	38.750	39.000	NT				
								Total returns using NM's forecast				14.87

(c) The number of trading days and ROI are positively related. The number of trading days for DELL when GP forecasts are used is 16 compared with only 9 when NM forecasts are used. Respectively, these numbers are 11 and 11 for CMB, 9 and 6 for GM, and 10 and 11 for WCOM. When viewed in light of the returns reported in Table 19.3, one can immediately see the effect of the increased number of trading days on ROI. trading!return on investment

(d) ROI is greater for relatively low-priced stocks. Given that these stocks trade at different prices (as the reported averages reveal) ROI on trading 1,000 shares at $40 each is much higher than trading 1,000 shares at $80 each for the same spread group given that the same spread applies for both. There would be no difference if the spread were set as a percentage of the stock price. This partially explains the relatively high ROI on trading DELL reported in Table 19.3.

Table 19.6. GM price-forecasts and trading decisions

Date	Actual-H	GP-H	Actual-L	GP-L	Spread	CP	OP	Trade	Buy @	Sell @	$ Profit	Return
7/9/99	68.750	68.921	68.125	66.682	2.239	66.125	68.125	Short	68.250	68.625	375	1.09
7/12/99	68.688	67.330	67.500	65.392	1.938	68.688	68.000	Short	67.375	68.500	1125	3.28
7/13/99	68.188	69.557	67.000	67.814	1.743	67.750	67.125	NT				
7/14/99	68.500	69.063	67.563	67.149	1.914	67.813	68.125	Short	67.625	68.375	750	2.19
7/15/99	68.688	69.034	67.875	67.096	1.938	68.063	68.000	Short	68.000	68.500	500	1.46
7/16/99	69.938	69.249	68.500	67.482	1.767	69.063	68.688	Short	68.625	69.125	500	1.45
7/19/99	69.438	70.316	67.938	68.383	1.932	68.313	69.063	NT				
7/20/99	69.250	69.478	66.438	67.457	2.021	66.500	68.313	Short	67.500	69.125	1625	4.70
7/21/99	67.000	67.775	66.125	66.190	1.585	66.500	67.000	NT				
7/22/99	66.125	67.779	64.813	65.993	1.786	65.688	65.500	NT				
7/23/99	66.188	66.406	64.750	64.473	1.933	65.625	66.125	Short	64.875	66.000	1125	3.41
7/26/99	66.125	66.498	64.813	64.665	1.833	65.938	65.375	Long	65.000	66.000	1000	3.08
7/27/99	66.438	66.917	65.813	65.194	1.723	66.125	66.000	Short	66.000	66.375	375	1.13
7/28/99	66.313	67.153	65.563	65.214	1.939	65.750	66.063	NT				
7/29/99	64.875	66.661	63.063	65.162	1.499	63.938	64.750	NT				
7/30/99	64.125	64.874	60.375	63.114	1.761	61.125	64.125	NT				
8/2/99	62.563	62.231	61.250	60.356	1.875	61.875	61.375	NT				
8/3/99	63.500	62.833	62.250	61.244	1.589	63.063	62.375	NT				
8/4/99	63.125	63.719	62.188	62.300	1.419	62.813	62.813	NT				
8/5/99	62.938	63.754	60.563	62.180	1.574	61.000	62.875	NT				
								Total returns using GP's forecast				21.80
Date	Actual-H	GP-H	Actual-L	GP-L	Spread	CP	OP	Trade	Buy @	Sell @	$ Profit	Return
7/9/99	68.750	70.125	68.125	67.000	3.125	66.125	68.125	NT				
7/12/99	68.688	68.750	67.500	68.125	0.625	68.688	68.000	NT				
7/13/99	68.188	68.688	67.000	67.500	1.188	67.750	67.125	NT				
7/14/99	68.500	68.188	67.563	67.000	1.188	67.813	68.125	NT				
7/15/99	68.688	68.500	67.875	67.563	0.937	68.063	68.000	NT				
7/16/99	69.938	68.688	68.500	67.875	0.813	69.063	68.688	NT				
7/19/99	69.438	69.938	67.938	68.500	1.438	68.313	69.063	NT				
7/20/99	69.250	69.438	66.438	67.938	1.500	66.500	68.313	Short	68.000	69.125	1125	3.25
7/21/99	67.000	69.250	66.125	66.438	2.812	66.500	67.000	Long	66.250	66.625	375	1.13
7/22/99	66.125	67.000	64.813	66.125	0.875	65.688	65.500	NT				
7/23/99	66.188	66.125	64.750	64.813	1.312	65.625	66.125	Short	65.000	66.000	1000	3.03
7/26/99	66.125	66.188	64.813	64.750	1.438	65.938	65.375	Short	65.000	66.000	1000	3.03
7/27/99	66.438	66.125	65.813	64.813	1.312	66.125	66.000	Short	66.000	66.000	0	0.00
7/28/99	66.313	66.438	65.563	65.813	0.625	65.750	66.063	NT				
7/29/99	64.875	66.313	63.063	65.563	0.750	63.938	64.750	NT				
7/30/99	64.125	64.875	60.375	63.063	1.812	61.125	64.125	NT				
8/2/99	62.563	64.125	61.250	60.375	3.750	61.875	61.375	NT				
8/3/99	63.500	62.563	62.250	61.250	1.313	63.063	62.375	NT				
8/4/99	63.125	63.500	62.188	62.250	1.250	62.813	62.813	Short	62.375	63.000	625	1.98
8/5/99	62.938	63.125	60.563	62.188	0.937	61.000	62.875	NT				
								Total returns using NM's forecast				12.43

19.6 Remarks

This chapter contained stock trading strategy and showed how GP's ability to evolve regression models to predict complex time series and produce superior forecasts can boost profits made by implementing that strategy. The strategy shows how trading into and out of the market the same day can reduce overnight investment risk which may result in higher profits when compared with a buy and hold traditional trading strategy. The reader should quickly realize that investment trading is a risky endeavor and using SDTS does not guarantee profits. Using such strategy assumes that the investor is experienced, willing to take losses, and fully understands the rules and guidelines outlined. It should also be clear that the work presented in this chapter is in an early stage. SDTS should first be tested on a much larger number of shares and for a reasonably long period of time. Such strategy would be more reliable to implement if it were tested for no less than a year. Therefore, the results presented here are only an invitation for further research to prove that it does indeed produce extra ordinary economic profits.

Table 19.7. WCOM price-forecasts and trading decisions

Date	Actual-H	GP-H	Actual-L	GP-L	Spread	CP	OP	Trade	Buy @	Sell @	$ Profit	Return
7/9/99	90.500	91.235	88.938	88.462	2.773	89.438	90.000	Short	89.000	90.375	1375	3.04
7/12/99	91.375	91.048	89.438	88.285	2.763	90.750	90.000	Long	89.500	91.000	1500	3.35
7/13/99	90.688	92.084	89.250	88.987	3.097	90.313	90.125	Long	89.375	90.500	1125	2.52
7/14/99	90.500	91.615	89.250	89.025	2.590	89.750	90.375	NT				
7/15/99	90.500	91.236	88.750	88.404	2.832	89.438	90.375	Short	88.875	90.375	1500	3.32
7/16/99	90.625	90.805	89.750	88.170	2.635	90.125	89.750	NT				
7/19/99	91.000	91.220	89.438	89.004	2.217	90.438	90.063	Long	89.500	90.875	1375	3.07
7/20/99	90.250	91.504	88.063	88.855	2.648	88.125	90.000	Short	89.000	90.125	1125	2.50
7/21/99	89.813	89.517	87.750	86.867	2.650	89.063	88.094	Short	87.875	89.500	1625	3.63
7/22/99	89.125	90.552	86.125	87.626	2.926	86.313	89.125	NT				
7/23/99	87.750	87.993	86.250	85.082	2.911	87.563	86.688	NT				
7/26/99	87.125	89.159	85.938	86.467	2.693	86.188	87.000	NT				
7/27/99	87.750	87.857	86.063	85.020	2.836	87.688	86.938	NT				
7/28/99	87.625	89.157	85.500	86.242	2.916	86.688	87.563	NT				
7/29/99	85.688	88.377	83.938	85.573	2.805	84.125	85.625	NT				
7/30/99	84.500	85.845	82.375	83.114	2.730	82.500	84.250	NT				
8/2/99	84.313	83.930	82.438	81.545	2.384	82.500	82.875	Short	82.500	83.875	1375	3.28
8/3/99	83.563	84.321	81.750	81.871	2.450	82.813	83.375	Short	82.000	83.375	1375	3.30
8/4/99	83.875	84.965	80.625	81.633	3.332	80.813	83.125	NT				
8/5/99	83.188	82.297	79.813	79.528	2.769	83.188	81.063	Long	80.000	82.250	2250	5.63
								Total returns using GP's forecast				33.63

Date	Actual-H	GP-H	Actual-L	GP-L	Spread	CP	OP	Trade	Buy @	Sell @	$ Profit	Return
7/9/99	90.500	90.063	88.938	85.688	4.375	89.438	90.000	Short	89.125	89.875	750	1.67
7/12/99	91.375	90.500	89.438	88.938	1.562	90.750	90.000	Short	89.375	90.375	1000	2.21
7/13/99	90.688	91.375	89.250	89.438	1.937	90.313	90.125	Long	89.375	90.500	1125	2.52
7/14/99	90.500	90.688	89.250	89.250	1.438	89.750	90.375	Short	89.375	90.375	1000	2.21
7/15/99	90.500	90.500	88.750	89.250	1.250	89.438	90.375	Short	88.875	90.375	1500	3.32
7/16/99	90.625	90.500	89.750	88.750	1.750	90.125	89.469	NT				
7/19/99	91.000	90.625	89.438	89.750	0.875	90.438	90.063	NT				
7/20/99	90.250	91.000	88.063	89.438	1.562	88.125	90.000	Short	89.000	90.125	1125	2.50
7/21/99	89.813	90.250	87.750	88.063	2.187	89.063	88.094	NT				
7/22/99	89.125	89.813	86.125	87.750	2.063	86.313	89.125	NT				
7/23/99	87.750	89.125	86.250	86.125	3.000	87.563	86.688	NT				
7/26/99	87.125	87.750	85.938	86.250	1.500	86.188	87.000	NT				
7/27/99	87.750	87.125	86.063	85.938	1.187	87.688	86.938	NT				
7/28/99	87.625	87.750	85.500	86.063	1.687	86.688	87.563	Short	86.375	87.500	1125	2.57
7/29/99	85.688	87.625	83.938	85.500	2.125	84.125	85.625	Long	85.750	85.500	-250	-0.58
7/30/99	84.500	85.688	82.375	83.938	1.750	82.500	84.250	NT				
8/2/99	84.313	84.500	82.438	82.375	2.125	82.500	82.875	NT				
8/3/99	83.563	84.313	81.750	82.438	1.875	82.813	83.375	Long	82.000	83.375	1375	3.35
8/4/99	83.875	83.563	80.625	81.750	1.813	80.813	83.125	Short	81.750	83.375	1625	3.90
8/5/99	83.188	83.875	79.813	80.625	3.250	83.188	81.063	Long	80.000	83.000	3000	7.50
								Total returns using NM's forecast				31.17

References

1. Andrew M., Prager R. (1994) Genetic Programming for the Acquisition of double auction market strategies, In: Kinnear K. Jr., (Eds.) Advances in Genetic Programming, Cambridge, MA: The MIT Press, 355–368
2. Bhattacharyya S., Pietet O., Zumbach G. (1998) Representational Semantics for Genetic Programming Based Learning in High-Frequency Financial Data. In: Koza J., Banzhaf W., Chellapilla K., Deb K., Dorigo M., Fogel D., Garzon M., Goldberg D., Iba H., Riolo R. (Eds.) Genetic Programming 1998: Proceedings of the Third Annual Conference, San Francisco, CA: Morgan Kaufmann, 11–16
3. Brock W., Lakonishok J., LeBaron B. (1992) Simple Technical Trading Rules and the Stochastic Properties of Stock Returns. Journal of Finance **47**, 1731–64
4. Chen S -H (1996) Genetic Programming, Predictability, and Stock Market Efficiency. In: Vlacic L. Nguyen T., Cecez-Kecmanovic D. (Eds.) Modelling and Control of National and Regional Economies, Oxford, Great Britain: Pergamon Press, 283–288
5. Chen S -H (1998) Hedging Derivative Securities with Genetic Programming. In: Application of Machine Learning and Data Mining in Finance: Workshop

at ECML-98, Dorint-Parkhotel, Chemnitz, Germany, 24 April 1998. ECML-98 Workshop 6.
6. Chen S -H, Ni C -C (1997) Evolutionary Artificial Neural Networks and Genetic Programming: A Comparative Study Based on Financial Data. Forthcoming in: Smith, G. D. (Eds.), Artificial Neural Networks and Genetic Algorithms, Vienna: Springer-Verlag, 397-400.
7. Chen S -H, Yeh C -H (1995). Predicting Stock Returns with Genetic Programming: Do the Short-term Nonlinear Regularities Exist? In: Fisher D. (Ed.) Proceedings of the Fifth International Workshop on Artificial Intelligence and Statistics, Ft. Lauderdale, Florida, 95–101
8. Chen S -H and Yeh, C -H (1996). Genetic Programming and the Efficient Market Hypothesis. In: Koza J., Goldberg D., Fogel D., Riolo R. (Eds.). Genetic Programming 1996: Proceedings of the First Annual Conference. Cambridge, MA: The MIT Press, 45–53
9. Chen S -H and Yeh C-H (1997a). Using genetic programming to model volatility in financial time series. In Koza, J. Deb, K., Dorigo, M., Fogel, D., Garzon, M., Iba, H. and Riolo, R. (editors), Genetic Programming 1997: Proceedings of the Second Annual Conference. Stanford University, CA: Morgan Kaufmann, 58-63.
10. Chen, S -H, Yeh, C -H (1997b). Toward a Computable Approach to the Efficient Market Hypothesis: An Application of Genetic Programming. Journal of Economic Dynamics and Control **21**, 1043–1063
11. Chen S., Yeh C., and Lee W. (1998). Option Pricing with Genetic Programming. In: Koza J., Banzhaf W., Chellapilla K., Deb K., Dorigo M., Fogel D., Garzon M., Goldberg D., Iba H., Riolo R. (Eds.) Genetic Programming 1998: Proceedings of the Third Annual Conference, San Francisco, CA: Morgan Kaufmann, 32–37
12. Chidambaran N., Lee C., Trigueros J. (1998) An Adaptive Evolutionary Approach to Option Pricing via Genetic Programming. In: Koza J., Banzhaf W., Chellapilla K., Deb K., Dorigo M., Fogel D., Garzon M., Goldberg D., Iba H., Riolo R. (Eds.) Genetic Programming 1998: Proceedings of the Third Annual Conference, San Francisco, CA: Morgan Kaufmann, 187–192
13. Efron, B. (1982). The Jackknife, the Bootstrap, and Other Resampling Plans. Philadelphia: Society for Industrial and Applied Mathematics.
14. Eglit J. T. (1994) Trend Prediction in Financial Time Series. In: Koza J. (Eds.) Genetic Algorithms at Stanford, Stanford Bookstore, Stanford, CA, 31–40
15. Fama E. (1965) The Behavior of Stock Market Prices. Journal of Finance **38**, 34–105
16. Fama E., French K. R. (1988) Dividend Yields and Expected Stock Returns. Journal of Financial Economics **22**, 3-25.
17. Faraway J., Chatfield C. (1998) Time Series Forecasting with Neural Networks: A Comparative Study Using the Airline Data, Applied Statistics **47**, 231–250
18. Hsieh D. (1989) Testing for Nonlinear Dependence in Daily Foreign Exchange Rates. Journal of Business **62**, 339–368
19. Kaboudan M. (1998a) Forecasting Stock Returns Using Genetic Programming in C++. In: Cook D. (Ed.) FLAIRS Proceedings of the Eleventh International Florida Artificial Intelligence Research Symposium Conference, Menlo Park, CA: AAAI Press, 73–77
20. Kaboudan M. (1998b) Statistical Properties of Time-series-complexity Measure Applied to Stock Returns. Computational Economics **11**, 167–187

21. Kaboudan M. (1999) A Measure of Time Series Predictability Using Genetic Programming Applied to Stock Returns. The Journal of Forecasting **18**, 345-357.
22. Kaboudan M., Vance M. (1998) Statistical Evaluation of Symbolic Regression Forecasting of Time-series. In: Proceedings of the International Federation of Automatic Control Symposium on Computation in Economics, Finance and Engineering: Economic Systems, Cambridge, UK, June 1998
23. Kendall M. G. (1953). The Analysis of Economic Time-series. Part I. Prices. Journal of the Royal Statistical Society **96**, 11–25
24. Koza J. (1992) Genetic Programming. Cambridge, MA: The MIT Press
25. Lehman B. (1990) Fads, Martingales, and Market Efficiency. Quarterly Journal of Economics, CV, 1–28
26. Lo A., MacKinlay C. (1988) Stock Prices Do Not Follow Random Walks: Evidence From A Simple Specification Test. Review of Financial Studies **1**, 41–66
27. Lo A., MacKinlay C. (1999) A Non-Random Walk Down Wall Street, Princeton, NJ: Princeton University Press
28. Neely C., Weller P., Dittmar R. (1996) Is Technical Analysis in the Foreign Exchange Market Profitable? A Genetic Programming Approach. Research Division Working Papers, The Federal Reserve Bank of St. Louis, 96-006B
29. Oussaidene M., Chopard B., Pictet O., Tomassini M. (1996) Parallel Genetic Programming: An Application to Trading Models Evolution. In: Koza J., Goldberg D., Fogel D., Riolo R. (Eds.) Genetic Programming 1996: Proceedings of the First Annual Conference. Cambridge, MA: The MIT Press, 357–362
30. Scheinkman J., LeBaron B. (1989) Nonlinear Dynamics and Stock Returns. Journal of Finance **62**, 311–337
31. Singleton A. (1995) Genetic programming with C++, public domain genetic programming package, Dublin, NH: Creation Mechanics: http://www.cen.uiuc.edu/~klipp/research/ssprob.cxx.
32. Warren M. A. (1994) Stock Price Prediction Using Genetic Programming. In: Koza J. (Ed.) Genetic Algorithms at Stanford 1994, Stanford, CA. Stanford Bookstore, 180–184

Appendix A

These are selected best-fit equations representing daily highs:

CMB 7/9/99:

HP = (ln(sqrt(exp(cos((ln(sin(sqrt((sqrt((x35/sqrt(sin(x17))))/
sqrt(sin(x17))))))/sin(x1))))))+(sqrt(exp(cos(sin(exp(exp((ln
(sin(sqrt((sqrt((x32/sqrt(sin(x17))))/sqrt(sqrt(sin(x17)))))))/
sqrt(sin(x17)))))))))-exp(sqrt(sin(x17))))).

CMB 8/2/99:

HP = (sqrt(sqrt(exp(sqrt(sqrt(sqrt(sqrt(sqrt(sqrt(sin(sqrt(sin(((sqrt
(sqrt(exp(sin((sqrt(sin(exp(sin(x1))))*x34)))))+sqrt(sin(exp(sin(x1)))))
*x34))))))))))))+ sqrt(sqrt(cos(cos(x1))))).

Dell 7/21/99:

HP = ((cos(sqrt(sqrt(((sqrt(x2)*cos(sqrt(sqrt(((cos(x10)*cos(sin(sqrt(cos
(x1)))))/x2)))))/ln(x26)))))/(sqrt(((cos(x20)*cos(sqrt(cos(x21))))/
sqrt((cos(x8)+sqrt(x32)))))*x10))+ x8).

Dell 8/4/99:

HP=(cos(cos((sqrt((x14*sqrt(x14)))/exp(exp(exp(cos(ln(cos((sqrt((
x14*sqrt(x16)))/exp(cos(cos(cos(ln(cos(sqrt(cos(x1)))))))))))))))))
+exp(cos(cos(sqrt((x14*sqrt(x30))))))).

GM 7/16/99:

HP = (cos(sqrt(exp(ln(sqrt(sin(sin(sin(sin(sin(sin(sin(sin(sin(
(sin(x1)/cos((sin(x1)/cos(sin(x1))))))))))))))))))+(sqrt(x1)+cos
((cos(cos(sqrt(cos(sqrt((sin(x1)/ cos(sin(x1))))))))+ exp(sin(x1)))))).

GM 7/29/99:

HP = ((exp(((sin(x26) / exp(sin(exp(exp(cos(x1)))))) / exp(sin(exp
(cos((sin(x26)/ cos(x1))))))))* exp(cos(x1)))+ x24).

WCOM 7/12/99:

HP = (exp(((exp(sin(sin(sqrt(x1))))/cos(sin(exp(sin(ln(x1))))))/
cos(sin((x22/(exp(sin(ln(x1)))/ cos(sin((x18/x1)))))))))+ x1).

WCOM 7/27/99:

HP = (x6+exp(cos((x1/ln(((sqrt(ln(ln(x12)))+(sqrt((x1+x1))/x1))
/ ln(((sin(ln(x9)) +(sqrt((x1+x1))/ x1))* sqrt(ln(x22)))))))))).

These are selected best-fit equations daily lows:

CMB 7/20/99:

LP = (sqrt(ln(cos(ln(cos(ln((x3+sin((cos(ln((x3+sin((ln(cos
((x1+sin((sqrt((ln(x1)+ x23))+ x23)))))* exp(ln((x1+x1))))))))
* x1))))))))+x1).

CMB 7/28/99:

LP = exp(ln((cos(((sin(sqrt((x13*sqrt((x13*(x13*sin(sqrt((x13*
sin((x13* sqrt((x13 / sin(x32))))))))))))))+ x1)/ cos(x32)))-(x4/68)))).

Dell 7/14/99:

LP = (ln(ln(ln(cos(cos(ln(cos(sqrt(exp(ln(ln(ln(cos(cos(ln(
ln(cos(ln(exp(ln((exp(cos(ln(x1)))/(ln(cos(sqrt(exp(ln((sin
(cos(x1))*ln(cos(x13)))))))) + x13)))))))))))))))))))))))+ x22).

Dell 7/26/99:

LP = (122*(x1/(x27+(x21/(42/(x21/(41 /(x21 /(x24+sin((x35*(x1/
(x8+(x17+ln(((39 / (x9+sin((x17-x24))))- 109)))))))))))))))).

GM 7/13/99:

LP = (x24*((x24*((x2*(x1/(x27+sin(cos(sin(sqrt(x11))))))) /
(x28+sqrt(sin((x27* sqrt(sin((x22*(x1/(x27+sin(x32)))))))))))/
(x28+sqrt(sin((x24*(x1/(x27+ sin(x32))))))))).
GM 8/3/99:
LP = (exp(ln(sin(sin(exp(ln(sin(sin(exp(ln(sin(sin(exp(ln(sin(sin
(exp(ln(cos(exp(sqrt(sin(sqrt((sin(sqrt(ln((x19/sin(x1)))))+
cos(ln(cos(ln(x1)))))))))))))))))))))))))))) + (x13+sin(ln(x1)))).
WCOM 7/23/99:
LP = (sqrt(ln(cos(ln(((exp((exp(x31)/((ln((((((exp(x1)/x1)+ln((x8*x24)))
+(x28+cos(x28)))+ln(ln((x8*x1))))+(x28+cos(x28))))/x11)+(x8*x1))))/x1)
+cos(x28))))))+ x1).
WCOM 8/5/99:
LP = ((ln(cos(ln((x16+exp(exp(cos((sqrt(cos(sqrt((x26*cos((x26*
sqrt(cos(sqrt((x26*sqrt((x26*cos((x9*x1))))))))))))))-x1)))))))))
-sqrt((x26*cos((x26*x1)))))+ x14).

20 Option Pricing Via Genetic Programming*

Nemmara Chidambaran[1], Joaquin Triqueros[2], and Chi-Wen Jevons Lee[2]

[1] New York University, New York, NY 10012, USA
chiddi@stern.nyu.edu
[2] Tulane University, New Orleans, LA 70118, USA

Abstract. We propose a methodology of Genetic Programming to approximate the relationship between the option price, its contract terms and the properties of the underlying stock price. An important advantage of the Genetic Programming approach is that we can incorporate currently known formulas, such as the Black-Scholes model, in the search for the best approximation to the true pricing formula. Using Monte Carlo simulations, we show that the Genetic Programming model approximates the true solution better than the Black-Scholes model when stock prices follow a jump-diffusion process. We also show that the Genetic Programming model outperforms various other models when pricing options in the real world. Other advantages of the Genetic Programming approach include its low demand for data, and its computational speed.

20.1 Introduction

The Black-Scholes model is a landmark in contingent claim pricing theory and has found wide acceptance in financial markets. The search for a better option pricing model continues, however, as the Black-Scholes model was derived under strict assumptions that do not hold in the real world and model prices exhibit systematic biases from observed option prices. We propose the Genetic Programming methodology for better approximating the elusive relationship between option prices, option contract terms, and properties of the underlying stock price.

Many researchers have attempted to explain the systematic biases of the Black-Scholes model as an artifact of its assumptions [13]. The most often challenged assumption is the normality of stock returns.[1] References [20] and [2] propose a Poisson jump-diffusion returns processes. References [11]

* Published previously in: Computational Finance – Proceedings of the Sixth International Conference, Leonard N. Stern School of Business, January 1999. MIT Press, Cambridge, MA

[1] Normality of stock returns has been repeatedly rejected. Indeed, [16] have ranked candidates for return distributions and found normality to be the least likely. Their rankings are: 1) Intertemporal dependence models (ARCH, GARCH), 2) Student t, 3) Generalized mixture of normal distributions, 4) Poisson jump, and 5) Stationary normal.

and [3] advocate GARCH([7]) processes. It is not possible to derive closed-form analytical solutions in all cases, in which cases numerical solutions are used.

The difficulty in finding an analytical closed-form parametric solution has also led to non-parametric approaches. Reference [22] suggests that we examine option data for the implied binomial tree to be used for pricing options. Reference [10] use a quasi-analytic approximation based on Monte Carlo simulation. Reference [14] build a numerical pricing model using neural networks. We apply Koza's [17] Genetic Programming to develop an adaptive evolutionary model of option pricing. This method is well suited to the task and can operate on small data sets, circumventing the large data requirement of the neural network approach noted by [14].

The philosophy underlying Genetic Programming is to replicate the stochastic process by which genetic traits evolve in offspring through a random combination of the genes of the parents, in the biological world. A random selection of functions of the option contract terms and basic statistical properties of the underlying stock price will have among them some elements that will ultimately make up the true option pricing formula. By selectively "breeding" the functions, these elements will be passed onto future generations of functions that price options more accurately.

Since it is impossible to ex-ante determine which element is the best, we instead focus on parent-selection, that is, the method of selecting equations to serve as parents for the next generation. Equations are chosen probabilisitically based on the pricing errors of the functions. We examine six alternative parent-selection methods: Best, Fitness, Fitness-overselection, Random, Tournament with 4 individuals and Tournament with 7 individuals. We find that the Fitness-overselection method seems to offer the best results for option pricing. We also explore the effect of varying other model parameters, such as the properties of the data set required to train the genetic programs, on model efficiency.

An important advantage of the Genetic Programming approach over other numerical techniques is its ability to incorporate known approximate solution into the initial "gene pool" to be used in evolving future generations, for example, we can include the Black-Scholes model. We illustrate how this approach quickly adapts the Black-Scholes model to a jump-diffusion process, where the Black-Scholes assumption of returns normality does not hold and for pricing options in the real world. We find that the Genetic Programming formulae beats the Black-Scholes equation in 9 out of 10 runs when the underlying stock prices are generated by a jump-diffusion process and in 10 out of 10 runs when we apply the analysis to the S&P Index options. The method also outperforms the Black-Scholes model for four out of five equity stock options in our sample.

20.2 Genetic Programming - A Brief Overview

Genetic Programming is a technique that applies the Darwinian theory of evolution to develop efficient computer programs.[2] We use a variant of Genetic Programming called Genetic Regression, where the desired program is a function that relates a set of inputs such as stock price, option exercise price, etc., to one output, the option price.

20.2.1 Basic Approach

The set of data on which the program operates to determine the relationship between input parameters and the options price is called the training set. The set of data on which the resulting formula is tested is called the test set. The procedure of the basic approach is as follows.

- Given a training set of matched inputs and outputs, a set of possible formulas is randomly generated. The formulas are represented as trees and we allow a maximum tree depth of 17.[3] Each formula is an individual and the set of individuals is called the population. The size of the population is held constant and is a control variable for optimizing the genetic program.
- Every individual in the population is evaluated to test whether it can accurately price options in the training data set. Based on individual fitness, a subset of the population is selected to act as the parents for the next generation of formulas.
- A pair of the parents generates a pair of offspring. Components of the parent formulas are crossed to generate offspring formula. A random point is selected in each parent tree. The sub-trees below that random point are switched between the two parent formulas. This operation creates a new pair of individuals, the offspring. It is possible that no crossover is performed and the parents themselves are placed in the new population (a clone). The process of selection and crossover is repeated until the new generation is completely populated.

[2] Genetic Programming is an offshoot of Genetic Algorithms. Genetic Algorithms have been used to successfully develop technical trading rules by [1] for the S&P 500 index and by [21] for foreign exchange markets. In a paper similar in spirit to our study, [15] uses Genetic Programming to value American put options. Genetic Programming has also been used in multi-agent financial markets by [18] and in multi-agent games by [12].

[3] A 17 deep tree is a popular number used to limit the size of tree sizes [17]. Practically, we chose the maximum depth size possible without running into excessive computer run times. Note that the Black-Scholes formula is represented by a tree of depth size 12. A depth size of 17, therefore, is large enough to accommodate complicated option pricing formulas and works well in practice.

- The steps above are repeated for a pre-specified number of times, or generations. Evolutionary pressure in the form of fitness-related selection combined with the crossover operator eventually produce populations of highly fit individuals. The best-fit individual is the solution to the option pricing problem.

20.2.2 Parent Selection Criteria

The method of selecting parents for the next generation can affect the efficiency of genetic programs. We analyze six different selection methods: Best, Fitness, Fitness-overselection, Random, Tournament with 4 individuals and Tournament with 7 individuals. These methods represent various attempts to preserve a degree of "randomnes" in the evolutionary process. In the Best method, individuals are ranked in terms of their fitness, ascending in order of the magnitude of their errors. The individuals with the smallest errors are thus picked to serve as parents of the next generation. In the Fitness method, individuals are selected randomly with a probability that is proportional to their fitness. In the Fitness-overselection method, 400 individuals are classified into two groups. Group 1 has 320 best-fit individuals and Group 2 has the remainder. Individuals are selected randomly with an 80% probability from Group 1 and a 20% probability from Group 2. In the Random method, the fitness of the individuals is completely ignored and parents are chosen at random from the existing population. Finally, in the Tournament method, n individuals are selected at random from the population and the best-fit individual is chosen to be a parent. We examine Tournament method with n=4 and n=7.

20.3 Performance Analysis in a Jump-Diffusion World

We first examine how well the Genetic Programming model can adapt and outperform the Black-Scholes model under controlled conditions. We choose a jump-diffusion world described by [20], since the closed form solution for the option prices in a jump-diffusion world is known and is a convenient benchmark. We can therefore measure the pricing errors of the Genetic Programming model and the Black-Scholes model.

The jump-diffusion process is a combination of a Geometric Brownian diffusion process and a Poisson jump process and can be written as:

$$\frac{dS(t)}{S(t)} = (\mu - \lambda k)dt + \sigma dW(t) + dq \tag{20.1}$$

where dq is the Poisson-lognormal jump process. The Poisson process determines when a jump occurs, with the jump size being lognormally distributed.

We simulate the price path of daily stock prices over a 24 month period with the initial price set at $S_0 = 50$. Each month is assumed to have 21 trading days. The diffusion parameters m (mean) and σ(standard deviation) were set at 10% and 20% respectively, and jump parameters k(jump size), λ (jump rate), and δ (standard deviation of the log-jumps), were set at $0.02, 25$, and 0.05 respectively. These values are well within the range estimated by stock price data. Thus 504 stock prices, $S(t)$, are simulated using random daily returns $z_t \sim N[(m - \sigma^2/2 - k)/252, \sigma/252]$ and $n(t) \sim Poisson[\lambda t]$ jumps, each of magnitude Y_j (where $lnY_j \sim N[ln(1 + k) - 0.5\delta^2, \delta])$ for each $t \in 1 \ldots 504$:

$$S(t) = S_0 e^{\sum_{i=1}^{t} z_t}\, Y(n(t)), \quad t = 1, \ldots, 4 \tag{20.2}$$

where, $Y(0) = 1$ and $Y(n(t)) = \prod_{i=1}^{n(t)} Y_i,\ n(t) > 0$

We use CBOE rules to create call options from the simulated stock price path. Figure 20.1 shows the distribution of option prices in a jump-diffusion world.

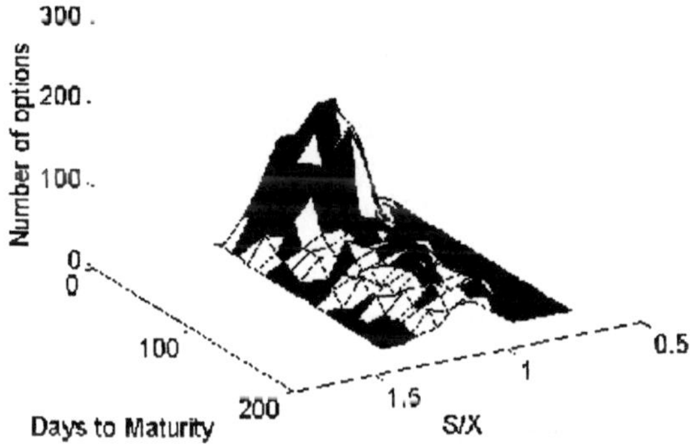

Fig. 20.1. Distribution of options simulated in the jump-diffusion world

Options are priced using [20] jump-diffusion formula, truncated at the point when the marginal contribution of additional terms is negligible.

$$F(S, X, r, \sigma, \tau, \lambda, k, \delta) = \sum_{n=0}^{\infty} \frac{e^{-\lambda'\tau}(\lambda'\tau)^n}{n!} C_{B-S}(S, X, r_n, v_n, \tau) \tag{20.3}$$

where,$\lambda' = \lambda(1 + k)$, $r_n = r - \lambda k + \frac{nln(1+k)}{\tau}$, $v_n = \sigma^2 + \frac{n\delta^2}{\tau}$, and τ is the option's time to maturity. C_{B-S} is the Black-Scholes option value given by,

$$C_{B-S}(S, X, r, \sigma, \tau) = SN(d_1) - Xe^{-rT}N(d_2) \tag{20.4}$$

where $d_1 = (ln[S/X] + (r + \sigma^2/2)T)/(\sigma\sqrt{T})$, $d_2 = d_1 - (\sigma\sqrt{T})$. $N[d_1]$ and $N[d_2]$ are the cumulative standard normal distribution values at d_1 and d_2.

Table 20.1 presents the set of operations and variables used to develop the Genetic Programs. Note that we include the Black-Scholes option value as a possible component of the tree. This serves as a good starting point and information from known analytical models can thus be used to find a better solution. We do correct for the volatility estimate an investor would have used based on a history of prices generated by a combination of the diffusion and jump processes. This reflects the approach of a naive investor who is unaware of the true nature of the returns' underlying process when using the Black-Scholes model to price the option. The estimated call option value with the modified Black-Scholes model is therefore([20]):

$$C_{AdjB-S} = C_{B-S}(S, X, r, \sigma_{adj} = \sqrt{\sigma^2 + \lambda\delta^2}, \tau) \tag{20.5}$$

Table 20.1. Genetic programming model specification in a jump-diffusion world - training variables.

Name	Source	Definition
S	Option Contract	Stock price
X	Option Contract	Exercise price
S/X	Option Contract	Option moneyness
τ	Option Contract	Time to maturity (years)
Max(S-X)	Boundary Condition	Intrinsic value Max (S-X,0)
Black-Scholes	Analytical Model	Black Scholes value
$+$	Standard arithmetic	Addition
$-$	Standard arithmetic	Subtraction
$*$	Standard arithmetic	Multiplication
%	Standard arithmetic	x%y = 1 , if y = 0 = x/y , otherwise
Exp	Black-Scholes	Exponent: $exp(x) = e^x$
plog	Black-Scholes	$plog(x) = ln(\|x\|)$
psqrt	Black-Scholes	$psqrt(x) = sqrt(\|x\|)$
Ncdf	Black-Scholes	Normal CDF

Table 20.2. Genetic programming model specification in a jump-diffusion world - size of training sets. For each training set (option pricing formula), the price path of a stock with beginning value $S_0 = 50$ was simulated through 24 21-day months. Options were created according to CBOE rules and valued using the Black Scholes formula. Each training set consisted of the daily values of these options.

Training Set	1	2	3	4	5	6	7	8	9	10
Data Points	311	350	364	308	288	420	318	387	409	319

Table 20.3. Genetic programming model specification in a jump-diffusion world - training parameters. Genetic programming algorithm training parameters used in the non-Black-Scholes world where stock prices are generated by a jump-diffusion process.

Fitness Criterion	Sum of absolute dollar and percentage errors
Population Size	5,000
Number of Generations	10

Table 20.2 and Table 20.3 present the size of the training sets used and the algorithm training criteria. We implement an additional step in setting the size of the population and the number of generations. Using an independent set of 25% of the options as a training set, we determine that a minimum population size of 5000 functions and 10 generations is needed to get a formula that outperforms the Black-Scholes model. We use these parameters on ten new 25% subsets of the options created to develop the genetic program for option pricing in a jump-diffusion world.

The formulas generated by the genetic program are adaptations of the Black-Scholes model, for example, one of the formulas generated is:

$$C(S,X,\tau) = \sqrt{C_{B-S} * \left[0.11734 + \sqrt{0.95461 * C_{B-S} * (C_{B-S} + \tau)}\right]} \quad (20.6)$$

where, C_{AdjB-S} is the adjusted Black-Scholes model given in 20.5 and τ is the option's time to maturity.

We examine the performance of our genetic program on ten out-of-sample test sets of option data. The Fitness-overselection method gives the lowest pricing errors and beats the modified Black-Scholes model in each of the ten out-of-sample tests. The next best performance is the Tournament method with n=7. The Genetic Programming model based on Fitness-overselection clearly outperforms the Black-Scholes model in out-of-sample tests.[4] The Genetic Programming formula performs better than the Black-Scholes model for each of the 10 out-of-sample test sets.

An important criticism usually leveled at complex numerical methodologies is, Can the method perform any better than a simple linear regression model? We therefore run single-stage and two-stage linear regression with and without Black-Scholes model as an independent variable. The two-stage model represents separate equations for in-the-money and out-of-the-money options.

[4] The only measure in which the original Black-Scholes model ever beats the Genetic Programming formula is in the training-set sum of percentage errors, and this occurs for only 1 out of the 10 Genetic Programming formulas. However, the error is large enough to blow up the average percentage error. We attribute this fluke to our decision to ignore during training all percentage errors for options worth less than \$0.01.

Table 20.4. Average absolute pricing errors of the genetic programming models (GP), the Black-Scholes model (BS), and linear models in a jump-diffusion world. Pricing errors are presented for six Genetic Programming formulas using alternate methods for generating new populations from the previous generation and for four linear models that are a function of the initial stock price, exercise price, and time to maturity. Each cell in the table presents the average pricing errors over ten sets of stock and option prices and for the entire sample of options generated in each set. Parameter values used to generate stock price and options data and the Genetic Programming parameters are given in Table 20.1.

	GP	B-S	1-stage, with B-S	1-stage, no B-S	2-stage, with B-S	2-stage, no B-S
Best	0.0655	0.0888	0.0350	0.9357	0.0158	0.4049
Fitness	0.0517	0.0888	0.0350	0.9357	0.0158	0.4049
Fitness-Overselection	0.0393	0.0888	0.0350	0.9357	0.0158	0.4049
Random	0.0704	0.0888	0.0350	0.9357	0.0158	0.4049
Tournament, Size = 4	0.0534	0.0888	0.0350	0.9357	0.0158	0.4049
Tournament, Size = 7	0.0464	0.0888	0.0350	0.9357	0.0158	0.4049

Table 20.4 presents the pricing errors for the Genetic Programming formulas, the Black-Scholes equation, and for the linear models. The absolute and average errors for the Genetic Programming formulas are the average pricing error for all options which is again averaged across the ten Genetic Program runs. Results are presented for all six parent-selection methods considered. The modified Black-Scholes equation has the largest errors compared to all other models. The linear models give very good results when we include the naive Black-Scholes model as an independent variable with the two-stage linear model giving the lowest errors. Among the Genetic Programming formulas, the Fitness-overselection parent selection method provides the smallest absolute pricing error and one of the smaller percentage pricing errors. The magnitudes are comparable to the two-stage linear model with Black-Scholes as an independent variable.[5]

We further evaluate the performance of the Genetic Programming formula by comparing its pricing errors with that of the Black-Scholes model and Neural Networks for options of various maturities and moneyness. The details

[5] The linear models that have the Black-Scholes model as an independent variable however have one major draw back – the partial derivatives of the pricing equation are equal to the Black-Scholes partial derivatives with a constant adjustment term. In a related paper (see [9]), we test the hedging effectiveness of the Genetic Program model and the Black-Scholes model by constructing a hedge portfolio of the Option, Stock, and Bonds. The hedging performance is calculated to be the deviation from zero in the portfolio value. Overall, the Genetic Program beats the Black-Scholes model in over 50% of the cases.

of the Neural Networks we use are reported in [9]. Results are reported for the network that gives the best results among the various normalization and initialization schemes considered.

Table 20.5. Mean absolute pricing errors of genetic programming models (GP), the Black-Scholes model (BS), and neural networks (NN) in a jump-diffusion world. The numbers in each cell are the average pricing errors from the models across 10 test sets. For each test set, the error value for each cell is calculated by taking the average pricing errors over five options. Rows in the table represent days-to-maturity and columns represent the degree-of-moneyness, S/X.

		Moneyness S/X					
Maturity		0.9	0.95	1	1.05	1.1	1.15
	BS=	0.05	0.08	0.03	0.03	0.01	0.00
5 days	GP=	0.03	0.02	0.12	0.06	0.01	0.01
	NN=	0.23	0.18	0.12	0.21	0.23	0.26
	BS=	0.08	0.09	0.00	0.04	0.01	0.00
10 days	GP=	0.03	0.02	0.08	0.07	0.03	0.02
	NN=	0.17	0.08	0.12	0.22	0.22	0.26
	BS=	0.13	0.12	0.07	0.02	0.00	0.01
30 days	GP=	0.03	0.03	0.02	0.03	0.05	0.04
	NN=	0.09	0.17	0.17	0.19	0.21	0.26
	BS=	0.15	0.14	0.10	0.06	0.03	0.01
45 days	GP=	0.04	0.04	0.03	0.02	0.03	0.04
	NN=	0.17	0.18	0.16	0.18	0.22	0.26
	BS=	0.16	0.15	0.12	0.09	0.05	0.03
60 days	GP=	0.04	0.05	0.03	0.03	0.03	0.04
	NN=	0.19	0.16	0.15	0.19	0.21	0.26
	BS=	0.19	0.18	0.16	0.13	0.10	0.07
90 days	GP=	0.06	0.06	0.05	0.05	0.05	0.06
	NN=	0.14	0.14	0.18	0.20	0.21	0.28

Table 20.5 shows the absolute pricing errors and the percentage pricing errors for the Genetic Programming formula developed with Fitness-overselection, for the Black-Scholes model, and for the best Neural Network, on an out-of-sample two-dimensional grid. Each cell in the table represents the average across ten out-of-sample test data sets and the value in each cell is the average over 5 options. Figure 20.2 plots the absolute pricing errors for the Black-Scholes equation and the Genetic Programming formulas.

The Genetic Programming formula tends to do better with in-the-money and short-maturity options whereas the Black-Scholes model seems to perform relatively better with out-of-the money and long-term options. This result is consistent with the notion that the jump term influences prices of short-maturity in-the-money options more, relative to long-term and out-of-the-money options. Genetic Programming beats Neural Networks over-

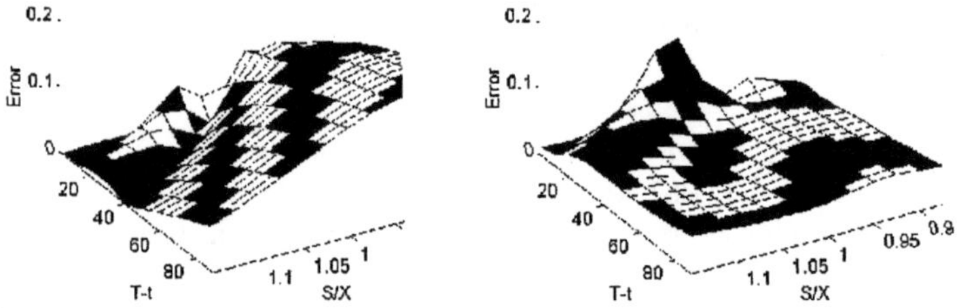

Fig. 20.2. Mean absolute pricing error for the Black-Scholes(left figure) and Genetic Programming (right figure) models.

all, however in 9 out of the 72 cases considered Neural Networks do show marginally lower errors.

To take advantage of Genetic Programming's ability to learn with small training sets [17] and reduce computational time, we tested its performance using random samples of 5% of the options generated in the simulation, without updating the algorithm parameters. We found that the training formulas with the smaller data sets resulted in only a minimal reduction in out-of-sample performance (see [9]). Our tests support the notion that Genetic Programming needs only small training sets in order to arrive at a good solution.

20.4 Application in the Real World

We next apply Genetic Programming to price real-world options data. Call options data for the S&P 500 Index and 5 different stocks were obtained from the Berkeley Options Data Base (BODB). BODB's data is time stamped to the nearest second and ensures a good match between the values of an option and its underlying asset.[6] Option prices are set to be the average of the bid and ask prices. The option's time to maturity is set to be the number of trading days between the trade date and the expiration date of the option. Interest rates from the term structure of zero-coupon treasuries([6]) were used to calculate the risk-free rate between two calendar dates.

[6] Raw BODB records are screened as follows. We do not include records from the first 2,500 seconds after 8:30 am or in the last 2,500 seconds before 3:00pm and required at least 300 seconds within a 1% deviation for the underlying index/equity price. We also reject data when the option bid-ask spread is more than \$0.25 or is more than 5% of option value. The first restriction eliminates artificial pricing that may occur due to the structure of the market at the beginning and the ending of the day. The second restriction is to allow the options market to adjust to changes in underlying asset value. The third restriction gives us a tighter handle on the option's equilibrium price.

We use a two step process to develop the Genetic Program. We first determine the optimal set of algorithm parameters using a training and validation step. That is, we vary algorithm parameters when training the program and test the performance on a validation data set of options prices from a later date. The parameters that give the best results in the validation step are the optimal algorithm parameters. These parameters are then used in the next stage when the genetic programs are developed. A separate training data set and out-of-sample data set of options prices are used for the subsequent training/test step.

Ten Genetic Programming formulas were developed using ten sets of training/validations sets and ten sets of training/test data sets. The training/validation data sets were created by randomly sampling April 3-4, 1995 screened S&P 500 Index Option data. The training/test data sets are separately created from April 6-10, 1995 screened S&P 500 Index Options data. All out-of-sample validation and test data occurred later in time than the training data. Training sets contained a mere 50 points each and training time do not exceed 3 minutes per formula.

The sets of operations, functions and variables allowed in our formulas are those used in the jump-diffusion world (given in Table 20.1), augmented by the risk-free rate and historical volatility. As in [14], we estimate the S&P 500 Index volatility by computing the standard deviation of the 60 most recent continuously compounded daily S&P 500 returns using 3.00pm (CST) prices. We adjusted for dividends by subtracting the present value of actual dividends between the record date t_0 and the option maturity date T.

The formulas generated by the genetic program for the index options were adaptations of the Black-Scholes model. For example, one formula is,

$$C(S, X, \tau) = C_{B-S} + 3\tau \qquad (20.7)$$

where, C_{B-S} is the Black-Scholes formula and τ is the option's time to maturity.

Table 20.6 presents the average absolute pricing error and the average percentage pricing error for the 10 Genetic Programming formulas and for the Black-Scholes model when pricing the S&P Index options. The out-of-sample performance of these Genetic Programming adaptations of Black-Scholes model is remarkable: 9 out of the 10 Genetic Programming formulae beat the Black-Scholes model in both average absolute and percentage pricing errors.

For testing the performance of Genetic Programming in pricing equity options, we choose five stocks that had options volume of at least 1500 contracts and which never paid cash dividends. The stocks are: Best Buy Company Inc., Broderbund Software Inc., CompUsa Inc., Digital Equipment Corporation, and Novellus Systems Inc. For each stock, we develop ten Genetic Programming formulas. The training/validation data sets are constructed using BODB records for April 3-4, 1995. The training/test data sets are constructed from options traded during the period April 6-13, 1995.

Table 20.6. This table shows the mean absolute pricing errors for ten Genetic Programming Formulas, the Black-Scholes model, and Neural Networks, on ten out-of-sample data sets of the S&P 500 Index (SPX) option and five equity options. Each formula came from a separate training set and was evaluated on a separate test set, each set having a sample of 50 options. The parameter search was performed using April 3-4 data to find algorithm parameters that give formulas with good out-of-sample performance. All training and test sets for SPX came from April 6-10 BODB data and those for five equity options came from April 6-13.

Run #	GP	B-S	NN	GP	B-S	NN
		S&P 500			Best Buy	
1	1.9555	3.2301	1.2371	0.0927	0.1181	0.0583
2	3.0323	4.6736	1.1773	0.0605	0.0859	0.0633
3	2.3617	3.3248	1.2008	0.1150	0.0794	0.0771
4	1.5177	3.2884	1.0456	0.0996	0.1019	0.0583
5	2.4803	3.5442	2.0528	0.1262	0.0868	0.0743
6	2.0220	3.0111	0.7804	0.1046	0.0895	0.0621
7	2.3358	3.1802	1.6938	0.1012	0.1145	0.0527
8	2.1766	3.3159	1.2111	0.1189	0.0786	0.0701
9	1.6305	3.2846	1.0720	0.0855	0.1011	0.0601
10	3.0803	4.1809	0.9376	0.1181	0.0780	0.0766
Average	2.2590	3.5030	1.2409	0.1022	0.0933	0.0653
		Broderbund			Comp USA	
1	0.1346	0.1346	0.7407	0.1475	0.1892	0.1640
2	0.1255	0.1255	0.6099	0.1919	0.1856	0.1396
3	0.1074	0.1074	0.5695	0.2145	0.1487	0.2774
4	0.1418	0.1415	0.5975	0.1622	0.1870	0.1395
5	0.1448	0.1448	0.6821	0.1592	0.1656	0.1620
6	0.1279	0.1279	0.6916	0.2133	0.1843	0.2652
7	0.1253	0.1253	0.6114	0.1665	0.1856	0.1211
8	0.1211	0.1211	0.5610	0.1502	0.1886	0.1583
9	0.1475	0.1475	0.7252	0.1796	0.1922	0.1453
10	0.1074	0.1158	0.6349	0.1519	0.1870	0.1724
Average	0.1283	0.1291	0.6424	0.1737	0.1814	0.1745
		DEC			Novellus	
1	0.1595	0.1595	0.4301	0.1423	0.2531	0.4756
2	0.1462	0.1462	0.4140	0.1976	0.2840	0.5227
3	0.1435	0.1435	0.4544	0.2249	0.2850	0.4574
4	0.1148	0.1148	0.4056	0.1817	0.2692	0.5035
5	0.0880	0.1223	0.4336	0.1882	0.2772	0.4485
6	0.1254	0.1254	0.4076	0.2049	0.2344	0.4389
7	0.1236	0.1236	0.4302	0.1668	0.3217	0.4507
8	0.1286	0.1286	0.3329	0.2006	0.1979	0.5819
9	0.1454	0.1454	0.4343	0.1953	0.3096	0.5859
10	0.1063	0.1063	0.4105	0.1617	0.2973	0.4420
Average	0.1281	0.1316	0.4153	0.1864	0.2729	0.4907

The formulas generated by the genetic program for the equity options were also adaptations of the Black-Scholes model. In most cases, the formulas are of the form,

$$C(S, X, \tau) = C_{B-s} + Constant \ * \ \tau \tag{20.8}$$

where, C_{B-S} is the Black-Scholes formula and τ is the option's time to maturity. The constant takes values from 1 to 4 depending on the stock underlying the option that is being priced.

Table 20.6 also presents the average absolute pricing errors and average percentage pricing errors for the ten Genetic Programming formulas, the Black Scholes formula, and for Neural Networks, when pricing equity options. When there is no difference between the errors for the Genetic Programming formula and the Black-Scholes model, it indicates that the Genetic Program converged on the Black-Scholes model. Except for Best Buy, the Genetic Programming method produced formulas that outperform the Black-Scholes model on average, though the results are not as strong as the case of S&P Index options. We attribute the results for Best Buy to the fact that the data set used for the parameter search is much smaller than the data sets used for the other four stocks. The resulting training and validation sets were thus not independent enough to yield an insight on satisfactory parameters.

Neural Network pricing errors, though higher, are comparable to those of Genetic Programming formulas for the S&P 500 index option. On the other hand, Neural Network pricing errors are a magnitude higher than that of Genetic Programming formulas when pricing equity options. Equity options are more thinly traded and there is less data available to train Neural Networks and Genetic Programs in comparison to the S&P 500 index option.

Note that for Broderbund and DEC, a majority of the Genetic Programs converge on the Black-Scholes model as the best possible pricing formula. This highlights the advantage of the Genetic Programming approach – it can easily converge on existing known models, if they are indeed the best solutions. By including known analytical solutions in the parameter set, we thus increase the efficiency of Genetic Programming by using it to improve on existing solutions.

20.5 Conclusion

In this paper we have developed a procedure to apply the principles of Genetic Programming to option pricing. Our results, from controlled simulations and real world data, are strongly encouraging and suggest that Genetic Programming work well in practice.

Genetic Programming is a non-parametric data driven approach and, using options data, extracts the implied pricing equation. Genetic Programming thus overcomes the need to make specific assumptions about the stock price process. Researchers have attributed the systematic biases in Black-Scholes prices to the assumption that stock prices follow a diffusion process. We show that Genetic Programming formulas beat the Black-Scholes model in 10 out of 10 cases in a simulation study where the underlying stock prices were generated using a jump-diffusion process. They work almost as well in pricing S&P Index options with genetic programs beating the Black-Scholes model in 9 out of 10 cases. For equity options, genetic programs beat or match the Black-Scholes model for 4 of the 5 stocks considered.

The Genetic Programming method requires less data than other numerical techniques such as Neural Networks ([14]). We show this by simulation studies that use smaller subsets of the data and by using both genetic programs and neural networks to price relatively thinly traded equity options. The time required to train and develop the genetic programming formulas is also relatively short.

Genetic Programming can incorporate known analytical approximations in the solution method. For example, we use the Black-Scholes model as a parameter in the genetic program to build a better option pricing model. The flexibility in adding terms to the parameter set used to develop the functional approximation can also be used to examine whether factors beyond those used in this study, for example, trading volume, skewness and kurtosis of returns, and inflation, are relevant to option pricing. Finally, since the Genetic Programming method is self-learning and self-improving, it is an ideal too for practitioners.

References

1. Allen F., Karjalainen R. (1999) Using Genetic Algorithms to Find Technical Trading Rules. Journal of Financial Economics **51(2)**, 245–71
2. Ball C. A., Torous W. N. (1985) On Jumps in Common Stock Prices and Their Impact on Call Option Pricing. Journal of Finance **40(1)**, 155–73
3. Ballie R., DeGennaro R. (1990) Stock Returns and Volatility. Journal of Financial and Quantitative Analysis **25(2)**, 203–14
4. Black F., Scholes M. (1972) The Valuation of Option Contracts and a Test of Market Efficiency. Journal of Finance **27(2)**, 399–417
5. Black F., Scholes M. (1973) The Pricing of Options and Corporate Liabilities. Journal of Political Economy **81(3)**, 637–54
6. Bliss R. (1997) Testing Term Structure Estimation Methods. In: Boyle P., Pennacchi G., Ritchken P. (Eds.) Advances in Futures and Options Research. Volume 9. Greenwich, Conn. and London: JAI Press, 197–231
7. Bollerslev T. (1986) Generalized Autoregressive Conditional Heteroskedasticity. Journal of Econometrics **31(3)**, 307–27
8. Chidambaran N. K., Lee C. W. J., Trigueros J. (1999) Option Pricing Via Genetic Programming. In: Abu-Mostafa Y. S., LeBaron B., Lo, A. W., Weigend A. S. (Eds.) Computational Finance – Proceedings of the Sixth International Conference, Leonard N. Stern School of Business, January 1999. Cambridge, MA: MIT Press
9. Chidambaran N. K., Lee C. W. J., Trigueros J. (1998) An Adaptive Evolutionary Approach to Option Pricing via Genetic Programming. Working paper, New York University

10. Chidambaran N. K., Figlewski S. (1995) Streamlining Monte Carlo Simulation with the Quasi-Analytic Method: Analysis of a Path–Dependent Option Strategy. Journal of Derivatives
11. French K. R., Schwert G. W., Stambaugh R. F. (1987) Expected Stock Returns and Volatility. Journal of Financial Economics **19(1)**, 3–29
12. Ho T. H. (1996) Finite Automata Play Repeated Prisoner's Dilemma with Information Processing Costs. Journal of Economic Dynamics and Control **20(1-3)**, 173–207
13. Hull J. (1997) Options, Futures, and Other Derivative Securities. 3rd edn. Prentice-Hall, Englewood Cliffs, New Jersey
14. Hutchinson J., Lo A., Poggio T. (1994) A Nonparametric approach to the Pricing and Hedging of Derivative Securities Via Learning Networks. Journal of Finance **49(3)**, 851–89
15. Keber C. (1998) Option Valuation with the Genetic Programming Approach. Working paper. University of Vienna
16. Kim D., Kon S. J. (1994) Alternative Models for the Conditional Heteroscedasticity of Stock Returns. The Journal of Business **67(4)**, 563–98
17. Koza J. R. (1992) Genetic Programming. MIT Press, Cambridge, Massachusetts
18. Lettau M. (1997) Explaining the Facts with Adaptive Agents. Journal of Economic Dynamics and Control **21(7)**, 1117–47
19. Merton R. C. (1973) Theory of Rational Option Pricing. Bell Journal of Economics **4(1)**, 141–83
20. Merton R. C. (1976) Option Pricing When Underlying Stock Returns Are Discontinuous. Journal of Financial Economics **3(1-2)**, 125–44
21. Neely C., Weller P., Dittmar R. (1997) Is Technical Analysis in the Foreign Exchange Market Profitable? A Genetic Programming Approach. Journal of Financial and Quantitative Analysis **32(4)**, 405–426
22. Rubinstein M. (1994) Implied Binomial Trees. Journal of Finance **49(3)**, 771–818
23. Trigueros J. (1997) A Nonparametric Approach to Pricing and Hedging Derivative Securities Via Genetic Regression. Proceedings of the Conference on Computational Intelligence for Financial Engineering

21 Evolutionary Computation in Option Pricing: Determining Implied Volatilities Based on American Put Options

Christian Keber

University of Vienna, Department of Business Administration, Austria
Christian.Keber@univie.ac.at

Abstract. In our investigation, Genetic Programming (GP) is used in the context of option pricing. Up to now very few publications exist in this field, the first papers, to our knowledge, come from Chen et al., Chidambaran et al., and Keber [13,14,24,25]. Currently, research activities are increasing in the "GP/Finance"–area, as, e.g., the publications of Allen and Karjalainen as well as Neely et al. show [2,31]. In the context of option valuation a lot of models are not amenable to an exact analytical solution. Such models must be solved either by using numerical procedures or analytical approximations. In our paper we derive analytical approximations for calculating implied volatilities based on American put options using Genetic Programming. Applying our approximations to experimental data sets we can show that the results obtained by our formulas are very close to the numerically calculated ones.

21.1 Introduction

In their seminal papers Black and Scholes [9] and Merton [30] derived an analytical exact solution for the valuation of European–style options on stocks paying no dividends during the time to expiration. Merton additionally showed that premature exercising of American call options on this type of stocks is never optimal and that the valuation can also be made using the Black/Scholes–Merton method. American put options on stocks paying no dividends during the time to expiration, on the other hand, have early exercise premiums implicitly embedded in their prices. Unlike the European–style option and American call option pricing problems, however, analytical exact pricing formulas for American put options have not as yet been derived. American put options are priced either by using numerical procedures or analytical approximations. The best known numerical procedures are the *lattice approach* of Cox, Ross, and Rubinstein [17] and the *finite difference method* of Brennan and Schwartz [10], and the most frequently quoted analytical approximations come from Johnson, Geske and Johnson,MacMillan, and Barone–Adesi and Whaley[23,20,28,4].

The Black/Scholes–Merton method is a – Nobel–prize awarded – breakthrough in the research area of contingent claim pricing. The five input parameters which are necessary to specify the pricing model, all are directly observable from market data except the volatility of the stock price. The volatility can be calculated either from historical data (*historical volatility*) or implicitly by using the current option price in the valuation model (*implied volatility*). The implied volatility represents investors' expectations on the option market and is therefore a better predictor of the future volatility than volatility predictors based on historical data (see, e.g., [6,15], and [27]).

Unfortunately, a closed–form solution for calculating the implied volatility is unknown. To obtain the implied volatility exactly numerical procedures – mainly the Newton–Raphson method and its variants – are required (see, e.g., [29]). However, calculating the exact implied volatility manually is error–prone and spreadsheet implementations are cumbersome. Therefore, analytical approximations are applied more often for determining the implied volatility (see, e.g., [7,8,11,12,16], and [24]). In doing so, researchers typically concentrate on implied volatilities based on European–style options (see, e.g., [18,19,32], and [33]) although options are frequently American style.

In this paper we derive *analytical approximations* for determining the *implied volatility* exclusively based on *American put options* on non–dividend paying stocks[1] using the *Genetic Programming* approach. Recently, Genetic Programming is used more and more in finance (see, e.g., [2,31,35]) especially in the context of option pricing. Chidambaran et al. [14], e.g., as well as Keber [25] derive approximations for calculating option prices and show that GP–models outperform various other models presented in the literature. Chen et al. [13] use the Genetic Programming approach for hedging derivative securities and Keber [24] shows that genetically determinded formulas for calculating the implied volatility based on the Black/Scholes–Merton model outperform the most frequently quoted analytical approximations for calculating implied volatilities. Genetic Programming is an approach designed to generate computer programs as well as formulas using a random oriented search technique based on principles of evolution and heredity. Hence, in our context a formula for determining the implied volatility based on American put options should itself evolve during the evolution process. Using experimental data sets we can show that the genetically derived formula for calculating implied volatilities based on American put options provides accurate approximation results.

In the second section we concentrate on implied volatility models. In the third section we focus on the concept of Genetic Programming. The fourth section presents the genetically derived approximation for calculating implied voltaltilities based on American put options. In the fifth section we show the experimental results. The paper concludes with a summary and a few closing remarks.

[1] An extension for the dividend paying case is straightforward.

21.2 The Implied Volatility Model

21.2.1 The Classical Approaches

The classical Black/Scholes–Merton option pricing model for an European call on a non–dividend paying stock is given as ([9,30])

$$c_0 = f(r, S_0, \sigma, T, X) = S_0 \Phi(d_1) - X e^{-rT} \Phi(d_2), \tag{21.1}$$

where

$$d_1 = \frac{\ln\left(\frac{S_0}{X}\right) + (r + \frac{\sigma^2}{2})T}{\sigma\sqrt{T}}, \qquad d_2 = d_1 - \sigma\sqrt{T}$$

and the parameters in the formula are defined as follows:

- c_0 the premium for the call option at $t = 0$.
- $\Phi(d)$ the cumulative normal distribution function evaluated at d.
- r the annual continuous risk–free interest rate in %.
- S_0 the stock price at $t = 0$.
- σ the annual volatility of the stock price in %.
- T the time to expiration in years.
- X the exercise price.

Using the relationship $c_0 = f(r, S_0, \sigma, T, X)$ given in (21.1), the exact implied volatility is denoted by σ_{BSM}, which for given values of c_0, r, S_0, T, and X satisfies

$$c_0 = f(r, S_0, \sigma_{\text{BSM}}, T, X). \tag{21.2}$$

A simple analytical approximation for calculating the implied volatility was presented by Brenner and Subrahmanyam [11]. Their formula is given as

$$\sigma \approx \sigma_{\text{BS}} = \sqrt{2\pi} \frac{1}{\sqrt{T}} \frac{c_0}{S_0} \tag{21.3}$$

and delivers accurate approximation results if the stock price is exactly equal to the discounted exercise price. However, stock prices are rarely exactly equal to a discounted exercise price, so that many more ambitious analytical approximations have been developed. The best–known of these come from Bharadia, Christofides, and Salkin [8] and Corrado and Miller [16].

Bharadia, Christofides, and Salkin's formula,

$$\sigma \approx \sigma_{\text{BCS}} = \frac{1}{\sqrt{T}} \left\{ \sqrt{2\pi} \left(\frac{C_0 - \frac{S_0 - X'}{2}}{S_0 + X'} \right) + \sqrt{2\pi \left(\frac{C_0 - \frac{S_0 - X'}{2}}{S_0 + X'} \right)^2 - \frac{2(S_0 - X') \ln\left(\frac{S_0}{X}\right)}{S_0 + X'}} \right\}, \tag{21.4}$$

is based on a quadratic approximation method and provides accurate approximations for the implied volatility for close–to–the money options. In (21.4) and hereafter X' denotes the discounted exercise price ($X' = X e^{-rT}$).

Corrado and Miller, however, showed that the accuracy of the quadratic formula in (21.4) can be significantly improved by minimizing its concavity. Accordingly, Corrado and Miller derived the approximation

$$\sigma \approx \sigma_{\mathrm{CM}} = \frac{1}{\sqrt{T}} \frac{\sqrt{2\pi}}{S_0 + X'} \left(C_0 - \frac{S_0 - X'}{2} + \sqrt{\left(C_0 - \frac{S_0 - X'}{2} \right)^2 - \frac{(S_0 - X')^2}{\pi}} \right) \tag{21.5}$$

which provides nearly perfect approximations for option maturities of three months or more and stock prices within ±10 percent of a discounted strike price. Accurate approximation results are also given for shorter maturities, i.e., one month, and stock prices within ±5 percent of a discounted strike price.

Keber [24] abandons traditional techniques in constructing approximations and uses the Genetic Programming approach to derive analytical approximations for determining the implied volatility based on the Black/Scholes–Merton model. Keber derived the approximation

$$\sigma \approx \sigma_{\mathrm{K}} = \frac{1}{\sqrt{T}} \frac{\sqrt{2\pi}}{\mathcal{S}_0 + \mathcal{X}'} \left\{ \mathcal{C}_0 - \frac{\mathcal{S}_0 - \mathcal{X}'}{2} + \frac{1}{\pi(\mathcal{C}_0^{\mathcal{C}_0} - r)} + \sqrt{\left(\mathcal{C}_0 - \frac{\mathcal{S}_0 - \mathcal{X}'}{2} + \frac{(\mathcal{S}_0 - T - \mathcal{X}')^2}{2\pi[2(\mathcal{S}_0 - \mathcal{X}) + \mathcal{X}']} \right)^2 - \frac{(\mathcal{S}_0 - \mathcal{X}')^2}{\pi}} \right\} \tag{21.6}$$

with $\mathcal{S}_0 = 240, \mathcal{X} = \mathcal{S}_0 X / S_0, \mathcal{X}' = \mathcal{X} e^{-rT}, \mathcal{C}_0 = \mathcal{S}_0 c_0 / S_0$, and stated that for option maturities of one month (three months or more) and stock prices within ±5 (±10) percent of a discounted strike price the approximation (21.6) outperforms Bharadia, Christofides, and Salkin's as well as Corrado and Miller's approximation formulas.

21.2.2 The Implied Volatility Based on American Puts

Although options are frequently American puts researchers typically rely on implied volatilities based on the Black/Scholes–Merton model (see, e.g., [18,19,32] and [33]). Since the American put option price is not less than the price of a corresponding European–style option, the implied volatility so obtained is biased upwards.

As mentioned in the introduction, American put options on stocks paying no dividends during the time to expiration have early exercise premiums implicitly embedded in their prices. Therefore, American put options are priced either by using numerical procedures or analytical approximations. Let

$$P_0 = g(r, S_0, \sigma, T, X) \tag{21.7}$$

be the price of an American put option on a stock paying no dividends during the time to expiration calculated by using the lattice approach [17], the finite difference method [10] or an analytical approximation (e.g., [4,20,23,25,28]). For given values of P_0, r, S_0, T, and X the exact implied volatility σ^* based on American put options satisfies

$$P_0 = g(r, S_0, \sigma^*, T, X). \tag{21.8}$$

Unfortunately, closed–form solutions as well as analytical approximations for solving (21.8) are unknown. The common numerical method used for solving (21.8) is the Newton–Raphson method. However, this can be very time–consuming when used in connection with the lattice approach or the finite difference method.

21.3 Genetic Programming

The Genetic Programming (GP) methodology comes from Koza [26] and is based on genetic algorithms which were originally presented by Holland [22]. Genetic Programming is an approach designed to generate computer programs as well as formulas using a random oriented search technique based on principles of natural evolution and heredity. The fundamental philosophy of Genetic Algorithms lies in the imitation of the strategy that determines the evolution process: The fittest survive and pass their genetic material, and thus their strengths, to the next generation. By imitating this process we get individuals which are ever more suited to their environment. In our GP–application we start with a defined programming language $\mathcal{L}$ (see, e.g., [1]). $\mathcal{L}$ is to be seen simply as the set of all analytical expressions which can be produced from a start symbol $\mathcal{S}$ under application of *substitution rules* $\mathcal{R}$ and a finite set or vocabulary of *terminal symbols* $\mathcal{T}$. Thus,

$$\mathcal{L} = \{c \mid \mathcal{S} \Longrightarrow c \ \wedge \ c \in \mathcal{T}^*\} \tag{21.9}$$

where $\mathcal{T}^*$ represents the set of all analytical expressions which can be produced from the symbols of the vocabulary $\mathcal{T}$. In the sense of Genetic Programming, $\mathcal{L}$ is the *search space* of all potential analytical expressions to be generated, $c \in \mathcal{L}$ is a *chromosome* or *individual*, and $\mathcal{P}_\tau \subset \mathcal{L}$ is the *population* of the τ–th generation ($\tau = 0, \ldots, \tau_{\max}$). Furthermore, the terminal symbols contained in c represent the *genes* of the chromosome c.

The genetic structure of an individual is evaluated by using the *fitness concept*. Fitness is here used as a measure of suitability of the individual to its environment. Each chromosome (analytical expression, computer program) $c \in \mathcal{L}$ can be seen as a function $c: \mathcal{E}_i \to \mathcal{A}_i$ because c transforms input data $\mathcal{E}_i$ into the solution or output data $\mathcal{A}_i$. Related to this, a fitness function $f: \mathcal{A} \to \mathbb{R}$ has to be defined in a way that it awards a higher fitness value to those individuals which represent a good solution to the task in hand, and a lower fitness value to less suitable individuals. The fitness is usually measured using a representative set of test records $\mathcal{E}_i$, for $i = 1, \ldots, n$. If these input data are processed using the program $c \in \mathcal{P}_\tau$, the result will be the output data $\mathcal{A}_i$, which can then be compared to the target output data $\mathcal{A}_i^S$. Using a deviation function $\Delta(\mathcal{A}_i, \mathcal{A}_i^S)$ in a way that it delivers higher values the more the output data differ from the target output data, the aggregated deviation is the so-called *raw fitness* f_r of the program c:

$$f_r(c) = \sum_{i=1}^{n} \Delta(\mathcal{A}_i, \mathcal{A}_i^S) \quad \text{with} \quad \mathcal{A}_i = c(\mathcal{E}_i). \tag{21.10}$$

The raw fitness function delivers smaller fitness values the better the individual c accomplishes its tasks. As in most applications, the raw fitness is modified in a way that the fitness values lie between zero and one, and larger fitness values represent better individuals (*adjusted fitness*).

A high fitness value expresses the ability of an individual to produce offspring more frequently (*fitness–proportional selection*), whereby the genetic material is passed on to the following generation (*reproduction/recombination*). The direct result of this is that genes which are advantageous to the individual in its environment are inherited more frequently than those from less well–suited individuals, so that in the course of evolution, populations arise which are ever more suited to their environment. Accordingly, a new population $\mathcal{P}_{\tau+1}$ is formed by selecting $|\mathcal{P}_\tau|$ individuals proportionally to their fitness and altering some survivors using *crossover* and *mutation*. By applying crossover – based on a given crossover rate p_c – we select pairs of formulas (individuals), swap randomly selected sub–expressions and get a new pair of offspring. By using mutation – based usually on a low mutation rate p_m – we alter formulas (chromosomes) by exchanging operators or operands (genes) randomly. Crossover as well as mutation can be interpreted as search operators. The crossover operator crosses between two individuals and enables the exchange of genetic material; the task of the mutation operator consists primarily of opposing a premature convergence on local optima. As formulated in Holland's *schema theorem* [22] – a proof is given in [21] – the performance of Genetic Programming can be measured as the ability of individuals to make themselves better suited to the environment (by fitness–proportional selection, reproduction, and recombination) with over–proportional speed relative to the number of generations.

21.4 Genetic Determination of Implied Volatilities

21.4.1 Implementation and Parameter Specification

By applying the Genetic Programming approach to derive approximations for calculating implied volatilities based on American put options we have to specify some parameters. In the fitness concept we used a randomly generated training sample of 1,000 American put options on non–dividend paying stocks described by the tuple $\langle P_0, r, S_0, \sigma, T, X\rangle$ which satisfies (21.7). Each option price P_0 was calculated by using the finite difference method[2] and served as the exact value of the American puts. The corresponding parameters were drawn randomly based on uniform distributions. In accordance with the literature as well as realistic circumstances we defined the domains shown in Table 21.1.

Table 21.1. Parameter domains used in the GP–application

Parameter	α	r	σ	T	X
Lower bound	0.9	2.00	10	0.00274	10.00
Upper bound	1.1	12.00	50	0.33333	200.00

In Table 21.1 the parameter α describes the moneyness of the option ($\alpha = S_0/X$). The α–domain is based on the consideration that option trading always starts near–the–money. Additionally, Barone–Adesi and Whaley's study [3] shows that the moneyness lies between 0.9 and 1.1 in approximately 67 % of cases in a sample of 697,733 stock option transactions. Furthermore, Stephan and Whaley [34] look at a sample of 950,346 stock option transactions and report that the moneyness is between 0.9 and 1.1 in approximately 78 % of cases.

In accordance with the Genetic Programming approach we transformed each tuple $\langle P_0, r, S_0, \sigma, T, X\rangle$ into an input data record $\mathcal{E}_i$ ($\widehat{=} \langle P_0, r, S_0, T, X\rangle$) and a corresponding target output data record $\mathcal{A}_i^S$ ($\widehat{=} \langle\sigma\rangle$), for $i = 1, \ldots, 1000$. For the fitness function we used the sum of the squared errors

$$f_r(c) = \sum_{i=1}^{1000} (\mathcal{A}_i - \mathcal{A}_i^S)^2$$

where $\mathcal{A}_i = c(\mathcal{E}_i)$ as described in the previous section. In the terminal symbol set we included the variables P_0, r, S_0, T, X, ephemeral random constants as the possible operands, and the commonly used mathematical operators $+$, $-$, $*$, $\div$, $\sqrt{x}$, $\ln(x)$, x^2, and x^y. Furthermore, we defined a population size of 50 individuals, a crossover rate $p_c = 0.9$, a mutation rate $p_m = 0.001$, and stopped the evolution process after 70,000 generations. With this combination of values, a balanced relationship between convergence speed, calculation effort and effectiveness of the genetic programming was obtained.

[2] The finite difference method was used with $\Delta t = (1/365)/5 = 0.00054795$, $\Delta S = 0.01$, and $S_{\max} = 2S_0$.

21.4.2 The Genetically Derived Analytical Approximation

Applying the Genetic Programming approach as described a raw fitness of 0.0105464 is obtained for the best individual after $\tau_{\max} = 70{,}000$ generations. The analytical expression of this individual represents an approximation σ_{proxy} for calculating the implied volatility based on American put options on non–dividend paying stocks:

$$\sigma \approx \sigma_{\text{proxy}} = \frac{a_0}{S_0 + \left[S_0 e^{\gamma} - (S_0 - X e^{\kappa})/2\right] e^{\nu}} \left(\frac{S_0 - X'}{2} + P_0 + \sqrt{P_0^2 + P_0\left(S_0 - X e^{-2r^2T^2}\right) - 2Tr^2} \right) \frac{1}{\sqrt{T}} \tag{21.11}$$

The parameters contained in (21.11) are as follows:

$$\gamma = rT \left[\frac{S_0 - X - r}{2P_0} - \frac{1}{\sqrt{T}} - 1 \right],$$

$$\kappa = r \left[\frac{a_1 \left(S_0 - X e^{r^2}\right)}{P_0} - \left(\frac{1}{2} + a_2 \frac{\sqrt{X} - \sqrt{r} + \sqrt{T} - P_0 - a_0}{S_0 T} \right) \right],$$

$$\nu = \frac{\lambda}{\mu} \frac{P_0}{S_0} - rT, \quad \lambda = P_0 + \frac{a_0}{S_0} + \frac{2r - T}{a_3^2}, \quad \mu = \frac{2 S_0^2 T}{S_0 + 2 P_0 \sqrt{a_0}},$$

$$a_0 = 2.5066282746, \quad a_1 = 0.4203491482,$$
$$a_2 = 0.2320584397, \quad a_3 = 0.3188095643.$$

The above formula is as derived by the Genetic Programming algorithm without possible simplification.

21.5 Experimental Results

To give a first impression of the accuracy of the genetically derived analytical approximation for calculating implied volatilities based on American put options on non–dividend paying stocks we use the data sets of Corrado and Miller [16] and Barone–Adesi and Whaley [5]. Furthermore, we use huge samples of randomly generated American put options (validation data sets). This ensures that the assessment of the approximations will be highly accurate.

For the *method–related* comparison we use implied volatilities based on the Black/Scholes–Merton model which are sometimes used to approximate implied volatilites based on American put options (see, e.g., [16]).

In the following Tables 21.2 and 21.3 the genetically derived approximation σ_{proxy} as well as the approximation σ_{BSM} are applied to the data sets of

Corrado and Miller [16] and Barone–Adesi and Whaley [5], respectively. σ denotes the exact implied volatilitiy, and at the end of each table the three error measures mean absolute deviation (MAD), mean squared deviation (MSD) and mean absolute percentage deviation (MAPD) are given for each approximation. The approximation results can be summarised as follows:

Table 21.2. Implied volatilities based on American put options. Approximations are calculated using (21.11) and (21.2) adapted for put options. Examples are based on data from Corrado and Miller [16, p. 601]

r	T	S_0	X	P_0	σ	σ_{proxy}	σ_{BSM}
4.00	$0.08\dot{3}$	398.67	390.00	2.976	15.00	15.183	15.077
4.00	$0.08\dot{3}$	398.67	400.00	6.987	15.00	14.669	15.219
4.00	$0.08\dot{3}$	398.67	410.00	13.327	15.00	14.838	15.673
4.00	0.250	396.02	390.00	7.629	15.00	14.785	15.299
4.00	0.250	396.02	400.00	12.268	15.00	14.522	15.534
4.00	0.250	396.02	410.00	18.334	15.00	14.453	15.981
MAD (in %)						0.319	0.464
MSD (in %)						0.124	0.307
MAPD (in %)						2.182	2.962

- When we look at the two data sets of Corrado and Miller as well as Barone–Adesi and Whaley it can be seen that σ_{BSM} delivers less accurate approximations having deviations to the value of around 11 percentage points. The mean absolute deviations are about 0.5 and 2.7 percentage points, respectively. As previously mentioned, the implied volatilities based on σ_{BSM} are biased upwards. In comparison, the genetically derived approximations are roughly speaking better than the σ_{BSM}–based ones because their deviations are at most about 3 percenatge points. The mean absolute deviations based on the approximation σ_{proxy} are about 0.3 and 0.8 percentage points, respectively.

The application results just shown cannot be used for general statements about the accuracy of the genetically derived approximation formula because the underlying data sets are too small. If we use the 1,000 data records used in the genetic programming approach as a basis for a general assessment the problem arises that this data sample represents training data, and thus the assessment would be open to criticism for being a "self–fulfilling prophecy". Therefore, the definitive judgement of the approximation formula is to be made from the above mentioned validation data sets which have a much larger scope and are independent of the training data set, too.

The first validation data set (VDS–I) consists of 37,000 randomly generated American put options and the domains of the corresponding parameters

Table 21.3. Implied volatilities based on American put options. Approximations are calculated using (21.11) and (21.2) adapted for put options. Examples are based on data from Barone–Adesi and Whaley [5, p. 315]

r	T	S_0	X	P_0	σ	σ_{proxy}	σ_{BSM}
8.00	0.25	90.00	100.00	10.035	20.00	20.190	26.931
8.00	0.25	100.00	100.00	3.222	20.00	19.712	20.955
8.00	0.25	110.00	100.00	0.664	20.00	18.756	20.223
12.00	0.25	90.00	100.00	10.000	20.00	22.936	31.197
12.00	0.25	100.00	100.00	2.922	20.00	20.124	21.573
12.00	0.25	110.00	100.00	0.554	20.00	16.780	20.370
8.00	0.25	90.00	100.00	12.563	40.00	40.143	42.316
8.00	0.25	100.00	100.00	7.105	40.00	39.957	40.864
8.00	0.25	110.00	100.00	3.697	40.00	39.835	40.390
8.00	0.50	90.00	100.00	10.289	20.00	20.413	26.042
8.00	0.50	100.00	100.00	4.188	20.00	20.102	21.517
8.00	0.50	110.00	100.00	1.409	20.00	19.577	20.525
MAD (in %)						0.774	2.742
MSD (in %)						1.668	17.680
MAPD (in %)						3.798	12.966

are shown in Table 21.4. The numerically exact put option price P_0 was calculated using the finite difference method[3] and satisfies (21.7).

Table 21.4. Parameter domains used in the validation data set

Parameter[4]	α	r	σ	T	X
Lower bound	0.9	2.00	10	0.00274	10.00
Upper bound	1.1	12.00	50	0.33333	100.00

Based on the validation data set VDS–I the graph in Fig. 21.1 shows the accuracy of the genetically derived approximation σ_{proxy} as well as the σ_{BSM}–approximation in terms of cumulated frequencies of the absolute deviations between the true implied volatility σ and the approximations σ_{proxy} and σ_{BSM}.

- If we focus on the 95 % probability we are able to say that the σ_{BSM}–approximation always causes absolute deviations less than about

[3] Here and hereafter, the finite difference method was used with $\Delta t = (1/365)/5 = 0.00054795$, $\Delta S = 0.01$, and $S_{\max} = 2S_0$.

[4] The domains of X and S_0 are less important because the option price P_0 is linear homogeneous in X and S_0.

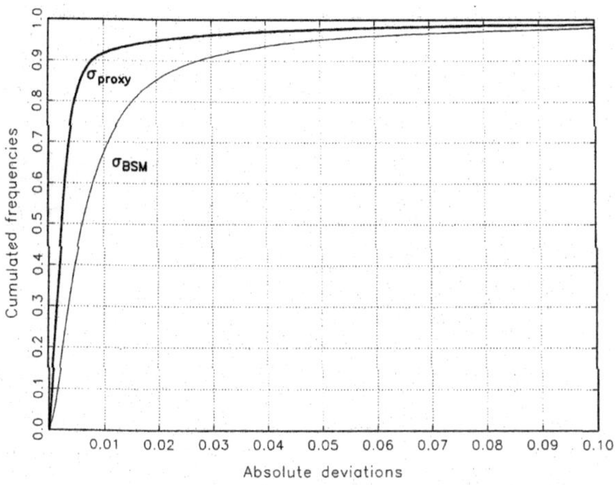

Fig. 21.1. Cumulated frequencies of the absolute deviations between the true implied volatility σ and the approximations (21.11) and (21.2) adapted for put options

4.6 percentage points. Focusing on the 90 % probability the σ_{BSM}–approximation always provides absolute deviations less than about 2.8 percentage points.

- In comparison, the genetically derived formula σ_{proxy} delivers better approximation than σ_{BSM}. The absolute deviations caused by the genetically derived formula are always less than about 2 percentage points – if we focus on the 95 % probability – and less than about 0.7 percentage points if we look at the 90 % probability.

Although the investigation just shown is very informative with respect to the accuracy of the approximations the analysis cannot be used for general statements about possible deviation patterns caused by the approximation formulas. Thus, we now investigate whether the approximation formulas cause systematic deviations. In doing so, we can reveal assumptions for using the formulas properly, too. For these purposes we calculate implied volatilities systematically with respect to the different parameters.

The graphs in Fig. 21.2 show for

- an annual continuous riskless interest rate of 5 %,
- a moneyness range of 0.9 to 1.1, and
- maturities between two weeks and four months

true implied volatilities ($\sigma = 20, 30, 40$, and 50 %, respectively) as well as their approximations ($\sigma_{proxy}, \sigma_{BSM}$). The horizontal shapes refer to the true implied volatilities, the dark surfaces represent the σ_{BSM}–approximations, and the bright ones belong to the genetically derived approximations (σ_{proxy}).

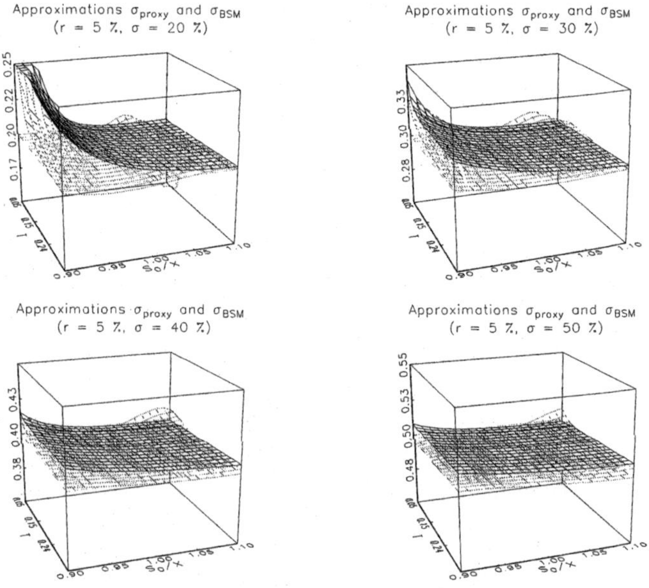

Fig. 21.2. True implied volatilities and approximation results based on (21.11) and (21.2) adapted for put options

The graphs in Fig. 21.3 are based on an annual continuous riskless interest rate of 7 %, the remaining parameters, however, are the same as the ones used in Fig. 21.2.

- When we look at the graphs in Figs. 21.2 and 21.3 we can identify for short maturity and (relatively deep) in–the money options that the deviations between the true implied volatility and the approximations σ_{BSM} and σ_{proxy} are the higher the smaller the true implied volatility. This can be seen clearly from each figure when the upper left graph is compared to the remaining threes.
- The sensitivity analysis with respect to the annual continuous riskless interest rate – based on the graphs in Figs. 21.2 and 21.3 as well as further investigations not shown here – can be summarized as follows: The deviations caused by the σ_{BSM}–approximation are the higher the more the annual continuous riskless interest rate increases. Although this is also true for the genetically derived approximation the dependencies are significantly smaller.
- As already mentioned, the implied volatilities based on σ_{BSM} are always biased upwards. The implied volatilities based on σ_{proxy} are also biased upwards except for short maturity and (relatively deep) out–

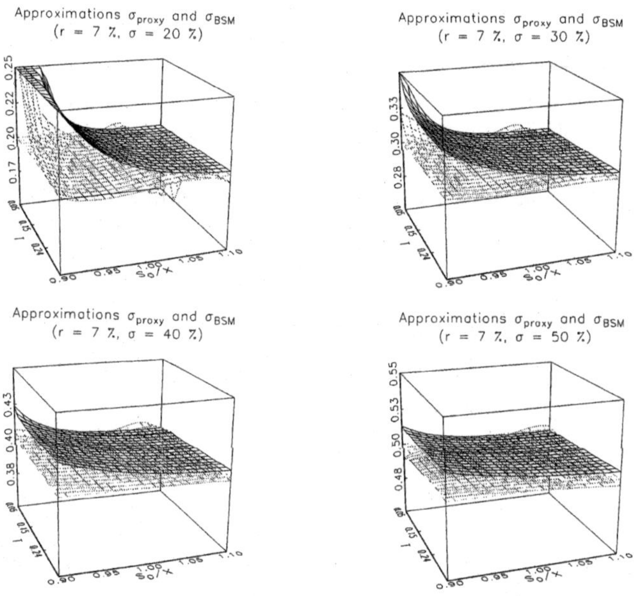

Fig. 21.3. True implied volatilities and approximation results based on (21.11) and (21.2) adapted for put options

of–the money options. However, it must be mentioned that the genetically derived formula generally causes smaller deviations than the σ_{BSM}–approximation.

The investigation just shown can be used to specify assumptions for using the genetically derived approximation properly. For calculating implied volatilities based on American put options on non dividend paying stocks, σ_{proxy} can be used to provide nearly perfect approximations

- for option maturities of about one and a half months or more and stock prices within ±10 percent of a strike price.
- For shorter maturities, say, one month, the approximations are also nearly perfect when the lower bound of the range is reduced to about −5 percent of a strike price.

What are the approximation results when the parameters used in the formulas are chosen in accordance with these assumptions? To show the accuracy of the genetically derived approximation as well as the σ_{BSM}–approximation we use a second validation data set (VDS–II) which is independent of the validation data set VDS–I. VDS–II consists of 20,000 randomly generated American put options on non dividend paying stocks. P_0 was calculated using the finite difference method and the corresponding parameters reflect option maturities of one month or more and stock prices within −5 and +10 percent of a strike price. The domains of the annual continuous riskless interest rate, the annual volatility, and the strike price are the same as the ones used in VDS–I.

The graph in Fig. 21.4 shows the accuracy of the approximations σ_{BSM} and σ_{proxy} applied to options whose parameters are in accordance with the VDS–II domains[5]. The accuracy is shown in terms of cumulated frequencies of the absolute deviations between the true implied volatility σ and the approximations σ_{proxy} and σ_{BSM}.

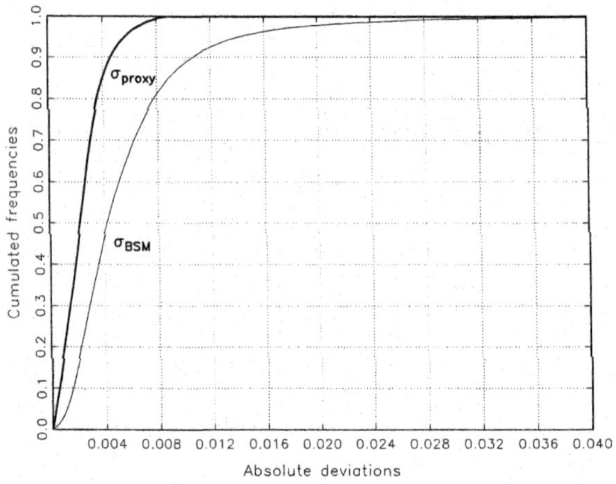

Fig. 21.4. Cumulated frequencies of the absolute deviations between the true implied volatility σ and the approximations (21.11) and (21.2) adapted for put options

- First of all it can be said with almost 100 % probability that the absolute deviations between the true implied volatilities and the genetically derived approximations are always less than about 0.8 percentage points. With respect to the σ_{BSM}–formula, however, deviations to this level are only found in about 81 % of cases.
- Focusing on the 95 and 90 % probabilities the σ_{BSM}–approximation always causes absolute deviations less than about 1.4 and 1.1 percentage points, respectively.
- In comparison, the genetically derived formula provides better approximations. If we focus on the 95 and 90 % probabilities we are able to say

[5] The accuracy of the approximation results with respect to option maturities of one and a half months or more and stock prices within ±10 percent of a strike price are similar to the one presented in Fig. 21.4, they are not shown.

that the absolute deviations caused by genetically derived formula are always less than about 0.6 and 0.5 percentage points, respectively.

The above analyses show that the genetically derived approximation σ_{proxy} for determining implied volatilities based on American put options on non–dividend paying stocks delivers better results, i.e., smaller deviations from the true implied volatility, than the σ_{BSM}–approximation. For a wide range of parameter values of the underlying option, the genetically derived formula provides accurate approximations. Nearly perfect approximations are given for option parameter values which are in accordance with specific assumptions. Restricting the formula to these assumptions, however, is consistent with current practice ([16]). Finally, the genetically derived formula has the additional advantage that it is a purely analytical approximation.

21.6 Concluding Remarks

In this paper we have shown that the Genetic Programming methodology can be used to derive accurate analytical approximations for determinig implied volatilites based on American put options on non–dividend paying stocks. Using sample data sets from the literature, huge validation data sets, and systematic sensitivity analyses it was possible to show that the genetically derived formula provides accurate approximations for a wide range of parameter values of the underlying option and nearly perfect approximations for parameter values following specific assumptions which are consistent with current practice.

We introduced Genetic Programming as a data–driven, non–parametric method for determining implied volatilities. In traditional parametric approaches, the derivation of a model is closely bounded to those parameters that model the stochastic price process of the underlying asset. Incorrect specifications for these parameters or the stochastic price process will lead to systematic errors in the valuation model. On the other hand, correct specifications lead to correct valuation models. A lot of such models, however, are not amenable to exact analytical solutions. In comparison, Genetic Programming methods need only a few assumptions with respect to the parameters mentioned and are rather insensitive to the effects of specification errors in these parameters. As any mathematical or statistical function can be included in the set of terminal symbols, Genetic Programming can be seen as a flexible instrument to derive valuation models for a broad variety of derivatives and associated stochastic price processes of the underlying asset. Exotic options and derivatives where the stochastic price process of the underlying asset is only given implicitly (e.g., by historical prices) should be mentioned here in particular as a field of application. If the stochastic price process of the underlying asset can be described correctly by appropriate parameters with the result that an analytically exact solution of the parametric model can be found, it is obvious that this solution is to be preferred to genetically derived one. However, is should be noted that such circumstances occur rarely enough, so that Genetic Programming can be seen as a reasonable alternative method to derive accurate analytical valuation models.

References

1. Aho A. V., Ullman J. D. (1972) The Theory of Parsing, Translation, and Compiling, Volume I: Parsing. Prentice–Hall, Englewood Cliffs
2. Allen F., Karjalainen R. (1999) Using Genetic Algorithms to Find Technical Trading Rules. Journal of Financial Economics **51**, 245–271
3. Barone–Adesi G., Whaley R. E. (1986) The Valuation of American Call Options and the Expected Ex–dividend Stock Price Decline. Journal of Financial Economics **17**, 91–111
4. Barone–Adesi G., Whaley R. E. (1987) Efficient Analytic Approximation of American Option Values. Journal of Finance **42**, 301–320
5. Barone–Adesi G., Whaley R. E. (1988) On the Valuation of American Put Options on Dividend–Paying Stocks. Advances in Futures and Options Research **3**, 1–13
6. Beckers S. (1981) Standard Deviations Implied in Option Prices as Predictors of Future Stock Price Volatility. Journal of Banking & Finance **5**, 363–381
7. Bharadia M. A. J., Christofides N., Salkin G. R. (1995) Computing the Black–Scholes Implied Volatility: Generalization of a Simple Formula. Advances in Futures and Options Research **8**, 15–29
8. Bharadia M. A. J., Christofides N., and Salkin G. R. (1996) A Quadratic Method for the Calculation of Implied Volatility Using the Garman–Kohlhagen Model. Financial Analysts Journal **52**, 61–64
9. Black F., Scholes M. (1973) The Pricing of Options and Corporate Liabilities. Journal of Political Economy **81**, 637–659
10. Brennan M., Schwartz E. (1977) The Valuation of American Put Options. Journal of Finance **32**, 449–462
11. Brenner M., Subrahmanyam M. G. (1988) A Simple Formula to Compute the Implied Standard Deviation. Financial Analysts Journal **44**, 80–83
12. Chance D.M. (1996) A Generalized Simple Formula to Compute the Implied Volatility. The Financial Review **31**, 859–867
13. Chen S. -H., Lee W. -C., Yeh C. -H. (1999) Hedging Derivative Securities with Genetic Programming. International Journal of Intelligent Systems in Accounting, Finance and Management, Vol. 8, No. 4, 237-251.
14. Chidambaran N. K., Lee C. W. J., Trigueros J. R. (2000) Option Pricing via Genetic Programming. In: Abu–Mostafa Y. S., LeBaron B., Lo A. W., Weigend A. S. (Eds.) Computational Finance – Proceedings of the Sixth International Conference. Leonard N. Stern School of Business, January 1999. The MIT Press, Cambridge, Massachusetts
15. Chiras D. P., Manaster S. (1978) The Information Content of Option Prices and a Test of Market Efficiency. Journal of Financial Economics **6**, 213–234
16. Corrado C.J., Miller Jr T. W. (1996) A Note on a Simple, Accurate Formula to Compute Implied Standard Deviations. Journal of Banking & Finance **20**, 595–603

17. Cox J., Ross S., Rubinstein M. (1979) Option Pricing: A Simplified Approach. Journal of Financial Economics **7**, 229–263
18. Day T. E., Lewis C. M. (1988) The Behaivor of the Volatility Implicit in the Prices of Stock Index Options. Journal of Financial Economics **22**, 103–122
19. Day T. E., Lewis C. M. (1992) Stock Market Volatility and the Information Content of Stock Index Options. Journal of Econometrics **52**, 267–287
20. Geske R., Johnson H. E. (1984) The American Put Option Valued Analytically. Journal of Finance **39**, 1511–1524
21. Goldberg D.E. (1989) Genetic Algorithms in Search, Optimization, and Machine Learning. Addison Wesley, Reading, Massachusetts
22. Holland J. H. (1975) Adaption in Natural and Artificial Systems. The University of Michigan Press, Ann Arbor
23. Johnson H.E. (1983) An Analytic Approximation for the American Put Price. Journal of Financial and Quantitative Analysis **18**, 141–148
24. Keber C. (1999) Genetically Derived Approximations for Determining the Implied Volatility. OR Spektrum **21**, 205–238
25. Keber C. (2000) Option Valuation with the Genetic Programming Approach. In: Abu–Mostafa Y. S., LeBaron B., Lo A. W., Weigend A. S. (Eds.) Computational Finance – Proceedings of the Sixth International Conference. Leonard N. Stern School of Business, January 1999. The MIT Press, Cambridge, Massachusetts
26. Koza J.R. (1992) Genetic Programming. On the Programming of Computers by Means of Natural Selection. The MIT Press, Cambridge, Massachusetts
27. Latane H. A., Rendleman Jr R. J. (1976) Standard Deviations of Stock Price Ratios Implied in Option Prices. Journal of Finance **31**, 369–381
28. MacMillan L. W. (1986) Analytic Approximation for the American Put Option. Advances in Futures and Options Research **1**, 119–139
29. Manaster S., Koehler G. (1982) The Calculation of Implied Variances from the Black–Scholes Model: A Note. Journal of Finance **37**, 227–230
30. Merton R. C. (1973) Theory of Rational Option Pricing. Bell Journal of Economics and Management Science **4**, 141–183
31. Neely C., Weller P., Dittmar R. (1997) Is Technical Analysis in the Foreign Exchange Market Profitable? A Genetic Programming Approach. Journal of Financial and Quantitative Analysis **32**, 405–426
32. Resnick B. G., Sheikh A. M., Song Y. S. (1993) Seasonality and Calculation of the Weighted Implied Standard Deviation. Journal of Financial and Quantitative Analysis **28**, 417–430
33. Schwert G. W. (1990) Stock Market Volatility and the Crash of '87. Review of Financial Studies **3**, 77–102
34. Stephan J. A., Whaley R. E. (1990) Intraday Price Changes and Trading Volume Relations in the Stock and Stock Option Markets. Journal of Finance **45**, 191–220
35. Szpiro G. G. (1997) A Search for Hidden Relationships: Data Mining with Genetic Algorithms. Computational Economics **10**, 267–277

Part V

Bibliography

22 Evolutionary Computation in Economics and Finance: A Bibliography

Shu-Heng Chen[1] and Tzu-Wen Kuo[2]

[1] AI-ECON Research Center, Department of Economics, National Chengchi University, Taipei, Taiwan, 11623
chchen@nccu.edu.tw
[2] AI-ECON Research Center, Department of Economics, National Chengchi University, Taipei, Taiwan, 11623
g7258502@grad.cc.nccu.edu.tw

Abstract. This chapter presents a bibliography on the application of evolutionary computation to economics and finance. Publications included in this bibliography are classified by application domain, published journal or conference proceedings. Information on some useful websites and software is also provided.

22.1 Introduction

While the goal of editing a single volume like this one is to give a comprehensive coverage of the field, there is a fundamental limit which one cannot transcend, and that is the size of the book. Normally speaking, 20 chapters already comes to the maximum capacity of a single volume. Anything beyond has to be reluctantly dropped out. Therefore, to compensate for what we might be missing, we add a chapter of a bibliography as a concluding chapter, which is composed of three parts: publications, useful web sites, and computer software. Complimentary guidance is also provided to enable readers to take full advantage of this bibliography.

22.2 Publications by Application Domains

More than 380 publications were produced in this field from 1986 to the early 2001.[1] Most of them are listed in the references. While many are *papers* from journals, proceedings, and chapters in books, *books* on this subject are also available([45], [54], [146], [200], [238], [324], [371]). One useful way to have a quick grasp of these publications is to divide them by sub-fields, as shown in Sect. 22.2.7.

Papers on general discussion of the relevance of evolutionary computation (**EC**) to economic modeling fall into the category of **General**. These papers

[1] While we have done our best to search for all possible publications, the list is by no means exhaustive. In particular, it is very difficult to get all publications in conference proceedings.

basically state why evolutionary computation can be a useful tool to model learning and adaptation or to capture the essence of bounded rationality in economics ([20]). Notions of artificial adaptive agents are well established in these papers ([199], [350].) The connection among evolutionary computation, evolutionary economics, and agent-based computational economics are well discussed in ([353], [67].)

22.2.1 Games and Industrial Organization

Games and Industrial Organization is one of the most active application domains of evolutionary computation. This is not surprising given the fact that this area suffers most from the criticism of *bounded rationality.* Among the various kinds of games studied in game theory, **repeated prisoner's dilemma (RPD)** draws most intensive attention from EC researchers. The RPD game is a very standard model in game theory. The issue addressed here is simple: *Under a conflicting situation, is cooperation possible and stable?* There is, of course, no simple answer even in a two-person case. But, using EC, the 28 papers cited in the references spell out the complicate determinants of this simple issue.

Important as they are, RPD games are not that realistic. **Oligopoly games**, a non-trivial extension of RPD games, bring us closer to the real world. Issues studied in oligopoly games are varied. Almost all standard models in industrial organization have been taken by researchers, which include Cournot competition, Bertrand competition, Stackelberg competition, the chain of monopolies model, competition with the capacity constraint, price wars, and predatory pricing. Other related issues such as price dispersion and product loyalty are also investigated.

Another non-trivial extension of RPD games to real situations is the **ultimatum games**, which may serve as the starting point to approach two more complicated and intriguing games, namely, **bargaining and auction games**. The latter is particularly important in the era of **Electronic Commerce** ([159]). Recently, applications of EC were extended to these games as well.

22.2.2 Macroeconomics

The earliest application of EC to macroeconomic modeling appeared in 1990, in the case of the Kiyotake-Wright model ([259]), which replicated the emergence of money from a barter economy. Throughout the decade, the study found fervent followers, among the more recent ones, [33], [337], and [156].

In addition to the Kiyotake-Wright model, there are two popular prototypes under intensive study. One is the *cobweb model*, and the other the *overlapping generation* (**OG**) *model.* The cobweb model is often taken by EC users as a standard model to study the stability of the *rational expectations equilibrium.* The OG model, on the other hand, is taken to exemplify how the *indeterminacy* (*equilibrium selection*) problem can be tackled by EC. There are three different types of the OG model to which EC is applied, namely, the Samuelson-Allais type, Kareken-Wallace type, and Azariadis-Drazen type. Application to the Samuelson-Allais type concerns the determination of inflation dynamics, whereas application to the Kareken-Wallace type concerns the determination of exchange rate dynamics.[2] As for the Azariadis-Drazen type, EC application involves the replication of the emergence of industrial revolution and economic growth ([16]), and in this bibliography we put it under the next sub-category: **Growth, Cycles and Transition**.

Applications collected in this sub-category are quite diverse. EC is used as a numerical technique to solve growth models ([43], [158]), a data mining tool for macroeconomic modeling of business cycles ([167], [2]), and a vehicle to simulate complex adaptive macroeconomic phenomena, such as innovation, industrial dynamics and economic transition.

22.2.3 Econometrics

In econometrics, while evolutionary computation is intensively applied to *parametric models*, the most challenging and promising direction is the applications to *non-parametric models*. All the papers listed under the category "**Model Discovery**" show how EC can trigger a global search in the immense space of non-parametric models. Among non-parametric models the one which receives overwhelming attention of econometricians is the class of *artificial neural nets* (ANNs). Technical issues involved in the selection of an ANN range from architecture, associated learning rules to weight determination. These issues make the optimal choice of an ANN extremely difficult, if not impossible. Evolutionary computation, as a tool for model discovery, provides a possible solution to these issues. This EC-oriented solution defines the research area "**evolutionary artificial neural nets**" (**EANNs**). Econometric applications of EANNs are listed under this category. However, unlike the conventional statistical foundation of econometrics, a well-established doctrine for the use of EC in econometric analysis is yet to be found. Recent efforts are devoted to building the foundation from *universal function approximations* ([203], [289], [290]) and *information-based complexity measure* ([40],[59],[141]). Related papers are listed under the category "**Foundations**".

To see how EC can effectively discover the underlying "*true*" (optimal) models, applications are directed to forecasting non-linear time series, in particular, the **Chaotic Time Series**. This is probably because when the series is deterministically chaotic, EC as a global search technique, has great potential to discover the underlying model, and hence can perfectly predict the future of the series ([105]). This type of applications also motivated a series

[2] Since applications devoted to the Kareken-Wallace type introduced agent-based foreign exchange markets, they are categorized as 5.(b)-ii in Sect. 22.2.7.

of fundamental issues. Is perfect prediction really possible? If it is, can we tell how hard (or how easy) it is to predict a series? Is there a measure of predictability? While these issue are by no means new, EC requires a new perspective to address them. Two papers are collected under the category "**Measures of Predictability**".

EC is also applied to more conventional econometric problems, such as "**Parameter Estimation**". Here, EC is taken as a numerical technique for non-linear optimization, such as minimizing non-linear least squares ([74], [194]) or maximizing non-linear likelihood ([251]). A comprehensive study can be found in [153]. EC is applied to other conventional issues as well, though in a non-standard way. For example, "**Hypothesis Testing**" was conducted as an implementation of the survival-of-the-fittest principle ([95]), and "**Structural Change**" was defined by a cognitive-theoretic approach built upon EC ([77]).

22.2.4 Financial Economics

With 148 related publications, financial economics is perhaps the most active research application domain of EC. Introductory or survey articles come under the category of "**General**". Others are divided into two sub-categories, *agent-based computational finance* (or *agent-based artificial financial markets*), and *financial engineering.* Of agent-based artificial financial markets, two types are extensively studied: *stock markets* and *foreign exchange markets.* Except for a couple of studies on commodity futures, there has been no research directed at artificial derivative markets. Since a basic element of all these markets is *double auction*, it would be useful to study artificial financial markets from this starting point. In fact, there are a few studies conducted on *agent-based double-auction markets.*

Application to financial engineering covers a variety of topics. Among them, "**Financial Forecasting**" is one of the most researched areas. However, despite numerous publications, subjects in these applications are limited. Efforts are predominantly devoted to stock prices and exchange rates, with a few exceptions on commercial banks deposit ([130]) and crude oil price ([223]). "**Trading strategies**" is another active arena. This application domain has gradually come to its mature stage in the sense that statistical techniques, such as bootstrap, Monte-Carol simulation, and asymptotic theory, are largely involved in the performance evaluation.[3] Other application areas include "**Investment, Portfolio Optimization and Risk Management**", "**Credit Scoring and Early Warning Systems**", "**Option Pricing**", and "**Volatility Modeling**". "**Cooperate Takeover** and "**Arbitrage**" are new directions for the financial application of EC.

Documentation of some of the commercial software to implement EC in financial engineering can be found under the category of **Software**.

[3] See Chapter 1, Sect.1.4.2.

22.2.5 Preferences, Risk and Uncertainty

Speaking of finance, one usually assumes that agents are *risk averse* by nature. In other words, risk preferences of agents are usually exogenously given. It is but a recent trend to endogenize preferences and study the preference formation process, and this is where EC comes into play. References [343], [248], [53], [201] and [346] all evidence that genetic algorithms can be used to simulate the formation process of risk preferences. References [343], [201], [53], and [346] show that *risk aversion* can emerge as a result of an evolutionary process. [248] shows that, under different settings, *risk embrace* can emerge as well. Whatever differences they may have, their explanation for the deviation of risk neutral is the same–when the length of evaluation cycle (time horizon) is too short, risk deviation can happen, be it risk aversion or risk embrace. Since time horizon can be empirically connected to institutional designs, these studies has given us a new perspective on risk attitude.

22.2.6 Others

Apart from the five major areas reviewed above, EC has recently made its way into **Labor Economics**, **Transportation and Regional Economics**, and **Experimental Economics**, the potentiality of which has yet to be realized.

We close this classification work with the category **Technical Issues**. As mentioned in the introductory chapter, EC involves many possible user-supplied procedures. The consequences of different choices have already received extensive treatment in the EC literature. Nonetheless, many of the established results may not be directly applicable when EC is further applied to economic and financial domain. As a result, economists and financial people have to develop their own treatment of those technical issues. Technical aspects of EC addressed here are the survival-of-the-fittest principle ([60]), election operator ([273]), designs of genetic operators ([169], [315]), selection schemes([171]), analysis of evolutionary dynamics ([82], [316], [306], [317]), architecture ([369]), and numerical and computational issues ([154], [30], [314], [302]).

22.2.7 Classification of Publications

Based on the description above, we classify the publications into ten fields. The number following each field title is the number of applications in each individual field.

1. **General: Learning, Adaptation, Bounded Rationality and Technical Issues** (17)
   [20], [21], [44], [49], [55], [67], [147], [160], [186], [190], [198], [199], [250], [266], [308], [350], [372].
2. **Games and Industrial Organizations** (100)

   (a) **General** (6)
   [3], [48], [258], [264], [305], [150]
   (b) **Repeated Prisoner's Dilemma** (28)
   [7], [23], [24], [25], [41], [42], [56], [143], [173], [174], [176], [184], [191], [196], [197], [207], [218], [232], [254], [262], [267], [277], [298], [329], [330], [361], [378], [380]
   (c) **Duopoly, Oligopoly and Market Games** (25)
   [29], [84], [85], [86], [88], [89], [90], [91], [131], [136], [142], [155], [231], [263], [265], [266], [278], [268], [271], [270], [276], [271], [272], [364], [365]
   (d) **Bargaining, Auction, and Ultimatum Games** (6)
   [5], [6], [148], [157], [159], [300]
   (e) **Electronic Commerce** (2)
   [299], [383]
   (f) **Electricity Market** (4)
   [28], [27], [241], [288]
   (g) **Cooperation, Coalition, and Coordination** (17)
   [15],[17], [12], [18], [123], [124], [125], [126], [161], [177], [208], [281], [284], [326], [361], [368], [379]
   (h) **Principal-Agent Games** (1)
   [370]
   (i) **Differential Games and International Relations** (5)
   [3], [230], [186], [188], [304]
   (j) **Stochastic Games** (1)
   [164]
   (k) **Firms, Trade Networks and Organizations** (5)
   [166], [274], [282], [348], [349]

3. Macroeconomics (40)
   (a) **General** (1)
   [8]
   (b) **The Kiyotake-Wright Model** (4)
   [33], [156], [259], [337]
   (c) **The Cobweb Model** (13)
   [9], [96], [99], [100], [104], [107], [108], [109], [110], [111], [145], [149], [179]
   (d) **Overlapping Generations Model** (7)
   [10], [13], [50], [51] [52], [114], [115]
   (e) **Growth, Cycles and Transition** (12)
      i. **Economic Growth** (3)
      [16], [43], [158]
      ii. **Innovation** (4)
      [47], [237], [239], [310]
      iii. **Economic Transition** (3)
      [110], [119], [193]

- iv. **Business Cycles** (2)
  [2], [167]
- (f) **Consumption Behavior** (1)
  [57]
- (g) **Large-Scale Models** (2)
  [34], [35]

4. **Econometrics** (65)
   - (a) **Model Discovery** (26)
     [202], [234], [235], [283], [312], [321], [328], [344]
     - i. **Evolutionary Artificial Neural Networks** (11)
       [87], [132], [151], [152], [172], [192], [260], [279], [325], [333], [331]
     - ii. **Foundation** (7)
       [40], [59], [141], [203], [224], [289], [290]
   - (b) **Forecasting Time Series** (23)
     [103], [162], [165], [178], [225], [275], [287], [313], [320], [340]
     - i. **Chaotic Time Series** (11)
       [175], [220], [280], [294], [295], [296], [297], [319], [345], [381], [384]
     - ii. **Measures of Predictability** (2)
       [105], [221]
   - (c) **Parameters Estimation** (6)
     [1], [74], [153], [194], [251], [332]
   - (d) **Hypothesis Testing** (4)
     [95], [97], [98], [102]
   - (e) **Structural Changes** (2)
     [77], [113]
   - (f) **Classification** (1)
     [163]
   - (g) **Data Pre-processing** (1)
     [183]
   - (h) **Agent-Based Foundations of Time Series** (2)
     [19], [70]
5. **Financial Economics** (148)
   - (a) **General** (8)
     [32], [62], [101], [140], [144], [226], [242], [336]
   - (b) **Artificial Financial Markets** (35)
     - i. **Artificial Stock Markets** (17)
       [22], [76], [116], [117], [118], [120], [121], [134], [135], [216], [217], [243], [244], [257], [307], [327], [347]
     - ii. **Artificial Foreign Exchange Markets** (8)
       [11], [19], [14], [210], [211], [212], [213], [256],
     - iii. **Artificial Future Markets** (5)
       [70], [107], [108], [109], [111]
     - iv. **Artificial Double Auction Markets** (5)
       [6], [64], [65], [148], [300]
   - (c) **Financial Engineering** (105)

i. **Financial Forecasting** (32)
[83], [87], [94], [97], [98], [102], [106], [132], [139], [182], [203], [204], [214], [219], [221], [222], [223], [236], [252], [255], [275], [279], [290], [318], [339], [340], [341], [359], [357], [358], [376], [377]
ii. **Trading Strategies** (28)
[4], [36], [37], [227], [46], [58], [63], [68], [69], [78], [80], [81], [92], [93], [122], [130], [180], [185], [189], [205], [240], [247], [285], [286], [292], [303], [309], [311]
iii. **Investment, Portfolio Optimization, and Risk Management** (16)
[26], [31], [38], [39], [195], [206], [233], [249], [253], [293], [301], [322], [334], [335], [367], [382]
iv. **Credit Scoring and Early Warning Indicators** (3)
[181], [209], [373]
v. **Option Pricing** (15)
[71], [72], [73], [75], [127], [128], [129], [133], [137], [138], [215], [228], [356], [360], [363]
vi. **Volatility** (5)
[61], [106], [112], [192], [229]
vii. **Arbitrage and Hedging** (2)
[261], [374]
viii. **Takeover** (1)
[291]
ix. **Software** (3)
[338], [354], [355]

6. **Preferences, Risk and Uncertainty** (9)
[53], [201], [245], [246], [248], [323], [343], [346],[366]
7. **Labor Economics** (3)
[351], [352], [362]
8. **Transportation and Regional Economics** (2)
[172], [375]
9. **Experimental Economics** (2)
[11], [156]
10. **Technical Issues** (18)
[30], [60], [66], [79], [82], [154], [168], [169], [170], [171], [273], [302], [306], [315], [316], [314], [317], [369]

22.3 Publications by Journals

For those who are interested in research on EC application and who would like to submit their own papers, we provide a list of journals which have published at least two papers on this kind of research, and we rank them in descending order. Top on the list is *Journal of Economic Dynamics and Control*, which has so far published a total of 28 papers in this area. Coming next is *Computational Economics* with 16 papers published, followed by *Journal of Computational Intelligence in Finance* (the former Neurovt$t Journal) and *Journal of Evolutionary Economics.* Publications of this research area are widely distributed in 57 journals.

(a) Journal of Economic Dynamics and Control (28)
(b) Computational Economics (16)
(c) Journal of Computational Intelligence in Finance (Neurove$t Journal) (11)
(d) Journal of Evolutionary Economics (9)
(e) IEEE Transactions on Evolutionary Computation (3)
(f) Journal of Economic Behavior and Organization (3)
(g) Macroeconomic Dynamics (3)
(h) BioSystems (2)
(i) Complexity (2)
(j) Evolutionary Computation (2)
(k) Financial Engineering News (2)
(l) Journal of Management and Economics (2)

22.4 Publications by Conference Proceedings

For those who would like to know more about this research area, here is a list of the conferences in which you may want to participate.

(a) GECCO (the former ICGA and GP) (43)
(b) CEC (the former ICEC and EP) (41)
(c) CIFER (9)
(d) IDEAL (8)
(e) CIEF (6)
(f) IFAC Computation in Economics, Finance and Engineering (6)
(g) Artificial Life (5)
(h) Computational Finance (5)
(i) Artificial Neural Networks and Genetic Algorithms (4)
(j) Asia-Pacific Conference on Genetic Algorithms and Applications (4)
(k) International Conference on Artificial Intelligence (4)
(l) Complex Systems: Complexity between the Ecos– From Ecology to Economics (3)
(m) IEEE International Conference on Intelligent Processing Systems (3)
(n) International Mendel Conference on Genetic Algorithms, Optimization Problems, Fuzzy Logic, Neural Networks, Rough Sets. (3)
(o) International Symposium on Dynamic Games and Applications (3)
(p) Simulating Social Phenomena (3)

The two major conference series are the *Genetic and Evolutionary Computation Conference* (**GECCO**) and *Congress on Evolutionary Computation* (**CEC**). GECCO integrated the former *international conference on genetic algorithms* (**ICGA**) and *international conference on genetic programming* (**GP**) in 1999. Coincidentally, the *conference on evolutionary programming* (**EP**) merged with the *international conference on evolutionary computation* (**ICEC**) in the same year and developed into the current CEC. These two series have made important contributions to the literature on EC application to economics and finance. Until 2001, there have been 43 publications in GECCO proceedings and 41 in CEC proceedings.

GECCO and CEC are general conferences that encompass a wide range of subjects, from engineering to art design. There are now a few series which are exclusively devoted to economics and finance, where one can find a lot of EC applications. Among them are *IEEE/IAFE Conference on Computational Intelligence for Financial Engineering* (**CIFER**), *Intelligent Data Engineering and Automated Learning* (**IDEAL**), *Computational Intelligence in Economics and Finance* (**CIEF**), and *IFAC Computation in Economics, Finance and Engineering.*

22.5 Useful Websites

For those who just entered this field and want to learn something basic, the following website should be useful. Many give good tutorial materials, and have established extensive linkages with other websites.

- The Hitch-Hiker's Guide to Evolutionary Computation (FAQ for comp.ai.genetic)
  http://www.cs.bham.ac.uk/Mirrors/ftp.de.uu.net/EC/clife/www/
- A WWW about Genetic Programming
  http://www.genetic-programming.org/
- Software in evolutionary computation
  ftp://ftp.cse.cuhk.edu.hk/pub/EC/FAQ/www/Q20.htm
- Centre for Research on Simulation in the Social Sciences, Department of Sociology, School of Human Sciences, University of Surrey, Guildford, England.
  http://www.soc.surrey.ac.uk/research/simsoc/cress.html
- Bibliography of the Evolutionary Models in Social Science (EMSS)
  http://www.soc.surrey.ac.uk/~scs1ec/emssbib.html
- Agent-Based Computational Economics maintained by Leigh Tesfatsion
  http://www.econ.iastate.edu/tesfatsi/ace.htm
- Computational Laboratories Group at Department of Economics, University of California, Santa Barbara
  http://www.econ.ucsb.edu/~cath/complab/index.html
- Agent-Based Computational Finance maintained by Blake LeBaron
  http://stanley.feldberg.brandeis.edu/~blebaron/acf

- Complexity of Cooperation Web Site maintained by Robert Axelrod
  http://pscs.physics.lsa.umich.edu/Software/ComplexCoop.html
- Santa Fe Institute
  http://www.santafe.edu/

22.6 Software

Writing EC software can be quite time-consuming. Fortunately, software to implement some standard procedures is available. Many of the programs can be directly downloaded from the website address given below. In addition, *Computational Intelligence in Economics and Finance* (**2(4)**) provides a list of GA software for finance (see ibid., pp.20-21).

Name	OS	Language	Contact
AGAC	Unix, DOS	ANCI C	http://www.aic.nrl.navy.mil/~spears/freeware.html
BUGS	X11	C	ftp://www.aic.nrl.navy.mil/pub/galist/src/BUGS.tar.Z
Ease	Unix	Tcl	http://ls11-www.cs.uni-dortmund.de/~joe/Ease/Ease.html
Evolver	PC	VB	http://www.palisade.com
FlexTool(GA)	PC	MATLAB	http://www.flextool.com
GAC	Unix	C	http://www.aic.nrl.navy.mil/~spears/freeware.html
GAL	PC	Lisp	http://www.aic.nrl.navy.mil/~spears/freeware.html
GAGA	Unix	C	ftp://ftp://cs.ucl.ac.uk/darpa/gaga.shar
GAGS	Unix, DOS	Perl	http://kal-el.ugr.es/GAGS
GAlib	Unix, DOS	C++	ftp://lancet.mit.edu/pub/ga/
GA Toolbox	Unix, PC	MATLAB	http://www.shef.ac.uk/uni/projects/gaipp/gatbx.html
GAucsd	Unix	C	http://www-cse.ucsd.edu/users/tkammeye/
GEA Toolbox	Unix, PC	MATLAB	http://www.geatbx.com/docu/fcnindex.html
GECO	Unix, MacOS	Lisp	ftp://www.aic.nrl.navy.mil/pub/galist/src/
Genesis	Unix, DOS	C	ftp://ftp.aic.nrl.navy.mil/pub/galist/src/
Genitor	Unix	C	ftp://ftp.cs.colostate.edu/pub/GENITOR.tar
LigGA	Unix, DOS	C	ftp://ftp.aic.nrl.navy.mil/pub/galist/src/
Matlab-GA	Unix, DOS	MATLAB	ftp://ftp.mathworks.com/pub/contrib/v4/optim/genetic
MicroGA	PC, MAC	C++	Emergent Behavior, 635 Wellsbury Way, Palo Alto CA 94306
NGO	PC		http://www.bio-comp.com
SGA-C	Unix	C	ftp://ftp.aic.nrl.navy.mil/pub/galist/src/

Table 22.1. Software in GA

Name	OS	Language	Contact
da Vinci	Unix	C	http://www.informatik.uni-bremen.de/~inform/forschung/daVinci/daVinci.html
Evolve	Windows	C++	http://www.digitalbiology.com/
lilGP	Unix, PC	C	http://GARAGe.cps.msu.edu/software/lil-gp/lilgp-index.html
Lisp-GP	Unix, PC	Lisp	http://www.genetic-programming.org/gplittlelisp.html

Table 22.2. Software in GP

References

1. Agapie A., Agapie A. (1996) Parameter Estimation in Time Series Forecasting Using Genetic Algorithms. In: Proceedings of the Fourth European Congress on Intelligent Techniques and Soft Computing, 2155–2158
2. Agapie A., Agapie A. (1997) Forecasting the Economic Cycles Based on an Extension of the Holt-Winters Model: A Genetic Algorithms Approach. In: Proceedings of the IEEE/IAFE 1997 Computational Intelligence for Financial Engineering. IEEE Press, 96–99

3. Alemdar N. M., Ozyildirim S. (1998) A Genetic Game of Trade, Growth and Externalities. Journal of Economic Dynamics and Control **22(6)**, 811–832
4. Allen F., Karjalainen R. (1999) Using Genetic Algorithms to Find Technical Trading Rules. Journal of Financial Economics **51(2)**, 245–271
5. Andreoni J., Miller J. H. (1995) Auctions with Artificial Adaptive Agents. Games and Economic Behavior **10**, 39–64
6. Andrew M., Prager R. (1994) Genetic Programming for the Acquisition of Double Auction Market strategies. In: Kinnear K. Jr. (Ed.), Advances in Genetic Programming. MIT Press, Cambridge, MA, 355–368
7. Angeline P. J. (1994) An Alternate Interpretation of the Iterated Prisoner's Dilemma and the Evolution of Nonmutual Cooperation. In: Brooks R., Maes P. (Eds.), Artificial Life IV. MIT Press, Cambridge, MA, 353–358
8. Arifovic J. (1991) Learning by Genetic Algorithms in Economic Environments. Ph.D. thesis, University of Chicago.
9. Arifovic J. (1994) Genetic Algorithms Learning and the Cobweb Model. Journal of Economic Dynamics and Control **18(1)**, 3–28
10. Arifovic J. (1995) Genetic Algorithms and Inflationary Economies. Journal of Monetary Economics **36(1)**, 219–243
11. Arifovic J. (1996) The Behavior of the Exchange Rate in the Genetic Algorithm and Experimental Economies. Journal of Political Economy **104(3)**, 510–541
12. Arifovic J. (1997) Strategic Uncertainty and the Genetic Algorithm Adaptation. In: Amman H., Rustem B., Whinston A. (Eds.), Computational Approaches to Economic Problems. Kluwer Academic Publishers, Dordrecht, 225–236
13. Arifovic J. (1998) Stability of Equilibria Under Genetic–Algorithm Adaptation: An Analysis. Macroeconomic Dynamics **2(1)**, 1–21
14. Arifovic J. (2001) Evolutionary Dynamics of Currency substitution. Journal of Economic Dynamics and Control **25**, 395–417
15. Arifovic J., Eaton B. C. (1995) Coordination via Genetic Learning. Computational Economics **8(3)**, 181–203
16. Arifovic J., Bullard J., Duffy J. (1997) The Transition from Stagnation to Growth: An Adaptive Learning Approach. Journal of Economic Growth **2(2)**, 185-209
17. Arifovic J., Eaton B. C., Morrison W. G. (1996) Drift and Selective Imitation. In: Angeline P., Back T., Fogel D. (Eds.), Evolutionary Programming: Proceeding of the Fifth Annual Conference on Evolutionary Programming. MIT Press, Cambridge, MA, 37–44
18. Arifovic J., Eaton B. C. (1998) The Evolution of Type Communication in a Sender/Receiver Game of Common Interest with Cheap Talk. Journal of Economic Dynamics and Control **22**(8-9), 1187–1207
19. Arifovic J., Gencay R. (2000) Statistical Properties of Genetic Learning in a Model of Exchange Rate. Journal of Economic Dynamics and Control **24**, 981–1005
20. Arthur W. B. (1992) On Learning and Adaptation in the Economy. SFI Working Paper, 92-07-038.
21. Arthur W. B. (1994) Inductive Reasoning and Bounded Rationality. American Economic Association Papers Proceedings **84**, 406–411
22. Arthur W. B., Holland J., LeBaron B., Palmer R., Tayler P. (1997) Asset Pricing under Endogenous Expectations in an Artificial Stock Market. In: Arthur W. B., Durlauf S., Lane D. (Eds.), The Economy as an Evolving Complex System II. Addison-Wesley, Reading, MA, 15–44

23. Ashlock D., Smucker M. D., Stanley E. A., Tesfatsion L. (1996) Preferential Partner Selection in an Evolutionary Study of Prisoner's Dilemma. BioSystems **37** (1–2), 99–125
24. Axelrod R. (1984) The Evolution of Cooperation. Basic Books, NY.
25. Axelrod R. (1987) The Evolution of Strategies in the Iterated Prisoner's Dilemma. In: Davis L. (Ed.), Genetic Algorithms and Simulated Annealing. Pitman, London, 32–41
26. Baglioni S., Sorbello D., Pereira C. da Costa, Tettamanzi A. G. B. (2000) Evolutionary Multiperiod Asset Allocation. In: Whitley D., Goldberg D., Cantú-Paz E., Spector L., Parmee I., Beyer H.-G. (Eds.), Proceedings of the Genetic and Evolutionary Computation Conference. Morgan Kaufmann, 597–604
27. Bagnall A. J. (2000) A Multi-Adaptive Agent Model of Generator Bidding in the UK Market in Electricity. In: Whitley D., Goldberg D., Cantú-Paz E., Spector L., Parmee I., Beyer H.-G. (Eds.), Proceedings of the Genetic and Evolutionary Computation Conference. Morgan Kaufmann, 605–612
28. Bagnall A. J. Smith G. D. (1999) Using an Adaptive Design to Bid in a Simplified Model of the UK Market in Electricity. In: Banzhaf W., Daida J., Eiben A. E., Garzon M. H., Honavar V., Jakiela M., Smith R. E. (Eds.), Proceedings of the Genetic and Evolutionary Computation Conference, Vol. 1. Morgan Kaufmann, 774
29. Baldassarre G. (1997) Neural Networks and Genetic Algorithms for the Simulation Models of Bounded Rationality Theory. An Application to Oligopolistic Markets. Rivista di Politica Economica **87(12)**, 107–145
30. Ball J. F., Dorsey R. E., Johnson J. D. (1997) Non-Linear Optimization on a Parallel Intel i860 RISC Based Architecture. Computational Economics **10(3)**, 279–294
31. Bao P., Chan H. (1998) Vector Quantization and Genetic Algorithm-Based Portfolio Theory Modeling. In: Xu L., Chan L. W., King I., Fu A. (Eds.), Intelligent Data Engineering and Learning: Perspectives on Financial Engineering and Data Mining. Springer-Verlag, Singapore, 149–156
32. Barrow D. (1992) Making Money with Genetic Algorithms. In: Proceedings of the Fifth European Seminar on Neural Networks and Genetic Algorithms. IBC International Service, London.
33. Bascl E. (1999) Learning by Imitation. Journal of Economic Dynamics and Control **23**, 1569–1585
34. Basu N., Pryor R. J. (1997) Growing a Market Economy. Sandia Report SAND-97-2093.
35. Basu N., Pryor R. J., Quint T. (1998) ASPEN: A Microsimulation Model of the Economy. Computational Economics **12(3)**, 223–241
36. Bauer R. J. Jr., Liepins G. E. (1992) Genetic Algorithms and Computerized Trading Strategies. In: O'leary D. E., Watkins R. R. (Eds.), Expert Systems in Finance. North Holland.
37. Bauer R. J. Jr. (1994a) Genetic Algorithms and Investment Strategies. John Wiley & Sons, New York.
38. Bauer R. J. Jr. (1994b) An Introduction to Genetic Algorithms: a Mutual Fund Screening Example. Neurove$t Journal **2(4)**, 16–19
39. Bauer R. J. Jr. (1995) Genetic Algorithms and the Management of Exchange Rate Risk. In: Biethahn J., Nissen V. (Eds.), Evolutionary Algorithms in Management Applications. Springer, Heidelberg and New York, 253–263

40. Bearse P. M., Bozdogan H., Schlottmann A. M. (1997) Empirical Econometric Modelling of Food Consumption Using a New Informational Complexity Approach. Journal of Applied Econometrics **12(5)**, 563–586
41. Beaufils B., Delahaye J.-P., Mathieu P. (1996) Our Meeting with Gradual: a Good Strategy for the Iterated Prisoner's Dilemma. Proceedings of the Fifth International Conference on Artificial Life, 159-166
42. Beaufils B., Delahaye J.-P., Mathieu P. (1998) Complete Classes of Strategies for the Classical Iterated Prisoner's Dilemma. In: Porto V. W., Saravanan N., Waagen D., Eiben A. E. (Eds.), Evolutionary Programming VII, Lecture Notes in Computer Science, Vol. 1447. Springer-Verlag, Berlin, 33–42
43. Beaumont P. M., Bradshaw P. T. (1995) A Distributed Parallel Genetic Algorithm for Solving Optimal Growth Models. Computational Economics **8(3)**, 159–179
44. Beckenbach F. (1999) Learning by Genetic Algorithms in Economics. In: Brenner T. (Ed.), Computational Techniques for Modelling Learning in Economics, The Series Advances in Computational Economics 11. Kluwer, Dordrecht.
45. Belttratti A., Margarita S., Terna P.(1996) Neural Networks for Economic and Financial Modeling. International Thomson Computer Press, London, UK.
46. Bhattacharyya S., Pictet O., Zumbach G. (1998) Representational Semantics for Genetic Programming Based Learning in High-Frequency Financial Data. In: Koza J. R., Banzhaf W., Chellapilla K., Deb K., Dorigo M., Fogel D. B., Garzon M. H., Goldberg D. E., Iba H., Riolo R. (Eds.), Genetic Programming 1998: Proceedings of the Third Annual Conference. Morgan Kaufmann, 11–16
47. Birchenhall C. R. (1995) Modular Technical Change and Genetic Algorithms. Computational Economics **8(3)**, 233–253
48. Birchenhall C. R. (1996) Evolutionary Games and Genetic Algorithms. In: Gilli M. (Ed.), Computational Economic Systems: Models, Methods & Econometrics, The Series Advances in Computational Economics 5. Kluwer Academic Publishers, Dordrecht, 3–23
49. Birchenhall C. R., Kastrinos N., Metcalfe S. (1997) Genetic Algorithms in Evolutionary Modelling. Journal of Evolutionary Economics **7(4)**, 375–393
50. Bullard J., Duffy J. (1998) A Model of Learning and Emulation with Artificial Adaptive Agents. Journal of Economic Dynamics and Control **22**, 179–207
51. Bullard J., Duffy J. (1998) Learning and the Stability of Cycles. Macroeconomic Dynamics **2(1)**, 22–48
52. Bullard J., Duffy J. (1999) Using Genetic Algorithms to Model the Evolution of Heterogeneous Beliefs. Computational Economics **13(1)**, 41–60
53. Cacho O., Simmons P. (1999) A Genetic Algorithm Approach to Farm Investment. Australian Journal of Agricultural and Resource Economics, **43(3)**, 305–322
54. Casti J. L. (1997) Would-Be Worlds. John Wiley, New York.
55. Chattoe E. (1997) Modelling Economic Interaction Using a Genetic Algorithm. In: Back T., Fogel D., Michalewicz Z. (Eds.) The Handbook of Evolutionary Computation. Oxford University Press/IOP Publishing, Section G7.1, 1–5
56. Chalk K., Smith G. D. (1997) Multi-Agent Classifier Systems and the Iterated Prisoner's Dilemma. In: Smith G. D., Steele N. C., Albrecht R. F. (Eds.), Artificial Neural Networks and Genetic Algorithms. Springer-Verlag, 615–618
57. Chattoe E., Gilbert N. (1997) A Simulation of Adaptation Mechanisms in Budgetary Decision Making. In: Conte R., Hegselmann R., Terna P. (Eds.), Simulating Social Phenomena. Springer-Verlag, 401–418

58. Chen J.-S., Deng S.-X., Lin P.-C. (2000) Generation of Trading Strategies Using Genetic Algorithms. In: Wang P. (Ed.), Proceedings of the Fifth Joint Conference on Information Sciences, Vol. II, 921–924
59. Chen S.-H. (1997a) Occam's Razor as an Emerging Property of ALIFE Economic Systems. In: Proceedings of the VII the Conference of International Association for the Development of Interdisciplinary Research on Learning: From Natural Principle to Artificial Methods, University of Geneva, 59–62
60. Chen S.-H. (1997b) On the Artificial Life of the General Economic System (I): the Role of Selection Pressure. In: Hara F., Yoshida K. (Eds.), Proceedings of International Symposium on System Life, 233–240
61. Chen S.-H. (1998a) Modeling Volatility with Genetic Programming: A First Report. Neural Network World **8(2)**, 181–190
62. Chen S.-H. (1998b) Evolutionary Computation in Financial Engineering: A Roadmap to GAs and GP. Financial Engineering News, Vol. 2, No. 4.
63. Chen S.-H. (1998c) Can We Believe that Genetic Algorithms Would Help without Actually Seeing Them Work in Financial Data Mining?: Part I, The Foundations. In: Xu L., Chan L. W., King I., Fu A. (Eds.), Intelligent Data Engineering and Learning: Perspectives on Financial Engineering and Data Mining. Springer-Verlag, Singapore, 81–87
64. Chen S.-H. (2000a) Toward an Agent-Based Computational Modeling of Bargaining Strategies in Double Auction Markets with Genetic Programming. In: Leung K.S., Chan L.-W., Meng H. (Eds.), Intelligent Data Engineering and Automated Learning- IDEAL 2000: Data Mining, Financial Engineering, and Intelligent Agents, Lecture Notes in Computer Sciences 1983. Springer, 517–531
65. Chen S.-H. (2000b) Evolving Bargaining Strategies with Genetic Programming: An Overview of "AIE-DA, Version 1". In: Proceedings of the Second Asia-Pacific Conference on Genetic Algorithms and Applications. Global Link Publishing Company, Hong Kong, 388–396
66. Chen S.-H. (2001a) Fundamental Issues in the Use of Genetic Programming in Agent-Based Computational Economics. In: Namatame A. (Ed.), Proceedings of the First International Workshop on Agent-based Approaches in Economic and Social Complex Systems, 175–185
67. Chen S.-H. (2001b) On the Relevance of Genetic Programming to Evolutionary Economics. In: Aruka Y. (Ed.), Evolutionary Controversies in Economics: A New Transdisciplinary Approach. Springer-Verlag, Tokyo, 135–150
68. Chen S.-H., Chen C.-F. (1995) Can GA-Based Technical Trading Rules Survive Well During the 1990-91 World-Wide Recession? Evaluation Based on the Crash of TAIEX and NIKKEI. In: Steele H. C., Yau O. (Eds.), Proceedings of International Conference on "Global Business in Transition: Prospects for the Twenty First Century", Vol. II. Center for International Business Studies (CIBS), Lingnan College, Hong Kong, 591–598
69. Chen S.-H., Chen C.-F. (1998) Can We Believe that Genetic Algorithms Would Help without Actually Seeing Them Work in Financial Data Mining?: Part II, Empirical Tests. In: Xu L., Chan L. W., King I., Fu A. (Eds.), Intelligent Data Engineering and Learning: Perspectives on Financial Engineering and Data Mining. Springer-Verlag, Singapore, 89–97
70. Chen S.-H., Kuo T.-W. (1999) Towards an Agent-Based Foundation of Financial Econometrics: An Approach Based on Genetic-Programming Artificial Markets. In: Banzhaf W., Daida J., Eiben A. E., Garzon M. H., Honavar V.,

Jakiela M., Smith R. E. (Eds.), Proceedings of the Genetic and Evolutionary Computation Conference, Vol. 2. Morgan Kaufmann, 966–973
71. Chen S.-H., Lee W.-C. (1997a) Option Pricing with Genetic Algorithms: A First Report. In: Tan C.-C. (Ed.), Proceedings of the Second Chinese Congress on Intelligent Control and Intelligent Automation. Xi'an Jiaotong University Press, Xi'an, China, 1682–1688
72. Chen S.-H., Lee. W.-C. (1997b) Option Pricing with Genetic Algorithms: The Case of European-Style Options. In: T. Back (Ed.), Proceedings of the Seventh International Conference on Genetic Algorithms. Morgan Kaufmann Publishers, San Francisco, CA, 704–711
73. Chen S.-H., Lee W.-C. (1997c) Option Pricing with Genetic Algorithms: Separating Out-of-the-Money from In-the-Money. In: Proceedings of the 1997 IEEE International Conference on Intelligent Processing Systems, Vol. 1. IEEE Press, 110–115
74. Chen S.-H., Lee W.-C. (1997d) Estimating the Effective Tax Function with Genetic Algorithms. In: Alander J. T. (Ed.), Proceedings of the Third Nordic Workshop on Genetic Algorithms and Their Applications, 275–290
75. Chen S.-H., Lee W.-C. (1997e) Option Pricing with Genetic Algorithms: A Second Report. In: The 1997 IEEE International Conference on Neural Networks Proceedings, 21–25
76. Chen S.-H., Liao C.-C. (2000) Price Discovery in Agent-Based Computational Modeling of Artificial Stock Markets. In: Proceedings of the Second Asia-Pacific Conference on Genetic Algorithms and Applications. Global Link Publishing Company, Hong Kong, 380–387
77. Chen S.-H., Lin W.-Y. (1997a) On the Cognitive System Which Can Detect and Adapt to Intrinsic Novelties. In: Graham J., Shin D.-G. (Eds.), Proceedings of the 6th ISCA International Conference on Intelligent Systems, 187–192
78. Chen S.-H., Lin W.-Y. (1997b) Financial Data Mining with Adaptive Genetic Algorithms. In: Philip T. (Ed.), Proceedings of the 10th International Conference on Computer Applications in Industry and Engineering, 154–159
79. Chen S.-H., Lin W.-Y. (1997c) Rethinking the Appeal of Evolution: Empirical Evidence from the Financial Application of Genetic Algorithms. In: Proceedings of 1997 Emerging Technologies Workshop (ET'97), University College London, London, U.K., 79–94
80. Chen S.-H., Lin W.-Y. (1998a) The Appeal of Evolution: The Case of the RGA-Based Portfolios. In: Debnath N. C. (Ed.), Proceedings of the ISCA 13th International Conference, 125–130
81. Chen S.-H., Lin W.-Y. (1998b) Two Ways to Improve Genetic Algorithms in Financial Data Mining: Sell Short with Recursive GAs. In: Proceedings of the Seventh International Conference on Information Processing and Management of Uncertainty in Knowledge-Based Systems, Vol. 2, 1090–1097
82. Chen S.-H., Lin W.-Y. (1998c) Rethinking the Appeal of Evolution: Empirical Evidences from the Financial Applications of Genetic Algorithms. Beijing Mathematics **4(2)**, 161–175
83. Chen S.-H., Lu C.-F. (1999) Would Evolutionary Computation Help in Designs of ANNs in Forecasting Exchange Rates? In: Proceedings of the 1999 Congress on Evolutionary Computation, Vol. 1. IEEE Press, 267–274
84. Chen S.-H., Ni C.-C. (1996a) Genetic Algorithm Learning and the Chain-Store Game. In: Proceedings of 1996 IEEE International Conference on Evolutionary Computation. IEEE Press, 480–484

85. Chen S.-H., Ni C.-C. (1996b) Understanding the Nature of Predatory Pricing in the Large-Scale Market Economy with Genetic Algorithms. In: Proceedings of the 1996 IEEE International Symposium on Intelligent Control. IEEE Press, 354–359
86. Chen S.-H., Ni C.-C. (1997a) Coevolutionary Instability in Games: An Analysis Based on Genetic Algorithms. In: Proceedings of 1997 IEEE International Conference on Evolutionary Computation. Institute of Electrical and Electronics Engineers, Piscataway, NJ, 703–708
87. Chen S.-H., Ni C.-C. (1997b) Evolutionary Artificial Neural Networks and Genetic Programming: A Comparative Study Based on Financial Data. In: Smith G. D., Steele N. C., Albrecht R. F. (Eds.), Artificial Neural Networks and Genetic Algorithms. Springer-Verlag, 397–400
88. Chen S.-H., Ni C.-C. (1998a) Simulating the Stability of Collusive Pricing of Oligopolists with Genetic Algorithms. In: John R. I. (Ed.), Recent Advances in Soft Computing, Vol. 1, 206–212
89. Chen S.-H., Ni C.-C. (1998b) Simulating the Adaptive Behaviour of Oligopolists: An Application of Genetic Algorithms. In: Neevel V. (Ed.), Proceedings of the 8th International Symposium on Dynamic Games and Applications, 153–157
90. Chen S.-H., Ni C.-C. (1999) Using Genetic Algorithms to Simulate the Evolution of an Oligopoly Game. In: McKay B., Yao X., Newton C. S., Kim J.-H., Furuhashi T. (Eds.), Simulated Evolution and Learning, Lecture Notes in Artificial Intelligence 1585. Springer, 293–300
91. Chen S.-H., Ni C.-C. (2000) Simulating the Ecology of Oligopolistic Competition with Genetic Algorithms. Knowledge and Information Systems **2**, 310–339
92. Chen S.-H., Tsao C.-Y. (1999a) Discovering Trading Rules with Genetic Algorithms: An Empirical Study Based on GARCH Time Series. In: Arabnia H. R. (Ed.), Proceedings of the International Conference on Artificial Intelligence, Vol. II. CSREA Press, 430–436
93. Chen S.-H., Taso C.-Y. (1999b) Genetic Algorithms and Trading Strategies: Evidences from the ARCH Nonlinear Time Series. In: Proceedings of the Eighth International Fuzzy Systems Association World Congress, Vol. 1, 315–319
94. Chen S.-H., Yeh C.-H. (1995a) Predicting Stock Returns with Genetic Programming: Do the Short-Term Nonlinear Regularities Exist? In: Fisher D. (Ed.), Proceedings of the Fifth International Workshop on Artificial Intelligence and Statistics, 95–101
95. Chen S.-H., Yeh C.-H. (1995b) On the Competitiveness of the Quantity Theory of Money: A Natural-Selection Test Based on Genetic Programming. In: Proceedings on the 11th International Conference on Advanced Science and Technology, 242–251
96. Chen S.-H., Yeh C.-H. (1995c) On the Coordination and Adaptability of the Large Economy: An Application of Genetic Programming to the Cobweb Model. In: Proceedings of the First International Conference on Applications of Dynamic Models to Economics, 121–159
97. Chen S.-H. Yeh C.-C. (1996a) Genetic Programming, Predictability, and Stock Market Efficiency. In: Vlacic L. Nguyen T., Cecez-Kecmanovic D. (Eds.), Modelling and Control of National and Regional Economies. Pergamon Press, Oxford, Great Britain, 283–288

98. Chen S.-H., Yeh C.-C. (1996b) Genetic Programming and the Efficient Market Hypothesis. In: Koza J., Goldberg D., Fogel D., Riolo R. (Eds.), Genetic Programming 1996: Proceedings of the First Annual Conference. MIT Press, Cambridge, MA, 45–53
99. Chen S.-H., Yeh C.-H. (1996c) Genetic Programming Learning and the Cobweb Model. In: Angeline P. (Ed.), Advances in Genetic Programming, Vol. 2, Chap. 22. MIT Press, Cambridge, MA, 443–466
100. Chen S.-H., Yeh C.-H. (1996d) Information Transmission, Market Efficiency and the Evolution of Information-Processing Technology. Journal of Technology Management, Vol. 1, No. 1, 23–41
101. Chen S.-H., Yeh C.-H. (1996e) Genetic Programming in Computable Financial Economics. In: Proceedings of the ISCA 11th Conference: Computers and Their Applications. ISCA Press, 135–138
102. Chen S.-H., Yeh C.-H. (1996f) Bridging the Gap between Nonlinearity Tests and the Efficient Market Hypothesis by Genetic Programming. In: Proceedings of the IEEE/IAFE 1996 Conference on Computational Intelligence for Financial Engineering. IEEE Press, 34–39
103. Chen S.-H., Yeh C.-H. (1996g) A Comparison of Forecast Accuracy between Genetic Programming and Other Forecastors: A Loss Differential Approach. In: Borrajo D., Isasi P. (Eds.), Proceedings of the First International Workshop on Machine Learning, Forecasting, and Optimization. Universidad Carlos III de Madrid Press, 39–52
104. Chen S.-H., Yeh C.-H. (1996h) On the Coordination and Adaptability of the Large Economy: An Application of Genetic Programming to the Cobweb Model. In: Preprints of 13th World Congress International Federation of Automatic Control, Vol. L, 279–284
105. Chen S.-H., Yeh C.-H. (1997a) Toward a Computable Approach to the Efficient Market Hypothesis: An Application of Genetic Programming. Journal of Economic Dynamics and Control **21(6)**, 1043–1063
106. Chen S.-H., Yeh C.-H. (1997b) Using Genetic Programming to Model Volatility in Financial Time Series. In: Koza J. R., Banzhaf W., Chellapilla K., Deb K., Dorigo M., Fogel D. B., Garzon M. H., Goldberg D. E., Iba H., Riolo R. (Eds.), Genetic Programming 1997: Proceedings of the Second Annual Conference. Morgan Kaufmann, 58–63
107. Chen S.-H., Yeh C.-H. (1997c) Speculative Trades and Financial Regulations: Simulation Based on Genetic Programming. In: Ghose A. (Ed.), Working Notes of The IJCAI-97: Workshop on Business Applications of AI. Fifteenth International Joint Conference on Artificial Intelligence, 1–8
108. Chen S.-H., Yeh C.-H. (1997d) Modeling Speculators with Genetic Programming. In: Angeline P., Reynolds R. G., McDonnell J. R., Eberhart R. (Eds.), Evolutionary Programming VI, Lecture Notes in Computer Science, Vol. 1213. Springer-Verlag, Berlin, 137–147
109. Chen S.-H., Yeh C.-H. (1997e) Speculative Trades and Financial Regulations: Simulations Based on Genetic Programming. In: Proceedings of the IEEE/IAFE 1997 Computational Intelligence for Financial Engineering. IEEE Press, 123–129
110. Chen S.-H., Yeh C.-H. (1997f) Simulating Economic Transition Processes by Genetic Programming. In: Kulikowski R., Nahorski Z., Owsinski J. W. (Eds.),

In: Proceedings of the International Conference on Transition to Advanced Market Institutions and Economies: Systems and Operations Research Challenges, 87–93

111. Chen S.-H., Yeh C.-H. (1997g) Trading Restrictions, Speculative Trades and Price Volatility: An Application of Genetic Programming. In: Proceedings of the 3rd International Mendel Conference on Genetic Algorithms, Optimization Problems, Fuzzy Logic, Neural Networks, Rough Sets, 31–37
112. Chen S.-H., Yeh C.-H. (1997h) Using Genetic Programming to Model Volatility in Financial Time Series: The Case of Nikkei 225 and S&P 500. In: Proceedings of the 4th JAFEE International Conference on Investments and Derivatives, 288–306
113. Chen S.-H., Yeh C.-H. (1997i) Modelling Structural Changes with Genetic Programming: An Outline. In: Sydow A. (Ed.), Proceedings of 15th IMACS World Congress on Scientific Computation, Modelling and Applied Mathematics, Vol. 2: Numerical Mathematics. Wissenschaft & Technik Verlag, 621–626
114. Chen S.-H., Yeh C.-H. (1998) Genetic Programming in the Overlapping Generations Model: An Illustration with Dynamics of the Inflation Rate. In: Porto V. W., Saravanan N., Waagen D., Eiben A. E. (Eds.), Evolutionary Programming VII, Lecture Notes in Computer Science, Vol. 1447. Springer-Verlag, Berlin, 829–838
115. Chen S.-H., Yeh C.-H. (1999a) Modeling the Expectations of Inflation in the OLG Model with Genetic Programming. Soft Computing **3(2)**, 53–62
116. Chen S.-H., Yeh C.-H. (1999b) Genetic Programming in the Agent-Based Artificial Stock Market. In: Proceedings of the 1999 Congress on Evolutionary Computation, Vol. 2. IEEE Press, 834–841
117. Chen S.-H., Yeh C.-H. (1999c) On the Consequence of "Following the Herd": Evidence from the Artificial Stock Market. In: Arabnia H. R. (Ed.), Proceedings of the International Conference on Artificial Intelligence, Vol. II. CSREA Press, 388–394
118. Chen S.-H., Yeh C.-H. (2000a) On the Role of Intensive Search in Stock Markets: Simulations Based on Agent-Based Computational Modeling of Artificial Stock Markets. In: Proceedings of the Second Asia-Pacific Conference on Genetic Algorithms and Applications. Global Link Publishing Company, Hong Kong, 397–402
119. Chen S.-H., Yeh C.-H. (2000b) Simulating Economic Transition Processes by Genetic Programming. Annals of Operation Research **97**, 265–286
120. Chen S.-H., Yeh C.-H. (2001a) Evolving Traders and the Business School with Genetic Programming: A new Architecture of the Agent-Based Artificial Stock Market. Journal of Economic Dynamics and Control **25**, 363–393
121. Chen S.-H., Yeh C.-H. (2001b) On the Emergent Properties of Artificial Stock Markets: The Efficient Market Hypothesis and the Rational Expectations Hypothesis. Forthcoming in Journal of Economic Behavior and Organization.
122. Chen S.-H., Chen C.-F., Tan C.-W. (1998) Toward an Effective Implementation of Genetic Algorithms in Financial Data Mining: Retraining Plus Validating. In: Xu L., Chan L. W., King I., Fu A. (Eds.), Intelligent Data Engineering and Learning: Perspectives on Financial Engineering and Data Mining. Springer-Verlag, Singapore, 99–105
123. Chen S.-H., Duffy J., Yeh C.-H. (1996a) Equilibrium Selection Using Genetic Programming. In: Amari S., Xu L., Chan L., King I., Leung K. (Eds.), Progress

in Neural Information Processing, Vol. 2. Springer-Verlag, Singapore, 1341–1346

124. Chen S.-H., Duffy J., Yeh C.-H. (1996b) Genetic Programming in the Coordination Game with a Chaotic Best-Response Function. In: Angeline P., Back T., Fogel D. (Eds.), Evolutionary Programming: Proceeding of the Fifth Annual Conference on Evolutionary Programming. MIT Press, Cambridge, MA, 277–286
125. Chen S.-H., Duffy J., Yeh C.-H. (1997) Genetic Programming in the Coordination Game with a Chaotic Best-Response Function. In: Fogel L., Angeline P., Back T. (Eds.), Evolutionary Programming V. MIT Press, 277–286
126. Chen S.-H., Duffy J., Yeh C.-H. (2001) Equilibrium Selection via Adaptation: Using Genetic Programming to Model Learning in a Co-Ordination Game. Forthcoming in Electronic Journal of Evolutionary Modeling and Economic Dynamics.
127. Chen S.-H., Lee W.-C., Yeh C.-H. (1998a) Pricing Financial Derivatives with Genetic Programming. In: Li G., Tang W., Chen M., Cui D. (Eds.), Systems Science and Its Applications. Tianjin People's Publishing House, 144–156
128. Chen S.-H., Lee W.-C., Yeh C.-H. (1998b) Hedging Securities with Genetic Programming. In: Nakhaeizadeh G., Steurer E. (Eds.), ECML'98 Workshop Notes: Application of Machine Learning and Data Mining in Finance, 140–151
129. Chen S.-H., Lee W.-C., Yeh C.-H. (1999) Hedging Derivative Securities with Genetic Programming. International Journal of Intelligent Systems in Accounting, Finance and Management **8(4)**, 237–251
130. Chen S.-H., Lin W.-Y., Tsao C.-Y. (1999) Genetic Algorithms, Trading Strategies and Stochastic Processes: Some New Evidence from Monte Carlo Simulations. In: Banzhaf W., Daida J., Eiben A. E., Garzon M. H., Honavar V., Jakiela M., Smith R. E. (Eds.), GECCO-99: Proceedings of the Genetic and Evolutionary Computation Conference. Morgan Kaufmann, 114–121
131. Chen S.-H., Ni C.-C., Feng S. (1997) Understanding the Nature of Predatory Pricing in Large-Scale Market Economy with Genetic Algorithms. Journal of Systems Engineering and Electronics **8(2)**, 33–44
132. Chen S.-H., Wang H.-S., Zhang B.-T. (1999) Forecasting High-Frequency Financial Time Series with Evolutionary Neural Trees: The Case of Hang-Seng Stock Index. In: Arabnia H. R. (Ed.), Proceedings of the International Conference on Artificial Intelligence, Vol. II. CSREA Press, 437–443
133. Chen S.-H., Yeh C.-H., Lee W.-C. (1998) Option Pricing with Genetic Programming. In: Koza J., Banzhaf W., Chellapilla K., Deb K., Dorigo M., Foegl D., Garson M., Goldberg D., Iba H., Riolo R. (Eds.), Proceedings of the Third Annual Genetic Programming Conference. Morgan Kaufmann Publishers, San Francisco, CA, 32–37
134. Chen S.-H., Yeh C.-H., Liao C.-C. (1999) Testing Rational Expectations Hypothesis with Agent-Based Models of Stock Markets. In: Arabnia H. R. (Ed.), Proceedings of the International Conference on Artificial Intelligence, Vol. II. CSREA Press, 381–387
135. Chen S.-H., Yeh C.-H., Liao C.-C. (2000) Testing for Granger Causality in the Stock-Price Volume Relation: A Perspective from the Agent-Based Model of Stock Markets. In: Wang P. (Ed.), Proceedings of the Fifth Joint Conference on Information Sciences, Vol. II, 950–956

136. Cheung Y., Bedingfield S., Huxford S. (1997) Monitoring and Interpreting Evolved Behavior in an Oligopoly. In: Proceedings of 1997 IEEE International Conference on Evolutionary Computation. Institute of Electrical and Electronics Engineers, Piscataway, NJ, 697–702
137. Chidambaran N., Lee C.-W. J., Trigueros J. (1998) An Adaptive Evolutionary Approach to Option Pricing via Genetic Programming. In: Koza J., Banzhaf W., Chellapilla K., Deb K., Dorigo M., Fogel D., Garzon M., Goldberg D., Iba H., Riolo R. (Eds.), Genetic Programming 1998: Proceedings of the Third Annual Conference. Morgan Kaufmann, San Francisco, CA, 187–192
138. Chidambaran N., Lee C.-W. J., Trigueros J. (1999) Option Pricing via Genetic Programming. In: Abu-Mostafa Y. S., LeBaron B., Lo A. W., Weigend A. S. (Eds.), Computational Finance – Proceedings of the Sixth International Conference. MIT Press, Cambridge, MA.
139. Chiraphadhanakful S., Dangprasert P., Avatchanakorn V. (1997) Genetic Algorithms in Forecasting Commercial Banks Deposit. In: Proceedings of the 1997 IEEE International Conference on Intelligent Processing Systems, Vol. 1. IEEE Press, 116–121
140. Colin A. M. (1994) Genetic Algorithms for Financial Modeling. In: Deboeck G. J. (Ed.), Trading on the Edge: Neural, Genetic, and Fuzzy Systems for Chaotic Financial Markets. John Wiley & Sons, 148–173
141. Costa P. (1997) A Methodology for the Analysis of Complex Systems Based on Qualitative Reasoning, Stochastic Complexity and Genetic Programming. In: Koza J .R. (Ed.), Late Breaking Papers at the Genetic Programming 1997 Conference, 35–41
142. Curzon Price T. (1997) Using Co-Evolutionary Programming to Simulate Strategic Behaviour in Markets. Journal of Evolutionary Economics **7**(3), 219–254
143. Darwen P. J., Yao X. (2001) Why More Choices Cause Less Cooperation in Iterated Prisoner's Dilemma. In: Proceedings of the 2001 Congress on Evolutionary Computation, Vol. 2. IEEE Press, 987–994
144. Davis L. (1994) Genetic Algorithms and Financial Applications. In: Deboeck G. J. (Ed.), Trading on the Edge: Neural, Genetic, and Fuzzy Systems for Chaotic Financial Markets. John Wiley & Sons, 133–147
145. Dawid H. (1996a) Learning of Cycles and Sunspot Equilibria by Genetic Algorithms. Journal of Evolutionary Economics **6(4)**, 361–373
146. Dawid H. (1996b) Adaptive Learning by Genetic Algorithms: Analytical Results and Applications to Economical Models. Springer, Heidelberg and New York.
147. Dawid H. (1996c) Algorithms as a Model of Adaptive Learning in Economic Systems. Central European Journal for Operations Research and Economics **4(1)**, 7–23
148. Dawid H. (1999) On the Convergence of Genetic Learning in a Double Auction Market. Journal of Economic Dynamics and Control **23**, 1544–1567
149. Dawid H., Kopel M. (1998) On Economic Applications of the Genetic Algorithm: A Model of the Cobweb Type. Journal of Evolutionary Economics **8(3)**, 297–315
150. Dawid H., Mehlmann A. (1996) Genetic Learning in Strategic form Games. Complexity **1(5)**, 51–59

151. Delgado A. et al. (1996) Hybrid System: Neural Networks and Genetic Algorithms Applied in Nonlinear Regression and Time Series Forecasting. In: Prat A. (Ed.), COMPSTAT: Proceedings in Computational Statistics. Physica, Heidelberg, 217–222
152. Dorsey R. E., Johnson J. D., Mayer W. J. (1994) A Genetic Algorithm for the Training of Feedforward Neural Networks. In: Johnson J. D., Whinston A. B. (Eds), Advances in Artificial Intelligence in Economics, Finance, and Management, Vol. 1. JAI Press, Greewich, CT, 93–111
153. Dorsey R. E., Mayer W. J. (1995a) Genetic Algorithms for Estimation Problems with Multiple Optima, Nondifferentiability, and Other Irregular Features. Journal of Business and Economic Statistics **13(1)**, 53–66
154. Dorsey R. E., Mayer W. J. (1995b) Optimization Using Genetic Algorithms. In: Johnson J. D., Whinston A. B. (Eds), Advances in Artificial Intelligence in Economics, Finance, and Management, Vol. 1. JAI Press, Greewich, CT, 69–91
155. Dosi G., Marengo L., Bassanini A., Valente M. (1999) Norms as Emergent Properties of Adaptive Learning: The Case of Economic Routines. Journal of Evolutionary Economics **9(1)**, 5–26
156. Duffy J. (2001) Learning to Speculate: Experiments with Artificial and Real Agents. Journal of Economic Dynamics and Control **25**, 295–319
157. Duffy J., Feltovich N. (1999) Does Observation of Others Affect Learning in Strategic Environments? An Experimental Study. International Journal of Game Theory **28**, 131–152
158. Duffy J., McNelis P. D. (2001) Approximating and Simulating the Stochastic Growth Model: Parameterized Expectations, Neural Networks, and the Genetic Algorithm. Journal of Economic Dynamics and Control **25(9)**, 1273–1303
159. Dworman G., Kimbrough S. O., Laing J. D. (1996) Bargaining by Artificial Agents in Two Coalition Games: A Study in Genetic Programming for Electronic Commerce. In: Koza J., Goldberg D., Fogel D., Riolo R. (Eds.), Genetic Programming 1996: Proceedings of the First Annual Conference. MIT Press, Cambridge, MA, 54–62
160. Edmonds B., Moss S. J. (1997) Modeling Bounded Rationality Using Evolutionary Techniques. In: AISB'97 Workshop on Evolutionary Computation, Lecture Notes in Computer Science 1305, 31–42
161. Edmonds B. (1998) Gossip, Sexual Recombination and the El Farol Bar: Modeling the Emergence of Heterogeneity. In: Computation in Economics, Finance and Engineering: Economic Systems. Society for Computational Economics, Cambridge, England.
162. Eglit J. T. (1994) Trend Prediction in Financial Time Series. In: Koza J. (Ed.), Genetic Algorithms at Stanford. Stanford Bookstore, Stanford, CA, 31–40
163. Eiben A. E., Koudijs A. E., Slisser F. (1998), Genetic Modelling of Customer Retention. In: Banzhaf W., Poli R., Schoenauer M., Fogarty T. C. (Eds.), Genetic Programming: Proceedings of the First European Workshop on Genetic Programming. Springer, 178–186
164. Eriksson A., Lindgren K. (2001) Evolution of Strategies in Repeated Stochastic Games with Full Information of the Payoff Matrix. In: Spector L., Goodman E. D., Wu A., Langdon W. B., Voigt H.-M., Gen M., Sen S., Dorigo M., Pezeshk S., Garzon M. H., Burke E. (Eds.), Proceedings of the Genetic and Evolutionary Computation Conference. Morgan Kaufmann, 853–859

165. Esparcia-Alcazar A. I., Sharman K. C. (1996) Some Applications of Genetic Programming in Digital Signal Processing. In: Koza J. (Ed.), Late Breaking Papers at the Genetic Programming 1996 Conference. Stanford Bookstore, Stanford, CA, 24–31
166. Eymann T., Padovan B., Schoder D. (1998) Artificial Coordination–Simulating Organizational Changes with Artificial Life Agents. In: Computation in Economics, Finance and Engineering: Economic Systems. Society for Computational Economics, Cambridge, England.
167. Farley A. M., Jones S. (1994) Using a Genetic Algorithm to Determine an Index of Leading Economic Indicators. Computational Economics **7(3)**, 163–173
168. Fernandez T., Evett M. (1997) Training Period Size and Evolved Trading Systems. In: Koza J. R., Banzhaf W., Chellapilla K., Deb K., Dorigo M., Fogel D. B., Garzon M. H., Goldberg D. E., Iba H., Riolo R. (Eds.), Genetic Programming 1997: Proceedings of the Second Annual Conference. Morgan Kaufmann, 95
169. Fernandez T., Evett M. (1998a) Numeric Mutation as an Improvement to Symbolic Regression in Genetic Programming. In: Porto V. W., Saravanan N., Waagen D., Eiben A. E. (Eds.), Evolutionary Programming VII, Lecture Notes in Computer Science, Vol. 1447. Springer-Verlag, Berlin, 251–260
170. Fernandez T., Evett M. (1998b) Numeric Mutation Improves the Discovery of Numeric Constants in Genetic Programming. In: Koza J., Banzhaf W., Chellapilla K., Deb K., Dorigo M., Foegl D., Garson M., Goldberg D., Iba H., Riolo R. (Eds.), Proceedings of the Third Annual Genetic Programming Conference. Morgan Kaufmann Publishers, San Francisco, CA, 66–71
171. Ficici S. G., Melnik O., Pollack J. (2000) A Game-Theoretic Investigation of Selection Methods Used in Evolutionary Algorithms. In: Congress on Evolutionary Computation 2000, Vol. 2. IEEE Press, 880–887
172. Fischer M. M., Leung Y. (1998) A Genetic-Algorithms Based Evolutionary Computational Neural Network for Modelling Spatial Interaction Data. Annals of Regional Science **32(3)**, 437–458
173. Fogel D. B. (1993) Evolving Behaviors in the Iterated Prisoner's Dilemma. Evolutionary Computation **1(1)**, 77–97
174. Fogel D. B. (1995) On the Relationship between the Duration of an Encounter and the Evolution of Cooperation in the Iterated Prisoner's Dilemma. Evolutionary Computation **3**, 349–363
175. Fogel D. B., Fogel L. J. (1996) Preliminary Experiments on Discriminating between Chaotic Signals and Noise Using Evolutionary Programming. In: Koza J., Goldberg D., Fogel D., Riolo R. (Eds.), Genetic Programming 1996: Proceedings of the First Annual Conference. MIT Press, Cambridge, MA, 512–521
176. Fogel D. B., Harrald P. G. (1994) Evolving Continuous Behaviors in the Iterated Prisoner's Dilemma. In: Sebald A., Fogel L. (Eds.), The Third Annual Conference on Evolutionary Programming. World Scientific, Singapore, 119–130
177. Fogel D. B., Chellapilla K., Angeline P. J. (1999) Inductive Reasoning and Bounded Rationality Reconsidered. IEEE Transactions on Evolutionary Computation **3(2)**, 142–146
178. Fong L. Y., Szeto K. Y. (2000) Rules Extraction in Short Memory Time Series Using Genetic Algorithms. Forthcoming in European Journal of Physics B.

179. Franke R. (1998) Coevolution and Stable Adjustments in the Cobweb Model. Journal of Evolutionary Economics **8(4)**, 383–406
180. Frick A., Herrmann R., Kreidler M., Narr A., Seese D. (1996) A Genetic-Based Approach for the Derivation of Trading Strategies on the German Stock Market. In: Amari S., Xu L., Chan L., King I., Leung K. (Eds.), Progress in Neural Information Processing, Vol. 2. Springer-Verlag, Singapore, 766–770
181. Fritz S., Hosemann D. (1998) Behavior Scoring for Deutsche Bank's German Corporates. In: Nakhaeizadeh G., Steurer E. (Eds.), Application of Machine Learning and Data Mining in Finance, 179–194
182. Fu K., Xu W. H. (1997) Training Neural Networks with Genetic Algorithms for Forecasting the Stock Price Index. In: Proceedings of the 1997 IEEE International Conference on Intelligent Processing Systems, Vol. 1. IEEE Press, 401–403
183. Fu T.-C., Chung F.-L., Ng V., Luk R. (2001) Evolutionary Segmentation of Financial Time Series into Subsequences. In: Proceedings of the 2001 Congress on Evolutionary Computation, Vol. 1. IEEE Press, 426–430
184. Fujiki C., Dickinson J. (1987) Using the Genetic Algorithm to Generate Lisp Source Code to Solve the Prisoner's Dilemma. In: Grefenstette J. J. (Ed.), Genetic Algorithms and their Applications, Proceedings of the 2nd. International Conference on Genetic Algorithms. Lawrence Erlbaum, Hillsdale, NJ, 236–240
185. Fyfe C., Marney J. P., Tarbert H. (1999) Technical Analysis Versus Market Efficiency: A Genetic Programming Approach. Applied Financial Economics **9**, 183–191
186. Gaivoronski A. A. (1999) Modeling of Complex Economic Systems with Agent Nets. In: Banzhaf W., Daida J., Eiben A. E., Garzon M. H., Honavar V., Jakiela M., Smith R. E. (Eds.), Proceedings of the Genetic and Evolutionary Computation Conference, Vol. 2. Morgan Kaufmann, 1265–1272
187. Gordillo F., Alcala I., Aracil J. (1999) A Tool for Solving Differential Games with Co-Evolutionary Algorithms. In: Banzhaf W., Daida J., Eiben A. E., Garzon M. H., Honavar V., Jakiela M., Smith R. E. (Eds.), Proceedings of the Genetic and Evolutionary Computation Conference, Vol. 2. Morgan Kaufmann, 1535–1542
188. Hackworth T. (1999) India and Pakistan, a Classic "Richardson" Arms Race: A Genetic Algorithm Approach. In: Banzhaf W., Daida J., Eiben A. E., Garzon M. H., Honavar V., Jakiela M., Smith R. E. (Eds.), Proceedings of the Genetic and Evolutionary Computation Conference, Vol. 2. Morgan Kaufmann, 1543–1550
189. Harland Z. (1999) Using Nonlinear Neurogenetic Models with Profit Related Objective Functions to Trade the US T-Bond Future. In: Abu-Mostafa Y. S., LeBaron B., Lo A. W., Weigend A. S. (Eds.), Computational Finance 1999. MIT Press, Cambridge, MA, 327–342
190. Harrald P. G. (1996) Evolutionary Algorithms and Economic Models: A View. In: Angeline P., Back T., Fogel D. (Eds.), Evolutionary Programming: Proceeding of the Fifth Annual Conference on Evolutionary Programming. MIT Press, Cambridge, MA, 3–8
191. Harrald P. G., Fogel D. B. (1996) Evolving Continuous Behaviors in the Iterated Prisoner's Dilemma. BioSystems **37(1-2)**, 135–145
192. Harrald P. G., Kamstra M. (1997) Evolving Artificial Neural Networks to Combine Financial Forecasts. IEEE Transactions on Evolutionary Computation **1(1)**, 40–52

193. Heymann D., Pearzzo R. P. J., Schuschny A. (1998) Learning and Contagion Effects in Transitions between Regimes: Some Schematic Multi-Agents Models. Journal of Management and Economics **2**. <http://www.econ.uba.ar/www/servicios/publicaciones/journal2/contents/contents.htm>
194. Hiden H., McKay B., Willis M., Montague G. (1998) Non-Linear Partial Least Squares Using Genetic Programming. In: Koza J., Banzhaf W., Chellapilla K., Deb K., Dorigo M., Foegl D., Garson M., Goldberg D., Iba H., Riolo R. (Eds.), Proceedings of the Third Annual Genetic Programming Conference. Morgan Kaufmann Publishers, San Francisco, CA, 128–135
195. Hiemstra Y. (1996) Applying Neural Networks and Genetic Algorithms to Tactical Asset Allocation. Neurove$t Journal **4(3)**, 8–15
196. Ho T. H. (1996) Finite Automata Play Repeated Prisoner's Dilemma with Information Processing Costs. Journal of Economic Dynamics and Control **20(1-3)**, 173–207
197. Hoffmann R., Waring N. G. (1997) Complexity Cost and Two Types of Noise in the Repeated Prisoner's Dilemma. In: Smith G. D., Steele N. C., Albrecht R. F. (Eds.), Artificial Neural Networks and Genetic Algorithms. Springer-Verlag, 619–623
198. Holland J. (1988) The Global Economy as An Adaptive Process. In: Arthur W. B., Durlauf S. N, Lane D. A. (Eds.), The Economy as an Evolving Complex System, SFI Studies in the Science of Complexity. Perseus Books Publishing, 117–124
199. Holland J., Miller J. (1991) Artificial Adaptive Agents in Economic Theory. American Economic Review **81(2)**, 365–370
200. Hillebrand E., Stender J. (Eds.) (1994) Many-Agent Simulation and Artificial Life. IOS Press, Amsterdam.
201. Huck S., Müller W., Strobel M. (1999) On the Emergence of Attitudes Towards Risk. In: Brenner T. (Ed.), Computational Techniques for Modelling Learning in Economics, The Series Advances in Computational Economics 11. Kluwer, Dordrecht, 123–144
202. Iba H., Kurita T., de Garis H., Sato T. (1993) System Identification Using Structured Genetic Algorithms. In: Forrest S. (Ed.), Proceedings of the Fifth International Conference on Genetic Algorithms. Morgan Kaufmann, San Mateo, CA, 279–286
203. Iba H., Nikolaev N. (2000) Genetic Programming Polynomial Models of Financial Data Series. In: Congress on Evolutionary Computation 2000, Vol. 2. IEEE Press, 1459–1466
204. Iba H., Sasaki T. (1999) Using Genetic Programming to Predict Financial Data. In: Angeline P. J., Michalewicz Z., Schoenauer M., Yao X., Zalzala A. (Eds.), Proceedings of the Congress on Evolutionary Computation, Vol. 1. IEEE Press, 244–251
205. Iglehart D. L., Voessner S. (1998) Optimization of a Trading System Using Global Search Techniques and Local Optimization. Journal of Computational Intelligence in Finance **6(6)**, 36–46
206. Irani Z., Sharif A. (1997) Genetic Algorithm Optimisation of Investment Justification Theory. In: Koza J. R. (Ed.), Late Breaking Papers at the Genetic Programming 1997 Conference, 87–92

207. Ishibuchi H., Nakari T., Nakashima T. (2000) Evolution of Strategies in Spatial IPD Games with Structured Demes. In: Whitley D., Goldberg D., Cantú-Paz E., Spector L., Parmee I., Beyer H.-G. (Eds.), Proceedings of the Genetic and Evolutionary Computation Conference. Morgan Kaufmann, 817–824
208. Ishibuchi H., Sakamoto R., Nakashima T. (2001) Effect of Localized Selection on the Evolution of Unplanned Coordination in a Market Selection Game. In: Proceedings of the 2001 Congress on Evolutionary Computation, Vol. 2. IEEE Press, 1011–1018
209. Ivanova P. J., Tagarev T. D., Hunter A. (1997) Selection of Indicators for Early Warning of Violent Political Conflicts by Genetic Algorithms. In: Proceedings of International Conference on Transition to Advanced Market Institutions and Economies, 180–183
210. Izumi K., Okatsu T. (1996) An Artificial Market Analysis of Exchange Rate Dynamics. In: Fogel L. J., Angeline P. J., Bäck T. (Eds.), Evolutionary Programming V. MIT Press, 27–36
211. Izumi K., Ueda K. (1998) Emergent Phenomena in a Foreign Exchange Market: Analysis Based on An Artificial Market Approach. In: Adami C., Belew R. K., Kitano H., Taylor C. E. (Eds.), Artificial Life VI. MIT Press, 398–402
212. Izumi K., Ueda K. (1999) Analysis of Dealers' Processing Financial News Based on An Artificial Market Approach. Journal of Computational Intelligence in Finance **7**, 23–33
213. Izumi K., Ueda K. (2000) Learning of Virtual Dealers in An Artificial Market: Comparison with Interview Data. In: Leung K. S., Chan L.-W., Meng H. (Eds.), Intelligent Data Engineering and Automated Learning- IDEAL 2000: Data Mining, Financial Engineering, and Intelligent Agents, Lecture Notes in Computer Sciences 1983. Springer, 511–516
214. Jang G., Lai F., Parng T. (1993) Intelligent Stock Trading Decision Support System Using Dual Adaptive-Structure Neural Networks. Journal of Information Science and Engineering **9**, 271–297
215. Jay White A. (1998) A Genetic Adaptive Neural Network Approach to Pricing Options: a Simulation Analysis. Journal of Computational Intelligence in Finance **6(5)**, 13–23
216. Joshi S., Bedau M. (1998) An Explanation of Generic Behavior in An Evolving Financial Market. In: Standish R., Henry B., Marks R., Stocker R., Green D., Keen S., Bossomaier T. (Eds.), Complex Systems'98. Complexity between the Ecos: From Ecology to Economics, 327–335
217. Joshi S., Parker J., Bedau M. A. (1999) Technical Trading Creates a Prisoner's Dilemma: Results from An Agent-Based Model. In: Abu-Mostafa Y. S., LeBaron B., Lo A. W., Weigend A. S. (Eds.), Computational Finance 1999. MIT Press, Cambridge, MA, 465–479
218. Julstrom B. A. (1996) Contest Length, Noise, and Reciprocal Altruism in the Population of a Genetic Algorithm for the Iterated Prisoner's Dilemma. In: Koza J. (Ed.), Late Breaking Papers at the Genetic Programming 1996 Conference. Stanford Bookstore, Stanford, CA, 88–93
219. Kaboudan M. A. (1998a) Forecasting Stock Returns Using Genetic Programming in C++. In: Cook D. (Ed.), FLAIRS Proceedings of the Eleventh International Florida Artificial Intelligence Research Symposium Conference. AAAI Press, Menlo Park, CA, 73–77

220. Kaboudan M. A. (1998b) A GP Approach to Distinguish Chaotic form Noisy Signals. In: Koza J., Banzhaf W., Chellapilla K., Deb K., Dorigo M., Foegl D., Garson M., Goldberg D., Iba H., Riolo R. (Eds.), Proceedings of the Third Annual Genetic Programming Conference. Morgan Kaufmann Publishers, San Francisco, CA, 187–191
221. Kaboudan M. A. (1999) A Measure of Time Series's Predictability Using Genetic Programming Applied to Stock Returns. Journal of Forecasting **18**, 345–357
222. Kaboudan M. A. (2000) Genetic Programming Prediction of Stock Prices. Computational Economics **16(3)**, 207–236
223. Kaboudan M. A. (2001a) Compumetric Forecasting of Crude Oil Prices. In: Proceedings of the 2001 Congress on Evolutionary Computation, Vol. 1. IEEE Press, 283–287
224. Kaboudan M. A. (2001b) Genetically Evolved Models and Normality of Their Fitted Residuals. Journal of Economic Dynamics and Control **25(11)**, 1719–1749
225. Kaboudan M. A., Vance M. (1998) Statistical Evaluation of Symbolic Regression Forecasting of Time-Series. In: Proceedings of the International Federation of Automatic Control Symposium on Computation in Economics, Finance and Engineering: Economic Systems. Cambridge, UK.
226. Kargupta H., Buescher K. (1996) The Gene Expression Messy Genetic Algorithm for Financial Application. In: Proceedings of the IEEE/IAFE 1996 Conference on Computational Intelligence for Financial Engineering. IEEE Press, 155–161
227. Katz J. D., McCormick D. L. (1994) Neurogenetics and Its Use in Trading System Development. Neurove$t Journal **2(4)**, 8–11
228. Keber C. (2000) Option Valuation with the Genetic Programming Approach. In: Abu–Mostafa Y. S., LeBaron B., Lo A. W., Weigend A. S. (Eds.), Computational Finance – Proceedings of the Sixth International Conference. MIT Press, Cambridge, MA, 689–703
229. Keber C., Schuster M. G. (2001) Evolutionary Computation and the Vega Risk of American Put Options. IEEE Transactions on Neural Networks **12(4)**, 704–715
230. Kim Y. W., Park D. J. (1996) Genetic Algorithm Approach for Obtaining Guidance Laws in Differential Game. In: Filar J. A., Gaitsgory V., Imado K. (Eds.), Proceedings of the Seventh International Symposium on Dynamic Games and Applications, Vol. 2, 461–468
231. Kirman A. P., Vriend N. (2001) Evolving Market Structure: An ACE Model of Price Dispersion and Loyalty. Journal of Economic Dynamics and Control **25(3-4)**, 459–502
232. Klos T. B. (1997) Spatially Coevolving Automata Play the Repeated Prisoner's Dilemma. In: Conte R., Hegselmann R., Terna P. (Eds.), Simulating Social Phenomena. Springer-Verlag, 153–159
233. Kosinski J. (1997) A Possibility of Application of Genetic Algorithms in Investment Project Cash-Flow Optimization. In: Proceedings of International Conference on Transition to Advanced Market Institutions and Economies, 234–239
234. Koza J. R. (1992) A Genetic Approach to Econometric Modeling. In: Bourgine P., Walliser B. (Eds.), Economics and Cognitive Science. Pergamon Press, 57–75

235. Koza J. R. (1995) Genetic Programming for Economic Modelling. In: Goonatilake S., Treleaven P. (Eds.), Intelligent Systems for Finance and Business. John Wiley & Sons, 251–269
236. Kuo R. J., Cheng C. H., Huang Y. C. (1999) Genetic Algorithm Initiated Fuzzy Neural Network with Application to Stock Market. In: Proceedings of the Eighth International Fuzzy Systems Association World Congress, Vol. 1, 372–376
237. Kwasnicki W., Kwasnicka H. (1992) Market, Innovation, Competition: An Evolutionary Model of Industrial Dynamics. Journal of Economic Behavior and Organization **19**, 343–368
238. Kwasnicki W. (1996) Knowledge, Innovation, and Economy. An Evolutionary Exploration. Edward Elgar, Cheltenham, UK.
239. Kwasnicki W., Kwasnicka H. (1997) Genetic Operators in An Evolutionary Model of Industrial Dynamics. In: Proceedings of the 3rd International Mendel Conference on Genetic Algorithms, Optimization Problems, Fuzzy Logic, Neural Networks, Rough Sets, 77–82
240. Lam S. S. (2001) A Genetic Fuzzy Expert System for Stock Market Timing. In: Proceedings of the 2001 Congress on Evolutionary Computation, Vol. 1. IEEE Press, 410–417
241. Lane D., Kroujiline A., Petrov V., Sheblé G. (2000) Electricity Market Power: Marginal Cost and Relative Capacity Effects. In: Congress on Evolutionary Computation 2000, Vol. 2. IEEE Press, 1048–1055
242. LeBaron B. (2000) Agent Based Computational Finance: Suggested Reading and Early Research. Journal of Economic Dynamics and Control **24**, 679–702
243. LeBaron B. (2001) Evolution and Time Horizons in An Agent Based Stock Market. Forthcoming in Macroeconomic Dynamics.
244. LeBaron B., Arthur W. B., Palmer R. (1999) Time Series Properties of An Artificial Stock Market. Journal of Economic Dynamics and Control **23**, 1487–1516
245. Lensberg T. (1997) A Genetic Programming Experiment on Investment Behavior under Knightian Uncertainty. In: Koza J. R., Banzhaf W., Chellapilla K., Deb K., Dorigo M., Fogel D. B., Garzon M. H., Goldberg D. E., Iba H., Riolo R. (Eds.), Genetic Programming 1997: Proceedings of the Second Annual Conference. Morgan Kaufmann, 231–239
246. Lensberg T. (1999) Investment Behavior under Knightian Uncertainty – An Evolutionary Approach. Journal of Economic Dynamics and Control **23**, 1587–1604
247. Lent B. (1994) Evolution of Trader Strategies Using Genetic Algorithms and Genetic Programming. In: Koza J. R. (Ed.), Genetic Algorithms at Stanford 1994. Standford Bookstore, Standford, CA, 87–98
248. Lettau M. (1997) Explaining the Facts with Adaptive Agents: the Case of Mutual Fund Flows. Journal of Economic Dynamics and Control **21(7)**, 1117–1147
249. Leinweber D., Arnott R. (1995) Quantitative and Computational Innovation in Investment Management. Journal of Portfolio Management **21(2)**, 8–15
250. Lindgren K. (1991) Evolutionary Phenomena in Simple Dynamics. In: Langton C., Taylor C., Farmer J., Rasmussen S., (Eds.), Artificial Life II, Vol. 10. Addison-Wesley, Reading, 295–324

251. Lis J., Czapkiewicz A. (1996) The Loghyperbolic Distribution Fitting by Genetic Algorithm. In: Angeline P., Back T., Fogel D. (Eds.), Evolutionary Programming: Proceeding of the Fifth Annual Conference on Evolutionary Programming. MIT Press, Cambridge, MA, 369–376
252. Liu Y., Yao X. (2001) Evolving Neural Networks for Hang Seng Stock Index Forecast. In: Proceedings of the 2001 Congress on Evolutionary Computation, Vol. 1. IEEE Press, 256–260
253. Loraschi A., Tettamanzi A. (1996) An Evolutionary Algorithm for Portfolio Selection within a Downside Framework. In: Dunis C. (Ed.), Forecasting Financial Markets: Exchange Rates, Interest Rates and Asset Management. John Wiley & Sons, 275–285
254. Macy M. (1996) Natural Selection and Social Learning in Prisoner's Dilemma: Co-Adaptation with Genetic Algorithms and Artificial Neural Networks. In: Liebrand W. B., Messick D. M. (Eds.), Frontiers in Social Dilemmas Research. Springer, Heidelberg and New York, 235–265
255. Mahfoud S., Mani G. (1996) Financial Forecasting Using Genetic Algorithms. Applied Artificial Intelligence **10**, 543–565
256. Marengo L., Tordjman H. (1996) Speculation, Heterogeneity and Learning: A Simulation Model of Exchange Rates Dynamics. Kyklos **49(3)**, 407–438
257. Margarita S., Beltratti A. (1993) Stock Prices and Volume in An Artificial Adaptive Stock Market. In: New Trends in Neural Computation: International Workshop on Artificial Networks. Springer Verlag, Berlin, 714–719
258. Marimon R. (1993) Adaptive Learning, Evolutionary Dynamics and Equilibrium Selection in Games. European Economic Review **37**, 603–611
259. Marimon R., McGrattan E., Sargent T. (1990) Money as a Medium of Exchange in An Economy with Artificially Intelligent Agents. Journal of Economic Dynamics and Control **14**, 329–374
260. Marin F. J., Sandoval F. (1997) Electric Load Forecasting with Genetic Neural Networks. In: Smith G. D., Steele N. C., Albrecht R. F. (Eds.), Artificial Neural Networks and Genetic Algorithms. Springer-Verlag, 49–52
261. Markose S., Tsang E., Er H., Salhi A. (2001) Evolutionary Arbitrage for FTSE-100 Index Options and Futures. In: Proceedings of the 2001 Congress on Evolutionary Computation, Vol. 1. IEEE Press, 275–282
262. Marks R. E. (1989a) Niche Strategies: the Prisoner's Dilemma Computer Tournaments Revisited. AGSM Working Paper 89–009. `<http://www.agsm.edu.au/~bobm/papers/niche.pdf>`
263. Marks R. E. (1989b) Breeding Optimal Strategies: Optimal Behavior for Oligopolists. In: Schaffer J. D. (Ed.), Proceedings of the Third International Conference on Genetic Algorithms. Morgan Kaufmann, San Mateo, CA, 198–207
264. Marks R. E. (1992a) Repeated Games and Finite Automata. In: Creedy J., Borland J., Eichberger J. (Eds), Recent Developments in Game Theory. Edward Elgar, Aldershot, 43–64
265. Marks R. E. (1992) Breeding Optimal Strategies: Optimal Behavior for Oligopolists. Journal of Evolutionary Economics **2(1)**, 17–28
266. Marks R. E. (1998) Evolved Perception and Behaviour in Oligopolies. Journal of Economic Dynamics and Control **22**(8–9), 1209–1233
267. Marks R. E., Schnabl H. (1999) Genetic Algorithms and Neural Networks: A Comparison Based on the Repeated Prisoner's Dilemma. In: Brenner T. (Ed.),

Computational Techniques for Modelling Learning in Economics, The Series Advances in Computational Economics 11. Kluwer, Dordrecht, 197–219
268. Marks R. E., Midgley D. F., Cooper L. G. (1995) Adaptive Behavior in An Oligopoly. In: Biethahn J., Nissen V. (Eds.), Evolutionary Algorithms in Management Applications. Springer-Verlag, Berlin, 225–239
269. Marks R. E., Midgley D. F., Cooper L. G. (1998) Refining the Breeding of Hybrid Strategies. Australian Graduate School of Management Working Paper 98–017, Sydney. `<http://www.agsm.edu.au/~bobm/papers/wp98017.html>`
270. Marks R. E., Midgley D. F., Cooper L. G. (2000) Breeding Better Strategies in Oligopolistic Price Wars. Paper submitted to the IEEE Transactions on Evolutionary Computation, special issue on Agent-Based Modelling of Evolutionary Economic Systems, INSEAD Working Paper 2000/65/MKT. `<http://www.agsm.edu.au/~bobm/papers/ieee-marks.html>`
271. Marks R. E., Midgley D. F., Cooper L. G., Shiraz G. M. (1998) The Complexity of Competitive Marketing Strategies. In: Standish R., Henry B., Marks R., Stocker R., Green D., Keen S., Bossomaier T. (Eds.) Complex Systems'98. Complexity between the Ecos: From Ecology to Economics, 336–345
272. Marks R. E., Midgley D. F., Cooper L. G., Shiraz G. M. (1999) Coevolution with Genetic Algorithms: Repeated Differential Oligopolies. In: Banzhaf W., Daida J., Eiben A. E., Garzon M. H., Honavar V., Jakiela M., Smith R. E. (Eds.), Proceedings of the Genetic and Evolutionary Computation Conference, Vol. 2. Morgan Kaufmann, 1609–1615
273. McCain R. A. (1994) Genetic Algorithms, Teleological Conservatism, and the Emergence of Optimal Demand Relations: the Case of Stable Preferences. Computational Economics **7(3)**, 187–202
274. McFadzean D., Tesfatsion L. (1999) A C++ Platform for the Evolution of Trade Networks. Computational Economics **14**, 109–134
275. Mehta K., Bhattacharyya S. (1999) Combining Rules Learnt Using Genetic Algorithms for Financial Forecasting. In: Proceedings of the 1999 Congress on Evolutionary Computation. IEEE Press, 1245–1252
276. Midgley D. F., Marks R. E., Cooper L. G. (1997) Breeding Competitive Strategies. Management Science **43 (3)**, 257–275
277. Miller J. H. (1996) The Coevolution of Automata in the Repeated Prisoner's Dilemma. Journal of Economic Behavior and Organization **29(1)**, 87–112
278. Miller J. H., Shubik M. (1994) Some Dynamics of a Strategic Market Game with a Large Number of Agents. Journal of Economics **60(1)**, 1–28
279. Muhammad A. King G. A. (1997) Foreign Exchange Market Forecasting Using Evolutionary Fuzzy Networks. In: Proceedings of the IEEE/IAFE 1997 Computational Intelligence for Financial Engineering. IEEE Press, 213–219
280. Mulloy B. S., Riolo R. L., Savit R. S. (1996) Dynamics of Genetic Programming and Chaotic Time Series Prediction. In: Koza J., Goldberg D., Fogel D., Riolo R. (Eds.), Genetic Programming 1996: Proceedings of the First Annual Conference. MIT Press, Cambridge, MA, 166–174
281. Mundhe M., Sen S. (2000) Evolving Agent Societies that Avoid Social Dilemmas. In: Whitley D., Goldberg D., Cantú-Paz E., Spector L., Parmee I., Beyer H.-G. (Eds.), Proceedings of the Genetic and Evolutionary Computation Conference. Morgan Kaufmann, 809–816
282. Muzyka D. F., De Koning A. J. (1996) Towards a Theoretical Model for Adaptive Entrepreneurial Organizations Using Genetic Algorithms. INSEAD Working Paper: 96/85/ENT

283. Mydlowec W. (1997) Discovery by Genetic Programming of Empirical Macroeconomic Models. In: Koza J. R. (Ed.), Genetic Algorithms and Genetic Programming at Stanford 1997. Stanford Bookstore, 168–177
284. Naing T. T., Mutoh A., Inuzuka N., Itoh H. (2000) A Framework in Which Rational Agents Yield Communal Profit. In: Proceedings of the Second Asia-Pacific Conference on Genetic Algorithms and Applications. Global Link Publishing Company, Hong Kong, 350–358
285. Neely C., Weller P., Dittmar R. (1997) Is Technical Analysis in the Foreign Exchange Market Profitable? A Genetic Programming Approach. Journal of Financial and Quantitative Analysis **32(4)**, 405–426
286. Neely C. J., Weller P. A. (1999) Technical Trading Rules in the European Monetary System. Journal of International Money and Finance **18(3)**, 429–458
287. Neubauer A. (1997) Prediction of Nonlinear and Nonstationary Time Series Using Self-Adaptive Evolutionary Strategies with Individual Memory. In: Back T. (Ed.), Proceedings of the Seventh International Conference on Genetic Algorithms. Morgan Kaufmann Publishers, San Francisco, CA, 727–734
288. Nicolaisen J., Smith M., Petrov V., Tesfatsion L. (2000) Concentration and Capacity Effects on Electricity Market Power. In: Congress on Evolutionary Computation 2000, Vol. 2. IEEE Press, 1041–1047
289. Nikolaev N., Iba H. (2001a) Genetic Programming Using Chebishev Polynomials. In: Spector L., Goodman E. D., Wu A., Langdon W. B., Voigt H.-M., Gen M., Sen S., Dorigo M., Pezeshk S., Garzon M. H., Burke E. (Eds.), Proceedings of the Genetic and Evolutionary Computation Conference. Morgan Kaufmann, 89–96
290. Nikolaev N. Y., Iba H. (2001b) Genetic Programming of Polynomial Models for Financial Forecasting. Forthcoming in IEEE Transactions on Evolutionary Computation.
291. Noe T. H., Pi L. (2000) Learning Dynamics, Genetic Algorithms, and Corporate Takeovers. Journal of Economic Dynamics and Control **24 (2)**, 189–217
292. Noever D., Baskaran S. (1994) Genetic Algorithms Trading on the S&P 500. AI in Finance **1**, 41
293. Novkovic S. (1998) A Genetic Algorithm Simulation of a Transition Economy: An Application to Insider-Privatization in Croatia. Computational Economics **11(3)**, 221–243
294. Oakley E. H. N. (1994a) The Application of Genetic Programming to the Investigation of Short, Noisy, and Chaotic Data Series. In: Fogarty (Ed.), Evolutionary Programming, Lecture Notes in Computer Science 865. Springer-Verlag, Berlin, Germany, 320–332
295. Oakley E. H. N. (1994b) Two Scientific Applications of Genetic Programming: Stack Filters and Non-Linear Equation Fitting to Chaotic Data. In: Kinnear K. E. Jr. (Ed.) Advances in Genetic Programming. MIT Press, Cambridge, MA, 369–389
296. Oakley E. H. N. (1995) Genetic Programming as a Means of Assessing and Reflecting Chaos. In: Genetic Programming, AAAI-95 Fall Symposium Series, American Association for Artificial Intelligence, 68–72
297. Oakley E. H. N. (1996) Genetic Programming, the Reflection of Chaos, and the Bootstrap: Toward a Useful Test for Chaos. In: Koza J., Goldberg D., Fogel D., Riolo R. (Eds.), Genetic Programming 1996: Proceedings of the First Annual Conference. MIT Press, Cambridge, MA, 175–181

298. Oh J. C. (1999) Ostracism for Improving Cooperation in the Iterated Prisoner's Dilemma Game. In: Angeline P. J., Michalewicz Z., Schoenauer M., Yao X., Zalzala A. (Eds.), Proceedings of the Congress on Evolutionary Computation, Vol. 2. IEEE Press, 891-896
299. Oh J. C. (2000) Benefits of Clustering among the Internet Search Agents Caught in the N-Person Prisoner's Dilemma Game. In: Congress on Evolutionary Computation 2000, Vol. 2. IEEE Press, 864–871
300. Olsson L. (2000) Evolution of Bargaining Strategies for Continuous Double Auction Markets Using Genetic Programming. In: Wang P. (Ed.), Proceedings of the Fifth Joint Conference on Information Sciences, Vol. II, 961–964
301. O'Neill M., Brabazon T. (2001) Developing a Market Timing System Using Grammatical Evolution. In: Spector L., Goodman E. D., Wu A., Langdon W. B., Voigt H.-M., Gen M., Sen S., Dorigo M., Pezeshk S., Garzon M. H., Burke E. (Eds.), Proceedings of the Genetic and Evolutionary Computation Conference. Morgan Kaufmann, 1375–1381
302. Ostermark R. (1999) Solving Irregular Econometric and Mathematical Optimization Problems with a Genetic Hybrid Algorithm. Computational Economics **13(2)**, 103–115
303. Oussaidene M., Chopard B., Pictet O., Tomassini M. (1996) Parallel Genetic Programming: An Application to Trading Models Evolution. In: Koza J., Goldberg D., Fogel D., Riolo R. (Eds.), Genetic Programming 1996: Proceedings of the First Annual Conference. MIT Press, Cambridge, MA, 357–362
304. Ozyildirim S. (1996) Three-Country Trade Relations: A Discrete Dynamic Game Approach. Computers and Mathematics with Applications **32**, 43–56
305. Ozyildirim S. (1997) Computing Open-Loop Noncooperative Solution in Discrete Dynamic Games. Journal of Evolutionary Economics **7(1)**, 23–40
306. Ozyildirim S., Alemdar N. M. (2000) Learning the Optimum as a Nash Equilibrium. Journal of Economic Dynamics and Control **24(4)**, 483–499
307. Palmer R. G., Arthur W. B., Holland J. H., LeBaron B., Tayler P. (1994) Artificial Economic Life: A Simple Model of a Stockmarket. Physica D **75**, 264–274
308. Page S. E. (1996) Two Measures of Difficulty. Economic Theory **8(2)**, 321–346
309. Pereira R. (1997) Genetic Algorithms and Trading Rules. In: Creedy J., Martin V. L. (Eds.), Nonlinear Economic Models: Cross-sectional, Time Series and Neural Network Applications, 191–210
310. Phelan S. E. (1997) Innovation and Imitation as Competitive Strategies: Revisiting a Simulation Approach. In: Conte R., Hegselmann R., Terna P. (Eds.), Simulating Social Phenomena. Springer-Verlag, 385–400
311. Pictet O. V., Docorogna M. M., Chopard B., Shirru M. O. R., Tomassini M. (1995) Using Genetic Algorithms for Robust Optimization in Financial Applications. In: Parallel Problem Solving from Nature–Applications in Statistics and Economics Workshop, Germany.
312. Rao S. S., Chellapilla K. (1996) Evolving Reduced Parameter Bilinear Models for Time Series Prediction Using Fast Evolutionary Programming. In: Koza J., Goldberg D., Fogel D., Riolo R. (Eds.), Genetic Programming 1996: Proceedings of the First Annual Conference. MIT Press, Cambridge, MA, 528–535
313. Rao S. S., Birru H. K. Chellapilla K. (1999) Evolving Nonlinear Time-Series Models Using Evolutionary Programming. In: Proceedings of the 1999 Congress on Evolutionary Computation, Vol. 2. IEEE Press, 236–243

314. Reeves C. R. (1997) Genetic Algorithms for the Operations Researchers. INFORMS Journal on Computing **9(3)**, 231–250
315. Riechmann T. (1998) Learning How to Learn. Towards An Improved Mutation Operator within GA Learning Models. In: Computation in Economics, Finance and Engineering: Economic Systems. Society for Computational Economics, Cambridge, England.
316. Riechmann T. (1999) Learning and Behavioural Stability: An Economic Interpretation of Genetic Algorithms. Journal of Evolutionary Economics **9(2)**, 225–242
317. Riechmann T. (2001) Genetic Algorithm Learning and Evolutionary Games. Journal of Economic Dynamics and Control **25(6-7)**, 1019–1037
318. Robinson G., McIlroy P. (1995) Exploring Some Commercial Applications of Genetic Programming. In: Fogarty T. C. (Ed.), Evolutionary Computing. Lecture Notes in Computer Sciences. Springer-Verlag, Berlin.
319. Rodriguez-Vazquez K., Fleming P J. (1999) Genetic Programming for Dynamic Chaotic System Modelling. In: Proceedings of the 1999 Congress on Evolutionary Computation, Vol. 1. IEEE Press, 22–28
320. Rogers D. (1997) Evolutionary Statistics: Using a Genetic Algorithm and Model Reduction to Isolate Alternate Statistical Hypotheses of Experimental Data. In: Back T. (Ed.), Proceedings of the Seventh International Conference on Genetic Algorithms. Morgan Kaufmann Publishers, San Francisco, CA, 735–742
321. Rolf S., Sprave J., Urfer W. (1997) Model Identification and Parameter Estimation of ARMA Models by Means of Evolutionary Algorithms. In: Proceedings of the IEEE/IAFE 1997 Computational Intelligence for Financial Engineering. IEEE Press, 237–243
322. Rubinson T., Yager R. R. (1996) Fuzzy Logic and Genetic Algorithms for Financial Risk Management. In: Proceedings of the IEEE/IAFE 1996 Conference on Computational Intelligence for Financial Engineering. IEEE Press, 90–95
323. Rubinson T., Geotsi Georgetti (1996) Estimation of Subjective Preference Using Fuzzy Logic & Genetic Algorithms. In: Proceedings of the Sixth International Conference on Information Processing and Management of Uncertainty in Knowledge-Based Systems, Vol II, 781-786
324. Sargent T. J. (1993) Bounded Rationality in Macroeconomics. Oxford University Press, Oxford.
325. Sato Y., Nagaya S. (1996) Evolutionary Algorithms that Generate Recurrent Neural Networks for Learning Chaos Dynamics. In: Proceedings of 1996 IEEE International Conference on Evolutionary Computation. IEEE Press, 144-149
326. Sato H., Namatame A. (2001) Co-Evolution in Social Interactions. In: Proceedings of the 2001 Congress on Evolutionary Computation, Vol. 2. IEEE Press, 1109–1114
327. Sato H., Koyama Y., Kurumatani K., Shiozawa Y., Deguchi H. (2001) U-Mart: A Test Bed for Interdisciplinary Research into Agent-Based Artificial Markets. In: Aruka Y. (Ed.), Evolutionary Controversies in Economics: A New Transdisciplinary Approach. Springer-Verlag, Tokyo, 179–190
328. Schmertmann C. P. (1996) Functional Search in Economics Using Genetic Programming. Computational Economics **9(4)**, 275–298
329. Seo Y.-G., Cho S.-B., Yao X. (1999) Emergence of Cooperative Coalition in NIPD Game with Localization of Interaction and Learning. In: Angeline P. J.,

Michalewicz Z., Schoenauer M., Yao X., Zalzala A. (Eds.), Proceedings of the Congress on Evolutionary Computation, Vol. 1. IEEE Press, 877-884
330. Seredynski F., Grzenda M. (1997) Behavior of Heterogeneous Agents in the Prisoner's Dilemma. In: Proceedings of the 3rd International Mendel Conference on Genetic Algorithms, Optimization Problems, Fuzzy Logic, Neural Networks, Rough Sets, 256–261
331. Sexton R. S. (1998) Identifying Irrelevant Input Variables in Chaotic Time Series Problems: Using a Genetic Algorithm for Training Neural Networks. Journal of Computational Intelligence in Finance **6(5)**, 34–41
332. Sheta A. F., De Jong K. (1996) Parameter Estimation of Nonlinear Systems in Noisy Environments Using Genetic Algorithms. In: Proceedings of the 1996 IEEE International Symposium on Intelligent Control. IEEE Press, 360–365
333. Sheta A. F., De Jong K. (2000) Time-Series Forecasting Using GA-Tuned Radial Basis Functions. In: Wang P. (Ed.), Proceedings of the Fifth Joint Conference on Information Sciences, Vol. I, 1074–1077
334. Shoaf J., Foster J. (1996) A Genetic Algorithm Solution to the Efficient Set Problem: A Technique for Portfolio Selection Based on the Markowitz Model. In: Proceedings of the Decision Sciences Institute Annual Meeting.
335. Smith C. E. (1999) An Application of Genetic Programming to Investment System Optimization. In: Banzhaf W., Daida J., Eiben A. E., Garzon M. H., Honavar V., Jakiela M., Smith R. E. (Eds.), Proceedings of the Genetic and Evolutionary Computation Conference, Vol. 2. Morgan Kaufmann, 1798
336. Smith S. N. (1998) Trading Applications of Genetic Programming. Financial Engineering News **2(6)**.
337. Staudinger S. (1999) Money as Medium of Exchange – An Analysis with Genetic Algorithm. Pennsylvania Economic Review **8(1)**, 1–9
338. Swernofsky S. (1994) Product Review: MircoGA. Neurove$t Journal **2(4)**, 30–31
339. Szeto K. Y., Chan K. O., Cheung K. H. (1997) Application of Genetic Algorithms in Stock Market Prediction. In: Weigend A. S., Abu-Mostafa, Refenes A. P. N. (Eds.), Proceedings of the Fourth International Conference on Neural Networks in the Capital Markets: Progress in Neural Processing, World Scientific, 95–103
340. Szeto K. Y., Cheung K. H. (1998) Multiple Time Series Prediction Using Genetic Algorithms Optimizer. In: Xu L., Chan L. W., King I., Fu A. (Eds.), Intelligent Data Engineering and Learning: Perspectives on Financial Engineering and Data Mining. Springer-Verlag, Singapore, 127–133
341. Szeto K. Y., Luo P. X. (1999) Self-Organizing Behavior in Genetic Algorithm for the Forecasting of Financial Time Series. In: Proceedings of the International Conference on Forecasting Financial Markets.
342. Szeto K. Y., Luo P. X. (2000) How Adaptive Agents in Stock Market Perform in the Presence of Random News: A Genetic Algorithm Approach. In: Leung K. S., Chan L.-W., Meng H. (Eds.), Intelligent Data Engineering and Automated Learning- IDEAL 2000: Data Mining, Financial Engineering, and Intelligent Agents, Lecture Notes in Computer Sciences 1983. Springer, 505–510
343. Szpiro G. G. (1997a) The Emergence of Risk Aversion. Complexity **2**, 31–39
344. Szpiro G. G. (1997b) A Search for Hidden Relationships: Data Mining with Genetic Algorithms. Computational Economics **10(3)**, 267–277
345. Szpiro G. G. (1997c) Forecasting Chaotic Time Series with Genetic Algorithms. Physical Review E, 2557–2568

346. Szpiro G. G. (1999) Can Computers Have Sentiments? The Case of Risk Aversion and Utility for Wealth. In: Floreano D., Nicoud J-D, Mondada F. (Eds.), Advances in Artificial Life, Lecture Notes in Artificial Intelligence, Vol. 1674. Springer, Heidelberg, 365–376
347. Tayler P. (1995) Modelling Artificial Stock Markets Using Genetic Algorithms. In Goonatilake S., Treleaven P. (Eds.), Intelligent Systems for Finance and Business. Wiley, New York, NY, 271–287
348. Tesfatsion L. (1996) An Evolutionary Trade Network Game with Preferential Partner Selection. In: Angeline P., Back T., Fogel D. (Eds.), Evolutionary Programming: Proceeding of the Fifth Annual Conference on Evolutionary Programming. MIT Press, Cambridge, MA, 45-54
349. Tesfatsion L. (1997a) A Trade Network Game with Endogenous Partner Selection. In: Amman H. M., Rustem B., Whinston A. B. (Eds.), Computational Approaches to Economic Problems. Kluwer Academic Publishers, Dordrecht, 249–269
350. Tesfatsion L. (1997b) How Economists Can Get Alife. In: Arthur W. B., Durlauf S., Lane D. (Eds.), The Economy as an Evolving Complex System, II. Proceedings Volume XXVII, Santa Fe Institute Studies in the Sciences of Complexity. Addison-Wesley, Reading, MA, 533–564
351. Tesfatsion L. (1998) Preferential Partner Selection in Evolutionary Labor Markets: A Study in Agent-Based Computational Economics. In: Porto V. W., Saravanan N., Waagen D., Eiben A. E. (Eds.), Evolutionary Programming VII. Proceedings of the Seventh Annual Conference on Evolutionary Programming. Springer-Verlag, Berlin, 15–24
352. Tesfatsion L. (2000) Concentration, Capacity, and Market Power in An Evolutionary Labor Market. In: Congress on Evolutionary Computation 2000, Vol. 2. IEEE Press, 1033–1040
353. Tesfatsion L. (2001) Introduction to the Special Issue on Agent-Based Computational Economics. Journal of Economic Dynamics and Control **25**, 281–293
354. Thomason M. (1994) Product Review: Genetic Training Option. Neurove$t Journal **2(1)**, 25–27
355. Thomason M. (1996) Product Review: Neurogenetic Optimizer. Neurove$t Journal **4(3)**, 35–37
356. Trigueros J. (1997) A Nonparametric Approach to Pricing and Hedging Derivative Securities via Genetic Regression. In: Proceedings of the IEEE/IAFE 1997 Computational Intelligence for Financial Engineering. IEEE Press, 1–6
357. Tsang E., Bulter J. M., Li J. (1998) EDDIE Beats the Bookies. Journal of Software, Practice and Experience **28(10)**, 285–300
358. Tsang E., Li J. (2000) Combining Ordinal Financial Prediction with Genetic Programming. In: Loung K.S., Chan L.-W., Meng H. (Eds.), Intelligent Data Engineering and Automated Learning- IDEAL 2000: Data Mining, Financial Engineering, and Intelligent Agents, Lecture Notes in Computer Sciences 1983. Springer, 532–537
359. Tsang E., Li J., Markose S., Er H., Salhi A., Iori G. (2001) EDDIE in Financial Decision Making. Journal of Management and Economics **4**. <http://www.econ.uba.ar/www/servicios/publicaciones/journal4/contents/contents.htm>
360. Tsang R., Lajbcygier P. (1999) Optimization of Technical Trading Strategy Using Split Search Genetic Algorithms. In: Abu-Mostafa Y. S., LeBaron B.,

Lo A. W., Weigend A. S. (Eds.), Computational Finance 1999. MIT Press, Cambridge, MA, 369–386
361. Uno K., Sato H., Namatame A. (1999) Social Evolution by Imitation. In: McKay B., Sarker R., Yao X., Tsujimura Y., Namatame A., Gen M. (Eds.), Proceedings of the Third Australia-Japan Joint Workshop on Intelligent and Evolutionary Systems, 33–40
362. Unver M. U. (2001) Backward Unraveling over Time: The Evolution of Strategic Behavior in the Entry Level British Medical Labor Markets. Journal of Economic Dynamics and Control **25(6-7)**, 1039–1080
363. Vacca L. (1997) Managing Options Risk with Genetic Algorithms. In: Proceedings of the IEEE/IAFE 1997 Computational Intelligence for Financial Engineering. IEEE Press, 29–35
364. Vallee T., Basar T. (1998) Incentive Stackelberg Solutions and the Genetic Algorithm. In: Neevel V. (Ed.), Proceedings of the 8th International Symposium on Dynamic Games and Applications, 633–639
365. Vallee T., Basar T. (1999) Off-Line Computation of Stackelberg Solutions with the Genetic Algorithm. Computational Economics **13(3)**, 201–209
366. Varetto F. (1998) Genetic Algorithms Applications in the Analysis of Insolvency Risk. Journal of Banking and Finance **22**, 1421–1439
367. Vedarajan G., Chan L .C., Goldberg D. (1997) Investment Portfolio Optimization Using Genetic Algorithms. In: Koza J .R. (Ed.), Late Breaking Papers at the Genetic Programming 1997 Conference, 255–263
368. Vriend N. (1995) Self-Organization of Markets: An Example of a Computational Approach. Computational Economics **8**, 205–231
369. Vriend N. J. (2000) An Illustration of the Essential Difference between Individual and Social Learning, and Its Consequences for Computational Analyses. Journal of Economic Dynamics and Control **24**, 1–19
370. Vila X. (1997) Adaptive Artificial Agents Plays a Finitely Repeated Discrete Principal-Agent Game. In: Conte R., Hegselmann R., Terna P. (Eds.), Simulating Social Phenomena. Springer-Verlag, 437–456
371. Waldrop M. M. (1992) Complexity: The Emerging Science at the Edge of Order and Chaos. Simon and Schuster.
372. Walker R., Barrow D., Gerrets M., Haasdijk E. (1994) Genetic Algorithms in Business. In: Stender J., Hillebrand E., Kingdon J. (Eds.), Genetic Algorithms in Optimisation, Simulation and Modelling. IOS press.
373. Walker R. F., Haasdijk E. W., Gerrets M. C. (1995) Credit Evaluation Using a Genetic Algorithm. In Goonatilake S., Treleaven P. (Eds.), Intelligent Systems for Finance and Business. Wiley, New York, NY, 39–59
374. Wang J. (2000) Trading and Hedging in S&P 500 Spot and Futures Markets Using Genetic Programming. Journal of Futures Markets **20(10)**, 911–942
375. Wang T. D., Fyfe C. (2000) Simulating Responses to Traffic Jams. In: Wang P. (Ed.), Proceedings of the Fifth Joint Conference on Information Sciences, Vol. II, 986–989
376. Warren M. A. (1994) Stock Price Prediction Using Genetic Programming. In: Koza J. (Ed.), Genetic Algorithms at Stanford 1994. Stanford Bookstore, Stanford, CA, 180–184
377. Wong F. (1994) Neurogenetic Computing Technology. Neurove$t Journal **2(4)**, 12–15
378. Wu J., Axelrod R. (1995) How to Cope with Noise in the Iterated Prisoner's Dilemma. Journal of Conflict Resolution **39 (1)**, 183–189

379. Yamashita T., Suzuki K., Ohuchi A. (1998) Agent Based Iterated Multiple Lake Game with Local Governments. In: Standish R., Henry B., Marks R., Stocker R., Green D., Keen S., Bossomaier T. (Eds.), Complex Systems'98. Complexity between the Ecos: From Ecology to Economics, 376–387
380. Yao X., Darwen P. J. (1994) An Experimental Study of N-Person Iterated Prisoner's Dilemma Games. Informatica **18**, 435–450
381. Yao X., Liu Y. (1996) EPNet for Chaotic Time Series Prediction. In: Yao X., Kim J.-H., Furuhashi (Eds.), Simulated Evolution and Learning. Lecture Notes in Artificial Intelligence 1285. Springer, 146–156
382. Ye Z., Ren Q., Lin W. (1998) A Random Segment Searching Algorithm for Optimal Portfolio with Risk Control. In: Xu L., Chan L. W., King I., Fu A. (Eds.), Intelligent Data Engineering and Learning: Perspectives on Financial Engineering and Data Mining. Springer-Verlag, Singapore, 121–126
383. Zhu F., Guan S.-U. (2001) Towards Evolution of Software Agents in Electronic Commerce. In: Proceedings of the 2001 Congress on Evolutionary Computation, Vol. 2. IEEE Press, 1303–1308
384. Zulkernine M., Uozumi T., Ono K. (1997) Genetic Based Fuzzy Inference System: Function Approximation and Time Series Prediction. Proceedings of the Third Joint Conference of Information Sciences, Vol. 1, 85–88

Index

9 783790 814767

GPSR Compliance

The European Union's (EU) General Product Safety Regulation (GPSR) is a set of rules that requires consumer products to be safe and our obligations to ensure this.

If you have any concerns about our products, you can contact us on ProductSafety@springernature.com

In case Publisher is established outside the EU, the EU authorized representative is:

Springer Nature Customer Service Center GmbH
Europaplatz 3
69115 Heidelberg, Germany

Batch number: 10725800

Printed by Printforce, the Netherlands